T0338117

Modern Characterization of
Electromagnetic Systems and Its
Associated Metrology

Modern Characterization of Electromagnetic Systems and Its Associated Metrology

Tapan K. Sarkar[†]
Syracuse University
11 Wexford Road, Syracuse, New York 13214

Magdalena Salazar-Palma
Carlos III University of Madrid
Avda. de la Universidad 30, 28911 Leganés, Madrid, Spain

Ming Da Zhu
Xidian University
No. 2 South Taibai Road, Xi'an, Shaanxi, China

Heng Chen
Syracuse University
211 Lafayette Rd. Room 425, Syracuse, NY, USA

IEEE PRESS

WILEY

The right of Tapan K. Sarkar, Magdalena Salazar-Palma, Ming Da Zhu, Heng Chen to be identified as the authors of the editorial material in this work has been asserted in accordance with law.

Registered Office
John Wiley & Sons, Inc., 111 River Street, Hoboken, NJ 07030, USA

Editorial Office
111 River Street, Hoboken, NJ 07030, USA

For details of our global editorial offices, customer services, and more information about Wiley products visit us at www.wiley.com.

Wiley also publishes its books in a variety of electronic formats and by print-on-demand. Some content that appears in standard print versions of this book may not be available in other formats.

Library of Congress Cataloging-in-Publication Data

Names: Sarkar, Tapan (Tapan K.), editor. | Salazar-Palma, Magdalena, editor. | Zhu, Ming Da, editor. | Chen, Heng, editor.
Title: Modern characterization of electromagnetic systems and its associated metrology / edited by Tapan K. Sarkar, Magdalena Salazar-Palma, Ming Da Zhu, Heng Chen.
Description: Hoboken, NJ : Wiley, 2020. | Includes bibliographical references and index.
Identifiers: LCCN 2020008264 (print) | LCCN 2020008265 (ebook) | ISBN 9781119076469 (hardback) | ISBN 9781119076544 (adobe pdf) | ISBN 9781119076537 (epub)
Subjects: LCSH: Electromagnetism–Mathematics. | Electromagnetic waves–Measurement.
Classification: LCC QC760 .M53 2020 (print) | LCC QC760 (ebook) | DDC 537/.12–dc23
LC record available at https://lccn.loc.gov/2020008264
LC ebook record available at https://lccn.loc.gov/2020008265

Cover Design: Wiley
Cover Image: © zf L/Getty Images

Set in 10/12pt Warnock by Straive, Pondicherry, India

SKY10028212_071721

Contents

Preface   *xiii*
Acknowledgments   *xxi*
Tribute to Tapan K. Sarkar – Magdalena Salazar Palma, Ming Da Zhu, and Heng Chen   *xxiii*

1   Mathematical Principles Related to Modern System Analysis   *1*
Summary   *1*
1.1   Introduction   *1*
1.2   Reduced-Rank Modelling: Bias Versus Variance Tradeoff   *3*
1.3   An Introduction to Singular Value Decomposition (SVD) and the Theory of Total Least Squares (TLS)   *6*
1.3.1   Singular Value Decomposition   *6*
1.3.2   The Theory of Total Least Squares   *15*
1.4   Conclusion   *19*
References   *20*

2   Matrix Pencil Method (MPM)   *21*
Summary   *21*
2.1   Introduction   *21*
2.2   Development of the Matrix Pencil Method for Noise Contaminated Data   *24*
2.2.1   Procedure for Interpolating or Extrapolating the System Response Using the Matrix Pencil Method   *26*
2.2.2   Illustrations Using Numerical Data   *26*
2.2.2.1   Example 1   *26*
2.2.2.2   Example 2   *29*
2.3   Applications of the MPM for Evaluation of the Characteristic Impedance of a Transmission Line   *32*
2.4   Application of MPM for the Computation of the S-Parameters Without any A Priori Knowledge of the Characteristic Impedance   *37*
2.5   Improving the Resolution of Network Analyzer Measurements Using MPM   *44*

2.6   Minimization of Multipath Effects Using MPM in Antenna
       Measurements Performed in Non-Anechoic Environments   *57*
2.6.1 Application of a FFT-Based Method to Process the Data   *61*
2.6.2 Application of MPM to Process the Data   *64*
2.6.3 Performance of FFT and MPM Applied to Measured Data   *67*
2.7   Application of the MPM for a Single Estimate of the SEM-Poles
       When Utilizing Waveforms from Multiple Look Directions   *74*
2.8   Direction of Arrival (DOA) Estimation Along with Their
       Frequency of Operation Using MPM   *81*
2.9   Efficient Computation of the Oscillatory Functional Variation in the
       Tails of the Sommerfeld Integrals Using MPM   *85*
2.10   Identification of Multiple Objects Operating in Free Space Through
       Their SEM Pole Locations Using MPM   *91*
2.11   Other Miscellaneous Applications of MPM   *95*
2.12   Conclusion   *95*
       Appendix 2A   Computer Codes for Implementing MPM   *96*
       References   *99*

3   The Cauchy Method   *107*
       Summary   *107*
3.1   Introduction   *107*
3.2   Procedure for Interpolating or Extrapolating the System Response
       Using the Cauchy Method   *112*
3.3   Examples to Estimate the System Response Using the Cauchy
       Method   *112*
3.3.1 Example 1   *112*
3.3.2 Example 2   *116*
3.3.3 Example 3   *118*
3.4   Illustration of Extrapolation by the Cauchy Method   *120*
3.4.1 Extending the Efficiency of the Moment Method Through
       Extrapolation by the Cauchy Method   *120*
3.4.2 Interpolating Results for Optical Computations   *123*
3.4.3 Application to Filter Analysis   *125*
3.4.4 Broadband Device Characterization Using Few Parameters   *127*
3.5   Effect of Noise Contaminating the Data and Its Impact on the
       Performance of the Cauchy Method   *130*
3.5.1 Perturbation of Invariant Subspaces   *130*
3.5.2 Perturbation of the Solution of the Cauchy Method Due
       to Additive Noise   *131*
3.5.3 Numerical Example   *136*
3.6   Generating High Resolution Wideband Response from Sparse and
       Incomplete Amplitude-Only Data   *138*

3.6.1 Development of the Interpolatory Cauchy Method for
       Amplitude-Only Data *139*
3.6.2 Interpolating High Resolution Amplitude Response *142*
3.7   Generation of the Non-minimum Phase Response from
       Amplitude-Only Data Using the Cauchy Method *148*
3.7.1 Generation of the Non-minimum Phase *149*
3.7.2 Illustration Through Numerical Examples *151*
3.8   Development of an Adaptive Cauchy Method *158*
3.8.1 Introduction *158*
3.8.2 Adaptive Interpolation Algorithm *159*
3.8.3 Illustration Using Numerical Examples *160*
3.8.4 Summary *171*
3.9   Efficient Characterization of a Filter *172*
3.10   Extraction of Resonant Frequencies of an Object from Frequency
       Domain Data *176*
3.11   Conclusion *180*
       Appendix 3A   MATLAB Codes for the Cauchy Method *181*
       References *187*

4   Applications of the Hilbert Transform – A Nonparametric Method for
       Interpolation/Extrapolation of Data *191*
       Summary *191*
4.1   Introduction *192*
4.2   Consequence of Causality and Its Relationship to the Hilbert
       Transform *194*
4.3   Properties of the Hilbert Transform *195*
4.4   Relationship Between the Hilbert and the Fourier Transforms
       for the Analog and the Discrete Cases *199*
4.5   Methodology to Extrapolate/Interpolate Data in the Frequency
       Domain Using a Nonparametric Methodology *200*
4.6   Interpolating Missing Data *203*
4.7   Application of the Hilbert Transform for Efficient Computation
       of the Spectrum for Nonuniformly Spaced Data *213*
4.7.1 Formulation of the Least Square Method *217*
4.7.2 Hilbert Transform Relationship *221*
4.7.3 Magnitude Estimation *223*
4.8   Conclusion *229*
       References *229*

5   The Source Reconstruction Method *235*
       Summary *235*
5.1   Introduction *236*
5.2   An Overview of the Source Reconstruction Method (SRM) *238*

5.3   Mathematical Formulation for the Integral Equations   *239*
5.4   Near-Field to Far-Field Transformation Using an Equivalent
        Magnetic Current Approach   *240*
5.4.1   Description of the Proposed Methodology   *241*
5.4.2   Solution of the Integral Equation for the Magnetic Current   *245*
5.4.3   Numerical Results Utilizing the Magnetic Current   *249*
5.4.4   Summary   *268*
5.5   Near-Field to Near/Far-Field Transformation for Arbitrary
        Near-Field Geometry Utilizing an Equivalent Electric Current   *276*
5.5.1   Description of the Proposed Methodology   *278*
5.5.2   Numerical Results Using an Equivalent Electric Current   *281*
5.5.3   Summary   *286*
5.6   Evaluating Near-Field Radiation Patterns of Commercial
        Antennas   *297*
5.6.1   Background   *297*
5.6.2   Formulation of the Problem   *301*
5.6.3   Results for the Near-field To Far-field Transformation   *304*
5.6.3.1 A Base Station Antenna   *304*
5.6.3.2 NF to FF Transformation of a Pyramidal Horn Antenna   *307*
5.6.3.3 Reference Volume of a Base Station Antenna for Human
        Exposure to EM Fields   *310*
5.6.4   Summary   *311*
5.7   Conclusions   *313*
        References   *314*

6   Planar Near-Field to Far-Field Transformation Using a Single Moving
        Probe and a Fixed Probe Arrays   *319*
        Summary   *319*
6.1   Introduction   *320*
6.2   Theory   *322*
6.3   Integral Equation Formulation   *323*
6.4   Formulation of the Matrix Equation   *325*
6.5   Use of an Magnetic Dipole Array as Equivalent Sources   *328*
6.6   Sample Numerical Results   *329*
6.7   Summary   *337*
6.8   Differences between Conventional Modal Expansion and the
        Equivalent Source Method for Planar Near-Field to Far-Field
        Transformation   *337*
6.8.1   Introduction   *337*
6.8.2   Modal Expansion Method   *339*
6.8.3   Integral Equation Approach   *341*
6.8.4   Numerical Examples   *344*
6.8.5   Summary   *351*

6.9 A Direct Optimization Approach for Source Reconstruction
and NF-FF Transformation Using Amplitude-Only Data *352*

6.9.1 Background *352*

6.9.2 Equivalent Current Representation *354*

6.9.3 Optimization of a Cost Function *356*

6.9.4 Numerical Simulation *357*

6.9.5 Results Obtained Utilizing Experimental Data *358*

6.9.6 Summary *359*

6.10 Use of Computational Electromagnetics to Enhance the Accuracy
and Efficiency of Antenna Pattern Measurements Using an
Array of Dipole Probes *361*

6.10.1 Introduction *362*

6.10.2 Development of the Proposed Methodology *363*

6.10.3 Philosophy of the Computational Methodology *363*

6.10.4 Formulation of the Integral Equations *365*

6.10.5 Solution of the Integro-Differential Equations *367*

6.10.6 Sample Numerical Results *369*

6.10.6.1 Example 1 *369*

6.10.6.2 Example 2 *373*

6.10.6.3 Example 3 *377*

6.10.6.4 Example 4 *379*

6.10.7 Summary *384*

6.11 A Fast and Efficient Method for Determining the Far Field
Patterns Along the Principal Planes Using a Rectangular
Probe Array *384*

6.11.1 Introduction *385*

6.11.2 Description of the Proposed Methodology *385*

6.11.3 Sample Numerical Results *387*

6.11.3.1 Example 1 *387*

6.11.3.2 Example 2 *393*

6.11.3.3 Example 3 *397*

6.11.3.4 Example 4 *401*

6.11.4 Summary *406*

6.12 The Influence of the Size of Square Dipole Probe Array
Measurement on the Accuracy of NF-FF Pattern *406*

6.12.1 Illustration of the Proposed Methodology Utilizing Sample
Numerical Results *407*

6.12.1.1 Example 1 *407*

6.12.1.2 Example 2 *411*

6.12.1.3 Example 3 *416*

6.12.1.4 Example 4 *419*

6.12.2 Summary *428*

6.13 Use of a Fixed Probe Array Measuring Amplitude-Only
Near-Field Data for Calculating the Far-Field *428*

6.13.1   Proposed Methodology *429*
6.13.2   Sample Numerical Results *430*
6.13.2.1 Example 1 *430*
6.13.2.2 Example 2 *434*
6.13.2.3 Example 3 *437*
6.13.2.4 Example 4 *437*
6.13.3   Summary *441*
6.14   Probe Correction for Use with Electrically Large Probes *442*
6.14.1   Development of the Proposed Methodology *443*
6.14.2   Formulation of the Solution Methodology *446*
6.14.3   Sample Numerical Results *447*
6.15   Conclusions *449*
          References *449*

7   Spherical Near-Field to Far-Field Transformation *453*
          Summary *453*
7.1   An Analytical Spherical Near-Field to Far-Field
          Transformation *453*
7.1.1   Introduction *453*
7.1.2   An Analytical Spherical Near-Field to Far-Field
          Transformation *454*
7.1.3   Numerical Simulations *464*
7.1.3.1   Synthetic Data *464*
7.1.3.2   Experimental Data *465*
7.1.4   Summary *468*
7.2   Radial Field Retrieval in Spherical Scanning for Current
          Reconstruction and NF–FF Transformation *468*
7.2.1   Background *468*
7.2.2   An Equivalent Current Reconstruction from Spherical
          Measurement Plane *470*
7.2.3   The Radial Electric Field Retrieval Algorithm *472*
7.2.4   Results Obtained Using This Formulation *473*
7.2.4.1   Simulated Data *473*
7.2.4.2   Using Measured Data *475*
7.3   Conclusion *482*
          Appendix 7A A Fortran Based Computer Program for Transforming
          Spherical Near-Field to Far-Field *483*
          References *489*

8   Deconvolving Measured Electromagnetic Responses *491*
          Summary *491*
8.1   Introduction *491*
8.2   The Conjugate Gradient Method with Fast Fourier Transform for
          Computational Efficiency *495*

8.2.1   Theory   *495*
8.2.2   Numerical Results   *498*
8.3   Total Least Squares Approach Utilizing Singular Value
           Decomposition   *501*
8.3.1   Theory   *501*
8.3.2   Total Least Squares (TLS)   *502*
8.3.3   Numerical Results   *506*
8.4   Conclusion   *516*
           References   *516*

9   Performance of Different Functionals for Interpolation/
           Extrapolation of Near/Far-Field Data   *519*
           Summary   *519*
9.1   Background   *520*
9.2   Approximating a Frequency Domain Response by Chebyshev
           Polynomials   *521*
9.3   The Cauchy Method Based on Gegenbauer Polynomials   *531*
9.3.1   Numerical Results and Discussion   *537*
9.3.1.1   Example of a Horn Antenna   *537*
9.3.1.2   Example of a 2-element Microstrip Patch Array   *539*
9.3.1.3   Example of a Parabolic Antenna   *541*
9.4   Near-Field to Far-Field Transformation of a Zenith-Directed
           Parabolic Reflector Using the Ordinary Cauchy Method   *543*
9.5   Near-Field to Far-Field Transformation of a Rotated Parabolic
           Reflector Using the Ordinary Cauchy Method   *552*
9.6   Near-Field to Far-Field Transformation of a Zenith-Directed
           Parabolic Reflector Using the Matrix Pencil Method   *558*
9.7   Near-Field to Far-Field Transformation of a Rotated Parabolic
           Reflector Using the Matrix Pencil Method   *564*
9.8   Conclusion   *569*
           References   *569*

10   Retrieval of Free Space Radiation Patterns from Measured Data
           in a Non-Anechoic Environment   *573*
           Summary   *573*
10.1   Problem Background   *573*
10.2   Review of Pattern Reconstruction Methodologies   *575*
10.3   Deconvolution Method for Radiation Pattern Reconstruction   *578*
10.3.1   Equations and Derivation   *578*
10.3.2   Steps Required to Implement the Proposed Methodology   *584*
10.3.3   Processing of the Data   *585*
10.3.4   Simulation Examples   *587*
10.3.4.1   Example I: One PEC Plate Serves as a Reflector   *587*
10.3.4.2   Example II: Two PEC Plates Now Serve as Reflectors   *594*

10.3.4.3   Example III: Four Connected PEC Plates Serve as Reflectors   *598*
10.3.4.4   Example IV: Use of a Parabolic Reflector Antenna as the AUT   *604*
10.3.5   Discussions on the Deconvolution Method for Radiation Pattern
            Reconstruction   *608*
10.4   Effect of Different Types of Probe Antennas   *608*
10.4.1   Numerical Examples   *608*
10.4.1.1   Example I: Use of a Yagi Antenna as the Probe   *608*
10.4.1.2   Example II: Use of a Parabolic Reflector Antenna as the Probe   *612*
10.4.1.3   Example III: Use of a Dipole Antenna as the Probe   *613*
10.5   Effect of Different Antenna Size   *619*
10.6   Effect of Using Different Sizes of PEC Plates   *626*
10.7   Extension of the Deconvolution Method to Three-Dimensional
            Pattern Reconstruction   *632*
10.7.1   Mathematical Characterization of the Methodology   *632*
10.7.2   Steps Summarizing for the Methodology   *635*
10.7.3   Processing the Data   *636*
10.7.4   Results for Simulation Examples   *638*
10.7.4.1   Example I: Four Wide PEC Plates Serve as Reflectors   *640*
10.7.4.2   Example II: Four PEC Plates and the Ground Serve as
            Reflectors   *643*
10.7.4.3   Example III: Six Plates Forming an Unclosed Contour Serve as
            Reflectors   *651*
10.7.4.4   Example IV: Antenna Measurement in a Closed PEC Box   *659*
10.7.4.5   Example V: Six Dielectric Plates Forming a Closed Contour
            Simulating a Room   *662*
10.8   Conclusion   *673*
            Appendix A: Data Mapping Using the Conversion between the
            Spherical Coordinate System and the Cartesian Coordinate
            System   *675*
            Appendix B: Description of the 2D-FFT during the Data
            Processing   *677*
            References   *680*

Index   *683*

Preface

The area of electromagnetics is an evolutionary one. In the earlier days the analysis in this area was limited to 11 separable coordinate systems for the solution of Helmholtz equations. The eleven coordinate systems are rectangular, circular cylinder, elliptic cylinder, parabolic cylinder, spherical, conical, parabolic, prolate spheroidal, oblate spheroidal, ellipsoidal and paraboloidal coordinates. However, Laplace's equation is separable in 13 coordinate systems, the additional two being the bispherical and the toroidal coordinate systems. Outside these coordinate systems it was not possible to develop a solution for electromagnetic problems in the earlier days. However, with the advent of numerical methods this situation changed and it was possible to solve real practical problems in any system. This development took place in two distinct stages and was primarily addressed by Prof. Roger F. Harrington. In the first phase he proposed to develop the solution of an electromagnetic field problem in terms of unknown currents, both electric and magnetic and not fields by placing some equivalent currents to represent the actual sources so that these currents produce exactly the same desired fields in each region. From these currents he computed the electric and the magnetic vector potentials in any coordinate system. In the integral representation of the potentials in terms of the unknown currents, the free space Green's function was used which simplified the formulation considerably as no complicated form of the Green's function for any complicated environment was necessary. From the potentials, the fields, both electric and magnetic, were developed by invoking the Maxwell-Hertz-Heaviside equations. This made the mathematical analysis quite analytic and simplified many of the complexities related to the complicated Green's theorem. This was the main theme in his book "Time Harmonic Electromagnetic Fields", McGraw Hill, 1961. At the end of this book he tried to develop a variational form for all these concepts so that a numerical technique can be applied and one can solve any electromagnetic boundary value problem of interest. This theme was further developed in the second stage through his second classic book "Field Computations by Moment Methods", Macmillan Company, 1968. In the second book he illustrated how to solve a general electromagnetic field problem. This gradual development took almost half a century to mature. In the experimental realm, unfortunately, no such progress has been made. This may be partially due to decisions taken by the past leadership of the IEEE Antennas and Propagation Society (AP-S) who first essentially disassociated measurements from their primary focus leading antenna measurement practitioners to form the Antenna Measurements Techniques Association (AMTA) as an organization different from IEEE AP-S. And later on even the numerical techniques part was not considered in the main theme of the IEEE Antennas and Propagation Society leading to the formation of the Applied Computational Electromagnetic Society (ACES). However, in recent times these shortcomings of the past decisions of the AP-S leadership have been addressed.

The objective of this book is to advance the state of the art of antenna measurements and not being limited to the situation that measurements can be made in one of the separable coordinate systems just like the state of electromagnetics over half a century ago. We propose to carry out this transformation in the realm of measurement first by trying to find a set of equivalent currents just like we do in theory and then solve for these unknown currents using the Maxwell-Hertz-Heaviside equations via the Method of Moments popularized by Prof. Harrington. Since the expressions between the measured fields and the unknown currents are analytic and related by Maxwell-Hertz-Heaviside equations, the measurements can be carried out in any arbitrary geometry and not just limited to the planar, cylindrical or spherical geometries. The advantage of this new methodology as presented in this book through the topic "Source Reconstruction Method" is that the measurement of the fields need not be done using a Nyquist sampling criteria which opens up new avenues particularly in the very high frequency regime of the electromagnetic spectrum where it might be difficult to take measurement samples half a wavelength apart. Secondly as will be illustrated these measurement samples need not even be performed in any specified plane. Also because of the analytical relationship between the sources that generate the fields and the fields themselves it is possible to go beyond the Raleigh resolution limit and achieve super resolution in the diagnosis of radiating structures. In the Raleigh limit the resolution is limited by the uncertainty principle and that is determined by the length of the aperture whose Fourier transform we are looking at whereas in the super resolution system there is no such restriction. Another objective of this book is to outline a very simple procedure to recover the non-minimum phase of any electromagnetic system using amplitude-only data. This simple procedure is based on the principle of causality which results in the Hilbert transform relationship between the real and the imaginary parts of a transfer function of any linear time invariant system. The philosophy of model order reduction can also be implemented using the concepts of total least squares along with the singular value decomposition. This makes the ill-posed deconvolution problem quite stable numerically. Finally, it is shown how to interpolate and extrapolate measured data including filling up the gap of missing measured near/far-field data.

The book contains ten chapters. In Chapter 1, the mathematical preliminaries are described. In the mathematical field of numerical analysis, model order reduction is the key to processing measured data. This also enables us to interpolate and extrapolate measured data. The philosophy of model order reduction is outlined in this chapter along with the concepts of total least squares and singular value decomposition.

In Chapter 2, we present the matrix pencil method (MPM) which is a methodology to approximate a given data set by a sum of complex exponentials. The objective is to interpolate and extrapolate data and also to extract certain parameters so as to compress the data set. First the methodology is presented followed by some application in electromagnetic system characterization. The applications involve using this methodology to deembed device characteristics and obtain accurate and high resolution characterization, enhance network analyzer measurements when not enough physical bandwidth is available for measurements, minimize unwanted reflections in antenna measurements and, when performing system characterization in a non-anechoic environment, to extract a single set of exponents representing the resonant frequency of an object when data from multiple look angles are given and compute directions of arrival estimation of signals along with their frequencies of operation. This method can also be used to speed up the calculation of the tails encountered in the evaluation of the Sommerfeld integrals and in multiple target characterization in free space from the scattered data using their characteristic external resonance which are popularly known as the singularity expansion method (SEM) poles. References to other applications, including multipath characterization of a propagating wave, characterization of the quality of power systems, in waveform analysis and imaging and speeding up computations in a time domain electromagnetic simulation. A computer program implementing the matrix pencil method is given in the appendix so that it can easily be implemented in practice.

In numerical analysis, interpolation is a method of estimating unknown data within the range of known data from the available information. Extrapolation is also the process of approximating unknown data outside the range of the known available data. In Chapter 3, we are going to look at the concept of the Cauchy method for the interpolation and extrapolation of both measured and numerically simulated data. The Cauchy method can deal with extending the efficiency of the moment method through frequency extrapolation. Interpolating results for optical computations, generation of pass band using stop band data and vice versa, efficient broadband device characterization, effect of noise on the performance of the Cauchy method and for applications to extrapolating amplitude-only data for the far-field or RCS interpolation/extrapolation. Using this method to generate the non-minimum phase response from amplitude-only data, and adaptive interpolation for sparsely sampled data is also illustrated. In addition, it has been applied to characterization of filters and extracting resonant frequencies of objects using frequency domain data. Other applications include non-destructive evaluation of fruit status of maturity and quality of fruit juices, RCS applications and to multidimensional extrapolation. A computer program implementing the Cauchy method has been provided in the Appendix again for ease of understanding.

The previous two chapters discussed the parametric methods in the context of the principle of analytic continuation and provided its relationship to reduced rank modelling using the total least squares based singular value decomposition methodology. The problem with a parametric method is that the quality of the solution is determined by the choice of the basis functions and use of unsuitable basis functions generate bad solutions. A priori it is quite difficult to recognize what are good basis functions and what are bad basis functions even though methodologies exist in theory on how to choose good ones. The advantage of the nonparametric methods presented in Chapter 4 is that no such choices of the basis functions need to be made as the solution procedure by itself develops the nature of the solution and no a priori information is necessary. This is accomplished through the use of the Hilbert transform which exploits one of the fundamental properties of nature and that is causality. The Hilbert transform illustrates that the real and imaginary parts of any nonminimum phase transfer function for a causal system satisfy this relationship. In addition, some parametrization can also be made of this procedure which can enable one to generate a nonminimum phase function from its amplitude response and from that generate the phase response. This enables one to compute the time domain response of the system using amplitude only data barring a time delay in the response. This delay uncertainty is removed in holography as in such a procedure an amplitude and phase information is measured for a specific look angle thus eliminating the phase ambiguity. An overview of the technique along with examples are presented to illustrate this methodology. The Hilbert transform can also be used to speed up the spectral analysis of nonuniformly spaced data samples. Therefore, in this section a novel least squares methodology is applied to a finite data set using the principle of spectral estimation. This can be applied for the analysis of the far-field pattern collected from unevenly spaced antennas. The advantage of using a non-uniformly sampled data is that it is not necessary to satisfy the Nyquist sampling criterion as long as the average value of the sampling rate is less than the Nyquist rate. Accurate and efficient computation of the spectrum using a least squares method applied to a finite unevenly spaced data is also studied.

In Chapter 5, the source reconstruction method (SRM) is presented. It is a recent technique developed for antenna diagnostics and for carrying out near-field (NF) to far-field (FF) transformation. The SRM is based on the application of the electromagnetic Equivalence Principle, in which one establishes an equivalent current distribution that radiates the same fields as the actual currents induced in the antenna under test (AUT). The knowledge of the equivalent currents allows the determination of the antenna radiating elements, as well as the prediction of the AUT-radiated fields outside the equivalent currents domain. The unique feature of the novel methodology presented is that it can resolve equivalent currents that are smaller than half a wavelength in size, thus providing super-resolution. Furthermore, the measurement field samples can be taken at spacing greater than half a wavelength, thus going beyond the classical sampling criteria. These two distinctive features are possible due to the choice of a model-based parameter estimation methodology where the unknown sources are approximated by a basis in the computational Method of Moment (MoM) context and, secondly, through the use of the analytic free space Green's function. The latter condition also guarantees the invertibility of the electric field operator and provides a stable solution for the currents even when evanescent waves are present in the measurements. In addition, the use of the singular value decomposition in the solution of the matrix equations provides the user with a quantitative tool to assess the quality and the quantity of the measured data. Alternatively, the use of the iterative conjugate gradient (CG) method in solving the ill-conditioned matrix equations for the equivalent currents can also be implemented. Two different methods are presented in this section. One that deals with the equivalent magnetic current and the second that deals with the equivalent electric current. If the formulation is sound, then either of the methodologies will provide the same far-field when using the same near-field data. Examples are presented to illustrate the applicability and accuracy of the proposed methodology using either of the equivalent currents and applied to experimental data. This methodology is then used for near-field to near/far-field transformations for arbitrary near-field geometry to evaluate the safe distance for commercial antennas.

In Chapter 6, a fast and accurate method is presented for computing far-field antenna patterns from planar near-field measurements. The method utilizes near-field data to determine equivalent magnetic current sources over a fictitious planar surface that encompasses the antenna, and these currents are used to ascertain the far fields. Under certain approximations, the currents should produce the correct far fields in all regions in front of the antenna regardless of the geometry over which the near-field measurements are made. An electric field integral equation (EFIE) is developed to relate the near fields to the equivalent magnetic currents. Method of moments (MOM) procedure is used to transform the integral equation into a matrix one. The matrix equation is solved using the iterative conjugate gradient method (CGM), and in the case of a rectangular matrix, a least-squares solution can still be found using this approach for the magnetic currents without explicitly computing the normal form of the equations. Near-field to far-field transformation for planar scanning may be efficiently performed under certain conditions by exploiting the block Toeplitz structure of the matrix and using the conjugate gradient method (CGM) and the fast Fourier transform (FFT), thereby drastically reducing computation time and storage requirements. Numerical results are presented for several antenna configurations by extrapolating the far fields using synthetic and experimental near-field data. It is also illustrated that a single moving probe can be replaced by an array of probes to compute the equivalent magnetic currents on the surface enclosing the AUT in a single snapshot rather than tediously moving a single probe over the antenna under test to measure its near-fields. It is demonstrated that in this methodology a probe correction even when using an array of dipole probes is not necessary. The accuracy of this methodology is studied as a function of the size of the equivalent surface placed in front of the antenna under test and the error in the estimation of the far-field along with the possibility of using a rectangular probe array which can efficiently and accurately provide the patterns in the principal planes. This can also be used when amplitude-only data are collected using an array of probes. Finally, it is shown that the probe correction can be useful when the size of the probes is that of a resonant antenna and it is shown then how to carry it out.

In Chapter 7, two methods for spherical near-field to far-field transformation are presented. The first methodology is an exact explicit analytical formulation for transforming near-field data generated over a spherical surface to the far-field radiation pattern. The results are validated with experimental data. A computer program involving this method is provided at the end of the chapter. The second method presents the equivalent source formulation through the SRM described earlier so that it can be deployed to the spherical scanning case where one component of the field is missing from the measurements. Again the methodology is validated using other techniques and also with experimental data.

Two deconvolution techniques are presented in Chapter 8 to illustrate how the ill-posed deconvolution problem has been regularized. Depending on the nature of the regularization utilized which is based on the given data one can obtain a reasonably good approximate solution. The two techniques presented here have built in self-regularizing schemes. This implies that the regularization process, which depends highly on the data, can be automated as the solution procedure continues. The first method is based on solving the ill-posed deconvolution problem by the iterative conjugate gradient method. The second method uses the method of total least squares implemented through the singular value decomposition (SVD) technique. The methods have been applied to measured data to illustrate the nature of their performance.

Chapter 9 discusses the use of the Chebyshev polynomials for approximating functional variations arising in electromagnetics as it has some band-limited properties not available in other polynomials. Next, the Cauchy method based on Gegenbauer polynomials for antenna near-field extrapolation and the far-field estimation is illustrated. Due to various physical limitations, there are often missing gaps in the antenna near-field measurements. However, the missing data is indispensable if we want to accurately evaluate the complete far-field pattern by using the near-field to far-field transformations. To address this problem, an extrapolation method based on the Cauchy method is proposed to reconstruct the missing part of the antenna near-field measurements. As the near-field data in this section are obtained on a spherical measurement surface, the far field of the antenna is calculated by the spherical near-field to far-field transformation with the extrapolated data. Some numerical results are given to demonstrate the applicability of the proposed scheme in antenna near-field extrapolation and far-field estimation. In addition, the performance of the Gegenbauer polynomials are compared with that of the normal Cauchy method using Polynomial expansion and the Matrix Pencil Method for using simulated missing near-field data from a parabolic reflector antenna.

Typically, antenna pattern measurements are carried out in an anechoic chamber. However, a good anechoic chamber is very expensive to construct. Previous researches have attempted to compensate for the effects of extraneous fields measured in a non-anechoic environment to obtain a free space radiation pattern that would be measured in an anechoic chamber. Chapter 10 illustrates a deconvolution methodology which allows the antenna measurement under a non-anechoic test environment and retrieves the free space radiation pattern of an antenna through this measured data; thus allowing for easier and more affordable antenna measurements. This is obtained by modelling the extraneous fields as the system impulse response of the test environment and utilizing a reference antenna to extract the impulse response of the environment which is used to remove the extraneous fields for a desired antenna measured under the same environment and retrieve the ideal pattern. The advantage of this process is that it does not require calculating the time delay to gate out the reflections; therefore, it is independent of the bandwidth of the antenna and the measurement system, and there is no requirement for prior knowledge of the test environment.

This book is intended for engineers, researchers and educators who are planning to work in the field of electromagnetic system characterization and also deal with their measurement techniques and philosophy. The prerequisite to follow the materials of the book is a basic undergraduate course in the area of dynamic electromagnetic theory including antenna theory and linear algebra. Every attempt has been made to guarantee the accuracy of the content of the book. We would however appreciate readers bringing to our attention any errors that may have appeared in the final version. Errors and/or any comments may be emailed to one of the authors, at salazar@tsc.uc3m.es, mingda.zhu@live.com, hchen43@syr.edu.

Acknowledgments

Grateful acknowledgement is made to Prof. Pramod Varshney, Mr. Peter Zaehringer and Ms. Marilyn Polosky of the CASE Center of Syracuse University for providing facilities to make this book possible. Thanks are also due to Prof. Jae Oh, Ms. Laura Lawson and Ms. Rebecca Noble of the Department of Electrical Engineering and Computer Science of Syracuse University for providing additional support. Thanks to Michael James Rice, Systems administrator for the College of Engineering and Computer Science for providing information technology support in preparing the manuscript. Also thanks are due to Mr. Brett Kurzman for patiently waiting for us to finish the book.

Tapan K. Sarkar
Magdalena Salazar Palma (salazar@tsc.uc3m.es)
Ming-da Zhu (mingda.zhu@live.com)
Heng Chen (hchen43@syr.edu)
Syracuse, New York

Tribute to Tapan K. Sarkar
by Magdalena Salazar Palma, Ming Da Zhu, and Heng Chen

Professor Tapan K. Sarkar, PhD, passed away on 12 March 2021. The review of the proofs of this book is probably the last task he was able to accomplish. Thus, for us, his coauthors, this book will be always cherished and valued as his last gift to the scientific community.

Dr. Sarkar was born in Kolkata, India, in August 1948. He obtained his Bachelor of Technology (BT) degree from the Indian Institute of Technology (IIT), Kharagpur, India, in 1969, the Master of Science in Engineering (MSCE) degree from the University of New Brunswick, Fredericton, NB, Canada, in 1971, and the Master of Science (MS) and Doctoral (PhD) degrees from Syracuse University, Syracuse, NY, USA, in 1975. He joined the faculty of the Electrical Engineering and Computer Science Department at Syracuse University in 1979 and became Full Professor in 1985. Prior to that, he was with the Technical Appliance Corporation (TACO) Division of the General Instruments Corporation (1975–1976). He was also a Research Fellow at the Gordon McKay Laboratory for Applied Sciences, Harvard University, Cambridge, MA, USA (1977–1978), and was faculty member at the Rochester Institute of Technology, Rochester, NY, USA (1976–1985). Professor Sarkar received the Doctor Honoris Causa degree from Université Blaise Pascal, Clermont Ferrand, France (1998), from Polytechnic University of Madrid, Madrid, Spain (2004), and from Aalto University, Helsinki, Finland (2012). He was now emeritus professor at Syracuse University. Professor Sarkar was a professional engineer registered in New York, USA, and the president of OHRN Enterprises, Inc., a small business founded in 1986 and incorporated in the State of New York, USA, performing research for government, private, and foreign organizations in system analysis.

Dr. Sarkar was a giant in the field of electromagnetics, a phenomenal researcher and teacher who also provided an invaluable service to the scientific community in so many aspects.

Dr. Sarkar research interests focused on numerical solutions to operator equations arising in electromagnetics and signal processing with application to electromagnetic systems analysis and design and with particular attention to building solutions that would be appropriate and scalable for practical adoption by industry. Among his many contributions together with his students and coworkers, it may be mentioned the development of the generalized pencil-of-function (GPOF) method, also known as matrix pencil method, for signal estimation with complex exponentials. Based on Dr. Sarkar group's work on the original pencil-of-function method, the technique is used in electromagnetic analyses of layered structures, antenna analysis, and radar signal processing. He is also coauthor of the general purpose electromagnetic solver HOBBIES (Higher Order Basis Based Integral Equation Solver). The list of Professor Sarkar's original and substantive contributions to the field of computational electromagnetics and antenna theory is quite long. Just to name a few, these include methods of evaluating the Sommerfeld integrals, the already mentioned matrix pencil method for approximating a function by a sum of complex exponentials, the conjugate gradient method and fast Fourier transform method for the efficient numerical solution of integral equations having convolutional kernels, the introduction of higher order basis functions in the numerical solution of integral equations using the method of moments, the solution of time domain problems using the associated Laguerre functions as basis functions, the application of the Cauchy method to the generation of accurate broadband information from narrowband data, broadband antenna design and analysis, and near-field to far-field transformation, and many more. Dr. Sarkar's work has modernized many systems that include wireless signal propagation, has made possible the design of antennas considering the effects of the platforms where they are deployed for the current and next generations of airborne surveillance system, and has developed adaptive methodologies that made performance of adaptive systems possible in real time. His advanced computational techniques have been implemented for parallel processing on super computers for fast and efficient solution of extremely large electromagnetic field problems. He has also developed antenna systems and processing for ultrawideband applications. He applied photoconductive switching techniques for generation of kilovolts amplitude electrical pulses of subnanosecond duration with applications in many fields including low probability intercept radar systems. It is remarkable that Dr. Sarkar has been able to keep innovating for such a sustained period of time throughout his career. Professor Sarkar has authored or coauthored more than 380 journal articles, innumerable contributions for conferences and symposia, 16 books and 32 book chapters, with 24 549 citations and h-index of 74 (Google Scholar). In the past, he was listed among the ISI Highly Cited

Researchers and in Guide 2 Research, Top H-index for Antennas and Propagation. He was the principal investigator of numerous research projects and contracts (some of them of multimillion USD), including setting up a parallel supercomputer center at NAVAIR (Naval Air Systems Command), USA. He worked also for foreign governmental organizations and institutions.

His research received a number of awards and recognitions: the Best Solution Award at the Rome Air Development Center (RADC) Spectral Estimation Workshop, May 1971, the Best Paper Award, *IEEE Transactions on Electromagnetic Compatibility*, October 1979, the College of Engineering Research Award, Syracuse University, 1996, the Best Paper Award, National Radar Conference, 1997, the Chancellor's Citation for Excellence in Research, Syracuse University, 1998. In 1992, he was elevated to IEEE Fellow, "for contributions to iterative solutions of numerical models in electromagnetic theory." As mentioned before, he received three Honorary Doctorate degrees from three European Universities. He also received the medal of the Friend of the City of Clermont-Ferrand, Clermont-Ferrand, France, in 2000. He was the 2020 recipient of the IEEE Electromagnetics Award, the highest technical recognition by IEEE in the field of electromagnetics, "for contributions to the efficient and accurate solution of computational electromagnetic problems in frequency and time domain, and for research in adaptive antennas."

Dr. Sarkar had a strong background in mathematics and physics, which gave him the ability to address any topic in the field with an original, unique, but always scientifically sound point of view. He had the ability to make complex problems, easy, and to draw practical conclusions. He admired, recognized, and respected the pioneers in our field but also the many individuals whose contributions made the body of electromagnetics. As mentioned before, recently, he was selected as the 2020 IEEE Electromagnetics Award recipient, a recognition that made him happy and humble.

Dr. Sarkar had a passion to learn. One of his mottos coming from Sri Ramakrishna was "As long as I live, so long do I learn." He exemplified it all along his life. He was constantly reading. And he was passionate about teaching others, starting with his students, both during his lessons and his weekly research meetings, which he conducted in a strong way asking hard work and dedication from his students, but at the same time offering insight on possible new approaches. He has advised students and postdocs from all over the world, who in many cases are now professors themselves in their countries of origin. He established collaborations with many of them and whenever possible he will visit them and their families.

Dr. Sarkar imposing scientific and physical stature, loud voice (no microphone needed!), were proverbial in the Applied Electromagnetics and Antennas and Propagation scientific community. Many of his students and colleagues have experienced also his generosity and kindness, which complemented his personality. He took a personal interest on daily-life aspects of his students, postdocs, and invited scholars and engineers, always going the extra mile for them.

It should be mentioned here that writing books, which compiled Dr. Sarkar's team research results on a given topic, is also a consequence of his generosity. He tried to facilitate the learning process of researchers by offering them the opportunity of finding all the necessary information in just one place, together in many cases with the relevant software. He was never scared of the tiring and time-consuming effort that writing a book or a book contribution implies. He will also negotiate with the editor the lowest possible price of the book.

Dr. Sarkar impact goes far beyond his scientific achievements and his work as teacher and research advisor. He was passionate about serving the scientific community. He devoted a huge amount of his time to The Institute of Electrical and Electronics Engineers (IEEE), and the IEEE Technical Societies relevant to him, mainly the Antennas and Propagation Society (AP-S). Besides serving as reviewer for many journals, he was associate editor of several of them. He was also AP-S Administrative Committee (AdCom) member, AP-S President, AP-S distinguished lecturer, chair of AP-S Membership and Geographic Activities (MGA) committee, chair of the AP-S Special Interest Group on Humanitarian Technologies (SIGHT), AP-S AdCom Honorary member, and other positions. He also served as member of the IEEE Technical Activities Board (TAB), and some of its committees, member of IEEE Fellow Committee and other IEEE committees. More recently, he was elected as IEEE Vice President for Publications Services and Products, and Chair of IEEE Publications Services and Products Board (PSPB). Many of these positions were elected positions, testament to his credibility and the immense respect in which the IEEE, AP-S, and scientific community held him. His tenacious character served IEEE well. Once he was convinced a change was needed, he would "fight" until the end in order to achieve the intended goals. Dr. Sarkar was the driving force behind the creation of several new AP-S publications and other journals in cooperation between AP-S and other IEEE Technical Societies. He also had a key influence in expanding the international reach of AP-S and its activities, through tireless visits and meetings around the globe, organizing events, and generating substantial growth of the AP-S membership and the AP-S local chapters over the past few years. Professor Sarkar also had a major role in strengthening and supporting two international conferences sponsored by the AP-S. He also held numerous leadership and editorial positions in other organizations, e.g. he was on the board of directors of the Applied Computational Electromagnetics Society (ACES) and was later the vice president of ACES. As mentioned before, Professor Sarkar was elected as 2020 IEEE Vice President for Publications Services and Products, IEEE Director, and IEEE Board of Directors member. He was full of energy and enthusiasm to continue to contribute to the improvement of IEEE and its service to Humanity and he started as soon as the election results were communicated in November 2019. Sadly, from April 2020, he was unable to continue his work because of his health condition.

Dr. Sarkar's kindness and genuine soft-heartedness – often expressed in the most outspoken and emphatic ways – will never be forgotten. His technical accomplishments are numerous and are what he is most well known for in our scientific community. He had passion and dedication for pursuing his visions for the advancement of science in our field and for building bridges between our technical know-how and societally impactful applications. He made many enormous positive impacts not only on the technical societies in which he was member but also on so many of us as individuals. He was a good friend who would support you without any fear while also giving you his opinion even in a strong way when needed. His energy was contagious. He recognized that leadership is most effective when building by consensus: through talking, listening, and being willing to adjust course when called for. He was one of the key leaders of the AP-S and promoted a global mindset, paving the way for substantial and unprecedented global reach for the AP-S. He facilitated the creation of chapters, meetings, and conferences across the world; he visited and recruited members from all corners of the Earth. He was generous with his time and his ideas; he was also one of the most effective advocates (and implementers) of diversity, equity, and inclusion in our community. He was supportive of good ideas, seeking them wherever and from whomever they originated. He would energetically recruit you to implement those ideas regardless of whether you were a student or a decorated senior colleague. He will be missed, but his contributions will endure.

It is also proverbial among those who knew Professor Sarkar well his love for history, archeology, nature, botany, wild life, and animals in general. The world was too small for him. He travelled as much as he could, always learning from different cultures, making new friends, visiting old ones. Colleagues who had the chance to travel with him enjoyed immensely his enthusiasm, lively conversation, and eagerness to explore new places and people, and the warm welcome of his hosts.

He dearly loved his family enjoying immensely his trips back to Kolkata and other Indian cities. He had many pets (dogs, a variety of birds) and a terrace garden that he always enjoyed and personally contributed to improve.

Dr. Sarkar was the best of friends for his colleagues, coworkers, local Bengali community members, and students, always interested in any aspect of their life. At times, he did not speak much, but he will show his kindness and friendship in so many other ways.

It is incredibly difficult to accept Dr. Sarkar's departure for those of us who knew him well and loved and respected him as a friend, a colleague, a mentor, and a leader.

These words are intended as a tribute to Dr. Sarkar and as a thank-you note to him for everything he gave us: his invaluable scientific contributions, his mentorship, his leadership, and his friendship!!! He will be missed most dearly...

Magdalena Salazar Palma
Ming Da Zhu
Heng Chen

1

Mathematical Principles Related to Modern System Analysis

Summary

In the mathematical field of numerical analysis, model order reduction is the key to processing measured data. This also enables us to interpolate and extrapolate measured data. The philosophy of model order reduction is outlined in this chapter along with the concepts of total least squares and singular value decomposition.

1.1 Introduction

In mathematical physics many problems are characterized by a second order partial differential equation for a function as

$$Au_{xx} + 2Bu_{xy} + Cu_{yy} + Du_x + Eu_y + F = f(x, y), \qquad (1.1)$$

and $u(x, y)$ is the function to be solved for a given excitation $f(x, y)$, where

$$u_{xx} = \partial^2 u/\partial x^2;$$
$$u_{xy} = \partial^2 u/(\partial x\, \partial y);$$

$$u_{yy} = \partial^2 u/\partial y^2;$$
$$u_x = \partial u/\partial x; \qquad (1.2)$$
$$u_y = \partial u/\partial y$$

When $B^2 - AC < 0$ and assuming $u_{xy} = u_{yx}$, then (1.1) is called an elliptic partial differential equation. These classes of problems arise in the solution of boundary value problems. In this case, the solution $u(x, y)$ is known only over a boundary {or equivalently a contour $B(x, y)$} and the goal is to continue the given solution $u(x, y)$ from the boundary to the entire region of the real plane $\Re(x, y)$.

Modern Characterization of Electromagnetic Systems and Its Associated Metrology, First Edition.
Tapan K. Sarkar, Magdalena Salazar-Palma, Ming Da Zhu, and Heng Chen.
© 2021 John Wiley & Sons, Inc. Published 2021 by John Wiley & Sons, Inc.

When $B^2 - AC = 0$ we obtain a parabolic partial differential equation for (1.1), which arises in the solution of the diffusion equation or an acoustic propagation in the ocean. Such applications are characterized by the term initial value problems. The solution is given for the initial condition $u(x, y = 0)$ and the goal is to find the solution $u(x, y)$ for all values of x and y.

Finally when $B^2 - AC > 0$, we obtain a hyperbolic partial differential equation. This type of equation arises from the solution of the wave equation. The characteristic of the wave equation is that if a disturbance is made in the initial data, then not every point of space feels the disturbance at once. The disturbance has a finite propagation speed. This feature makes it distinct from the elliptic and parabolic partial differential equations when a disturbance of the initial data is felt at once by all points in the domain. Even though these equations have significantly different mathematical properties, the solution methodology, just like for every numerical method in solution of an operator equation, is essentially the same, by exploiting the principle of analytic continuation.

The solution u of these equations is made in a straight forward fashion by assuming: it to be of the form

$$u(x, y) = \sum_i \alpha_i \, \phi_i(x, y), \tag{1.3}$$

where $\phi_i(x, y)$ are some known basis functions; and the final solution is to be composed of these functions multiplied by some constants α_i which are the unknowns to be determined using the specific given boundary conditions. The solution procedure then translates the solution of a functional equation to the solution of a matrix equation, the solution of these unknown constants is much easier to address. The methodology starts by substituting (1.3) into (1.1) and then solving for the unknown coefficients α_i from the boundary conditions for the problem if the equations are in the differential form or by integrating if it is an integral equation. Then once the unknown coefficients α_i are determined, the general solution for the problem can be obtained using (1.3).

A question that is now raised is: what is the optimum way to choose the known basis functions ϕ_i as the quality of the final solution depends on the choice of ϕ_i? It is well known in the numerical community that the best choices of the basis functions are the eigenfunctions of the operator that characterizes the system. Since in most examples one is dealing with a real life system, then the operators, in general, are linear time invariant (LTI) and have a bounded input and bounded output (BIBO) response resulting in a second-order differential equation, which is the case for Maxwell's equations. In the general case, the eigenfunctions of these operators are the complex exponentials, and in the transformed domain, they form a ratio of two rational polynomials. Therefore, our goal is to fit the given data for a LTI system either by a sum of complex exponentials or in the transformed domain approximate it by a ratio of polynomials. Next, it is illustrated how the eigenfunctions are used through a bias-variance tradeoff in reduced rank modelling [1, 2].

1.2 Reduced-Rank Modelling: Bias Versus Variance Tradeoff

An important problem in statistical processing of waveforms is that of feature selection, which refers to a transformation whereby a data space is transformed into a feature space that, in theory, has exactly the same dimensions as that of the original space [2]. However, in practical problems, it may be desirable and often necessary to design a transformation in such a way that the data vector can be represented by a reduced number of "effective" features and yet retain most of the intrinsic information content of the input data. In other words, the data vector undergoes a dimensionality reduction [1, 2]. Here, the same principle is applied by attempting to fit an infinite-dimensional space given by (1.3) to a finite-dimensional space of dimension p.

An important problem in this estimation of the proper rank is very important. First if the rank is underestimated then a unique solution is not possible. If on the other hand the estimated rank is too large the system equations involved in the parameter estimation problem can become very ill-conditioned, leading to inaccurate or completely erroneous results if straight forward LU-decomposition is used to solve for the parameters. Since, it is rarely a "crisp" number that evolves from the solution procedure determining the proper rank requires some analysis of the data and its effective noise level. An approach that uses eigenvalue analysis and singular value decomposition for estimating the effective rank of given data is outlined here.

As an example, consider an M-dimensional data vector $u(n)$ representing a particular realization of a wide-sense stationary process. (Stationarity refers to time invariance of some, or, all of the statistics of a random process, such as mean, autocorrelation, nth-order distribution. A random process $X(t)$ [or $X(n)$] is said to be strict sense stationary (SSS) if all its finite order distributions are time invariant, i.e., the joint cumulative density functions (cdfs), or probability density functions (pdfs), of $X(t_1), X(t_2), ..., X(t_k)$ and $X(t_1 + \tau), X(t_2 + \tau), ..., X(t_k + \tau)$ are the same for all k, all $t_1, t_2, ..., t_k$, and all time shifts τ. So for a SSS process, the first-order distribution is independent of t, and the second-order distribution — the distribution of any two samples $X(t_1)$ and $X(t_2)$ — depends only on $\tau = t_2 - t_1$ To see this, note that from the definition of stationarity, for any t, the joint distribution of $X(t_1)$ and $X(t_2)$ is the same as the joint distribution of $X\{t_+ + (t - t_1)\} = X(t)$ and $X\{t_2 + (t - t_1)\} = X\{t + (t_2 - t)\}$. An independent and identically distributed (IID) random processes are SSS. A random walk and

Poisson processes are not SSS. The Gauss-Markov process (as we defined it) is not SSS. However, if we set X_1 to the steady state distribution of X_n, it becomes SSS. A random process $X(t)$ is said to be wide-sense stationary (WSS) if its mean, i.e., $\varepsilon(X(t)) = \mu$, is independent of t, and its autocorrelation functions $RX(t_1, t_2)$ is a function only of the time difference $t_2 - t_1$ and are time invariant. Also $\varepsilon[X(t)^2] < \infty$ (technical condition) is necessary, where ε represents the expected value in a statistical sense. Since $RX(t_1, t_2) = RX(t_2, t_1)$, for any wide sense stationary process $X(t)$, $RX(t_1, t_2)$ is a function only of $|t_2 - t_1|$. Clearly a SSS implies a WSS. The converse is not necessarily true. The necessary and sufficient conditions for a function to be an autocorrelation function for a WSS process is that it be real, even, and nonnegative definite By nonnegative definite we mean that for any n, any $t_1, t_2, ..., t_n$ and any real vector $a = (a_1, ..., a_n)$, and $X(n)$; $a_i a_j R(t_i - t_j) \geq 0$. The power spectral density (psd) $SX(f)$ of a WSS random process $X(t)$ is the Fourier transform of $RX(\tau)$, i.e., $SX(f) = \Im\{RX(\tau)\} = \int_{-\infty}^{\infty} RX(\tau) \exp(-j2\pi\tau)\ d\tau$.

For a discrete time process X_n, the power spectral density is the discrete-time Fourier transform (DTFT) of the sequence $RX(n)$: $SX(f) = \sum_{n=-\infty}^{\infty} RX(n) \exp(-j2\pi nf)$. Therefore $RX(\tau)$ (or $RX(n)$) can be recovered from $SX(f)$ by taking the inverse Fourier transform or inverse DTFT.

In summary, WSS is a less restrictive stationary process and uses a somewhat weaker type of stationarity. It is based on requiring the mean to be a constant in time and the covariance sequence to depend only on the separation in time between the two samples. The final goal in model order reduction of a WSS is to transform the M-dimensional vector to a p-dimensional vector, where $p < M$. This transformation is carried out using the Karhunen-Loeve expansion [2]. The data vector is expanded in terms of q_i, the eigenvectors of the correlation matrix $[R]$, defined by

$$[R] = \varepsilon\left[u(n)u^H(n)\right] \tag{1.4}$$

and the superscript H represents the conjugate transpose of $u(n)$. Therefore, one obtains

$$u(n) = \sum_{i=1}^{M} c_i(n)q_i \tag{1.5}$$

so that

$$[R]\, q_i = \lambda_i q_i, \tag{1.6}$$

where $\{\lambda_i\}$ are the eigenvalues of the correlation matrix, $\{q_i\}$ represent the eigenvectors of the matrix R, and $\{c_i(n)\}$ are the coefficients defined by

$$c_i(n) = q_i^H u(n) \text{ for } i = 1, 2, ..., M. \tag{1.7}$$

To obtain a reduced rank approximation $\hat{u}(n)$ of $u(n)$, one needs to write

$$\hat{u}(n) = \sum_{i=1}^{p} c_i(n) q_i \tag{1.8}$$

where $p < M$. The reconstruction error Ξ is then defined as

$$\Xi(n) = u(n) - \hat{u}(n) = \sum_{i=p+1}^{M} \lambda_i. \tag{1.9}$$

Hence the approximation will be good if the remaining eigenvalues $\lambda_{p+1}, \ldots \lambda_M$ are all very small.

Now to illustrate the implications of a low rank model [2], consider that the data vector $u(n)$ is corrupted by the noise $v(n)$. Then the data $y(n)$ is represented by

$$y(n) = u(n) + v(n). \tag{1.10}$$

Since the data and the noise are uncorrelated,

$$\varepsilon\left[u(n)v^H(n)\right] = [0] \text{ and } \varepsilon\left[v(n)v^H(n)\right] = \sigma^2[\mathrm{I}], \tag{1.11}$$

where $[0]$ and $[\mathrm{I}]$ are the null and identity matrices, respectively, and the variance of the noise at each element is σ^2. The mean squared error now in a noisy environment is

$$\Xi_o = \varepsilon\left[\left\|y(n) - u(n)\right\|^2\right] = \varepsilon\left[\left\|v(n)v^H(n)\right\|\right] = \sum_{i=1}^{M}\left|v(n)\right|^2 = M\sigma^2. \tag{1.12}$$

Now to make a low-rank approximation in a noisy environment, define the approximated data vector by

$$r(n) = \hat{u}(n) + \hat{v}(n) = \sum_{i=1}^{p} c_i(n) q_i + v_i(n). \tag{1.13}$$

In this case, the reconstruction error for the reduced-rank model is given by

$$\Xi_{rr} = \varepsilon\left[\left\|r(n) - \hat{u}(n)\right\|^2\right] = \sum_{i=p+1}^{M} \lambda_i + p\,\sigma^2. \tag{1.14}$$

This equation implies that the mean squared error Ξ_{rr} in the low-rank approximation is smaller than the mean squared error Ξ_o to the original data vector without any approximation, if the first term in the summation is small. So low-rank modelling provides some advantages provided

$$\sum_{i=p+1}^{M} \lambda_i < (M - p)\sigma^2, \tag{1.15}$$

which illustrates the result of a *bias-variance* trade off. In particular, it illustrates that using a low-rank model for representing the data vector $u(n)$ incurs a bias through the p terms of the basis vector. Interestingly enough, introducing this bias is done knowingly in return for a reduction in variance, namely the part of the mean squared error due to the additive noise vector $v(n)$. *This illustrates that the motivation for using a simpler model that may not exactly match the underlying physics responsible for generating the data vector u(n), hence the bias, but the model is less susceptible to noise, hence a reduction in variance* [1, 2].

We now use this principle in the interpolation/extrapolation of various system responses. Since the data are from a linear time invariant (LTI) system that has a bounded input and a bounded output and satisfy a second-order partial differential equation, the associated time-domain eigenvectors are sums of complex exponentials and in the transformed frequency domain are ratios of two polynomials. As discussed, these eigenvectors form the optimal basis in representing the given data and hence can also be used for interpolation/extrapolation of a given data set. Consequently, we will use either of these two models to fit the data as seems appropriate. To this effect, we present the Matrix Pencil Method (MP) which approximates the data by a sum of complex exponentials and in the transformed domain by the Cauchy Method (CM) which fits the data by a ratio of two rational polynomials. In applying these two techniques it is necessary to be familiar two other topics which are the singular value decomposition and the total least squares which are discussed next.

1.3 An Introduction to Singular Value Decomposition (SVD) and the Theory of Total Least Squares (TLS)

1.3.1 Singular Value Decomposition

As has been described in [https://davetang.org/file/Singular_Value_Decomposition_Tutorial.pdf] *"Singular value decomposition (SVD) can be looked at from three mutually compatible points of view. On the one hand, we can see it as a method for transforming correlated variables into a set of uncorrelated ones that better expose the various relationships among the original data items. At the same time, SVD is a method for identifying and ordering the dimensions along which data points exhibit the most variation. This ties in to the third way of viewing SVD, which is that once we have identified where the most variation is, it's possible to find the best approximation of the original data points using fewer dimensions. Hence, SVD can be seen as a method for data reduction.* We shall illustrate this last point with an example later on.

First, we will introduce a critical component in solving a Total Least Squares problem called the Singular Value Decomposition (SVD). The singular value decomposition is one of the most important concepts in linear algebra [2].

To start, we first need to understand what eigenvectors and eigenvalues are as related to a dynamic system. If we multiply a vector x by a matrix $[A]$, we will get a new vector, Ax. The next equation shows the simple equation relating a matrix $[A]$ and an eigenvector x to an eigenvalue λ (just a number) and the original x.

$$Ax = \lambda x \tag{1.16}$$

$[A]$ is assumed to be a square matrix, x is the eigenvector, and λ is a value called the eigenvalue. Normally when any vector x is multiplied by any matrix $[A]$ a new vector results with components pointing in different directions than the original x. However, eigenvectors are special vectors that come out in the same direction even after they are multiplied by the matrix $[A]$. From (1.16) we can see that when one multiplies an eigenvector by $[A]$, the new vector Ax is just the eigenvalue λ times the original x. This eigenvalue determines whether the vector x is shrunk, stretched, reversed, or unchanged. Eigenvectors and eigenvalues play crucial roles in linear algebra ranging from simplifying matrix algebra such as taking the 500th power of $[A]$ to solving differential equations. To take the 500th power of $[A]$, one only needs to find the eigenvalues and eigenvectors of $[A]$ and take the 500th power of the eigenvalues. The eigenvectors will not change direction and the multiplication of the 500th power of the eigenvalues and the eigenvectors will result in $[A]^{500}$. As we will see in later sections, the eigenvalues can also provide important parameters of a system transfer function such as the poles.

One way to characterize and extract the eigenvalues of a matrix $[A]$ is to diagonalize it. Diagonalizing a matrix not only provides a quick way to extract eigenvalues but important parameters such as the rank and dimension of a matrix can be found easily once a matrix is diagonalized. To diagonalize matrix $[A]$, the eigenvalues of $[A]$ must first be placed in a diagonal matrix, $[\Lambda]$. This is completed by forming an eigenvector matrix $[S]$ with the eigenvectors of $[A]$ put into the columns of $[S]$ and multiplying as such

$$[S]^{-1}[A][S] = [\Lambda] = \begin{bmatrix} \lambda_1 & & \\ & \ddots & \\ & & \lambda_n \end{bmatrix} \tag{1.17}$$

(1.17) can now be rearranged and $[A]$ can also be written as

$$[A] = [S][\Lambda][S]^{-1} \tag{1.18}$$

We start to encounter problems when matrices are not only square but also rectangular. Previously we assumed that $[A]$ was an n by n square matrix. Now we will assume $[A]$ is any m by n rectangular matrix. We would still like to simplify the matrix or "diagonalize" it but using $[S]^{-1}[A][S]$ is no longer ideal for a few reasons; the eigenvectors of $[S]$ are not always orthogonal, there are sometimes not enough eigenvectors, and using $Ax = \lambda x$ requires $[A]$ to be a square matrix.

However, this problem can be solved with the singular value decomposition but of course at a cost. The SVD of [A] results in the following

$$[A]_{(m \times n)} = [U][\Sigma][V]^T = \begin{bmatrix} u_1 & \cdots & u_r \end{bmatrix}_{(m \times r)} \begin{bmatrix} \sigma_1 & & \\ & \ddots & \\ & & \sigma_r \end{bmatrix}_{(r \times r)} \begin{bmatrix} v_1 & \cdots & v_r \end{bmatrix}_{(n \times r)}$$

(1.19)

where m is the number of rows of [A], n is the number of columns of [A], and r is the rank of [A]. The SVD of [A], which can now be rectangular or square, will have two sets of singular vectors, u's and v's. The u's are the eigenvectors of [A][A]∗ and the v's are the eigenvectors of $[A]^T[A]$. [U] and [V] are also unitary matrices which means that $[U]*[U] = [I]$ and $[V]*[V] = [I]$. In other words, they are orthogonal where ∗ denoted complex conjugate transpose. The σ's are the singular values which also so happen to be the square roots of the eigenvalues of both $[A][A]^*$ and $[A]^*[A]$. We are not totally finished however because [U] and [V] are not square matrices. While (1.19) is the diagonalization of [A], the matrix equation is technically not valid since we cannot multiply these rectangular matrices of different sizes. To make them square we will need $n - r$ more v's and $m - r$ more u's. We can get these required u's and v's from the nullspace $N(A)$ and the left nullspace $N(A*)$. Once the new u's and v's are added, the matrices are square and [A] will still equal $[U][\Sigma][V]^T$. The true SVD of [A] will now be

$$[A]_{(m \times n)} = [U][\Sigma][V]^T =$$

$$\begin{bmatrix} u_1 & \cdots & u_r & \cdots & u_m \end{bmatrix}_{(m \times m)} \begin{bmatrix} \sigma_1 & & \\ & \ddots & \\ & & \sigma_r \\ & & & 0 \end{bmatrix}_{(m \times n)} \begin{bmatrix} v_1 & \cdots & v_r & \cdots & v_n \end{bmatrix}_{(n \times n)}$$

(1.20)

The new singular value matrix [Σ] is the same matrix as the old $r \times r$ matrix but with $m - r$ new rows of zero and $n - r$ columns of new zero added. The theory of total least squares (TLS) heavily utilizes the SVD as will be seen in the next section.

As an example consider the letter **X**. We now discretize the image on a 20×20 grid as seen in Figure 1.1. We place a 1 where there is no portion of the letter **X** and a zero where there is some portion of the letter. This results in the following matrix (1.21) representing the letter **X** digitally on a 20×20 grid of the matrix *A*. Now if we ask the question "What is the information content in the matrix **X** consisting of 1's and 0's? Clearly one would not require 20×20 = 400 pieces of information to represent **X** in (1.21) as there is a lot of redundancy in the system. This is where the SVD comes in and addresses this issue in quite a satisfactory way. If one performs a SVD of the matrix A given by (1.21), one will find that diagonal $\sum$ matrix in (1.20) has a lot of zero entries. In fact only seven of the singular values are not close to zero as presented in Table 1.1. This implies that

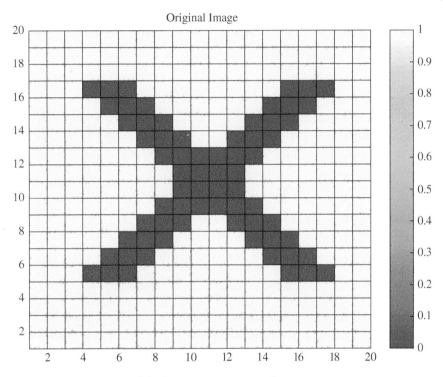

Figure 1.1 Discretization of the letter X on a 20×20 grid.

Table 1.1 List of all the singular values of the matrix **A**.

16.9798798959208
4.57981833410903
3.55675452579199
2.11591593148044
1.74248432449203
1.43326538670857
0.700598651862344
$8.40316310453450 \times 10^{-16}$
$2.67120960711993 \times 10^{-16}$
$1.98439889201509 \times 10^{-16}$
$1.14953653060279 \times 10^{-16}$
$4.74795444425376 \times 10^{-17}$
$1.98894713470634 \times 10^{-17}$
$1.64625309682086 \times 10^{-18}$
$5.03269559457945 \times 10^{-32}$
$1.17624286081108 \times 10^{-32}$
$4.98861204114981e \times 10^{-33}$
$4.16133005156854 \times 10^{-49}$
$1.35279056788548 \times 10^{-80}$
$1.24082758075064 \times 10^{-112}$

the information content of the picture in Figure 1.1 is small. The SVD can also be used to reconstruct the original data matrix with reduced information without sacrificing any accuracy.

$$A = \begin{bmatrix}
1 & 1 & 1 & 1 & 1 & 1 & 1 & 1 & 1 & 1 & 1 & 1 & 1 & 1 & 1 & 1 & 1 & 1 \\
1 & 1 & 1 & 1 & 1 & 1 & 1 & 1 & 1 & 1 & 1 & 1 & 1 & 1 & 1 & 1 & 1 & 1 \\
1 & 1 & 1 & 1 & 1 & 1 & 1 & 1 & 1 & 1 & 1 & 1 & 1 & 1 & 1 & 1 & 1 & 1 \\
1 & 1 & 1 & 1 & 1 & 1 & 1 & 1 & 1 & 1 & 1 & 1 & 1 & 1 & 1 & 1 & 1 & 1 \\
1 & 1 & 1 & 0 & 0 & 0 & 1 & 1 & 1 & 1 & 1 & 1 & 0 & 0 & 0 & 1 & 1 & 1 \\
1 & 1 & 1 & 1 & 0 & 0 & 0 & 1 & 1 & 1 & 1 & 1 & 1 & 0 & 0 & 0 & 1 & 1 & 1 \\
1 & 1 & 1 & 1 & 1 & 0 & 0 & 0 & 1 & 1 & 1 & 0 & 0 & 0 & 1 & 1 & 1 & 1 \\
1 & 1 & 1 & 1 & 1 & 1 & 0 & 0 & 0 & 1 & 1 & 0 & 0 & 0 & 1 & 1 & 1 & 1 \\
1 & 1 & 1 & 1 & 1 & 1 & 1 & 0 & 0 & 0 & 0 & 0 & 1 & 1 & 1 & 1 & 1 & 1 \\
1 & 1 & 1 & 1 & 1 & 1 & 1 & 0 & 0 & 0 & 0 & 1 & 1 & 1 & 1 & 1 & 1 & 1 \\
1 & 1 & 1 & 1 & 1 & 1 & 1 & 0 & 0 & 0 & 0 & 1 & 1 & 1 & 1 & 1 & 1 & 1 \\
1 & 1 & 1 & 1 & 1 & 1 & 1 & 0 & 0 & 0 & 0 & 0 & 1 & 1 & 1 & 1 & 1 & 1 \\
1 & 1 & 1 & 1 & 1 & 1 & 0 & 0 & 0 & 1 & 1 & 0 & 0 & 0 & 1 & 1 & 1 & 1 \\
1 & 1 & 1 & 1 & 1 & 0 & 0 & 0 & 1 & 1 & 1 & 0 & 0 & 0 & 1 & 1 & 1 & 1 \\
1 & 1 & 1 & 1 & 0 & 0 & 0 & 1 & 1 & 1 & 1 & 1 & 1 & 0 & 0 & 0 & 1 & 1 & 1 \\
1 & 1 & 1 & 0 & 0 & 0 & 1 & 1 & 1 & 1 & 1 & 1 & 1 & 0 & 0 & 0 & 1 & 1 & 1 \\
1 & 1 & 1 & 1 & 1 & 1 & 1 & 1 & 1 & 1 & 1 & 1 & 1 & 1 & 1 & 1 & 1 & 1 \\
1 & 1 & 1 & 1 & 1 & 1 & 1 & 1 & 1 & 1 & 1 & 1 & 1 & 1 & 1 & 1 & 1 & 1 \\
1 & 1 & 1 & 1 & 1 & 1 & 1 & 1 & 1 & 1 & 1 & 1 & 1 & 1 & 1 & 1 & 1 & 1 \\
1 & 1 & 1 & 1 & 1 & 1 & 1 & 1 & 1 & 1 & 1 & 1 & 1 & 1 & 1 & 1 & 1 & 1
\end{bmatrix} \tag{1.21}$$

To illustrate the point, we now perform a rank-1 approximation for the matrix and we will require

$$A \approx U_1 \sigma_1 V_1^* \tag{1.22}$$

where U_1 is a column matrix of size 20×1 and so is V_1. The rank one approximation of A is seen in Figure 1.2.

The advantage of the SVD is that an error in the reconstruction of the image can be predicted without actually knowing the actual solution. This is accomplished by looking at the second largest singular value. The result is not good and we did not expect it to be. So now if we perform a Rank-2 reconstruction for the image, it will be given by

$$A \approx \sigma_1 U_1 V_1^* + \sigma_2 U_2 V_2^* \tag{1.23}$$

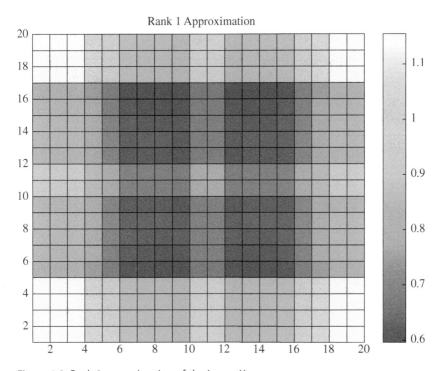

Figure 1.2 Rank-1 approximation of the image X.

where now two of the right and the left singular values are required. In this case the reank-2 approximation will be given by Figure 1.3. In this case we have captured the largest two singular values as seen from Table 1.1.

Again as we study how the picture evolves as we take more and more on the singular values and the vectors the accuracy in the reconstruction increases. For example, Figure 1.4 provides the rank-3 reconstruction, Figure 1.5 provides the rank-4 reconstruction, Figure 1.6 provides the rank-5 reconstruction, Figure 1.7 provides the rank-6 reconstruction. It is seen with each higher rank approximation the reconstructed picture resembles the actual one. The error in the approximation decreases as the magnitudes of the neglected singular values become small. This is the greatest advantage of the Singular Value Decomposition over say for example, the Tikhonov regularization as the SVD provides an estimate about the error in the reconstruction even though the actual solution is unknown.

Finally, it is seen that there are seven large singular values and after that they become mostly zeros. Therefore, we should achieve a perfect reconstruction with Rank-7 which is seen in Figure 1.8 and going to rank-8 which is presented in Figure 1.9, will not make any difference in the quality of the image. This is

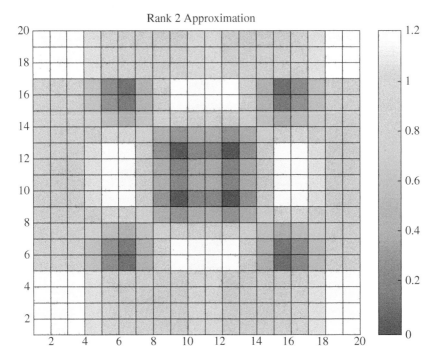

Figure 1.3 Rank-2 approximation of the image X.

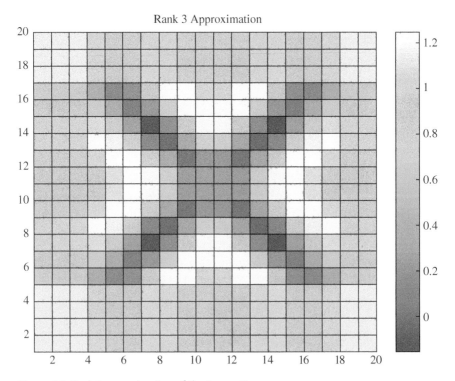

Figure 1.4 Rank-3 approximation of the image X.

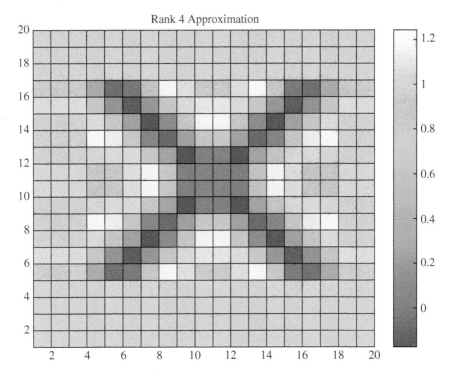

Figure 1.5 Rank-4 approximation of the image X.

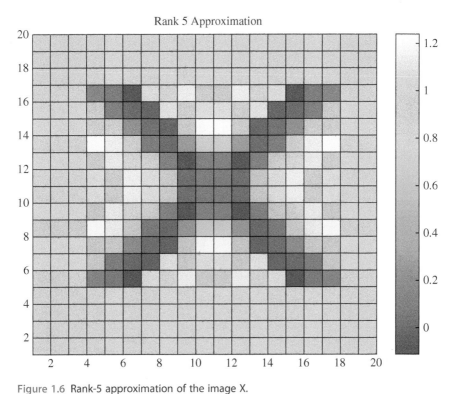

Figure 1.6 Rank-5 approximation of the image X.

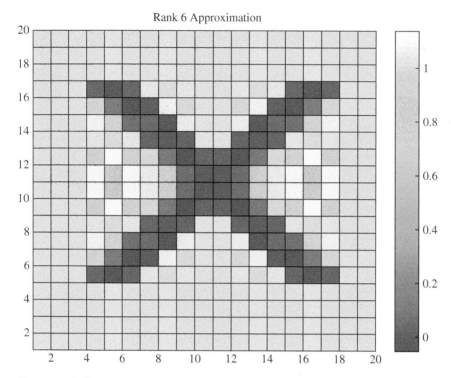

Figure 1.7 Rank-6 approximation of the image X.

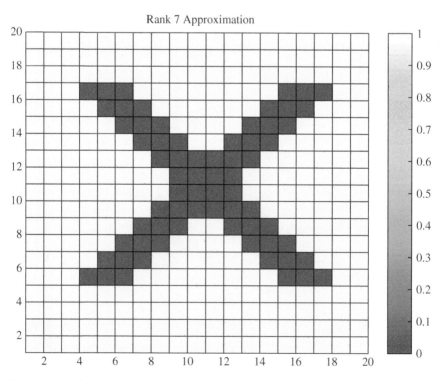

Figure 1.8 Rank-7 approximation of the image X.

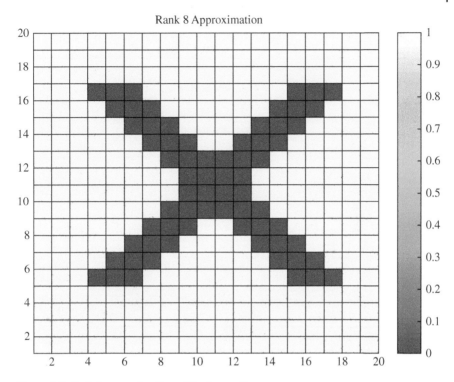

Figure 1.9 Rank-8 approximation of the image X.

now illustrated next through the mean squared error between the true solution and the approximate ones.

In Figure 1.10, the mean squared error between the actual picture and the approximate ones is presented. It is seen, as expected the error is given by this data

$$(0.32, 0.22, 0.16, 0.09, 0.06, 0.02, 0.00, 0.00, \ldots\ldots\ldots). \qquad (1.24)$$

This is a very desirable property for the SVD. Next, the principles of total least squares is presented.

1.3.2 The Theory of Total Least Squares

The method of total least squares (TLS) is a linear parameter estimation technique and is used in wide variety of disciplines such as signal processing, general engineering, statistics, physics, and the like. We start out with a set of m measured data points $\{(x_1, y_1), \ldots, (x_m, y_m)\}$, and a set of n linear coefficients $(a_1, \ldots, a_n)$ that describe a model, $\hat{y}(x; a)$ where $m > n$ [3, 4]. The objective of total least

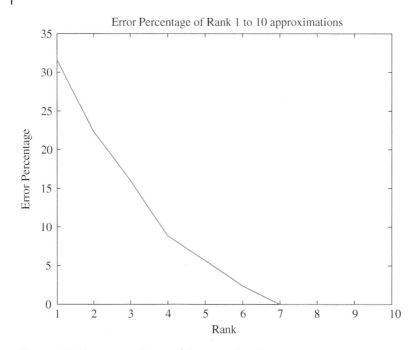

Figure 1.10 Mean squared error of the approximation.

squares is to find the linear coefficients that best approximate the model in the scenario that there is missing data or errors in the measurements. We can describe the approximation by a simple linear expression

$$y \approx Xa \tag{1.25}$$

Since $m > n$, there are more equations than unknowns and therefore (1.25) has an overdetermined set of equations. Typically, an overdetermined system of equation is best solved by the ordinary least squares where the unknown is given by

$$a = \left(X^*X\right)^{-1}X^*y \tag{1.26}$$

where X^* represents the complex conjugate transpose of the matrix X. The least squares can take into account if there are some uncertainties like noise in y as it is a least squares fit to it. However, if there is uncertainty in the elements of the matrix X then the ordinary least squares cannot address it. This is where the total Least squares come in. In the total least squares the matrix equation (1.25) is cast into a different form where uncertainty in the elements of both the matrix X and y can be taken into account.

$$\left[X \vdots y\right] \begin{bmatrix} a \\ -1 \end{bmatrix} = \left[X \vdots y\right][b] = [0] \tag{1.27}$$

In this form one is solving for the solution to the composite matrix by searching for the eigenvector/singular vector corresponding to the zero eigen/singular value. If the matrix X is rectangular then the eigenvalue concept does not apply and one needs to deal with the singular vectors and the singular values.

The best approximation according to total least squares is that minimizes the norm of the difference between the approximated data and the model $\hat{y}(x;a)$ as well as the independent variables X. Considering the errors of the measured data vector, y, and the independent variables, X, (1.25) can be re-written as

$$y + \tilde{y} = \left[X + \tilde{X}\right]a \tag{1.28}$$

where $\tilde{y}$ and $\tilde{X}$ are the errors in both the dependent variable measurements and independent variable measurements, respectively. We then want to approximate in a way that minimizes these errors in the dependent and independent variables. This can be expressed by,

$$min \left\| \left[\tilde{X}\,\tilde{y}\right] \right\|_F^2 \tag{1.29}$$

where $\left[\tilde{X}\,\tilde{y}\right]$ is an augmented matrix with the columns of error matrix $\tilde{X}$ concatenated with the error vector $\tilde{y}$. The operator $\|\bullet\|_F$ represents the Frobenius norm of the augmented matrix. The Frobenius norm is defined as the square root of the sum of the absolute squares of all of the elements in a matrix. This can be expressed in equation form as the following, where A is any matrix,

$$\|A\|_F^2 = \sum_{i=1}^{m}\sum_{j=1}^{n} A_{ij}^2 = trace\left(A^T A\right) = \sum_{i=1}^{n} \sigma_i^2 \tag{1.30}$$

and where σ_i is the i-th singular value of matrix A.

We will now bring the right-hand side of (1.28) over to the left side of the equation and equate it to zero as such

$$\left[X + \tilde{X}; y + \tilde{y}\right] \begin{bmatrix} a \\ -1 \end{bmatrix} = 0 \tag{1.31}$$

If the concatenated matrix $[X\,y]$ has a rank of $n + 1$, the $n + 1$ columns of the matrix are linearly independent and the $n + 1$, m-dimensional columns of the matrix span the same n-dimensional space as X. In order to have a unique solution for the coefficients, a, the matrix $[X + \tilde{X}; y + \tilde{y}]$ must have n linearly independent columns. However, this matrix has $n + 1$ columns in total and therefore is rank is deficient by 1. We then must find the smallest matrix $[\tilde{X}\,\tilde{y}]$ that changes matrix $[X\,y]$ with a rank of $n + 1$, to a matrix $\{[X\,y] + [\widetilde{X}\,\widetilde{y}]\}$ with a rank n. According to the Eckart-Young-Mirsky theorem we can achieve this by defining $\{[X\,y] + [\widetilde{X}\,\widetilde{y}]\}$ as the best rank-n approximation to $[X\,y]$ and by eliminating the last singular value of $[X\,y]$ which contains the least amount of system information and provides a unique solution. The Eckart–Young–Mirsky theorem (https://en.wikipedia.org/wiki/Low-rank_approximation) states a low-rank approximation is a minimization problem, in which the cost function measures the fit between a given matrix (the data) and an approximating matrix (the optimization variable), subject to a constraint that the approximating matrix has reduced rank. To illustrate how this is accomplished, we take the SVD of $[X\,y]$ as follows

$$[X \quad y] = \begin{bmatrix} U_x & u_y \end{bmatrix} \begin{bmatrix} \Sigma_x & \\ & \sigma_y \end{bmatrix} \begin{bmatrix} V_{xx} & V_{xy} \\ V_{yx} & V_{yy} \end{bmatrix}^T \tag{1.32}$$

where U_x has n columns, u_y is a column vector, Σ_x contains the n largest singular values diagonally, σ_y is the smallest singular value, V_{xx} is a $n \times n$ matrix, and v_{yy} is scalar. Let us multiple both sides by matrix V.

$$[X \quad y] \begin{bmatrix} V_{xx} & V_{xy} \\ V_{yx} & V_{yy} \end{bmatrix} = \begin{bmatrix} U_x & u_y \end{bmatrix} \begin{bmatrix} \Sigma_x & \\ & \sigma_y \end{bmatrix} \tag{1.33}$$

Next, we will equate just the last columns of the matrix multiplication occurring in (1.32).

$$[X \quad y] \begin{bmatrix} V_{xy} \\ V_{yy} \end{bmatrix} = u_y \sigma_y \tag{1.34}$$

From the Eckart-Young theorem, we know that $\{[X\,y] + [\widetilde{X}\,\widetilde{y}]\}$ is the closest rank-n approximation to $[X\,y]$. Matrix $\{[X\,y] + [\widetilde{X}\,\widetilde{y}]\}$ has the same singular vectors contained in Σ_x above with σ_y equal to zero. We can then write the SVD of $\{[X\,y] + [\widetilde{X}\,\widetilde{y}]\}$ as such

$$\begin{bmatrix} X + \widetilde{X}; & y + \widetilde{y} \end{bmatrix} = \begin{bmatrix} U_x & u_y \end{bmatrix} \begin{bmatrix} \Sigma_x & \\ & 0 \end{bmatrix} \begin{bmatrix} V_{xx} & V_{xy} \\ V_{yx} & V_{yy} \end{bmatrix}^T \tag{1.35}$$

To obtain $[\widetilde{X}\,\widetilde{y}]$ we must solve the following

$$\begin{bmatrix} \widetilde{X} & \widetilde{y} \end{bmatrix} = [X \quad y] + \begin{bmatrix} \widetilde{X} & \widetilde{y} \end{bmatrix} - [X \quad y] \tag{1.36}$$

(1.36) can be solved by first using (1.32) and (1.35) which results in

$$\begin{bmatrix} \widetilde{X} & \widetilde{y} \end{bmatrix} = -\begin{bmatrix} U_x & u_y \end{bmatrix} \begin{bmatrix} 0 & \\ & \sigma_y \end{bmatrix} \begin{bmatrix} V_{xx} & V_{xy} \\ V_{yx} & V_{yy} \end{bmatrix}^T$$

$$= -\begin{bmatrix} 0 & u_y\sigma_y \end{bmatrix} \begin{bmatrix} V_{xx} & V_{xy} \\ V_{yx} & V_{yy} \end{bmatrix}^T = -u_y\sigma_y \begin{bmatrix} V_{xy} \\ V_{yy} \end{bmatrix}^T \tag{1.37}$$

Then, from (1.34) we can rewrite (1.37) as

$$\begin{bmatrix} \widetilde{X} & \widetilde{y} \end{bmatrix} = -\begin{bmatrix} X & y \end{bmatrix} \begin{bmatrix} V_{xy} \\ V_{yy} \end{bmatrix} \begin{bmatrix} V_{xy} \\ V_{yy} \end{bmatrix}^T \tag{1.38}$$

Finally, $\{[X\ y] + [\widetilde{X}\ \widetilde{y}]\}$ can be defined as

$$\begin{bmatrix} X + \widetilde{X}; & y + \widetilde{y} \end{bmatrix} = \begin{bmatrix} X & y \end{bmatrix} - \begin{bmatrix} X & y \end{bmatrix} \begin{bmatrix} V_{xy} \\ V_{yy} \end{bmatrix} \begin{bmatrix} V_{xy} \\ V_{yy} \end{bmatrix}^T \tag{1.39}$$

After multiplying each term in (1.39) by $\begin{bmatrix} V_{xy} \\ V_{yy} \end{bmatrix}$ we get the following

$$\begin{bmatrix} X + \widetilde{X}; & y + \widetilde{y} \end{bmatrix} \begin{bmatrix} V_{xy} \\ V_{yy} \end{bmatrix} = \begin{bmatrix} X & y \end{bmatrix} \begin{bmatrix} V_{xy} \\ V_{yy} \end{bmatrix} - \begin{bmatrix} X & y \end{bmatrix} \begin{bmatrix} V_{xy} \\ V_{yy} \end{bmatrix} \tag{1.40}$$

The right-hand side cancels and we are left with

$$\begin{bmatrix} X + \widetilde{X}; & y + \widetilde{y} \end{bmatrix} \begin{bmatrix} V_{xy} \\ V_{yy} \end{bmatrix} = 0 \tag{1.41}$$

From (1.31) and (1.41) we can solve for the model coefficient a as

$$a = -V_{xy}V_{yy}^{-1} \tag{1.42}$$

The vector V_{xy} is the first n elements of the $n+1$-th columns of the right singular matrix V, of $[X\ y]$ and v_{yy} is the $n + 1$-th element of the $n + 1$ columns of V. The best approximation of the model is then given by

$$\hat{y} = \begin{bmatrix} X + \widetilde{X} \end{bmatrix} a \tag{1.43}$$

This completes the total least squares solution.

1.4 Conclusion

This first chapter provides the mathematical fundamentals that will be utilized later. The principles presented are the basis of low-rank modelling, singular value decomposition and the method of total least squares.

References

1 L. I. Scharf and D. Tufts, "Rank Reduction for Modeling Stationary Signals," *IEEE Transactions on Acoustics, Speech, and Signal Processing*, Vol. ASSP-35, pp. 350–355, 1987.
2 S. Haykin, *Adaptive Filter Theory*, Prentice Hall, Upper Saddle River, NJ, Third Edition, 1996.
3 G. Strang, *Introduction to Linear Algebra*, Cambridge Press, Wellesley, MA, Fifth Edition, 2016.
4 G. Golub and C. Van Loan, *Matrix Computations*, Johns Hopkins Studies in the Mathematical Sciences, The Johns Hopkins University Press, Baltimore, MD, 2013.

2

Matrix Pencil Method (MPM)

Summary

In this chapter, we present the matrix pencil method (MPM) which is a methodology to approximate a given data set by a sum of complex exponentials. The objective is to interpolate and extrapolate data and also to extract certain parameters so as to compress the data set. First the methodology is presented followed by some application in electromagnetic system characterization. The applications involve using this methodology to de-embed device characteristics and obtain accurate and high resolution characterization, enhance network analyzer measurement data when not enough bandwidth is available for analysis of that data, minimization of unwanted reflections in antenna characterization when performing measurements in a non-anechoic environment, to extract a single set of exponents representing the resonant frequencies of an object when data from multiple look angles is given, for direction of arrival estimations of signals along with their frequency of operation, to speed up the calculation of the tails encountered in the Sommerfeld integrals which are important expressions for characterizing propagation over an imperfect ground plane and for multiple target characterization operating in free space using their scattered data to compute their characteristic external resonance. References to other applications including multipath characterization of a propagating wave, characterization of the quality of power systems, in waveform analysis and imaging and speeding up analysis of FDTD computations followed by some conclusions. A computer program implementing the MPM is given in the appendix followed by a list of references.

2.1 Introduction

An electromagnetic (EM) transient waveform can be described by

$$y(t) = x(t) + n(t) \approx \sum_{i=1}^{M} R_i \, e^{s_i t} + n(t), \quad 0 \le t \le T, \tag{2.1}$$

Modern Characterization of Electromagnetic Systems and Its Associated Metrology, First Edition.
Tapan K. Sarkar, Magdalena Salazar-Palma, Ming Da Zhu, and Heng Chen.
© 2021 John Wiley & Sons, Inc. Published 2021 by John Wiley & Sons, Inc.

where $y(t)$ is the measurement data involving the waveform of interest $x(t)$ with an additive noise component $n(t)$. The problem is to estimate the parameters (M, R_i, s_i) and use them to interpolate or extrapolate the available measurement data. Such a model is valid because the system generating the data set can be treated as a linear time-invariant (LTI) system [1]. It is well known that for an LTI system, the eigenfunctions of the operator are decaying exponentials expressed as $e^{s_i t}$ where s_i are the complex valued poles or characteristic exponents of the system. After sampling the data, the time variable, t is replaced by kT_s, where T_s is the sampling period. The series of (2.1) can be rewritten as [2–6]

$$y(kT_s) = x(kT_s) + n(kT_s) \approx \sum_{i=1}^{M} R_i z_i^k + n(kT_s) \text{ for } k = 0, 1, \cdots, N - 1$$

(2.2)

$$\text{with} \quad z_i = e^{s_i T_s} = e^{(\alpha_i + j\omega_i)T_s}, \quad \text{for } i = 1, 2, \cdots, M, \quad \text{where} \quad (2.3)$$

$y(t)$ = observed time response,
$x(t)$ = waveform of interest,
$n(t)$ = noise in the system,
R_i = residue or complex amplitudes of the ith pole,
$s_i = \alpha_i + j\omega_i$ (ith pole of the system),
α_i = negative damping factor of the ith pole,
ω_i = angular frequency of the ith pole,
N = number of data samples,
M = number of poles approximating the sequence.
$j = \sqrt{-1}$.

The transient response from a structure can be characterized by the best estimates of M, R_i, and z_i using the MPM [2–6], especially for the case of noise contaminated data resulting from numerical errors and random noise. For noiseless data, one can define the $(N - L) \times (L + 1)$ data matrix $[Y]$ as

$$[Y] = \begin{bmatrix} y_0 & y_1 & \cdots & y_L \\ y_1 & y_2 & \cdots & y_{L+1} \\ \vdots & \vdots & \ddots & \vdots \\ y_{N-L-1} & y_{N-L} & \cdots & y_{N-1} \end{bmatrix}_{(N-L) \times (L+1)}$$

(2.4)

One can also define $[Y]$ as

$$[Y] = [c_1, Y_1] = [Y_2, c_{L+1}],$$

(2.5)

where c_i represents the ith column of matrix $[Y]$. These matrices $[Y_1]$ and $[Y_2]$ can be written as [2–6]

$$[Y_1] = [Z_1][R][Z_0][Z_2], \tag{2.6}$$

$$[Y_2] = [Z_1][R][Z_2], \tag{2.7}$$

where

$$[Z_1] = \begin{bmatrix} 1 & 1 & \cdots & 1 \\ z_1 & z_2 & \cdots & z_M \\ \vdots & \vdots & \ddots & \vdots \\ z_1^{(N-L-1)} & z_2^{(N-L-1)} & \cdots & z_M^{(N-L-1)} \end{bmatrix}_{(N-L) \times M}, \tag{2.8}$$

$$[Z_2] = \begin{bmatrix} 1 & z_1 & \cdots & z_1^{L-1} \\ 1 & z_2 & \cdots & z_2^{L-1} \\ \vdots & \vdots & \ddots & \vdots \\ 1 & z_M & \cdots & z_M^{L-1} \end{bmatrix}_{M \times L}, \tag{2.9}$$

$$[Z_0] = \begin{bmatrix} z_1 & 0 & \cdots & 0 \\ 0 & z_2 & \cdots & 0 \\ \vdots & \vdots & \ddots & \vdots \\ 0 & 0 & \cdots & z_M \end{bmatrix}_{M \times M}, \tag{2.10}$$

$$[R] = \begin{bmatrix} R_1 & 0 & \cdots & 0 \\ 0 & R_2 & \cdots & 0 \\ \vdots & \vdots & \ddots & \vdots \\ 0 & 0 & \cdots & R_M \end{bmatrix}_{M \times M}. \tag{2.11}$$

Now, consider the following matrix pencil

$$[Y_1] - \lambda[Y_2] = [Z_1][R]\{[Z_0] - \lambda[I]\}[Z_2], \tag{2.12}$$

provided $M \leq L \leq N - M$. The matrix $[Y_1] - \lambda[Y_2]$ has rank M. However, if $\lambda = z_i$, $i = 1, 2, \ldots, M$, the ith row of $[Z_0] - \lambda[I]$ is zero, and the rank of $[Z_0] - \lambda[I]$ is $M - 1$. Here $[I]$ is the identity matrix. Therefore, the matrix pencil $[Y_1] - \lambda[Y_2]$ will also be reduced in rank to $M - 1$. It implies that z_i are the generalized eigenvalues of the matrix pair $\{[Y_1], [Y_2]\}$. Therefore,

$$[Y_1][\psi_i] = z_i[Y_2][\psi_i], \tag{2.13}$$

where ψ_i is the generalized eigenvector corresponding to z_i. In an equivalent form,

$$\left\{[Y_2]^\dagger[Y_1] - z_i[I]\right\}[\psi_i] = 0, \tag{2.14}$$

where $[Y_2]^\dagger$ is the Moore-Penrose pseudo-inverse of $[Y_2]$, i.e., $[Y_2]^\dagger = \{[Y]^H[Y]\}^{-1}$ $[Y]^H$. Here, the superscript H is the complex conjugate transpose of a matrix. From (2.14), one can obtain z_i from the eigenvalues of $[Y_2]^\dagger[Y_1]$. Hence, for the MPM, the poles are computed directly as a one-step process. Once the poles or the exponents z_i are obtained the residues at the poles can be obtained from

$$[X] = [Z][A]$$

where

$$[X] = \begin{bmatrix} y(0) \\ y(1) \\ \vdots \\ y(N) \end{bmatrix}_{(N+1)\times 1} = \begin{bmatrix} 1 & 1 & \cdots & 1 \\ z_1 & z_2 & \cdots & z_M \\ z_1^2 & z_2^2 & \cdots & z_M^2 \\ \vdots & \vdots & & \vdots \\ z_1^N & z_2^N & \cdots & z_M^N \end{bmatrix}_{(N+1)\times M} \times \begin{bmatrix} R(1) \\ R(2) \\ \vdots \\ R(M) \end{bmatrix}_{M\times 1} = [Z][A]$$

$$\tag{2.15}$$

The various residues at the poles can now be computed from the least squares solution of (2.15) which yields

$$[A] = [Z]^\dagger[X] = \left\{[Z]^H[Z]\right\}^{-1}[Z]^H[X]. \tag{2.16}$$

2.2 Development of the Matrix Pencil Method for Noise Contaminated Data

However, when the data is contaminated by noise a slightly different procedure needs to be carried out. In this case, for efficient noise filtering, the pencil parameter L is chosen between $N/3$ to $N/2$. Then we introduce the singular value decomposition which has been described in Chapter 1. Define the Singular Value Decomposition (SVD) [6–8] of a matrix $[Y]$ to be

$$[Y] = [U][\Sigma][V]^H. \tag{2.17}$$

The matrices $[U]$ and $[V]$ are $(N-L)\times(N-L)$ and $(L+1)\times(L+1)$ unitary matrices, respectively. A unitary matrix is one whose inverse is the conjugate transpose of itself, i.e., $[U][U]^H = [I] = [U]^H[U]$, where $[I]$ is the identity matrix.

The matrix $[\Sigma]$ is a $(N - L) \times (L + 1)$ diagonal matrix with the singular values of matrix $[Y]$ in descending order. If the given data $y(kT_s)$ were noise free, matrix $[Y]$ would have exactly M nonzero singular values.

However, due to the presence of noise, the zero singular values are perturbed, yielding several small nonzero singular values. This error due to the noise can be suppressed by eliminating these spurious singular values from matrix $[\Sigma]$. To that end, define $[\Sigma']$ as a $M \times M$ diagonal matrix with the M largest singular values of $[Y]$ on its main diagonal. Furthermore, define $[U']$ and $[V']$ as submatrices of $[U]$ and $[V]$ corresponding to these singular values:

$$[U'] = [U(:, 1:M)]; \tag{2.18}$$

$$[V'] = [V(:, 1:M)]; \tag{2.19}$$

$$[\Sigma'] = [\Sigma(1:M, 1:M)]; \tag{2.20}$$

$$[Y'] = [U'][\Sigma'][V']^H \tag{2.21}$$

Therefore, using matrix $[Y']$ instead of matrix $[Y]$ in (2.5) results in filtering the noise in both $[Y_1]$ and $[Y_2]$. From (2.5) and (2.18)-(2.21),

$$[Y_1] = [U'][\Sigma']\left[V_1'\right]^H, \tag{2.22}$$

$$[Y_2] = [U'][\Sigma']\left[V_2'\right]^H, \tag{2.23}$$

where $\left[V_1'\right]$ and $\left[V_2'\right]$ are equal to $[V']$ without its first and the last rows, respectively. Using (2.22)-(2.23), the poles associated with the data (eigenvalues of $[Y_2]^\dagger[Y_1]$) are given by the nonzero eigenvalues of

$$\left\{\left[V_2'\right]\right\}^\dagger\left[V_1'\right]. \tag{2.24}$$

The number of modes/poles of the data is M and are chosen by observing the ratio of the various singular values to the largest one as defined by the SVD and is given by the integer value of the ratio

$$\frac{\sigma_R}{\sigma_{\max}} \approx 10^{-w}, \tag{2.25}$$

where w is the number of accurate significant decimal digits associated with the given data. Based on w, one can determine the proper values of M with the desired precision [2–6]. If the smallest singular value is larger than the round-off error of the given data, then the only possible solution is to acquire more data. At this stage the tradeoff between bias and variance as illustrated in Chapter 1 is made to select the system order M. Using this better choice of M, one can evaluate the poles $\{z_i\}$ and the amplitudes $\{R_i\}$. Hence, for the MPM, the residues and poles are obtained from the noise contaminated data using the SVD and the Total Least Squares (TLS) method [6–8]. In summary, there are five steps to be performed to approximate the given data by a sum of complex exponentials computed using the MPM.

The MPM Approach is computationally very efficient and quite robust to noise in the data [2–6]. It has been shown that, of all the techniques available in the published literature to achieve that goal, the variance in the estimate of the poles given by the MPM comes close to the Cramer-Rao bound [4], i.e., the procedure is very robust to noise (numerical or otherwise) in the data.

2.2.1 Procedure for Interpolating or Extrapolating the System Response Using the Matrix Pencil Method

Step 1. Choose the number of time sample points (N).

Step 2. Construct the $(N - L) \times (L + 1)$ matrix $[Y]$ from the selected time sampled data. Here the pencil parameter L is chosen to be $N/2 - 1$ for efficient noise filtering.

Step 3. Select the value for M (number of singular values) satisfying (2.25) before applying the TLS method.

Step 4. Generate matrices $\left[V_2'\right]$ and $\left[V_1'\right]$ from $[V]$ as submatrices of $[V]$ corresponding to M. Then, compute the poles and residues of the waveform from (2.22) - (2.24) followed by (2.15) and (2.16).

Step 5. Generate the approximation to the data using the estimated poles and residues of the signal from (2.2).

2.2.2 Illustrations Using Numerical Data

2.2.2.1 Example 1

Consider approximation of the waveform for different time samples:

$$y(t) = e^{-0.02\pi t} \sin(0.2\pi t) + e^{-0.035\pi t} \sin(0.35\pi t), \quad \text{for } t = 0, 0.5, 1.0, ..., 100.$$

The number of data samples is 201. In this case, $\alpha_1 = -0.02\pi$, $\omega_1 = 0.2\pi$, $\alpha_2 = -0.035\pi$, $\omega_2 = 0.35\pi$. In order to estimate the poles and residues of the waveform and then to interpolate or extrapolate from a small number of the data samples is illustrated by the five steps just outlined.

Step 1. Choose 14 samples from the 201 data samples ($N=14$).

$$t = 0, 2.500, 5.000, 7.500, 10.000, 12.500, 15.000, 17.500, 20.000, 22.500,$$
$$25.000, 27.500, 30.000, 32.500$$

$$y(t) = 0, 1.1453, -0.4081, -0.2192, -0.3330, 0.6897, -0.1359, -0.2772, -0.0000,$$
$$0.2110, 0.0453, -0.2226, 0.0369, 0.1038$$

Step 2. Construct the $(N - L) \times (L + 1)$ matrix $[Y]$ from the selected time sampling data. For efficient noise filtering, set the pencil parameter to be 6. Consequently, the size of $[Y]$ is $(N - L) \times (L + 1) = 8 \times 7$.

$$
[Y] = \begin{bmatrix}
0 & 1.1453 & -0.4081 & -0.2192 & -0.3330 & 0.6897 & -0.1359 \\
1.1453 & -0.4081 & -0.2192 & -0.3330 & 0.6897 & -0.1359 & -0.2772 \\
-0.4081 & -0.2192 & -0.3330 & 0.6897 & -0.1359 & -0.2772 & -0.0000 \\
-0.2192 & -0.3330 & 0.6897 & -0.1359 & -0.2772 & -0.0000 & 0.2110 \\
0.3330 & 0.6897 & -0.1359 & -0.2772 & -0.0000 & 0.2110 & 0.0453 \\
0.6897 & -0.1359 & -0.2772 & -0.0000 & 0.2110 & 0.0453 & -0.2226 \\
-0.1359 & -0.2772 & -0.0000 & 0.2110 & 0.0453 & -0.2226 & 0.0369 \\
-0.2772 & -0.0000 & 0.2110 & 0.0453 & -0.2226 & 0.0369 & 0.1038
\end{bmatrix}
$$

Step 3. The SVD of matrix $[Y]$ provides the optimal number of singular values (M). Figure 2.1 shows the normalized singular values (dB scale) from the singular values of $[Y]$. From (2.25), if the number of accurate significant decimal digits w is 1, M is 4 as shown in Figure 2.1.

Step 4. For $M = 4$, the matrices $\left[V_1' \right]$ and $\left[V_2' \right]$ from $[V']$ as submatrices of $[V]$ are

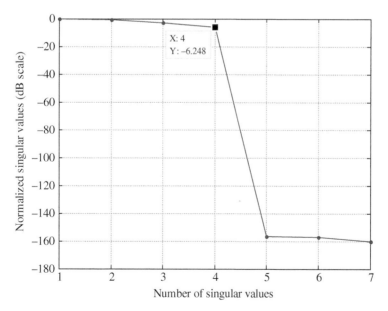

Figure 2.1 Normalized singular values from the SVD of matrix $[Y]$.

$$
[V'] = \begin{bmatrix} -0.6228 & 0.5077 & -0.1059 & 0.4279 \\ 0.6023 & 0.5516 & 0.0393 & -0.1783 \\ 0.0114 & -0.3679 & -0.7233 & 0.1140 \\ 0.0232 & -0.3379 & 0.6312 & 0.3722 \\ -0.4024 & 0.1759 & 0.0641 & -0.6691 \\ 0.2690 & 0.3397 & -0.1902 & 0.4108 \\ 0.1200 & -0.2052 & -0.1590 & -0.1314 \end{bmatrix}_{(7 \times 4)} \begin{matrix} [V_2'] \\ \\ \\ \\ \\ \\ [V_1'] \end{matrix}
$$

Then, one can compute z_i from the nonzero eigenvalues of $\{[V_2']\}^{\dagger}[V_1']$. From (2.12), the poles of the data can be obtained as

$$
s_i = \alpha_i + j\omega_i = \log_e(z_i)/T_s, \quad \text{for} \quad i = 1, 2, \cdots, M,
$$
$$
s_1 = -0.0628 + j0.6283; \quad s_2 = -0.0628 - j0.6283;
$$
$$
s_3 = -0.1099 + j1.0996; \quad s_4 = -0.1099 - j1.0996
$$

The poles are the complex conjugate pairs for s_1, s_2 and s_3, s_4. From the original data sequence, the exponents are seen to be $\alpha_1 = -0.0628$, $\omega_1 = 0.6283$, $\alpha_2 = -0.1099$, $\omega_2 = 1.0996$. Therefore, the computed exponents using the MPM are exactly the same as that of the original data sequence. One can calculate the residues corresponding to each pole as

$$
R_1 = -2.4712 \times 10^{-16} - j5.0000 \times 10^{-1} \simeq -j0.5 \; ;
$$
$$
R_2 = 1.8066 \times 10^{-17} + j5.0000 \times 10^{-1} \simeq j0.5,
$$
$$
R_3 = 5.0670 \times 10^{-16} - j5.0000 \times 10^{-1} \simeq -j0.5;
$$
$$
R_4 = -4.3417 \times 10^{-16} + j5.0000 \times 10^{-1} \simeq j0.5
$$

Now, one can check whether the estimated parameters from the poles and residues are the same for the data of this example. The data approximation problem can be reformulated as

$$
y(t) = e^{-0.02\pi t}\left(0.5je^{-0.2\pi t} - 0.5je^{0.2\pi t}\right) + e^{-0.035\pi t}\left(0.5je^{-0.35\pi t} - 0.5je^{0.35\pi t}\right)
$$
$$
= e^{-0.0628t}\left(0.5je^{-0.6283t} - 0.5je^{0.6283t}\right) + e^{-0.1100t}\left(0.5je^{-1.0996t} - 0.5je^{1.0996t}\right).
$$

On extracting the $\{R_i\}$ and $\{s_i\}$ from the preceding equation, the values seem to be virtually identical to the assumed values. Consequently, it is proved that one can extract accurately poles and residues of the data using the MPM.

Step 5. Based on the computed poles and residues using the MPM, one can estimate the poles for the data as shown in Figure 2.2. To evaluate the accuracy of

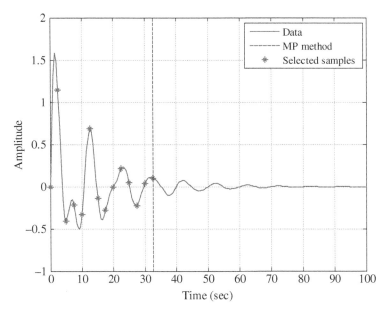

Figure 2.2 Comparison between the original data and the estimated data by applying the MPM to 14 selected samples.

the performance of the interpolation and extrapolation, one needs to compute the *estimated error* with the mean squared error (MSE) in the time domain. In this case, the estimated MSE is 1.9165×10^{-15}.

2.2.2.2 Example 2

Now add noise to the data of the previous example. White Gaussian noise is added to the data of the previous example 1 as shown in Figure 2.3. Here the signal-to-noise ratio for the data (SNR) is 15 dB. All conditions are the same as in *example 2.2.2.1* except for the added noise of 15 dB SNR. The same procedure applied to estimate the parameters for the poles and then to interpolate or extrapolate the given data as described before.

Step 1. Choose 100 samples of the data from t = 0 to 49.5 sec ($\Delta t = 0.5$ sec) to increase the assumed pencil parameter (*L*).

Step 2. Construct the $(N - L) \times (L + 1)$ matrix $[Y]$ from the selected time sampled data. For efficient noise filtering, the pencil parameter is chosen to be 49. The size of the matrix $[Y]$ is $(N - L) \times (L + 1) = 51 \times 50$.

Step 3. The SVD of matrix $[Y]$ will provide the optimal number of singular values (*M*). Figure 2.4(a) shows the normalized singular values (dB scale) from the SVD of matrix $[Y]$. From (2.25), if the number of accurate significant

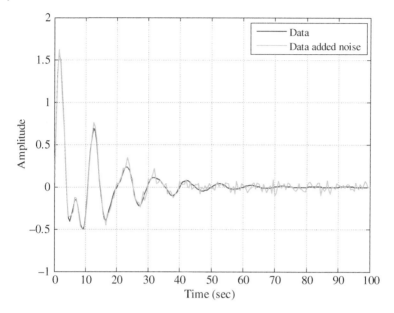

Figure 2.3 Comparison between the original data and the data with added white Gaussian noise.

decimal digits (w) is 1, the number of singular values (M) is 4 as shown in Figure 2.4(a).

Step 4. One can obtain matrices $[V'_1]$ and $[V'_2]$ from $[V']$ as submatrices of $[V]$ corresponding to M. Then, one can compute z_i from the nonzero eigenvalues of $\{[V'_2]\}^{\dagger}[V'_1]$. From (2.24), poles related to the data can be obtained as

$$s_1 = -6.1271 \times 10^{-2} + j6.2896 \times 10^{-1}; \quad s_2 = -6.1271 \times 10^{-2} - j6.2896 \times 10^{-1}$$
$$s_3 = -1.0737 \times 10^{-1} + j1.0915 \; ; \quad s_4 = -1.0737 \times 10^{-1} - j1.0915$$

The poles for the approximation of the data occurs in complex conjugate pairs. Also, one can calculate the residues corresponding to each pole from (2.3). From the noise free data, the original poles are computed as $\alpha_1 = 0.0628$, $\omega_1 = 0.6283$, $\alpha_2 = -0.1099$, $\omega_2 = 1.0996$. Due to the effects of the noise, the computed poles in this case using the MPM is slightly different. Then the relative error (E_r) for the variable α and ω is now defined as

$$E_r = \frac{|\hat{x} - x|}{x} \times 100\% \tag{2.26}$$

The deviations are tabulated in table 2.1. It is seen from the table that the MPM yields a solution with an approximation error of less than 3% in the

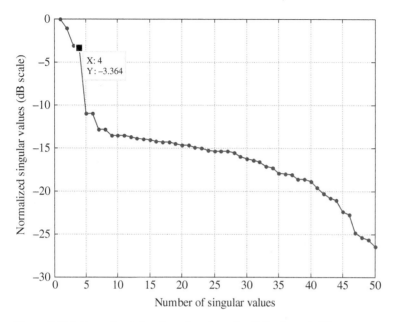

Figure 2.4(a) Normalized singular values from the SVD of matrix [Y].

Table 2.1 Percentage Errors occurring in the estimates using the MPM for noise contaminated data.

	α_1	ω_1	α_2	ω_2
Original data (SNR=∞)	−0.06283	0.62832	−0.10996	1.0996
Noisy data (SNR=15 dB)	−0.06127	0.62896	−0.10737	1.0915
E_r	2.5 %	0.1 %	2.4 %	0.7 %

solution when the SNR associated with the data is 15 dB. This implies that the noise is approximately six times smaller than the signal strength of interest. One can also calculate the residues corresponding to each poles from (2.3) as

$$R_1 = -0.0205 - j\,0.5054 \simeq -j\,0.5; \quad R_2 = -0.0205 + j\,0.5054 \simeq j\,0.5$$
$$R_3 = 0.0299 - j\,0.4992 \simeq -j\,0.5; \quad R_4 = 0.0299 + j\,0.4992 \simeq j\,0.5$$

Step 5. From the computed poles and residues, one can estimate the error in the extrapolated/interpolated data as shown in Figure 2.4(b). The estimated MSE is 0.03615.

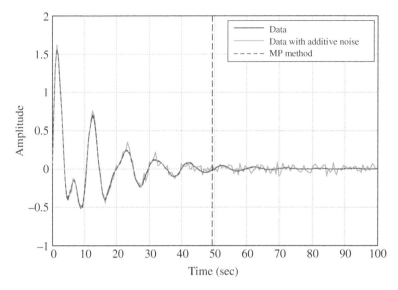

Figure 2.4(b) Comparison between the original data, the data with additive noise, and the estimated data from the MP method using 100 selected samples.

Next, the various applications where the MPM can be and has been applied with great success are presented.

2.3 Applications of the MPM for Evaluation of the Characteristic Impedance of a Transmission Line

As a first example, the MPM has been successfully applied to high-resolution de-embedding procedures using band-limited data [9–18] prevalent in most microwave and antenna measurements. Primarily, there are two parameters in certainty which always exist in a de-embedding operation but is absent in the presented procedure. The first is what is the loss both conducting and radiating involved in shifting the plane of reference on the transmission line. The second is the computation of the S-parameters without the value of the characteristic impedance which requires one to evaluate the voltage on the line which itself is quite strange. Here the S-parameters are computed directly from the current distribution on the structure where there is no uncertainties.

In characterizing any circuit one requires two independent variables. Typically one variable is the current and the other is the voltage. In general, the current on the transmission line of an integrated circuit can be uniquely determined and can be computed quite accurately. However, for the definition of the voltage there is some arbitrariness, particularly when frequencies become higher. This is because at high frequencies the voltage is dependent on the path of integration of the electric field. In addition, there are different definitions for the characteristic impedance. There are four different definitions for the characteristic impedance Z_0, in terms of the power, voltage and current and their particular combinations as illustrated through:

$$Z_0^{(\varepsilon_{r,\text{eff}})} = \frac{Z_0^{\varepsilon_0}}{\sqrt{\varepsilon_{r,\text{eff}}}} \quad Z_0^{(PI)} = \frac{2P_0}{I_0^2} \quad Z_0^{(PV)} = \frac{V_0^2}{2P_0} \quad Z_0^{(VI)} = \frac{V_0}{I_0} \quad (2.27)$$

where P_0 represents the transported power, V_0 is the voltage defined by integrating the electric field along a path on the section of the feed line of the device and I_0 is the current flowing in the feed line, and $Z_0^{\varepsilon_0} = \sqrt{L/C}$ is the characteristic impedance for the static case in vacuum.

As an example, from [13], consider the computation of the characteristic impedance of a microstrip line of width w and height h from the ground plane, with the ratio $w/h = 1$. The conductor thickness in all the examples has been assumed to be negligible. This line is situated on a dielectric slab of $\varepsilon_r = 9.7$. Therefore, h is also the thickness of the dielectric. Figure 2.5 shows Z_0 for this microstrip line as a function of h/λ_0, where λ_0 is the free space wavelength. In

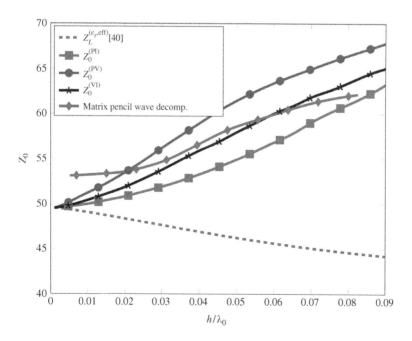

Figure 2.5 Z_0 as a function of h/λ_0 using various definitions for the characteristic impedance.

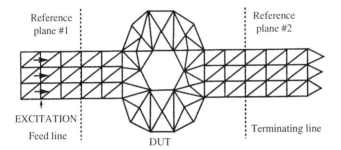

Figure 2.6 De-embedding of a device under test.

this analysis, for computational purposes, consider a finite microstrip line of length $100w$. The line is divided into 100 segments resulting in 200 triangular patches as shown in Figure 2.6. The first non-boundary edge at the end of the line is excited by the usual delta gap voltage source as used in any computational tools. The voltage and the current distribution on the line are obtained by solving the electric field integral equation using the appropriate Green's functions for the layered dielectric slab [9–18]. Then the current distribution on the transmission line feeding the structure is decomposed into various components of the travelling current waves. This is achieved by applying the MPM, primarily to the current distribution on the line as that is the only parameter which is not affected as the frequency of operation increases.

The purpose of the MPM is to approximate the current distribution on the line by a sum of complex exponentials. Therefore, the goal is to approximate the current by

$$I(x) = \sum_{m=1}^{N} A_m e^{\gamma_m x} = \sum_{m=1}^{N} A_m e^{(\alpha_m + j\beta_m)x} \quad \text{for } x > 0 \tag{2.28}$$

where γ_m are the complex propagation constants and A_m are complex amplitudes. Here, N, the number of complex exponentials is also an unknown to be solved for in addition to the set of γ_m and A_m. The philosophy in the application of the MPM is equivalent to numerically separating the incident wave, the reflected wave, the evanescent waves, the complex waves and/or the higher order modes that are propagating through the reference plane at $x = 0$ as illustrated in Figure 2.6.

The plane $x = 0$ is the end of the line away from the excitation. It is important to state that in all the simulations presented in this chapter, the line is left open at both ends and not terminated in the characteristic impedance as its true value cannot be estimated. In addition, in this approach there is no need to terminate the line. The excitation is located at the other end, at a distance approximately $99w$ from the open end. The value of N is determined from the singular value decomposition of the data. The details are available in [13]. The characteristic impedance is then obtained by dividing the forward incident voltage wave by the current wave (or with a change of sign, if backward travelling waves are used). The results are compared with that of Hoffman [19]. The quasi static solution for the characteristic impedance is obtained to be 49.8 Ω. There is a difference of 4% between this quasi-static solution and the solution obtained by the technique just described at a frequency f_0 such that $h/\lambda_0 = 0.005$. In Figure 2.5, the dashed line corresponds to the effective dielectric constant methodology [13, 19], resulting in an estimation of the impedance that diverges from the other methods when increasing the dielectric thickness. Three of these other methods are the power-current, power-voltage, and voltage-current relationships (2.27). The last line in Figure 2.1 corresponds to the methodology here outlined consisting on the decomposition into voltage and current waves as in (2.28) the computation of their ratio. The point is that depending on the chosen definition for the characteristic impedance the results are different. Hence, it would be useful to develop a procedure for network characterization that does not require the computation of the characteristic impedance.

Therefore, it becomes quite confusing in applying the classical S-parameters to study lossy transmission lines or study radiation from transmission line devices as the computation of the power is not unique. Secondly, the classical S-parameters of a device are computed quite far away from the device so that the evanescent and the higher order modes have died down and only the dominant mode is prevalent. Then a de-embedding is done by shifting the reference planes to the device of interest. Again, it becomes difficult to perform a de-embedding if there is any loss, ohmic or otherwise, associated with these transmission line sections. And finally, for the evaluation of the conventional S-parameters, the line should be terminated by its characteristic impedance which is difficult to do in a numerical simulation: the knowledge of the characteristic impedance itself can be a difficult task! It is for these reasons we seek an alternate procedure where the device can be characterized directly at the reference planes of interest and without terminating the lines in any form. We now illustrate this alternate methodology.

In addition, there is another problem that one has to address when the reference planes are not shifted, namely, that the dominant mode has to be isolated from the higher order modes.

As a second example, consider the analysis of a microstrip line at a fixed frequency, so that the width of the line is $w/h = 0.02$, and the dielectric constant of the substrate is $\varepsilon_r = 16$. We consider a line of length $100w$, excited by a unit voltage source at the first non-boundary edge. Once the current distribution on the

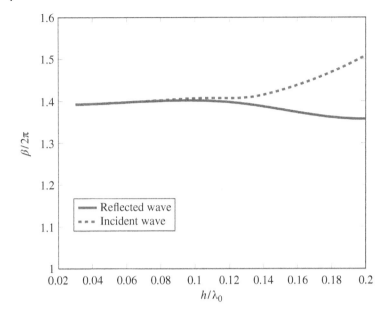

Figure 2.7 Plot of the propagation constant β for the incident and reflected waves as a function of h/λ.

structure is solved for, the matrix pencil method is used to fit the current. The two dominant values of the exponent corresponding to the propagation constant of the incident and the reflected waves are analyzed on the microstrip line as functions of h/λ. It is seen in Figure 2.7 that for $h/\lambda > 0.12$, the propagation constant of the incident and the reflected waves tend to differ. This indicates that for $h/\lambda > 0.12$, the microstrip line is no longer suitable as a transmission line, since there may be significant dispersion as the incident and the reflected waves cannot be clearly defined. This illustrates that by decomposing the total current into various waves, one can obtain a physical insight into the problem and can guarantee that the S-parameters are indeed defined with respect to the fundamental mode.

In summary, without a unique stable value for the characteristic impedance, it is difficult to compute the S-parameters. And when they are computed, it is important to be sure that they are defined with respect to the dominant mode, with the incident and the reflected wave having the same propagation constants but with a different sign. The question now is, can one compute the S-parameters in an alternate way without any knowledge of the characteristic impedance? In the next section a procedure is described that can achieve that goal. Also in this procedure there is no need to shift the reference planes and thereby remove the uncertainties associated with the losses.

2.4 Application of MPM for the Computation of the S-Parameters Without any A Priori Knowledge of the Characteristic Impedance

The characterization of the S-parameters for complex objects is considered next, where there can be radiative and ohmic losses in the structure and it may be difficult to augment the structure with a terminating set of characteristic impedances in a numerical simulation of a planar structure [13].

In this scheme, a port is selected which provides an excitation in the Method of Moments context of analysis of a microwave device. This is illustrated with some specific examples of finding the reflection coefficient of a 90° radial stub, and the radiation from a rectangular and a mitered bend. The reason behind these examples is that the radiation loss from these devices is extremely small and therefore very accurate characterization of the S-parameters are necessary. Also, because of the radiation losses and the presence of the higher order modes, it is difficult to transform the reference planes as is done in a de-embedding procedure. The question that is also addressed is how much is the radiation loss lower for a metered bend than for a rectangular bend. If that loss is really quantifiably small then it is a wastage of valuable resources to reproduce mitered bend in the fabrication process of a system.

As an example, consider the analysis of radiation from a 90° radial stub as shown in Figure 2.8, which is drawn not to scale [20]. The computational procedure for the radial stub constructed using a microstrip line is used to illustrate the novel method of computing the S-parameters without any knowledge of the characteristic impedance of the line. The radial stub shown in Figure 2.8 is connected to a microstrip line. As in the previous examples, zero-thickness conductors are considered, and the lines are left open in the simulations. The 90° radial stub is divided into a number of patches. The stub has a radius $r = 0.889$ mm, and is connected to a transmission line of width $w = 0.254$ mm. The structure is situated over a dielectric slab of relative permittivity $\varepsilon_r = 12$ and thickness $h = 0.635$ mm. The frequency of operation is 30 GHz. The length of the line connected to the radial stub is considered to be $2.0\lambda_0$ long, and it is divided into 100 subsections and into 200 triangles. At a non-boundary edge away from the stub (typically one to two wavelengths away from the discontinuity of interest, to the left in Figure 2.8: it is important to note that this figure is not to scale) a voltage of 1 V is applied to excite the structure as required in a numerical procedure [13]. The stub itself is divided into 80 triangular patches (not shown in Figure 2.8). Utilizing the conventional electric field integral equation and the classical Sommerfeld integrals, which take into account the effects of the infinite ground and dielectric slab, the unknown amplitudes of the currents flowing across non-boundary edges can be solved by the usual method of moments matrix [9–18].

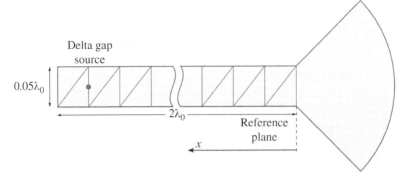

Figure 2.8 Characterization of a 90° radial stub with $x = 0$ starting from the right near the flare.

Once the current distribution is solved for on the entire structure, the MPM is then applied to fit the current distribution on the transmission line as a sum of exponentials starting from the plane $x = 0$. The reference plane is thus defined at the junction between the radial stub and the microstrip line. As in the previous example, the incident wave, the reflected wave, the evanescent waves, the complex waves and the higher order modes propagating through the reference plane $x = 0$, that are separated by the application of the MPM are shown in Figure 2.9, where the abscissa represents the distance in free space wave-lengths along the transmission line away from the radial stub. Only one wavelength spatial distance from the radial stub is represented in the figure. On the other hand, the ordinate represents current, barring a constant factor w, the width of the line in terms of λ_0. Not only the incident and the reflected waves are represented, but also the higher-order modes, although they are much weaker. In particular, the first higher order mode, which is an evanescent wave, and two other higher order modes plotted in this figure are almost negligible in magnitude. Possibly, some of the exponentials may be due to modeling error, finite discretizations, or truncation or representation of numerical errors. Those spurious solutions/modes would have a small amplitude A_m, as all the physical higher order modes. The amount of radiation coming out from the structure is also small, making the separation of the various effects tricky. However, if all the characterizations are based on the incident and reflected current amplitudes defined at the plane of interest, the results will be accurate, since their amplitudes are several orders of magnitude above the other mentioned waves and artifacts.

Now, the incident and the reflected current wave amplitudes can be directly determined from the S-parameters defined by the pseudo wave scattering parameters. For example, for a one-port network, the reflection coefficient is defined by $S_{p,\,11} = \Gamma_p = b_p/a_p$ where a_p and b_p are proportional to the incident and

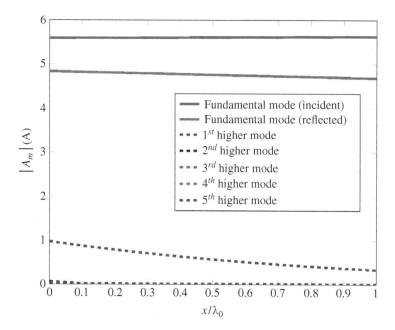

Figure 2.9 Mode decomposition shown in equation (2.28) of the current in the microstrip feed line of the 90° radial stub (Figure 2.8) with MPM.

reflected current waves, respectively. From their amplitudes, $|S_{p, 11}|$ is computed so that $1 - |S_{p, 11}|^2$ provides the radiated power, which is estimated to be 24.7% of the incident one, as compared to 21%, evaluated in [20] by integrating the entire radiated fields in 3D which is a herculean task. Since now the S-parameters are defined at the plane of interest, there is no need to shift the reference planes, making it unnecessary to have a numerical value of the characteristic impedance. Here, the S-parameters have been automatically normalized with respect to the characteristic impedance of the transmission line at the frequency for which the current distribution on the structure has been computed.

It is thus clear that utilizing the conventional concepts of "shifting of reference planes" may yield erroneous results when de-embedding the power radiated by the radial stub because the power in the higher-order propagating modes is not properly accounted for. There may also be problems in transferring the reference planes when the structure is lossy. Even though these modes are also radiating, the differentiation between the radiation from the discontinuity and from other source-induced terms become important.

In order to apply this technique for a two-port, nonreciprocal, nonsymmetric network, the procedure is as follows. Port 1 is excited first and the currents are computed on the structure. Based on this current distribution, the incident and the reflected waves are computed at each plane of reference. If $i_1^-, i_1^+, i_2^-, i_2^+$ are the incident and reflected current waves at the two reference planes then one would obtain

$$\begin{bmatrix} i_1^- \\ i_2^- \end{bmatrix} = \begin{bmatrix} S_{11} & S_{12} \\ S_{21} & S_{22} \end{bmatrix} \begin{bmatrix} i_1^+ \\ i_2^+ \end{bmatrix} \qquad (2.29)$$

Next, port 2 is excited and the current distribution on the entire structure is again solved for. The MPM is applied one more time to obtain the incident and reflected current waves at the second reference plane identified earlier. Let these values be $I_1^-, I_1^+, I_2^-, I_2^+$, related by

$$\begin{bmatrix} I_1^- \\ I_2^- \end{bmatrix} = \begin{bmatrix} S_{11} & S_{12} \\ S_{21} & S_{22} \end{bmatrix} \begin{bmatrix} I_1^+ \\ I_2^+ \end{bmatrix} \qquad (2.30)$$

Finally, the S-parameters are computed using both (2.29) and (2.30), as

$$\begin{bmatrix} S_{11} \\ S_{12} \\ S_{21} \\ S_{22} \end{bmatrix} = \begin{bmatrix} i_1^+ & i_2^+ & 0 & 0 \\ 0 & 0 & i_1^+ & i_2^+ \\ I_1^+ & I_2^+ & 0 & 0 \\ 0 & 0 & I_1^+ & I_2^+ \end{bmatrix} \begin{bmatrix} i_1^- \\ i_2^- \\ I_1^- \\ I_2^- \end{bmatrix} \qquad (2.31)$$

It should be noted that the described procedure is easily generalizable to the multiport case.

As an example, consider the radiation characterization of two-port devices, namely a right-angled bend and a mitered one. Both bends are connected to two transmission lines $1.95\lambda_0$ long and $0.05\lambda_0$ wide, λ_0 being the free space wavelength at the operating frequency. The conductor strips are situated on a dielectric slab of relative permittivity $\varepsilon_r = 2.33$ and thickness equal to $h = 0.017\lambda_0$. It should be noted that these dimensions correspond to a 50 Ω microstrip line operating at 10 GHz. The conductor thicknesses are neglected. The structure of the right-angled bend (Figure 2.10) is divided into 80 triangular patches, while the transmission line portions are divided into 39 triangular patches each. The structure is excited by a 1 V delta gap source at the first non-boundary edge of a triangular patch located away from the bend in one of the arms. First, the method of moments is used to solve for the current distribution on the structure. As in previous examples, both ends of the line are open, since this procedure does not require any particular form of termination by their characteristic impedance. It has been assumed that the transverse current distribution is constant, this being a reasonably good approximation for the analysis of wire-like structures where the width of the strip line is much smaller than a wavelength and $w/h > 1$. It has been shown in [18] that for $w/h > 1$, even though the transverse current goes to infinity at the edge, assuming it to be a constant has practically no effect on the longitudinal current amplitudes on

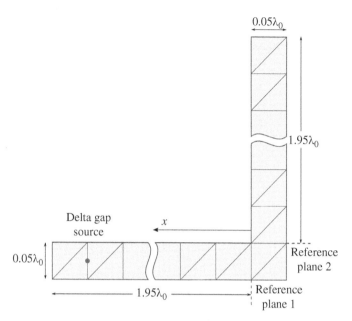

Figure 2.10 Geometry of a right angled bend.

the structure. However, the singular behavior of the current needs to be considered accurately when $w/h < 1$. Since in this case $w/h = 3$, assuming a constant transverse current distribution (and thereby neglecting the singular behavior at the edges), is a valid approach and has practically no effect on the solution.

Once the total current distribution on the structure is solved, the MPM is applied to approximate the complex current amplitudes on the straight edges parallel to the boundary by a sum of complex exponentials. The amplitudes of the exponentials, whose origins are located at the reference planes, are illustrated in Figure 2.11. It can be seen that only the incident and the reflected modes show non-negligible amplitudes. At the reference planes 1 and 2 they are $I_1^+ = 2.7169\angle107.80°$, $I_1^- = 2.6893\angle-103.33°$, $I_2^+ = 2.7368\angle94.64°$, and $I_2^- = 2.7547\angle-90.67°$. The conventional S-parameters are obtained by means of (2.30). Since the structure is symmetric and reciprocal we assume $S_{11} = S_{22}$ and $S_{21} = S_{12}$. In that case, the solution matrix will be of half the size, with only two unknowns.

Therefore, (2.31) becomes

$$\begin{bmatrix} S_{11} \\ S_{12} \end{bmatrix} = \begin{bmatrix} S_{22} \\ S_{21} \end{bmatrix} = \begin{bmatrix} I_1^+ & I_2^+ \\ I_2^+ & I_1^+ \end{bmatrix} \begin{bmatrix} I_1^- \\ I_2^- \end{bmatrix} = \begin{bmatrix} 0.0696\angle-115.3° \\ 0.9924\angle+157.7° \end{bmatrix} \quad (2.32)$$

Here symmetry has been used to reduce the four equations of (2.31) into two equations in (2.32). The relative power radiated from the bend is therefore given by $1 - |S_{11}|^2 - |S_{22}|^2 = 0.0102$, or 1.02% of the incident power. It should be

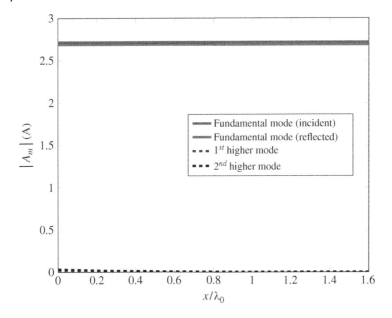

Figure 2.11 Mode decomposition shown of the current in the microstrip feed line of the right angle bend (Figure 2.10) with MPM.

remarked that all the computations involved in this prediction of radiation have been performed without any knowledge of the characteristic impedance of the transmission lines. Next a mitered bend is considered to study what is the nature of radiation from a mitered bend as opposed to a right angled bend.

Figure 2.12 shows the structure corresponding to a similar bend but mitered, including its dimensions. The metallic structure is subdivided into 70 triangular patches. As before, the two ends of the lines were left open and the transverse current distribution has been assumed to be constant. Also as before, the structure is excited by a 1 V source located at a non-boundary edge close to the end furthest from the bend. The current distribution on the structure is computed the same way as before, and then a sum of complex exponentials is used to fit the current distribution using the MPM.

The results of the approximation at the two reference planes are given by $I_1^+ = 2.7959\angle107.80°$, $I_1^- = 2.7689\angle-103.27°$, $I_2^+ = 2.7566\angle94.70°$, and $I_2^- = 2.7745\angle-90.61°$. From these values one can get the S-parameters as

$$\begin{bmatrix} S_{11} \\ S_{12} \end{bmatrix} = \begin{bmatrix} S_{22} \\ S_{21} \end{bmatrix} = \begin{bmatrix} I_1^+ & I_2^+ \\ I_2^+ & I_1^+ \end{bmatrix} \begin{bmatrix} I_1^- \\ I_2^- \end{bmatrix} = \begin{bmatrix} 0.0275\angle & 80.4° \\ 0.9947\angle163.5° \end{bmatrix} \tag{2.33}$$

Hence, one can compute the effective radiated power coming out from the mitered bend as $1 - |S_{11}|^2 - |S_{22}|^2 = 0.0098$, which implies that only 0.98% of the input power is radiated from the bend.

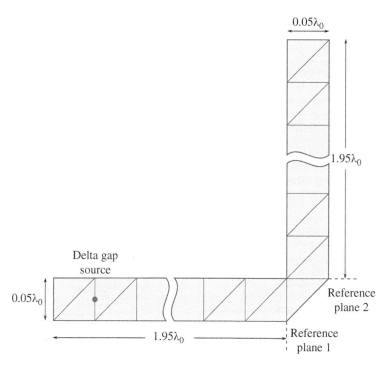

Figure 2.12 Geometry of a mitered bend.

This essentially confirms what we knew intuitively that a right angled bend not only reflects more energy but also radiates more than a mitered bend. Use of the MPM makes it possible to compare quantitatively the radiations from a right angled with a mitered bend by using the S-parameters directly and without having to transfer the reference planes as in a de-embedding procedure, and without the need to terminate the end sections of the line and most importantly without any *a priori* knowledge of the characteristic impedance of the line. However, the question still lingers on: Is it really necessary to consider mitered bend in a circuit as opposed to a rectangular bend solely based on the criterion of radiation as the differences between the two is merely 0.04%!

A novel de-embedding scheme utilizing the MPM has been described. By combining the electromagnetic analysis using an electromagnetic simulator [21] and the MPM it is possible to compute the S-parameters of any structure without knowing *a priori* the characteristic impedance of the transmission lines connected to it. In this approach only the current distribution on the transmission line needs to be known and no *a priori* information about the characteristic impedance matrix of the transmission lines is required. Since the MPM solves for the incident and reflected current waves, the S-parameters can be computed directly from these two without any need to have a numerical value of the characteristic impedance. The results are automatically normalized to the value of the characteristic impedance at which the current distribution on the structure has been computed. Also, there is no need to terminate the structures in their characteristic impedances to obtain the S-parameters. Numerical results have been presented to illustrate this technique.

2.5 Improving the Resolution of Network Analyzer Measurements Using MPM

As an example consider the Beatty standard, presented in Figure 2.13. There are two impedance step discontinuities in the standard. If the output port is terminated with 50 Ω and the standard is excited with an impulse at port 1, there will be reflections due to the discontinuities. From a theoretical point of view there will be four dominant reflections, as shown in Figure 2.14. The expected amplitude and time delay of the various reflections are given in Table 2.2.

Next we utilize the HP 8510B vector network analyzer and measure the S_{11} parameter from 45 MHz to 18 GHz using 801 data points. The measured magnitude and the phase responses are given in Figure 2.15. Utilizing the HP 8510B internal frequency to time domain conversion technique, one obtains the plot in Figure 2.16. The first three impulse like returns are quite clear and the fourth appears where it is marked by an arrow. Comparing the measured values with the theoretical values of Table 2.2 demonstrates a reasonable agreement. Next, the bandwidth of the sweep is reduced from 18 GHz to 2 GHz. In this case the magnitude and the phase response are plotted in Figure 2.17. If the inverse transform is taken with the HP 8510B internal mechanism, one obtains the time-domain response of Figure 2.18. Observe that, as expected, no information is available about the discontinuities [22].

The MPM is now applied to the same 2 GHz bandwidth data as plotted in Figure 2.17. The conventional FFT based approach will not provide any meaningful result as shown in Figure 2.18. The MPM technique illustrates that there are three dominant singular values from the application of the SVD. So the value for the parameter M is set to be equal to 3. The amplitude and location of the impulses are shown in Figure 2.19 Observe that the first three impulses have been identified and their positions quite accurately located, and that the fourth impulse is not visible. The MPM is now applied to the same 18 GHz bandwidth data as shown in Figure 2.15. These results are also superposed on Figure 2.19. Observe that, for this example, reducing the bandwidth from 18 GHz to 2 GHz had no visible impact on the time-domain resolution for the first three impulses. This is because a parametric technique such as the accuracy of MPM depends on the number of samples of data points rather than on the actual bandwidth of the measured data. This is the strength of the MPM over conventional FFT techniques.

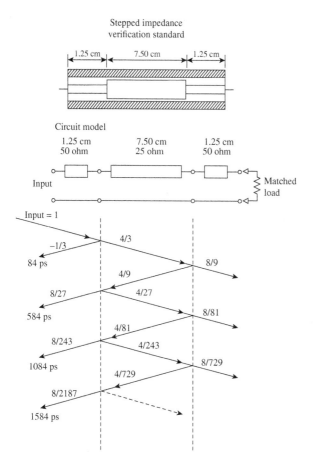

Figure 2.13 Multiple reflections from the Beatty standard terminated with a matched load.

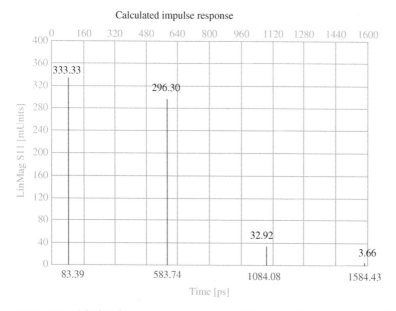

Figure 2.14 Calculated Impulse Response from (2.13) using MPM for the Beatty Standard.

Table 2.2 Calculated Impulse Response of a Beatty standard terminated with 50 Ω.

Impulse	Delay time [ps]	Amplitude
1.	83.39	0.333
2.	583.74	0.296
3.	1084.08	0.033
4.	1584.43	0.00366

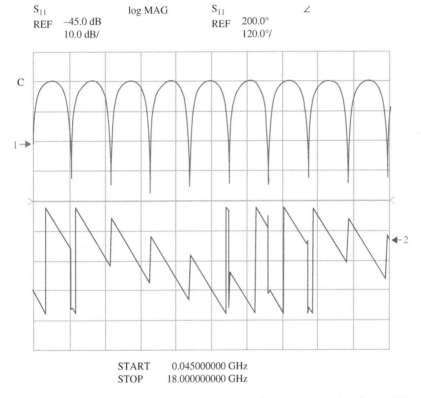

Figure 2.15 The Magnitude and the Phase Response of the Beatty Standard from 45 MHz to 18 GHz measured using the HP8510B network analyzer.

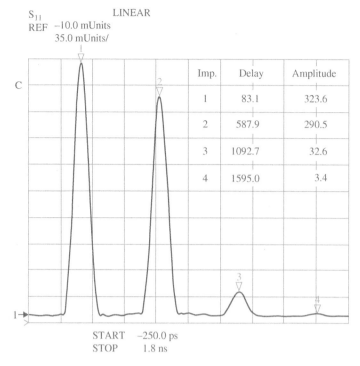

	Imp.	Delay	Amplitude
	1	83.1	323.6
	2	587.9	290.5
	3	1092.7	32.6
	4	1595.0	3.4

Figure 2.16 Time Domain Impulse Response using the standard built-in bandpass option on HP8510B with 18 GHz bandwidth.

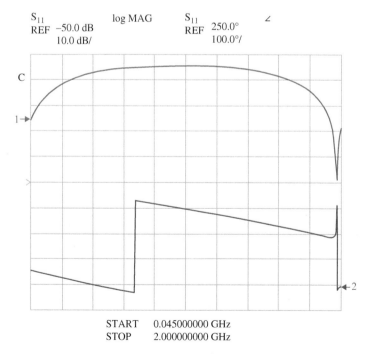

Figure 2.17 The Magnitude and Phase response of the Beatty standard measured from 45 MHz to 2 GHz.

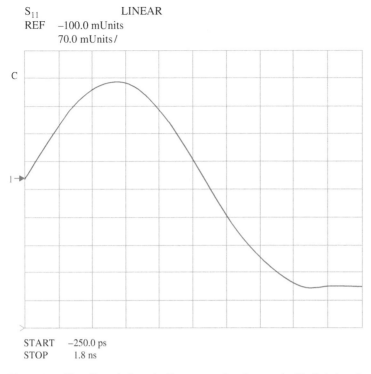

Figure 2.18 Time Domain Impulse Response using the standard built-in band-pass option on HP8510B with 2 GHz bandwidth.

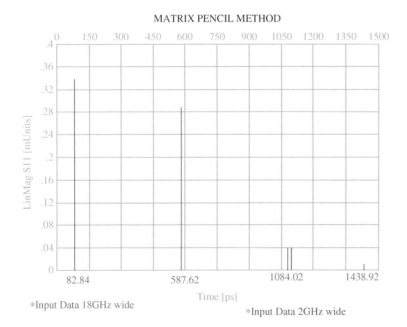

MATRIX PENCIL METHOD

*Input Data 18GHz wide

*Input Data 2GHz wide

Figure 2.19 Time Domain Impulse response obtained using the MPM with 2 GHz bandwidth (green) and 18 GHz (red) bandwidth.

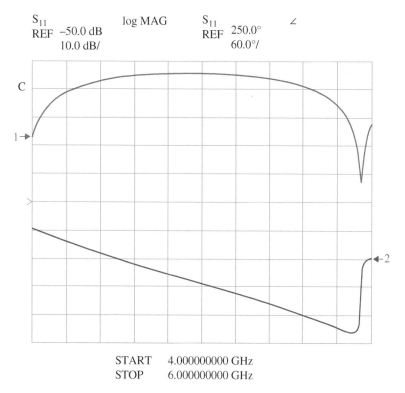

S_{11} log MAG S_{11} ∠
REF −50.0 dB   REF 250.0°
   10.0 dB/   60.0°/

START   4.000000000 GHz
STOP   6.000000000 GHz

Figure 2.20 The magnitude and the Phase response of the Beatty standard from 4-6 GHz bandwidth.

Next the frequency-domain response of the Beatty standard is generated over 4-6 GHz. The magnitude and the phase are plotted in Figure 2.20. The time domain response using the internal HP8510B standard band-pass option is shown in Figure 2.21. Observe that no useful information about the discontinuities can be extracted. However, the MPM technique applied to the same data generates three impulses, as shown in Figure 2.22. Observe that the fourth impulse is not visible for this bandwidth data. Time delay of the first three impulses is within 3.5% of the theoretical results, while the amplitude of the first impulse is off by 1.37%, the second by 2.5%, and the third by 6.67%. The CPU time needed to compute the time-domain response by using the MPM in Figure 2.19 is of the order of 4 s on an a HP9000/370 workstation.

The advantage of using the MPM over a Fourier based technique is now quite clear. The resolution for the MPM method is no longer limited by the bandwidth of the measurements but by the number of significant digits of the measured data. In contrast, the resolution of the Fourier based methods are limited

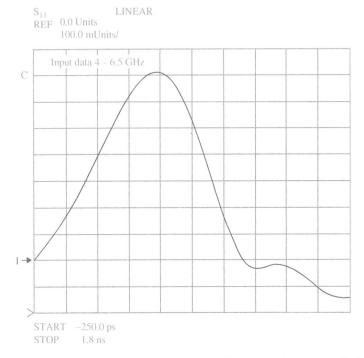

Figure 2.21 Time domain impulse response of the shorted Beatty standard using standard built-in bandpass option in HP8510B, with a bandwidth from 4-6 GHz.

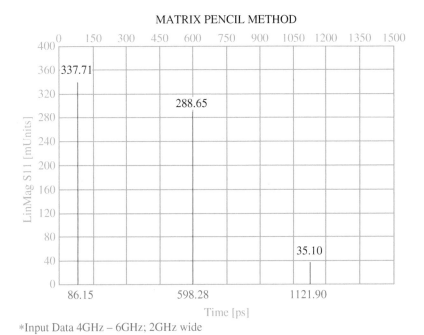

*Input Data 4GHz – 6GHz; 2GHz wide

Figure 2.22 Time Domain impulse Response obtained using the Matrix Pencil method with a bandwidth of 4-6 GHz.

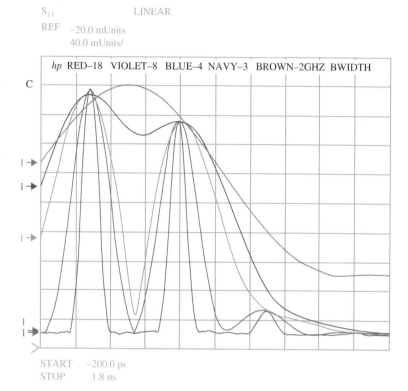

S$_{11}$ LINEAR
REF   −20.0 mUnits
      40.0 mUnits/

hp RED–18 VIOLET–8 BLUE–4 NAVY–3 BROWN–2GHZ BWIDTH

C

START   −200.0 ps
STOP   1.8 ns

Figure 2.23 Time Domain Response for the Fourier Technique Based method as a function of the Bandwidth of the available data.

by the bandwidth of the measurements as seen in Figure 2.23. The resolution in determining the impulses is clearly seen to be determined by the bandwidth of the measurements.

As an even more convincing example, consider the Beatty standard terminated with a short (Figure 2.24). The expected impulse response is given in Table 2.3 and in Figure 2.25. Utilizing the HP 8510B, the S$_{11}$ parameter is measured from 45 MHz to 18 GHz, using 801 data points. The magnitude and the phase response are given in Figure 2.26. Applying the HP 8510B internal band pass frequency to time conversion technique, one obtains the plot in Figure 2.27. Only impulses 1 and 3 (see Table 2.3) are clearly observable. Figure 2.27 shows the unwindowed time-domain impulse response, obtained from the same 18 GHz wide data set. Resolution is enhanced at the expense of the higher ringing and reduced dynamic range.

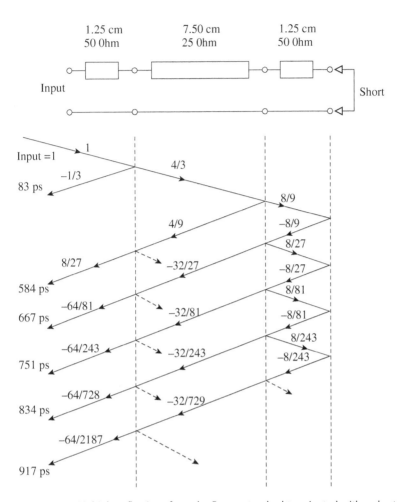

Figure 2.24 Multiple reflections from the Beatty standard terminated with a short.

Table 2.3 Calculated Impulse Response of a Beatty standard terminated with a Short.

Impulse	Delay time [ps]	Amplitude
1.	83.39	0.3333
2.	583.74	0.2963
3.	667.13	0.7901
4.	750.52	0.2634
5.	833.91	0.0878
6.	917.30	0.0293
7.	1000.69	0.0097
8.	1084.08	0.0297
9.	1167.47	0.1767
10.	1250.87	0.2045
11.	1334.26	0.1172
12.	1417.65	0.0488

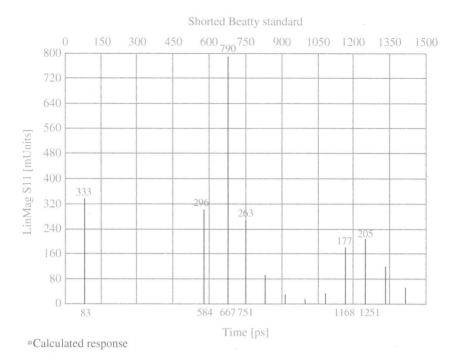

*Calculated response

Figure 2.25 Calculated Impulse Response of the Beatty Standard terminated with a short.

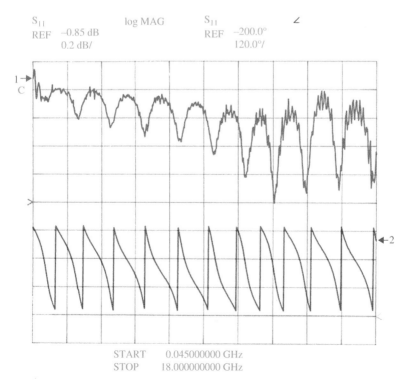

Figure 2.26 The Magnitude and the Phase Response of the Beatty Standard terminated in a short and S_{11} measured from 45 MHz to 18 GHz using the HP8510B network analyzer.

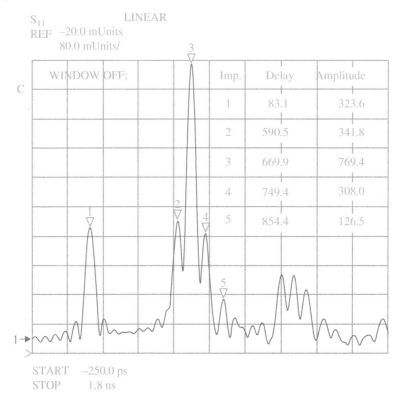

Figure 2.27 Time-domain impulse response using the internal built-in band-pass in the HP8510B, with 18 GHz bandwidth. Windowing option turned off.

Next, the bandwidth of the sweep is reduced to the 4-6.5 GHz range. In this case the magnitude and the phase response are plotted in Figure 2.28. If the inverse transform is taken with the internal built-in band-pass option in the HP 8510B, one obtains the time-domain response of Figure 2.29 which displays the response with the windows on. Observe that, as expected, no information is available about the discontinuities.

The MPM is now applied to the same 4-6.5 GHz data as plotted in Figure 2.28. The amplitude and location of the impulses are shown in Figure 2.30 along with the theoretically calculated response. If the MPM is applied to the 18 GHz bandwidth data of Figure 2.26, the impulse response of Figure 2.31 is obtained. Observe again that, for this example, reducing the bandwidth from 18 GHz to 2.5 GHz had no visible impact on the time-domain resolution (compare

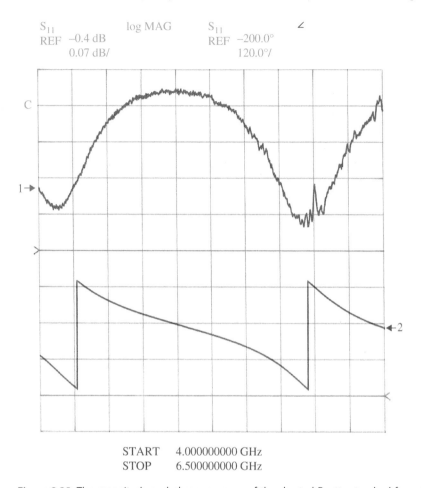

S_{11} log MAG S_{11} ∠
REF −0.4 dB   REF −200.0°
      0.07 dB/   120.0°/

START   4.000000000 GHz
STOP   6.500000000 GHz

Figure 2.28 The magnitude and phase response of the shorted Beatty standard from 4 to 6.5 GHz.

Figure 2.30 and Figure 2.31). From Figure 2.30 the time delay of the first four impulses is within 0.6% of the theoretical results, while the amplitude of the first impulse is off by 0.5%, the second by 3.1%, the third by 1.9%, and the fourth by 4.4%. For this particular example, which is an extreme case, involves solving for eigenvalues of a 400 by 400 matrix, and the elapsed CPU time on a laptop computer is of the order of a few minutes.

In summary, the MPM is used for extraction of the impulse response out of limited frequency bandwidth data. The examples show that a parametric

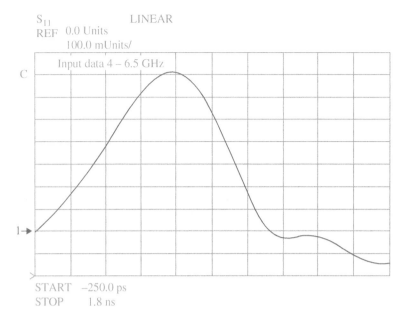

Figure 2.29 Time-domain impulse response of the shorted Beatty standard. using standard built-in band-pass option on HP 8510B, with a bandwidth of 4 to 6.5 GHz.

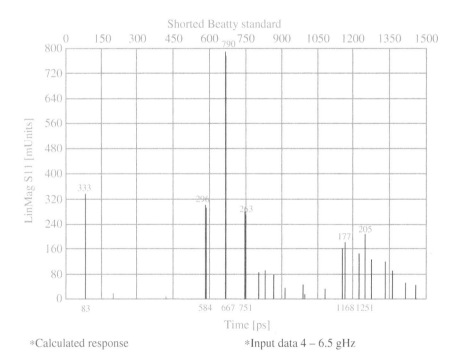

Figure 2.30 Time-domain impulse response of shorted Beatty standard obtained using the MPM, with a bandwidth of 4-6.5 GHz.

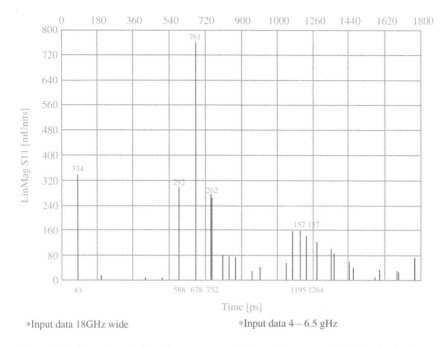

*Input data 18GHz wide   *Input data 4 – 6.5 gHz

Figure 2.31 Time-domain impulse response of shorted Beatty standard obtained using MPM, with a bandwidth of 18 GHz.

technique such as the MPM can provide accurate, reliable results with a high degree of resolution even when the FFT-based technique fails for lack of adequate bandwidth in the measurement. The method has wide application in antenna measurements, location of transmission line discontinuities, and radar cross-section measurements. It could be implemented as a standard firmware feature on advanced modern vector network analyzers.

2.6 Minimization of Multipath Effects Using MPM in Antenna Measurements Performed in Non-Anechoic Environments

In general, the measurement of antenna radiation patterns takes place in anechoic chambers, where one attempts to reduce as much as possible the reflection and diffraction contributions from walls and mechanical devices present inside the chamber. The reduction in the various scattered and diffracted fields is achieved by using appropriate absorbing materials. However, additional reflected radiation components from undesired elements will result in inaccuracies in the measured patterns. In those situations, an improvement of the measurements can only be achieved through the removal [23–26] or compensation [27–29] of the perturbations.

In this section, two techniques are considered and tested to obtain the radiation pattern of antennas from measurements carried out in reverberant or non-anechoic conditions [25]. For this purpose, a metal plate has been introduced inside an anechoic chamber, which will introduce contributions due to the reflected and diffracted components of the fields in the measurements. The goal of this section is to describe a methodology that will reduce these unwanted components, and that will eliminate them from the measurements in order to obtain a radiation pattern as similar as possible to the one obtained in an anechoic chamber.

The first technique to be described estimates the impulse response of the reverberant chamber from its frequency response by using the inverse Fourier transform. In the time domain, the direct contribution is detected and gated, eliminating the undesired echoes. Applying the Fourier transform to this new time response where only the direct contribution is present, the radiation pattern can be retrieved at the frequency of interest. The second technique is based on the MPM which is used to approximate signals into a sum of complex exponentials. The performance of the MPM is compared to that of an FFT based methodology which is limited by the Rayleigh limit. The Rayleigh limit states that to resolve signals at two frequencies f_1 and f_2 one needs a time record proportional to $\dfrac{1}{(f_2 - f_1)}$ with $f_2 > f_1$. In contrast, the MPM method is a model based parameter estimation technique and hence can perform super-resolution which means it can obtain a higher resolution than the Rayleigh limit. Here, the MPM method is applied to reconstruct the frequency response of antennas measured in a semi-anechoic chamber by identification and elimination of the reflected components.

The measurement system and the procedure used to obtain the data are described in the next section. The two techniques introduced here, based on the Fourier transform and MPM algorithms, are described in sections that follow, as well as their application to the data measured in the reverberant chamber. Sample processed results are presented and compared to the data measured in a fully anechoic chamber.

All the measurements presented in this section were obtained using the near-field/far-field spherical-range measurement system in an anechoic chamber (8 m × 5 m × 4.5 m) located in the ANTEM Lab (Antennas and Electromagnetic Emissions Measurement Laboratory) at the University of Oviedo, Spain designed to operate from 1 GHz to 40 GHz. Figure 2.32 and Figure 2.33 show both the antenna under test (AUT), mounted on a rollover azimuth-positioning system, and the probe. A simplified scheme of the whole measurement system is

Figure 2.32 An anechoic chamber in ANTEM-LAB, with the AUT on a rollover azimuth positioner.

Figure 2.33 An anechoic chamber in ANTEM-LAB, with the probe on a polarization positioner.

seen in Figure 2.34(a). For these experiments, the probe and the AUT were selected to be identical pyramidal-horn antennas, with an approximately constant gain of 20 dB in the frequency band between 17.7 GHz and 26.7 GHz. They were placed at a height of 2 m above ground level, and separated from each other by 5.4 m (396 wavelengths at the frequency of 22 GHz) during the whole

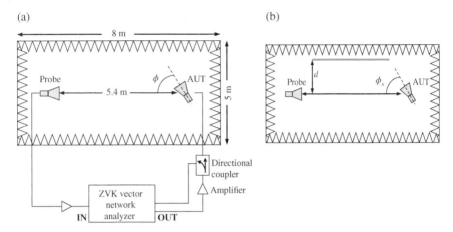

Figure 2.34 A top view of the antenna-measurement scheme in (a). The copper plate has been introduced into the anechoic chamber to model a reverberant chamber as seen in (b).

measurement campaign. The system controlling the positioners allowed the setting of the azimuth and roll orientation of the AUT, as well as the polarization of the probe, via software.

In order to compensate for the signal attenuation along the different RF coaxial cables, two amplifiers were used in the measurement system, one in the transmission path and another one in the reception section, both operating in the 18 GHz to 26 GHz band, with a typical gain of 28 dB. After the transmission amplifier, a directional coupler was inserted to get a reference of the signal radiated by the antenna. Finally, a Rohde-Schwarz ZVK vector network analyzer was used as transmitter-receiver equipment.

In this anechoic environment, the necessary reference data were taken: the azimuth (φ) radiation pattern of the AUT between 0° and 90° at the frequency of 22 GHz for an elevation angle (θ) equal to zero degrees. This information is used later to validate the results obtained by using both the FFT and MPM. Once the reference measurement was obtained, a 2 m × 1 m rectangular copper plate was placed inside the anechoic chamber, in order to simulate a partially reverberant chamber. This is shown in Figure 2.34(b), where the separation between the plate and the antennas line-of-sight (parameter d) could be configured for the different measurements, generating different scenarios or situations that had to be separately analyzed.

The $S_{21}(f)$ parameter of the system was measured in the frequency domain (18 to 26 GHz, with 22 GHz being the center frequency of the band) for each azimuth angle of radiation, φ_i, ranging from 0° to 90° in steps of 0.5°. Thus, for each step of the azimuth positioner, the channel transfer function around 22 GHz was measured. The process was repeated for each position of the copper plate, Considering that one of the two techniques is based on the Fourier transform, and bearing in mind the relationships that this algorithm establishes between the frequency domain and the time domain, the bandwidth (BW), the time resolution (Δt), and the step (Δf) of the frequency sweep must be carefully chosen, These parameters will be established as functions of the time delay (δt) from the arrival of the direct contribution until the arrival of the reflected or diffracted contributions (multipath components). This time delay, which is an indication of the minimum time resolution (Δt) required for processing in the time domain, will depend on the separation between the antennas and the location of the obstacles causing the undesirable received components.

Therefore, the minimum bandwidth required for the frequency sweep can be directly obtained as the reciprocal of this time delay, i.e., BW = 1/δt. However, it is advisable to use a higher time resolution to better distinguish the different contributions in the time domain. For this reason, a larger bandwidth, of approximately 10/δt, has been used in these measurements. On the other hand, the frequency step, Δf, is chosen as a function of the length of the timeresponse, again taking into account the distance between the antennas and the location of the main reflecting/diffracting elements located inside the chamber.

Once the parameters of the frequency sweep have been selected, the measurement procedure is straightforward and very fast As commented earlier, for each azimuth angle of the AUT, a measure of the S_{21} parameter, relating the input and output ports of the vector network analyzer, is done for the whole frequency range. The frequency sweep performed by the analyzer is fast, and it takes just a few seconds to obtain the response S_{21} (f, φ_i) for a large number of frequencies. After performing the whole azimuth sweep, the results are stored in a matrix, S_{21} (f, φ), which is the starting point for the analysis detailed in the following sections.

2.6.1 Application of a FFT-Based Method to Process the Data

It is well known that in an ideal situation, the only path that exists between the AUT and the probe is the direct propagation path. The radio channel in the frequency domain is then characterized by a constant amplitude response, independent of the frequency, and having a linear phase. However, considering multipath propagation, the above is not true anymore, and frequency-selective fading may appear, showing the influence of the echo contributions both in the amplitude and the phase of the channel frequency response. The measure of the channel frequency response is a process widely used to estimate the impulse response or time response of the multipath propagation channel [25, 26]. Once in the time domain, it is relatively easy to identify those multipath components and to estimate the channel impulse response for the anechoic conditions. This is the idea that inspired the technique that has been called the "FFT-Based Method."

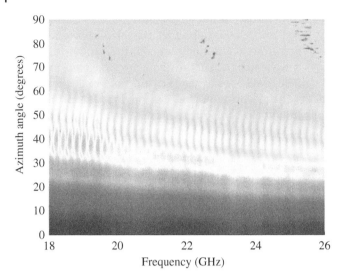

Figure 2.35(a) The steps followed in the FFT-Based Method: $S_{21}(f, \varphi)$, the channel frequency response matrix is from (18-26 GHz) for each azimuth angle between (0°-90°).

Figure 2.35 illustrates the different steps to be followed when applying this technique to the data previously measured for a given location of the copper plate. First, the measured coherent (amplitude/phase) frequency response, S_{21} (f, φ), shown in Figure 2.35(a), is windowed using a Hanning window (Figure 2.35(b)) before applying the inverse FFT to obtain the time-domain response of the system, S_{21} (t, φ). The windowing of the raw data turns out to have a slightly worse time resolution, which will now be the reciprocal of the bandwidth swept multiplied by the width of the window used. This can be particularly relevant in the case of a very short time delay between the main and the echo components. Bearing this in mind, as well as the level of the secondary lobes of the different windows commonly used, the Hanning window was finally chosen since it shows a good compromise between the reduction of the sidelobe level and the widening of the main lobe [30]. It can be observed in Figures 2.35(a) and 2.35(b) that the propagation channel between the two antennas is almost ideal for azimuth angles in the range between 0° and 20°. For that orientation of the AUT, the wave that impinges on the metal plate is radiated through a secondary lobe and, consequently, with a very low power level. However, for higher angles, the frequency-selective fading starts being evident, as the main beam of the AUT points at the reflection point on the metal plate.

Once in the time domain, Figure 2.35(c), the direct contribution is detected and gated, eliminating the delayed response due to the reflected and diffracted

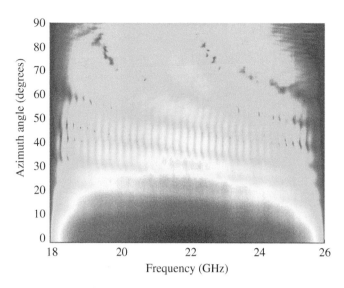

Figure 2.35(b) The steps followed in the FFT-Based Method: S_{21} (f, φ) matrix is multiplied appropriately in the frequency by the Hanning window.

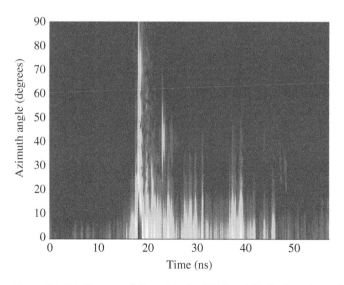

Figure 2.35(c) The steps followed in the FFT-Based Method: S_{21} (t, φ), the time response for each azimuth angle.

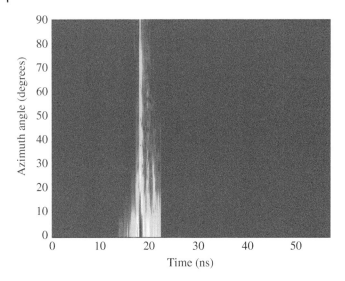

Figure 2.35(d) The steps followed in the FFT-Based Method: $S_{21}(t, \varphi)$, the time response after elimination of the delayed components.

components. This results in an impulse response $S'_{21}(t, \varphi)$, shown in Figure 2.35(d), comparable to the impulse response measured in an anechoic situation as opposed to measurements in a non-anechoic environment as seen in Figure 2.36.

Returning to the frequency domain by means of the application of the FFT algorithm to $S'_{21}(t, \varphi)$, the antenna radiation pattern at the center frequency. f_c, is obtained as $S_{21}(f_c, \varphi)$, This should be similar to the response obtained in a fully anechoic chamber. The comparison between the AUT pattern measured in this way and the reference measurement (in anechoic conditions) is shown in Figure 2.37.

2.6.2 Application of MPM to Process the Data

The second technique tested in this section is based on the MPM, which is used to obtain an approximation of the chamber response in terms of the $S_{21}(f)$ parameters approximated by a sum of complex exponential. This approximation is applied to the $S_{21}(f)$ measured data in the frequency domain so the complex exponentials can be directly related to the different contributions arriving at the probe (direct, reflected, and diffracted components) [25, 26]. The $S_{21}(f)$ response obtained for a given frequency range will be decomposed as

$$S_{21}(f) = \sum_{m=1}^{M} R_m \exp\left(s_m f\right) \tag{2.34}$$

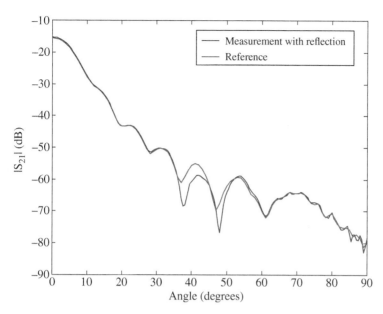

Figure 2.36 Example 1, *d* = 2.05 m: The radiation pattern measured in reverberant conditions.

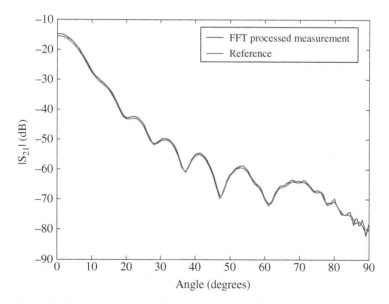

Figure 2.37 Example 1, *d* = 2.05 m: A comparison between the processed result using the Fourier-Transform Method and the reference measurement.

Once the $S_{21}(f)$ response is modeled as a combination of different complex exponentials, the next step is to determine which term corresponds to the direct contribution, so the other contributions (reflections, diffractions) can be suppressed from the measured data. The chamber response in anechoic conditions will then be obtained. The detailed procedure is explained next.

The procedure starts with the measured data, $S_{21}(f, \varphi)$. The measurements are performed for different azimuth angles φ_i and, for each azimuth angle, a set of data for different frequencies between f_0 and f_{N-1} is measured at intervals of Δf ($f_k = f_0 + k\Delta f$). The MPM is now applied to the measured data in order to model the frequency response of $S_{21}(f)$ for each one of the azimuth angles. Then, $S_{21}(f)$ for the azimuth angle φ_i can be expressed as a sum of complex exponentials

$$S_{21}\left(f_k\right)\Big|_{\phi_i} = \sum_{m=1}^{M} R_{m,i} \exp(s_{m,i} k\Delta f) = \sum_{m=1}^{M} R_{m,i} z_{m,i}^k \ \forall \ k = 0, \cdots, N-1 \tag{2.35}$$

where M is the number of complex exponentials used in (2.35), and $s_{m,i}$ and $R_{m,i}$ are the coefficients given by the MPM corresponding to the mth complex exponential term for the ith angle. The $s_{m,i}$ coefficients provide information about the propagation of each contribution in terms of attenuation and time delay, while the $R_{m,i}$ are the amplitudes of each exponential term. The procedure for obtaining those coefficients is as follows. First, a Hankel matrix, $[Y]$, is formed with the sample data $S_{21}\left(f_k\right)\Big|_{\phi_i}$ as

$$[Y] = \begin{bmatrix} S_{21}(f_0)\big|_{\phi_i} & S_{21}(f_1)\big|_{\phi_i} & \cdots & S_{21}(f_L)\big|_{\phi_i} \\ S_{21}(f_1)\big|_{\phi_i} & S_{21}(f_2)\big|_{\phi_i} & \cdots & S_{21}(f_{L+1})\big|_{\phi_i} \\ \vdots & \vdots & & \\ S_{21}(f_{N-L-1})\big|_{\phi_i} & S_{21}(f_{N-L})\big|_{\phi_i} & \cdots & S_{21}(f_{N-1})\big|_{\phi_i} \end{bmatrix}_{(N-L)\times(L+1)} \tag{2.36}$$

where L is referred to as the pencil parameter and H stands for the conjugate transpose of a matrix. Then, the $z_{m,i}$ terms are obtained as the eigenvalues of $[V_2]^\dagger[V_1]$, where the matrices $[V_1]$ and $[V_2]$ are obtained by truncating the M first rows of the $[V]$ matrix that results from the SVD of $[Y] = [U][\Sigma][V]^H$, and suppressing its first and last columns, respectively. Finally, the complex amplitude coefficients, $R_{m,i}$, are obtained by solving the following least-square problem:

$$S_{21}(f_k)\big|_{\phi_i} = \sum_{m=1}^{M} \left(\exp^{s_{m,i}\Delta f}\right)^k R_{m,i} \ \forall \ k = 0, \cdots, N-1 \tag{2.37}$$

Once the MPM is applied to the S_{21} data, an expression similar to Equation (2.35) is obtained for each measured azimuth angle. In order to reconstruct the radiation pattern of the AUT in anechoic conditions, it is necessary to determine the complex exponential that models the direct-contribution propagation for each azimuth angle. For that purpose, the information about the attenuation and delay of the propagation of the different contributions arriving at the probe, provided by the complex coefficients $s_{m,i}$, is used. Considering that the propagation delay of the direct contribution should remain constant for all the azimuth angles, this delay can be easily obtained for the azimuth angle $\varphi_0 = 0°$. For this azimuth angle, the direct-contribution amplitude ($R_{d,0}$ coefficient) must be greater than the amplitudes of the rest of the exponential terms $R_{i,0}$. Once the direct contribution term is identified for φ_0, the direct contributions for the rest of the azimuth angles are determined by comparing their propagation delay with that obtained at φ_0.

Then, the radiation pattern of the AUT at the central frequency $f_c = f_k$ {with $k = (N + 1)/2$} in full anechoic conditions can be approximated in terms of an amplitude and a complex exponential for each azimuth angle as

$$S_{21}(f_c, \phi) = R_{d,i} \exp(s_{d,i} \, c\Delta f)$$

Finally, some considerations for the practical implementation of the MPM are discussed. First, an order of $M = 3$ or $M = 4$ yields accurate modeling of the direct and reflected contributions, as well as other contributions due to diffractions in the plate edges and/or in the structures supporting the antennas. The frequency bandwidth needed for a good approximation by the MPM corresponds to one cycle of the interference between the direct ray and the reflected ray. This bandwidth can be directly related to the time delay of the reflected contribution by $1/\delta t$. For a given frequency f_c, the frequency sweep for the MPM should cover from $f_c - 1/(2 \, \delta t)$ to $f_c + 1/(2 \, \delta t)$.

2.6.3 Performance of FFT and MPM Applied to Measured Data

Two examples, corresponding to two different locations of the copper plate, are presented and analyzed in this section, in order to verify the accuracy of the procedures. The frequency of 22 GHz, as commented above, was chosen to obtain the radiation pattern of the horn antenna under test.

In the first example, the distance, d, from the plate to the antennas' line-of-sight was $d = 2.05$ m. In this situation, the main effects produced by the copper plate on the antenna pattern were expected when the azimuth position of the AUT ranged approximately from 30° to 50°, with the main beam of the antenna pointing directly at the plate, as had already been displayed in Figure 2.35(a). The time delay (δt) of the component reflected from the plate in relation to the time of arrival of the direct contribution can easily be estimated from the specular reflection, and was expected to be about 4.7 nsec leading to the choice of a bandwidth of 8 GHz, thus achieving an approximate time resolution of $\Delta t = \delta t/36$ (before windowing the frequency response). A frequency step, Δf, of 5 MHz provided a time response long enough to avoid aliasing among the most important contributions, giving a total of 1601 frequency samples for each azimuth position. Therefore, the size of the matrix, S_{21} (f, φ), measured for each situation to be 1601 rows by 801 columns.

Figure 2.36 compares the radiation pattern measured for this position of the metal plate with the reference pattern measured in the fully anechoic chamber. It can be observed that the presence of the plate certainly affected the radiation pattern, mainly for azimuth angles in the range from 30° to 50°, as had been foreseen. The time response of S_{21} (t), as a function of the azimuth angle is the one shown in Figure 2.35(c). The presence of the direct contribution can be noticed at all angles, with a propagation time of approximately 18 nsec and an amplitude that decreased as the azimuth angle increased, due to radiation by the lower secondary lobes. The delayed contribution due to reflection from the metal plate is noticeable at the estimated delay of about 4.7 nsec, with a higher amplitude in the range of angles already observed in Figure 2.36, i.e., between 30° and 50°. For the lowest azimuth angles, mainly between 0° and 10°, some other less important echoes appeared, which can be attributed to small reflections produced by the structures that supported the antennas, Following the FFT-Based Method, the elimination of all the contributions with a propagation time greater than 22 nsec guaranteed that all the effects due to the metallic plate disappeared. The length of the S'_{21} (t, φ) time response was then determined by detecting the direct component, and considering the widening of the time-domain impulses due to the Hanning window used. Transforming this modified time response (shown in Figure 2.35(d)) back to the frequency domain, and plotting the azimuth variation of S'_{21} at the frequency of 22 GHz, i.e., S'_{21} (22 GHz, φ), a radiation pattern similar to the one obtained in an anechoic chamber was achieved. This can be checked in Figure 2.37, which shows the comparison between the reference measurement taken under anechoic conditions and the result obtained when the measurement in reverberant conditions was processed with the method based on the FFT algorithm. The resemblance between both the curves is evident, and has been quantified by the mean value of the error (0.58 dB), defined as the absolute value of the difference between the curves, and the standard deviation of that error (0.30 dB). The maximum value of the error was 2.05 dB, corresponding to the highest azimuth angles, around 90°, where the received signal levels were much lower than those of the main lobe, and consequently a certain lack of precision can be attributed to the measurement equipment itself. These error values, together with those corresponding to the rest of the analyzed situations, are summarized in Table 2.4.

To apply the MPM to the measured data in this example, the first step was to determine the frequency sweep needed for this method, as discussed earlier. In

Table 2.4 The error levels obtained for the processed results using both techniques.

	Example 1		Example 2	
	FFT	MPM	FFT	MPM
Mean Error Level (dB)	0.58	0.49	0.62	0.56
STD Error Level (dB)	0.30	0.36	0.72	0.67
Maximum Error Value (dB)	2.05	1.87	3.91	3.03

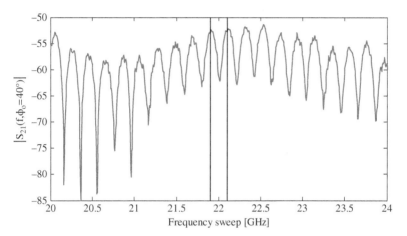

Figure 2.38 Bandwidth required by the MPM to carry out the processing is displayed between the two vertical lines whereas the bandwidth used for the FFT method is 2 GHz more in each end of the plot which has not been displayed.

this case, the estimated delay between the direct contribution and the reflected contribution was about 4.7 nsec, so the frequency bandwidth was established around 200 MHz. A total of 51 samples, from f_0 = 21.875 GHz to f_{N-1} = 22.125 GHz (as shown in Figure 2.38 in conjunction with the bandwidth used by the FFT based method) and an order of M = 3 were set as inputs to the MPM. The radiation pattern obtained by processing the S_{21} (f, φ) data using the MPM is shown in Figure 2.39. In this figure, a comparison between the reference radiation pattern and the radiation pattern reconstructed with the MPM is plotted. It can be seen that there is excellent agreement between both radiation patterns, the error being quantified by a mean of 0.49 dB, a standard deviation of 0.36 dB, and a maximum value of 1.87 dB as summarized in Table 2.4.

For the second example, the metal plate was located closer to the antennas, with d = 1 m. This situation presented a strong interaction between the antennas and the metal plate, noticeable over a wider range of azimuth angles, as shown in Figure 2.40. Figure 2.40 compares the AUT azimuthal patterns measured with

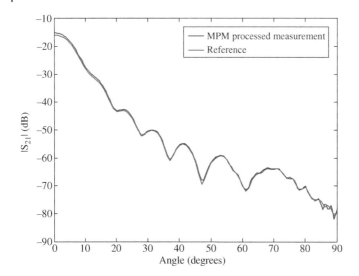

Figure 2.39 Example 1, d = 2.05 m: A comparison between the processed result using the MPM and the reference measurement.

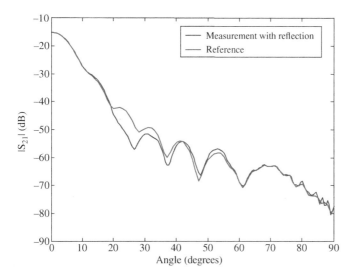

Figure 2.40 Example 2, d = 1 m: The radiation pattern measured in reverberant conditions.

and without the metal plate inside the anechoic chamber. Now, the range of azimuth angles where the pattern was evidently distorted extended from 20° to about 60°. In this situation, the estimated delay for the contribution reflected in the metal plate was 1.2 nsec. Therefore, using the same bandwidth and frequency sweep as in the previous example, a time resolution of $\delta t/10$ was obtained.

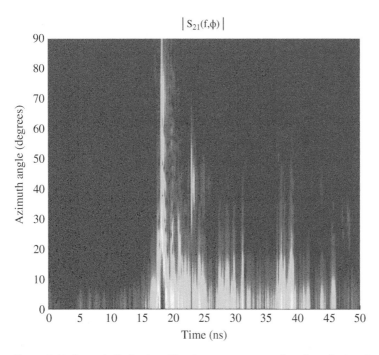

Figure 2.41 Example 2, *d* = 1 m: The time response as a function of azimuth angle.

Figure 2.41 shows the time response for this new situation, where the presence of the delayed component immediately after the main path is noticeable over a wide range of azimuth angles. In this case, when applying the method based on the Fourier transform algorithm, all the echoes with a delay greater than 1 nsec were eliminated. The FFT was performed on this modified time response, and the radiation pattern obtained at the frequency of 22 GHz again presented good agreement with the reference pattern obtained in anechoic conditions. The comparison between both patterns is presented in Figure 2.42. The mean error of the estimated pattern was 0.62 dB and the standard deviation of the error was 0.72 dB, the maximum error having reached a value of 3.91 dB.

Finally, the MPM was applied to the data measured in the second configuration. In this case, the beat frequency between the direct and reflected rays was lower than in the previous example, so the number of data samples needed for the Matrix-Pencil Method was larger. This meant that the bandwidth of S_{21} (f, φ) had to be increased up to 800 MHz, from $f_0 = 21.6$ GHz to $f_{N-1} = 22.4$ GHz, with a total of 151 samples (keeping Δf the same, used in the previous example) as seen in Figure 2.43 and comparing with the frequency samples required for a FFT based method. Figure 2.44 shows a comparison between the reconstructed

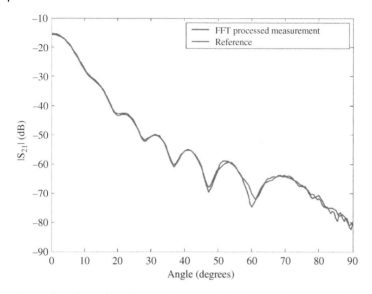

Figure 2.42 Example 2, d = 1 m: A comparison between the processed result using the Fourier-Transform Method and the reference measurement.

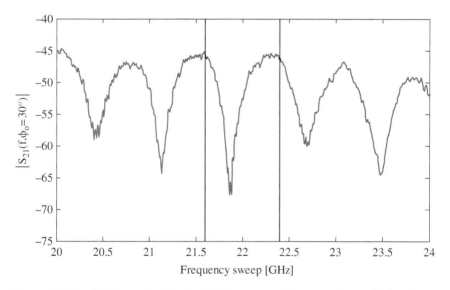

Figure 2.43 Bandwidth required by the MPM to carry out the processing is displayed between the two vertical lines whereas the bandwidth used for the FFT method is 2 GHz more in each end of the plot which has not been displayed.

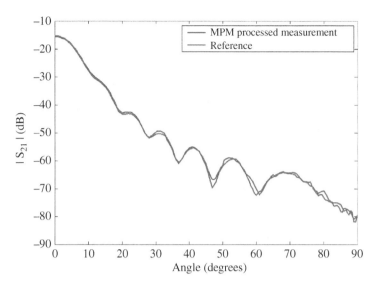

Figure 2.44 Example 2, *d* = 1 m: A comparison between the processed result using the MPM and the reference measurement.

radiation pattern and the reference pattern (measured in anechoic conditions), using three exponential terms (M = 3) to model the S_{21} (f, φ) for each azimuth angle. As a reference of the agreement between both results, the mean error level was 0.56 dB with a standard deviation of 0.67 dB, while the maximum error level was 3.3 dB as summarized in Table 2.4.

In summary, two different techniques have been used in this section to obtain the radiation pattern of antennas from measurements obtained in non-anechoic sites. One technique is based on the Fourier transform, and the other is based on the MPM. Different situations have been considered, checking how these techniques deal with echoes from nearby objects where the time delay between the direct contribution and the reflected/diffracted contributions is only a few nanoseconds. The accuracy of these techniques has been evaluated, showing examples of how the reflections and diffractions from metallic walls and objects can be suppressed. The comparison of the processed patterns with the patterns measured in an anechoic chamber has shown that the mean error of the estimations was less than 0.7 dB, and that both techniques exhibited similar behavior. Meanwhile, the maximum error found was 3.9 dB, appearing at azimuth angles where the level of the radiation pattern was below −70 dB. From the results presented, it is clear that for the parameters chosen both techniques showed very good agreement with the reference pattern measured in anechoic conditions, even in the scenario where the metal plate was separated by 1 m from the antennas. This last situation was more critical, considering that the time delay between the direct contribution and the reflected contribution was only 1.2 nsec.

Both the techniques require performing measurements in a frequency range, even though the MPM method requires a significantly less bandwidth than the FFT-based technique. Another important consideration is that the accuracy of the FFT technique is more dependent on the parameters (Δf and BW) that define the frequency sweep. The higher the bandwidth, the more accurate are the results obtained in the time domain, although it has been checked that a time resolution Δt (before windowing) of approximately δt/5 is enough to obtain satisfactory results. For that resolution, the error values found were very similar to those obtained for the highest resolutions analyzed. In Example 1, a mean error and standard deviation of 0.60 dB and 0.29 dB, respectively, were found, while the maximum value of the error was 1.79 dB. In Example 2, the mean error, standard deviation, and the maximum error were, respectively, 0.69 dB, 0.65 dB, and 3.12 dB. On the other hand, the MPM is quite dependent on the parameter, M, that defines the number of exponentials used in the expansion of Equation (2.35). The value of M may vary under noisy conditions and may affect the accuracy of the method.

Finally, some considerations must be taken into account for a practical application of these techniques. First, under the aim of developing an automatic algorithm for both techniques, a good characterization of the measurement chamber must be done. The main purpose is to determine or estimate the time delay, Δt, of the first echo arriving at the probe after the direct contribution, since this value will allow one to set up the different input parameters of both techniques (bandwidths and time resolution). The other considerations have to do with the characteristics of the different components of the measured process: depending on the amplifiers' bandwidths and/or the AUT and the frequency bands of the probe, some frequency sweeps cannot be performed under the same conditions, i.e., the components do not present constant gain in the selected band, which may result in an erroneous solution for the reconstruction of the pattern in an anechoic-chamber environment.

2.7 Application of the MPM for a Single Estimate of the SEM-Poles When Utilizing Waveforms from Multiple Look Directions

The singularity expansion method (SEM) [31, 32] provides the poles that are characteristics of the external resonances of an object and they remain invariant when the directions of the incident wave changes form snapshot to snapshot. In summary, when the transient responses from the object of interest whose SEM poles need to be found out is looked at from different angles both in azimuth and in elevation, the residues of the poles are angle dependent whereas the SEM poles modeling the time-domain waveforms are not angle dependent. In addition, for each look direction there are two possible polarizations. One could also use both polarizations to increase the number of waveforms available. Conventionally, to estimate the SEM poles from multiple look angle data one takes the average of all the various look directions waveforms and then obtains a single waveform. Then a sum of complex exponentials is used to fit the single waveform and an estimate of the SEM poles is obtained along with the averaged values of the residues. However, this is not a good approach if the signal-to-noise ratios of the different waveforms are different; namely in some of them the transient response dies down quite fast whereas in some of the responses it may continue to ring for a long time. Hence, taking an average of those two classes of waveshapes actually deteriorates the signal-to-noise ratio of the data. This is because by taking an average of the signal along with waveforms where the signal has died down may lead to an unnecessary contamination of the signal by noise. In this section, the MPM is applied to obtain a single estimate for the SEM poles utilizing simultaneously all the transient waveforms recorded from multiple look directions and without averaging them [33].

Let us denote the transient response of length $N + 1$ along a particular look direction k by the set $[Y_k]$. So that the column vector $[Y_k]$ is represented by

$$[Y_k]_{(N+1) \times 1} = \left[y_k(0); y_k(1); \dots; y_k(N) \right]^T_{1 \times (N+1)} \tag{2.38}$$

where T denotes the transpose of a matrix. The elements $y_k(j)$ represent the values of the transient response at the jth time sample, so that

$$y_k(j) = \sum_{i=1}^{M} A_k(i) \, \exp\left(s_i j \Delta t\right) \text{ for } j = 0, 1, 2, \cdots, N \tag{2.39}$$

and ΔT is the sampling time. Each transient response consists of the same M exponents s_i, which are to be solved for along their amplitudes $A_k(i)$ for a particular look direction k. Please note that M is also an unknown along with the exponents and their residues. The exponents s_i are look direction independent but not their residues $A_k(i)$. In the sampled domain, (2.39) can be rewritten as

$$y_k(j) = \sum_{i=1}^{M} A_k(i) \, z_i^j \text{ for } j = 0, 1, 2, \cdots, N \text{ with } z_i = \exp\left(s_i \Delta T\right) \tag{2.40}$$

It has been further assumed that all the waveforms from different look angles $k = 1, 2, \dots, K$, have been sampled uniformly at the same sampling rate ΔT and that each waveform contains the same number of samples $N + 1$.

Next, consider two matrices $[B_1]$ and $[B_2]$ defined as

$$[B_1]_{N \times K} = \begin{bmatrix} y_1(0) & y_2(0) & \cdots & y_K(0) \\ y_1(1) & y_2(1) & \cdots & y_K(1) \\ \vdots & & \vdots & \\ y_1(N-1) & y_2(N-1) & \cdots & y_K(N-1) \end{bmatrix}_{N \times K}$$

$$[B_2]_{N \times K} = \begin{bmatrix} y_1(1) & y_2(1) & \cdots & y_K(1) \\ y_1(2) & y_2(2) & \cdots & y_K(2) \\ \vdots & & \vdots & \\ y_1(N) & y_2(N) & \cdots & y_K(N) \end{bmatrix}_{N \times K}$$

Now it can be shown that the two matrices $[B_1]$ and $[B_2]$ can be decomposed into

$$[B_1]_{N \times K} = [Z_1]_{N \times M} [I]_{M \times M} [A]_{M \times K}$$
$$\text{and} \quad [B_2]_{N \times K} = [Z_1]_{N \times M} [Z_0]_{M \times M} [A]_{M \times K} \tag{2.41}$$

where

$$[Z_1]_{N \times M} = \begin{bmatrix} 1 & 1 & \cdots & 1 \\ z_1 & z_2 & \cdots & z_M \\ z_1^2 & z_2^2 & \cdots & z_M^2 \\ \vdots & \vdots & & \\ z_1^{N-1} & z_2^{N-1} & \cdots & z_M^{N-1} \end{bmatrix}_{N \times M} \qquad [Z_0]_{N \times M} = \begin{bmatrix} z_1 & & & O \\ & z_2 & & \\ & & \ddots & \\ O & & & z_M \end{bmatrix}_{M \times M}$$

$$[A]_{M \times K} = \begin{bmatrix} A_1(1) & A_2(1) & \cdots & A_k(2) \\ A_1(2) & A_2(2) & \cdots & A_k(2) \\ \vdots & & \vdots & \\ A_1(M) & A_2(M) & \cdots & A_k(M) \end{bmatrix}_{M \times K}$$

Now if the matrix pencil $[B_2] - \lambda [B_1]$ is considered, then one observes

$$[B_2] - \lambda [B_1] = [Z_1] \{ [Z_0] - \lambda[I] \} [A] \tag{2.42}$$

This matrix pencil becomes linearly dependent when λ is one of the system poles as then the rank of $\{ [Z_0] - \lambda[I] \}_{M \times M}$ is reduced by one as $\lambda = z_i$. Equation (2.42) can be transformed into a computationally palatable form by considering the ordinary eigenvalue problem in either of the following forms:

$$[B_2][B_1]^{\dagger} - \lambda\,[I] \quad \text{or} \quad [I] - \lambda[B_1][B_2]^{\dagger} \tag{2.43}$$

where the superscript $\dagger$ is the pseudo-inverse of the respective matrices. The pseudo-inverse is defined in terms of the singular value decompositions of the respective matrix. Let

$$[B_1]_{N \times K} = [U_1]_{N \times N} \begin{bmatrix} \sigma_1^2 & & & & \bigcirc \\ & \sigma_2^2 & & & \\ & & \ddots & & \\ & & & \sigma_K^2 & \\ \bigcirc & & & \vdots \end{bmatrix}_{N \times K} \qquad [V_1]_{K \times K}^{H} = [U_1]\left[\sum\right][V_1]^{H}$$

$$\tag{2.44}$$

where $[U_1]$ and $[V_1]$ are two orthogonal matrices, so that $[U_1]^{-1} = [U_1]^{H}$ and $[V_1]^{-1} = [V_1]^{H}$ and the superscript H denotes the conjugate transpose of a matrix. Here $[\sum]$ is a rectangular matrix whose diagonal elements are related to the singular values of $[B_1]$. In summary, one has the following relationships:

$$[B_1]_{N \times K}\,[\{V_c\}]_{K \times 1} = \sigma_c\,[\{U_c\}]_{N \times 1} \text{ for } c = 1,\ 2,\ldots,$$

$$[U_1]_{N \times N} = \left[\ \{u_1\}_{N \times 1} \vdots \{u_2\}_{N \times 1} \vdots \ \cdots \ \vdots \{u_N\}_{N \times 1}\ \right]_{N \times N}$$

$$[V_1]_{K \times K} = \left[\ \{v_1\}_{K \times 1} \vdots \{v_2\}_{K \times 1} \vdots \ \cdots \ \vdots \{v_K\}_{K \times 1}\ \right]_{K \times K}$$

Now the pseudo-inverse of $[B_1]$ can be computed from $[B_1]^{+} = [V_1]\,[\sum]^{-1}$

$$[U_1]^{H} \text{ with } [\sum]^{-1} = \begin{bmatrix} 1/\sigma_1^2 & & & & \bigcirc \\ & 1/\sigma_2^2 & & & \\ & & \ddots & & \\ & & & 1/\sigma_K^2 & \\ \bigcirc & & & \vdots \end{bmatrix}_{N \times K} . \text{ It is interesting to observe from}$$

(2.43) that the matrix pencil has a solution provided $K \geq M$, i.e., the multiple look directions must be greater than or equal to the number of poles of the system to be estimated. This can be a serious limitation in many cases as described in [33] as the number of exponents can be quite large for many practical systems and it may not be possible to provide as many sensors for each look directions.

Hence, this method is extended to the case where one may have $K < M$. If $K < M$, one can assume that, $N \gg K$ or M.

To deal with this more general situation consider the following two matrices $[D_1]$ and $[D_2]$. They are defined as

$[D_1]_{(L+1) \times K \cdot (N-L)}$

$$
= \begin{bmatrix}
y_1(0) & y_1(1) & \cdots & y_1(N-L-1) & y_2(0) & y_2(1) & \cdots y_2(N-L-1) & \cdots & y_K(0) & y_K(1) & \cdots & y_K(N-L-1) \\
y_1(1) & y_1(2) & \cdots & y_1(N-L) & y_2(1) & y_2(2) & \cdots & y_2(N-L) & \cdots & y_K(1) & y_K(2) & \cdots & y_K(N-L) \\
\vdots & \vdots & & \vdots & \vdots & \vdots & & \vdots & & \vdots & \vdots & & \vdots \\
\vdots & \vdots & & \vdots & \vdots & \vdots & & \vdots & & \vdots & \vdots & & \vdots \\
y_1(L) & y_1(L+1) & \cdots & y_1(N-1) & y_2(L) & y_2(L+1) & \cdots y_2(N-1) & \cdots & y_K(L) & y_K(L+1) & \cdots & y_K(N-1)
\end{bmatrix}_{(L+1) \times K \cdot (N-L)}
$$

$[D_2]_{(L+1) \times K \cdot (N-L)}$

$$
= \begin{bmatrix}
y_1(1) & y_1(2) & \cdots & y_1(N-L) & y_2(1) & y_2(2) & \cdots & y_2(N-L) & \cdots y_K(1) & y_K(2) & \cdots & y_K(N-L) \\
y_1(2) & y_1(3) & \cdots y_1(N-L+1) & y_2(2) & y_2(3) & \cdots & y_2(N-L+1) & \cdots y_K(2) & y_K(3) & \cdots & y_K(N-L+1) \\
\vdots & \vdots & & \vdots & \vdots & \vdots & & \vdots & & \vdots & \vdots & & \vdots \\
\vdots & \vdots & & \vdots & \vdots & \vdots & & \vdots & & \vdots & \vdots & & \vdots \\
y_1(L+1) & y_1(L+2) & \cdots & y_1(N) & y_2(L+1) & y_2(L+2) & \cdots & y_2(N) & \cdots y_K(L+1) & y_K(L+2) & \cdots & y_K(N)
\end{bmatrix}_{(L+1) \times K \cdot (N-L)}
$$

Next, it can be shown that the two matrices $[D_1]$ and $[D_2]$ can be factored into

$$
[D_1]_{(L+1) \times K \cdot (N-L)} = [P]_{(L+1) \times M} [I]_{M \times M} [R]_{M \times (K \cdot M)} [Q]_{(K \cdot M) \times K \cdot (N-L)} \quad \text{and}
$$

$$
[D_2]_{(L+1) \times K \cdot (N-L)} = [P]_{(L+1) \times M} [Z_0]_{M \times M} [R]_{M \times (K \cdot M)} [Q]_{(K \cdot M) \times K \cdot (N-L)}
$$

$$(2.45)$$

where

$$
[P]_{(L+1) \times M} = \begin{bmatrix}
1 & 1 & \cdots & 1 \\
z_1 & z_2 & \cdots & z_M \\
z_1^2 & z_2^2 & \cdots & z_M^2 \\
\vdots & \vdots & & \\
z_1^L & z_2^L & \cdots & z_M^L
\end{bmatrix}_{(L+1) \times M}
$$

$$
[R]_{M \times K \cdot M} = \begin{bmatrix}
A_1(1) & & O & \vdots A_2(1) & & O & \vdots & \cdots & \vdots A_K(1) & & O \\
& A_1(2) & & \vdots & A_2(2) & & \vdots & \cdots & \vdots & A_K(2) & \\
& & \ddots & \vdots & & \ddots & \vdots & \cdots & \vdots & & \ddots & \vdots \\
O & & A_1(M) & \vdots \; O & & A_2(M) & \vdots & \cdots & \vdots \; O & & A_K(M)
\end{bmatrix}_{M \times K \cdot M}
$$

$$[Q]_{(K \cdot M) \times [K \cdot (N-L)]} = \begin{bmatrix} [Q_1]_{M \times (N-L)} & & & O \\ & [Q_1]_{M \times (N-L)} & & \\ & & \ddots & \\ O & & & [Q_1]_{M \times (N-L)} \end{bmatrix}_{(K \cdot M) \times [K \cdot (N-L)]}$$

where

$$[Q_1]_{M \times (N-L)} = \begin{bmatrix} 1 & 1 & \cdots & 1 \\ z_1 & z_2 & \cdots & z_M \\ z_1^2 & z_2^2 & \cdots & z_M^2 \\ \vdots & \vdots & & \vdots \\ z_1^{N-L-1} & z_2^{N-L-1} & \cdots & z_M^{N-L-1} \end{bmatrix}^T_{(N-L) \times M}$$

In addition $[I]$ and $[Z_0]$ are two diagonal matrices, which have been defined by (2.41), respectively Now if one considers the matrix pencil $[D_2] - \lambda [D_1]$ then λ becomes one of the exponents.

In order to deal with noisy data, the formulation just described is made more robust to noise. One can now consider the composite matrix $[D]$ as

$$[D]_{(L+2) \times [K \cdot (N-L)]}$$

$$= \begin{bmatrix} y_1(0) & y_1(1) & \cdots & y_1(N-L-1) & y_2(0) & y_2(1) & \cdots & y_2(N-L-1) & \cdots & y_K(0) & y_K(1) & \cdots & y_K(N-L-1) \\ y_1(1) & y_1(2) & \cdots & y_1(N-L) & y_2(1) & y_2(2) & \cdots & y_2(N-L) & \cdots & y_K(1) & y_K(2) & \cdots & y_K(N-L) \\ \vdots & \vdots & & \vdots & \vdots & \vdots & & \vdots & & \vdots & \vdots & & \vdots \\ \vdots & \vdots & & \vdots & \vdots & \vdots & & \vdots & & \vdots & \vdots & & \vdots \\ y_1(L) & y_1(L+1) & \cdots & y_1(N-1) & y_2(L) & y_2(L+1) & \cdots & y_2(N-1) & \cdots & y_K(L) & y_K(L+1) & \cdots & y_K(N-1) \\ y_1(L+1) & y_1(L+2) & \cdots & y_1(N) & y_2(L+1) & y_2(L+2) & \cdots & y_2(N) & \cdots & y_K(L+1) & y_K(L+2) & \cdots & y_K(N) \end{bmatrix}_{(L+2) \times [K \cdot (N-L)]}$$

It is important to note that $[D_1]$ is obtained from $[D]$ by eliminating its last row and $[D_2]$ is obtained from $[D]$ by eliminating the first row. One can now perform a singular value decomposition of $[D]$ to obtain

$$[D]_{(L+2) \times [K \cdot (N-L)]} = [U]_{(L+2) \times (L+2)} \left[\sum\right]_{(L+2) \times [K \cdot (N-L)]}$$

$$[V]^H_{[K \cdot (N-L)] \times [K \cdot (N-L)]} \tag{2.46}$$

To combat the effects of noise and to determine the order M, a singular value filtering of $[\sum]$ is carried out by retaining its dominant singular values. The details are available in [4–8]. Also, it can be seen that the ordinary eigenvalue problem of $[D_2] - \lambda [D_1] = 0$ can be transformed into

$$[U_2]_{(L+1) \times (L+2)} \left[\sum\right]_{(L+2) \times [K \cdot (N-L)]}$$

$$[V]^H_{[K \cdot (N-L)] \times [K \cdot (N-L)]} = \lambda$$

$$\left[[U_1]_{(L+1) \times (L+2)} \left[\sum\right]_{(L+2) \times [K \cdot (N-L)]} [V]^H_{[K \cdot (N-L)] \times [K \cdot (N-L)]} \right]$$

where $[U_2]$ and $[U_1]$ are obtained from $[U]$ by eliminating the first and the last row, respectively. Then the exponents are obtained to form the solution of either of the following four ordinary eigenvalue problem:

$$[U_2]^H[U_2] - \lambda [U_2]^H[U_1] = 0; \qquad [U_2][U_2]^H - \lambda [U_1][U_2]^H = 0;$$
$$[U_2][U_1]^H - \lambda [U_1][U_1]^H = 0; \qquad [U_1]^H[U_2] - \lambda [U_1]^H[U_1] = 0$$

$$(2.47)$$

Once the exponents are obtained the residues at the poles can be computed from

$$[Y] = [Z][A]; \text{where} \qquad\qquad\qquad\qquad\qquad\qquad\qquad (2.48)$$

$$\begin{bmatrix} y_1(0) & y_2(0) & \cdots & y_K(0) \\ y_1(1) & y_2(1) & \cdots & y_K(1) \\ \vdots & \vdots & & \\ y_1(N) & y_2(N) & \cdots & y_K(N) \end{bmatrix}_{(N+1) \times K} = \begin{bmatrix} 1 & 1 & \cdots & 1 \\ z_1 & z_2 & \cdots & z_M \\ z_1^2 & z_2^2 & \cdots & z_M^2 \\ \vdots & \vdots & & \\ z_1^N & z_2^N & \cdots & z_M^N \end{bmatrix}_{(N+1) \times M}$$

$$\times \begin{bmatrix} A_1(1) & A_2(1) & \cdots & A_k(2) \\ A_1(2) & A_2(2) & \cdots & A_k(2) \\ \vdots & \vdots & & \\ A_1(M) & A_2(M) & \cdots & A_k(M) \end{bmatrix}_{M \times K}$$

The various residues can now be computed from the least squares solution of (2.48) as

$$[A] = [Z]^+[Y] = \left\{[Z]^H[Z]\right\}^{-1}[Z]^H[Y]. \qquad\qquad\qquad (2.49)$$

Numerical examples of how such problems can be handled is addressed in [33]. Application of MPM in extrapolating the time domain response has been presented in [34–37].

2.8 Direction of Arrival (DOA) Estimation Along with Their Frequency of Operation Using MPM

One of the applications of the MPM is to estimate the direction of arrival (DOA) of the signal along with its unknown frequency. In a typical DOA estimation problem, it is often assumed that all the signals are arriving at the antenna array at the same frequency which is assumed to be known. The antenna elements in the array are then placed half wavelength apart at the frequency of operation. However, in practice seldom all the signals arrive at the antenna array at a single pre-specified frequency, but at different frequencies. The question then is what to do when there are signals of multiple frequencies impinging at the array from different directions, which are unknown. This section presents an extension of the MPM to simultaneously estimate the DOA along with their operating frequency for each of the signals. This novel approach involves approximating the voltages that are induced in a three-dimensional (3-D) antenna array, by a sum of complex exponentials by jointly estimating the direction of arrival (both azimuth and elevation angles) along with the carrier frequencies of multiple far-field sources impinging on the array by using the three-dimensional MPM [38].

The MPM is a direct data domain method for approximating a function by a sum of complex exponentials in the presence of noise. The variances of the estimates computed by the MPM are quite close to the Cramer–Rao bound (CRB) [5]. Computer simulation results are provided to illustrate the performance of this novel technique.

The noise contaminated signal model, consisting of 3-D exponential sinusoids, is formulated as

$$\tilde{v}(a;b;c) = \sum_{i=1}^{I} M_i e^{j(\gamma_i + \frac{2\pi}{\lambda_i}\Delta_x \cos\phi_i \sin\theta_i a + \frac{2\pi}{\lambda_i}\Delta_y \sin\phi_i \sin\theta_i b + \frac{2\pi}{\lambda_i}\Delta_z \cos\theta_i c)} + w(a,b,c)$$

(2.50)

where $w(a, b, c)$ is a zero mean Gaussian white noise with variance κ. A three dimensional array of omni directional isotropic point sensors are considered in this section. The distance between the antenna elements all the three Cartesian directions are $\Delta_x = 0.5$ m, $\Delta_y = 0.5$ m, and $\Delta_z = 0.5$ m. The size of the antenna elements are $a = 1,..., A$, $b = 1,..., B$, and $c = 1,..., C$. It is assumed that there are three signals that are impinging on the array with amplitudes $M_1 = M_2 = M_3 = 1$.

Table 2.5 Summary of the Signal Parameters Incident on the Antenna Array.

	Signal I	Signal 2	Signal 3
Frequency	300 MHz	290 MHz	280 MHz
ϕ	30°	40°	50°
θ	45°	35°	25°

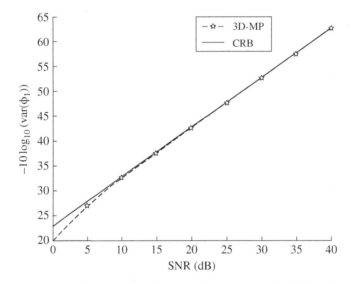

Figure 2.45 The variance - $10\log_{10}(\text{var}(\phi_1))$, 3-D MP and the CRB are plotted against the SNR.

Numerical examples illustrate the performance of the estimator in the presence of white Gaussian noise. The attributes of the signals are given in Table 2.5. The three signals are assumed to have a phase of $\gamma_i = 0$ degrees.

The number of the antenna elements along each Cartesian directions are, $A = B = C = 10$. The inverse of the sample variance of the estimates of ϕ_1 (azimuth angle), θ_1(elevation angle), λ_1(wavelength) is compared against the corresponding CRB versus signal-to-noise ratio (SNR) of the incoming signals and are plotted in Figures 2.45 to 2.47. Different values of SNR are plotted along the x-axis and the inverse of the variance of the estimated azimuth, elevation angles and wavelength are in logarithmic domain, $-10\log_{10}(\text{var}(\cdot))$ is shown along the y-axis.

The variances of the estimated values of elevation and azimuth angle and the wavelength of the first source plotted against SNR are shown below. The results

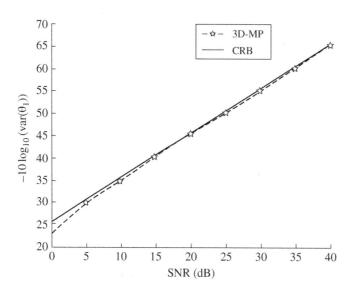

Figure 2.46 The variance - $10\log_{10}(\text{var}(\theta_1))$, 3-D MP and the CRB are plotted against the SNR.

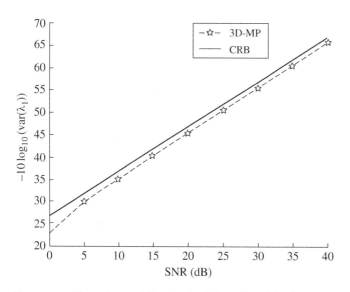

Figure 2.47 The variance - $10\log_{10}(\text{var}(\lambda_1))$, 3-D MP and the CRB are plotted against the SNR.

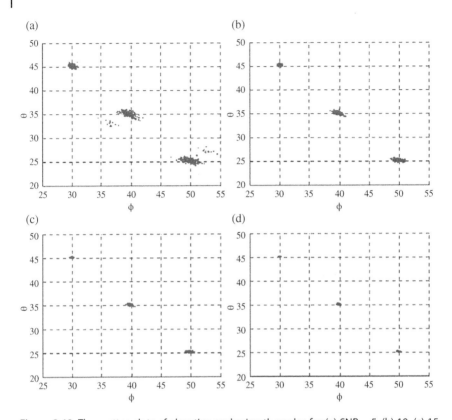

Figure 2.48 The scatter plots of elevation and azimuth angles for (a) SNR = 5, (b) 10, (c) 15, and (d) 25 dB.

are based on 1000 Monte Carlo simulations. Since the simulation results for other two signals have very similar characteristic, only the first signals results are provided.

The scatter plot of the estimated elevation and azimuth angles are shown in Figure 2.48 for different signal-to-noise (SNR) ratios of SNR = 5, 10, 15, and 25 dB. The results are based on 200 Monte Carlo simulations. As it is expected, when the SNR increases, the estimated values approach to its true values in the scatter plot.

The histogram of the estimated elevation and azimuth angles and the wavelength of the sources are shown in Figures 2.49–2.51 for different signal-to-noise ratios of SNR = 5, 10, 15, and 25 dB. The results are based on 1000 Monte Carlo simulations. As can be seen, for the increased SNR values, the estimated values approach to its true values in the histogram plot. Other examples are available in [39–51].

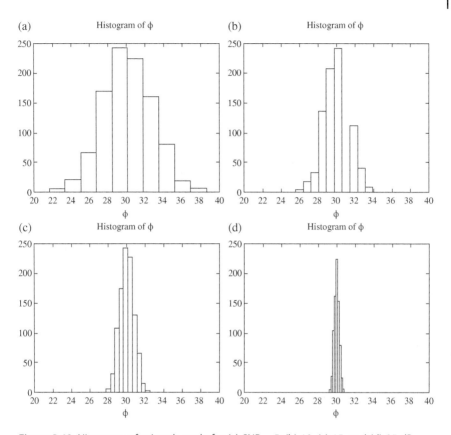

Figure 2.49 Histogram of azimuth angle for (a) SNR = 5, (b) 10, (c) 15, and (d) 25 dB.

2.9 Efficient Computation of the Oscillatory Functional Variation in the Tails of the Sommerfeld Integrals Using MPM

Another application of the MPM is to generate fast and accurate results in the evaluation of the oscillatory functional variation in the tails of the Sommerfeld integrals (SI) encountered in radiation and scattering from a two media problem. The oscillating infinite domain Sommerfeld integrals are difficult to integrate using a numerical procedure when dealing with structures in a layered media, even though several researchers have attempted to do that. Generally, integration along the real axis is used to compute the SI. However, significant computational effort is required to integrate the oscillating and slowly decaying function along the tail. Extrapolation methods are generally applied to

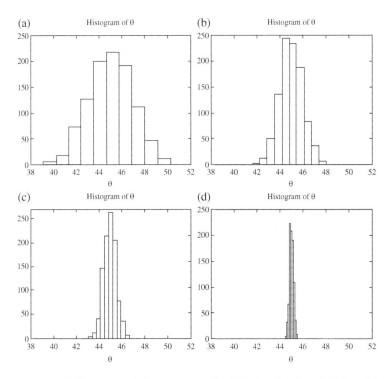

Figure 2.50 Histogram of elevation angle for (a) SNR = 5, (b) 10, (c) 15, and (d) 25 dB.

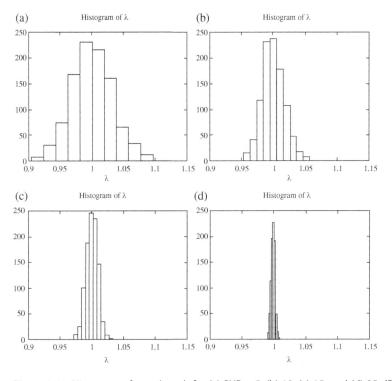

Figure 2.51 Histogram of wavelength for (a) SNR = 5, (b) 10, (c) 15, and (d) 25 dB.

accelerate the rate of convergence of these integrals. However, there are difficulties with the extrapolation methods, such as locations for the breakpoints and the like.

In this section, a simplified approach for accurate and efficient calculation of the integrals dealing with the oscillatory functional variations in the tails of the SI is presented. The oscillatory functional variation in the tail is approximated by a sum of finite (usually 10 to 20) complex exponentials using the MPM. The integral of the tail of the SI is then simply calculated by summing some complex numbers. No numerical integration is needed in this process, as the integrals approximated by complex exponentials can be performed analytically. Good accuracy is achieved with a small number of evaluations for the integral kernel (60 points for the MPM, as compared with hundreds or thousands of functional evaluations using the traditional extrapolation methods) along the tails of the SI [52, 53].

The simplest of the SI is the one encountered with the Sommerfeld identity. Suppose one has the following SI

$$\int_a^\infty \frac{e^{-jk_z|z|}}{jk_z} J_0(\lambda_\rho)\lambda d\lambda = \frac{e^{-jkr}}{r} - \int_0^a \frac{e^{-jk_z|z|}}{jk_z} J_0(\lambda_\rho)\lambda d\lambda, \qquad (2.51)$$

where $r = \sqrt{\rho^2 + z^2}$ and $k_z = \sqrt{k^2 - \lambda^2}$. $k = k_0\sqrt{\varepsilon}$ is the wave number in the media with dielectric constant ε, where k_0 is the wave number in free space. The dielectric constant is selected as $\varepsilon = 16 - j0.1$ in this case. There is a singularity at $\lambda = k_0\sqrt{\varepsilon}$. To avoid this singularity, the integration of the right hand side of (2.51) is computed along the path from 0 to a as shown in Figure 2.52. It is integrated to machine accuracy by using the adaptive Lobatto quadrature. The integration for the tail starts at a.

The most troublesome case is the one when $z = 0$, in which case the tail is oscillating and slowly decaying to 0 as $\lambda \to \infty$. The remaining parameters to be decided are the samples $\{f_p\}$ used as inputs for the MPM. Simulation shows that stable and accurate results can be obtained by equally sampling $f(\lambda)$ with $m = 6$ points per half period, and $K = 10$ half periods are good enough for the extrapolation from b to ∞. Hence the total number of evaluations of the

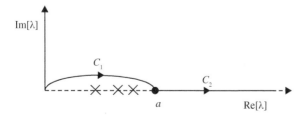

Figure 2.52 Path for typical Sommerfeld integration.

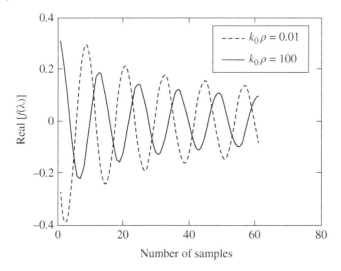

Figure 2.53 Functional samples of the tails of the Sommerfeld integral used in MPM.

integrand is $N = 60$. Figure 2.53 shows the real part of the 60 samples of the tail used in the MPM for $k_0\rho = 0.01$ and $k_0\rho = 100$. The two series of samples show the similar oscillating and slow decaying shape. For larger $k_0\rho$, the rate of decay is smaller. Note that the plots shown in Figure 2.53 are typical illustrations of the oscillatory functional variations occurring along the tails of the SI. Additional real forms of the SI for a stratified media due to horizontal or vertical sources have also been computed by using the MPM.

Simulation results are as shown in Figures 2.54 through 2.58 for different values of *tol* along $k_0\rho$ from 10^{-3} to 10^3. Note that $k_0 = 2\pi/\lambda_0$ where λ_0 is the wavelength in free space. Recall that the parameter *tol* is the tolerance, related to the ratio of the singular values in the MPM. Hence ρ varies from around $1.6 \times 10^{-4}\lambda_0$ to $160\lambda_0$. The integration results occurring on the left hand side of (2.51) are compared with the results obtained from the right hand side, and the normalized errors are plotted. Among the various traditional extrapolation methods, the weighted-average method (WAM) is one of the most versatile and efficient convergence accelerators for evaluation of the tails of the Sommerfeld integrals. For this reason, the asymptotic WAM is chosen to compare the performance to the proposed MPM. This parameter is also applied as a measure of relative error for adaptive Lobatto quadrature used for integration in each half period in the WAM. The number of half periods is set to 10 for both methods.

Figure 2.54 shows the normalized errors for the MPM and the WAM when $tol = 10^{-14}$. The performance of the WAM is better when $k_0\rho$ is small. However, when $k_0\rho$ is larger than 1 ($\rho > 0.16\lambda_0$), the MPM has better accuracy than the

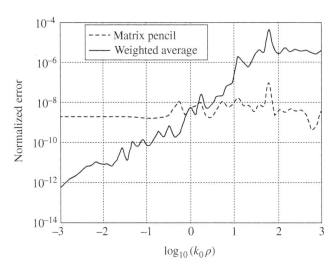

Figure 2.54 Normalized errors for the tail integration, $tol = 10^{-14}$.

WAM. Moreover, the MPM is more stable than WAM when $k_0\rho$ increases. The normalized error of MPM remains below 10^{-7} for all $k_0\rho$.

The computational CPU-times involved for both methods are shown in Figure 2.55. To calculate the integrals to $tol = 10^{-14}$, the MPM is generally 19 times faster than the WAM. An average of 17 exponentials are required to approximate the function $f(\lambda)$. The dominant computational time of WAM is spent on the evaluation of the integrand. Figure 2.56 shows that thousands of integrand evaluations are required for the WAM compared to a constant 60 evaluations for the MPM.

When tol becomes higher from 10^{-14} to 10^{-10}, the performance of the MPM remains consistent, as shown in Figure 2.57 and Figure 2.58. However, the performance of the WAM deteriorates. For $tol = 10^{-12}$ as in Figure 2.59, the accuracy of the MPM is better than the WAM for most $k_0\rho$, and the CPU-time for the WAM averages 8 times longer than the MPM. An average of 14.5 exponential functions are required to approximate $f(\lambda)$. For $tol = 10^{-10}$, the CPU-time used for the WAM is 2 times longer than MPM since less evaluations of the integrand is necessary, but the performance of the WAM is even worse. An average of 13 exponentials are required to approximate $f(\lambda)$ when $tol = 10^{-10}$. Based on the numerical results shown in Figures 2.54–2.59, one can observe that the MPM is generally more accurate than the WAM. The MPM has two additional advantages: greater robustness and faster computational time.

Other applications for efficient characterization of the Green's function is available in [54–57].

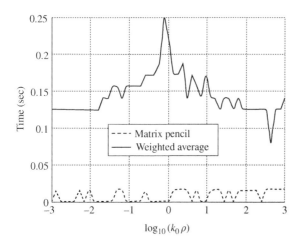

Figure 2.55 Comparison of CPU times for the two methods, $to1 = 10^{-14}$.

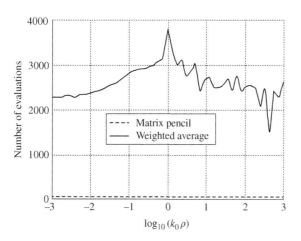

Figure 2.56 Number of functional evaluations for the two methods, $to1 = 10^{-14}$.

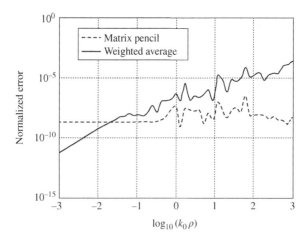

Figure 2.57 Normalized errors for the tail integration, $to1 = 10^{-12}$.

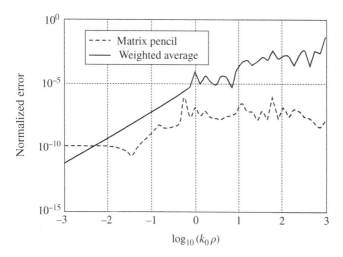

Figure 2.58 Normalized errors for the tail integration, *to1* $= 10^{-10}$.

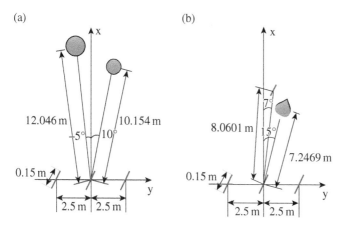

Figure 2.59 (a) Two sphere model. (b) One wire and one cone model.

2.10 Identification of Multiple Objects Operating in Free Space Through Their SEM Pole Locations Using MPM

Another application of the MPM is to identify multiple objects in free space using their natural resonant frequencies [58]. By observing the complete impulse response, the presence of multiple objects can be isolated in the time domain. Then the MPM can be applied to the late time response of this transient temporal impulse response. In the time domain it is relatively easy to locate the late time response. MPM approximates a time domain function by a sum of complex exponentials and this approximation is valid only for the late time response. One can also generate a library of poles associated with the objects using the frequency domain response (as will be illustrated in the next Chapter 3), and the actual poles computed using the time domain data, we illustrate that the correlation between the two pole sets obtained using (both time domain and frequency domain information) totally different methodologies provide a robust identification procedure. The poles using responses from data generated in different domains can be used for comparison purposes. In addition, one can find the coordinates of the unknown object using the Time-Difference-of-Arrival (TDOA) technique.

Two simulation examples are presented to illustrate the application of the proposed methodology for detection of multiple objects. The specification for each of the antenna for transmitter and receiver are 0.15 m in length and 1.5 mm radius. Spacing between the transmitter and the receiver is 2.5 m to fully minimize the effects of the antenna coupling. The transmitting antenna is excited by a voltage generator (delta-function generator) with a 1 V excitation. The response of the object is computed from 0.01 GHz to 5 GHz (sampling frequency $\Delta f = 0.01$ GHz), and the number of samples used is 500.

Figure 2.59 shows the Higher Order Basis Based Integral Equation Solver (HOBBIES) [21] simulation models for the case of two PEC spheres, one PEC cone and one PEC wire case. All are radiating in free space for these examples.

For the two PEC spheres model, the diameters of the first sphere and the second sphere are 0.1 m and 0.15 m, respectively. The location of the first sphere from the origin (the location of the transmitter) and in angle with respect to the normal joining the three antennas are 10.154 m and 10°. The coordinates of the second PEC sphere is 12.046 m radial distance and −5° azimuthal angle. For one PEC cone and one PEC wire model, the coordinates of the 0.1 m diameter and 0.1 m high PEC cone and the 0.1 m long, 1 mm radius PEC wire from the origin are 7.2469 m in radial distance, 15° in azimuthal angle and 8.0601 m in radial distance, 7° azimuthal angle, respectively as seen in Figure 2.59.

Tables 2.6 and 2.7 describe the actual vs. estimated coordinates of the unknown objects from the origin using the TDOA technique. Figures 2.60 and 2.61 compare the pole library using the Cauchy method with the computed

Table 2.6 Actual vs. Estimated Target Coordinates (Two Spheres).

	Sphere (D=0.1 m)		Sphere (D=0.15 m)	
	R(m)	Angle (°)	R(m)	Angle (°)
Actual target coordinates	10.104	10	11.971	−5
Estimated target coordinates	10.133	10.071	12.002	−4.9779

Table 2.7 Actual vs. Estimated Target Coordinates (a Cone and a Wire).

	Cone		Wire	
	R(m)	Angle (°)	R(m)	Angle (°)
Actual target coordinates	7.2469	15	8.0591	7
Estimated target coordinates	7.2642	15.036	8.0802	7.0419

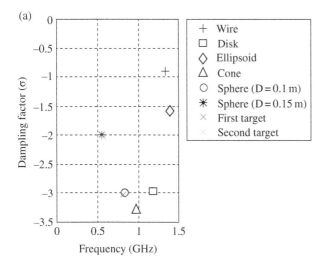

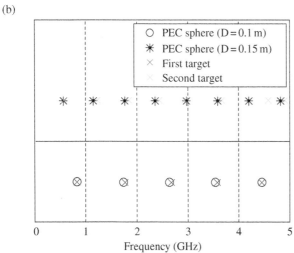

Figure 2.60 Pole Library vs. Computed poles of the unknown objects (two PEC spheres) using the MP method; (a) First order pole, (b) Resonant frequency.

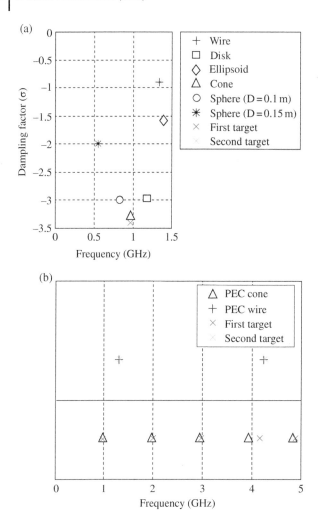

Figure 2.61 Pole Library vs. Computed poles of the unknown objects (one PEC cone and one PEC wire) using the MP method; (a) First order pole, (b) Resonant frequency.

poles of the unknown objects using the MPM. From Table 2.6 and Figure 2.60, one can locate the 0.1 m diameter PEC sphere with approximately a 99 % accuracy with 10.133 m radial distance, 10.071° azimuthal angle. The 0.15 m diameter PEC sphere is located with approximately a 95 % accuracy in 12.002 m radial distance and −4.9779° azimuthal angle. One also can identify the PEC cone (0.1 m diameter and 0.1 m height) with approximately a 97 % accuracy in 7.2642 m (*R*) with an angle of 15.036° and the PEC wire (0.1 m diameter, 1 mm radius) with approximately a 99 % accuracy in 8.0802 m (*R*), an angle 7.0419° as shown in Table 2.7 and Figure 2.61. Finally, one can detect and identify unknown multiple objects with high accuracy using the proposed methodology.

2.11   Other Miscellaneous Applications of MPM

This technique has also been used in system identification [59–67] and in characterization of medical signals [68] and in transient spectroscopy [69]. It has found applications in system modelling [70, 71], in signal analysis [72, 73] and to characterize the material parameters of a media including the dielectric constant and loss tangent in [74–76]. Further applications can be found in multipath characterization for a wave propagating in a medium [77, 78], in the characterization of the quality of power systems which in addition may also be affected by environmental effects [79–83], in waveform analysis and imaging [84–86] and in array characterization [87, 88]. Finally, it has been used in speeding up of FDTD in transmission line analysis [89].

2.12   Conclusion

In this chapter, the MPM is described from the development of the theory to its application through some examples. The basic assumption of the MPM is that for a LTI system its eigen functions are the exponential signal model with the poles and residues and the goal is to approximate the system responses by these eigen functions which is the mathematically the robust approximation possible. The main key to address this problem then is to estimate the poles and residues with high accuracy computed in an efficient way. The five steps to accomplish this goal have been introduced to solve this problem from selecting the number of samples to interpolate/extrapolate the data with the estimated poles and residues. This chapter also deals with data that are both noiseless and contaminated with noise. For the noiseless case, one can compute the poles and residues simultaneously to interpolate/extrapolate the data with high accuracy using only 14 sampling points. Furthermore, when the data is contaminated by additive noise with a noise level of approximately 15 dB SNR, it also have high performance to interpolate and extrapolate the data with 100 available data samples. As a result, the MPM is a high resolution method to interpolate/ extrapolate the data with accuracy and computational efficiency. The MPM is also more robust to noise present in the sampled data than any of the other cotemporary parameter estimation techniques as outlined in the references.

   Some applications of the MPM has been presented to illustrate the application of this novel methodology. Therefore, one can address many practical problems using the MPM. A computer code implementing the MPM using MATLAB is presented in the APPENDIX for an interested reader.

Appendix 2A   Computer Codes for Implementing MPM

MATLAB CODES

Example 2.1

```
%% Matrix Pencil method: Example 2.1
format short     % Format for displaying numbers
om1=0.2*pi;      % Angular frequency
om2=0.35*pi;
ap1=0.02*pi;     % Damping factor
ap2=0.035*pi;
t=0:0.5:100;
y=exp(-ap1.*t).*sin(om1*t)+exp(-ap2.*t).*sin(om2*t);

%% STEP 1. Choose the number of time sample points
Nt=t(1:5:floor(end/3)); % Selected sampling data (Time)
dt=Nt(2)-Nt(1);        % Delta_t
Ny=y(1:5:floor(end/3)); % Selected sampling data
(System response)
N=length(Ny);          % Number of samples
L=floor(N/2-1);        % Pencil parameter, between N/3 to N/2

%% STEP 2. Creating matrix Y
%% STEP 3. Select optimal number of sigular values (M)
%% STEP 4. Compute residues and poles
%% STEP 5. Generate signal using estimated poles and residues
[R,P,yr] = mpm_fn(Ny,L,dt,t);

MSE=norm(yr-y)/norm(y)   % Mean Square Errors
R     % Residues
P     % Poles
% Relative error
Er_ap1=abs((real(-P(1))-ap1))/ap1*100
Er_w1=abs((imag(P(1))-om1))/om1*100
Er_ap2=abs((real(-P(3))-ap2))/ap2*100
Er_w2=abs((imag(P(3))-om2))/om2*100

% Plot for the comparison between original data and the
estimated data
```

```
figure(2);
plot(t,real(y),'b-','LineWidth',1.2); hold on; grid on;
plot(t,real(yr),'r--','LineWidth',1.2);
plot(Nt,real(Ny),'r*');
legend('Data','MP Method','Selected samples');
plot([Nt(end) Nt(end)],[-1 2],'k--');
xlabel('Time (sec)','fontsize',11)
ylabel('Amplitude','fontsize',11)
hold off


function [R,P,yr] = mpm_fn(Ny,L,dt,t)

% Function of the Matrix Pencil Method
% R:   Residues
% P:   Poles
% yr:     Estimated data
% Ny:     Selected sampling data (System response)
% L:   Pencil parameter
% dt:     Delta_t
% t:   Time data

%% STEP 2. Creating matrix Y
N=length(Ny);        % Sampling number
for k=1:N-L
    Y(k,1:L+1)=Ny(k:L+k);    % (N-L) by (L+1) matrix
end

%% STEP 3. Select optimal number of sigular values (M)
[U,S,V]=svd(Y);        % Singular value decomposition
N_SV=10*log10(diag(S)./diag(S(1,1)));  % Normalized
singular values
figure(1)        % Plot normalized singular values
plot(1:length(N_SV),N_SV,'b-',1:length(N_SV),
N_SV,'b.','MarkerSize',11); grid on;
xlabel('Number of Singular values', 'fontsize',11);
ylabel('Normalized singular values (dB scale)',
'fontsize',11);

% M is chosen by the number of accurate significant decimal
digits
M = input('Please input the number of singular values:
');
```

```
%% STEP 4. Compute residues and poles
UU=U(:,1:M);
VV=V(:,1:M);
SS=S(1:M,1:M);
YY=UU*SS*ctranspose(VV);
V2=VV(1:end-1,:);
V1=VV(2:end,:);
Y1=UU*SS*ctranspose(V1);
Y2=UU*SS*ctranspose(V2);

[A,B]=eig(pinv(V2)*V1);
Z0=B(1:M,1:M);
zi=diag(Z0);    % M by M matrix
P=log(zi)/dt    % Poles

for k=1:N
    ZI(k,1:M)=(zi.^(k-1));    % (N-L) by (L+1) matrix
end
R=(pinv(ZI)*transpose(Ny));    % Residues

%% STEP 5. Generate signal using estimated poles and
residues
yr=0;
for k=1:M;
yr=yr+R(k).*(exp(P(k)*t));    % Estimated response
end
```

Example 2.2
```
%% Matrix Pencil method: Example 2.1
format short e    % Format for displaying numbers
om1=0.2*pi; % Angular frequency
om2=0.35*pi;
ap1=0.02*pi;        % Damping factor
ap2=0.035*pi;
t=0:0.5:100;
y=exp(-ap1.*t).*sin(om1*t)+exp(-ap2.*t).*sin(om2*t);
yy=awgn(y,15,'measured');  % Added white Gaussian noise

%% STEP 1. Choose the number of time sample points
Nt=t(1:1:100);          % Selected sampling data (Time)
dt=Nt(2)-Nt(1);         % Delta_t
Ny=yy(1:1:100);         % Selected sampling data (System
response)
```

```
N=length(Ny);      % Number of samples
L=floor(N/2-1);        % Pencil parameter, between N/3 to
N/2


%% STEP 2. Creating matrix Y
%% STEP 3. Select optimal number of sigular values (M)
%% STEP 4. Compute residues and poles
%% STEP 5. Generate signal using estimated poles and
residues
[R,P,yr] = mpm_fn(Ny,L,dt,t);


MSE=norm(yr-y)/norm(y)  % Mean Square Errors
R     % Residues
P     % Poles
% Relative error
Er_ap1=abs((real(-P(1))-ap1))/ap1*100
Er_w1=abs((imag(P(1))-om1))/om1*100
Er_ap2=abs((real(-P(3))-ap2))/ap2*100
Er_w2=abs((imag(P(3))-om2))/om2*100


% Plot for the comparison between original data and the
estimated data
figure(2);
plot(t,real(y),'b-','LineWidth',1.2); hold on; grid on;
plot(t,real(yy),'g-','LineWidth',1.2);
legend('Data','Data added noise')
xlabel('Time (sec)','fontsize',11)
ylabel('Amplitude','fontsize',11); hold off
figure(3);
plot(t,real(y),'b-','LineWidth',1.2); hold on; grid on;
plot(t,real(yy),'g-','LineWidth',1.2);
plot(t,real(yr),'r--','LineWidth',1.2);
legend('Data','Data with additive noise','MP Method')
plot([Nt(end) Nt(end)],[-1 2],'k--');
xlabel('Time (sec)','fontsize',11)
ylabel('Amplitude','fontsize',11); hold off
```

References

1 T. K. Sarkar, M. Salazar-Palma, and E. L. Mokole, "Application of the Principle of Analytic Continuation to Interpolate/Extrapolate System Responses Resulting in Reduced Computations – Part A: Parametric Methods," *IEEE Journal on*

Multiscale and Multiphysics Computational Techniques, Vol. 1, pp. 48–59, 2016.

2 Y. Hua, *On Techniques for Estimating Parameters of Exponentially Damped/Undamped Sinusoids in Noise*, Ph.D. Dissertation, Syracuse University, New York, August 1988.

3 Y. Hua and T. K. Sarkar, "Generalized Pencil-of-Function Method for Extracting Poles of an EM System from Its Transient Response," *IEEE Transactions on Antennas and Propagation*, Vol. 37, No. 2, pp. 229–234, 1989

4 Y. Hua and T. K. Sarkar, "Matrix Pencil Method for Estimating Parameters of Exponentially Damped/Undamped Sinusoids in Noise," *IEEE Transactions on Acoustics, Speech, and Signal Processing*, Vol. 38, No. 5, pp. 814–824, 1990.

5 Y. Hua and T. K. Sarkar, "On SVD for Estimating Generalized Eigenvalues of Singular Matrix Pencil in Noise," *IEEE Transactions on Signal Processing*, Vol. 39, No. 4, pp. 892–900, 1991.

6 T. K. Sarkar and O. M. Pereira-Filho, "Using the Matrix Pencil Method to Estimate the Parameters of a Sum of Complex Exponentials," *IEEE Antennas & Propagation Magazine*, Vol. 37, No.1, pp. 48–55, 1995.

7 G. Strang, *Introduction to Linear Algebra*, Cambridge Press, Wellesley, MA, Fifth Edition, 2016.

8 G. Golub and C. Van Loan, *Matrix Computations*, Johns Hopkins Studies in the Mathematical Sciences, The Johns Hopkins University Press, Baltimore, MD, 2013.

9 T. K. Sarkar, Z. A. Maricevic, and M. Kahrizi, "An Accurate De-Embedding Procedure for Characterizing Discontinuities," *International Journal of Microwave and Millimeter-Wave Computer-Aided Engineering*, Vol. 2, No. 3, pp. 135–143, 1992.

10 T. K. Sarkar, P. Midya, Z. A. Maricevic, M. Kahrizi, S. M. Rao, and A. R. Djordjevic, "Analysis of Arbitrarily Shaped Microstrip Patch Antennas Using the Sommerfeld Formulation," *International Journal of Microwave and Millimeter-Wave Computer-Aided Engineering*, Vol. 2, No. 3, pp. 168–178, 1992.

11 T. R. Arabi, A. T. Murphy, T. K. Sarkar, R. F. Harrington, and A. R. Djordjevic, "Analysis of Arbitrarily Oriented Microstrip Lines Utilizing a Quasi-Dynamic Approach," *IEEE Transactions on Microwave Theory and Techniques*, Vol. 39, No. 1, pp. 75–82, 1991.

12 T. K. Sarkar, Z. A. Maricevic, M. Kahrizi, and A. R. Djordjevic, "Frequency Dependent Characterization of Radiation from an Open End Microstrip Line," *Archiv fur Elektronik und Ubertragungstechnik*, Vol. 48, pp. 101–107, 1994.

13 S. Llorente-Romano, A. Garca-Lampérez, T. K. Sarkar, and M. Salazar-Palma, "An Exposition on the Choice of the Proper S Parameters in Characterizing Devices Including Transmission Lines with Complex Reference Impedances and a General Methodology for Computing Them," *IEEE Antennas and Propagation Magazine*, Vol. 55, No. 4, pp. 94–112, 2013.

14 O. M. C. P. Filho and T. K. Sarkar, "Full-Wave Analysis of MICS in Multilayer Dielectric Media in a Rectangular Waveguide," *IEEE Transactions on Microwave Theory and Techniques*, Vol. 48, No. 10, pp. 1611–1622, 2000.

15 M. Kahrizi, T. K. Sarkar, and Z. A. Maricevic, "Space Domain Approach for the Analysis of Printed Circuits," *IEEE Transactions on Microwave Theory and Techniques*, Vol. 42, No. 3, pp. 450–457, 1994.

16 M. Kahrizi, T. K. Sarkar, and Z. A. Maricevic, "Dynamic Analysis of a Microstrip Line Over a Perforated Ground Plane," *IEEE Transactions on Microwave Theory and Techniques*, Vol. 42, No. 5, pp. 820–825, 1994.

17 M. Kahrizi, T. K. Sarkar, and Z. A. Maricevic, "Analysis of a Wide Radiating Slot in the Ground Plane of a Microstrip Line," *IEEE Transactions on Microwave Theory and Techniques*, Vol. 41, No. 1, pp. 29–37, 1993.

18 T. K. Sarkar and M. Salazar, "An Alternate Interpretation of Complex Modes in Closed Perfectly Conducting (Lossless) Structures," *Archiv fur Elektronik und Ubertragungstechnik*, Vol. 48, No. 3, pp. 123–129, 1994.

19 R. K. Hoffmann, *Handbook of Microwave Integrated Circuits*, Artech House, Inc., Norwood, MA, 1987.

20 W. P. Harokopus, Jr., L. P. B. Katehi, W. Y. Ali-Ahmad, and G. M. Rebeiz, "Surface Wave Excitation from Open Microstrip Discontinuities," *IEEE Transactions on Microwave Theory and Techniques*, Vol. 39, No. 7, pp. 1098–1107, 1991.

21 Y. Zhang, T. K. Sarkar, X. Zhao, D. Garcia-Donoro, W. Zhao, M. Salazar-Palma, and S. Ting, *Higher Order Basis Based Integral Equation Solver [HOBBIES]*, John Wiley & Sons, Hoboken, NJ, 2012.

22 Z. A. Maricevic, T. K. Sarkar, Y. Hua, and A. R. Djordjevic, "Time-Domain Measurements with the Hewlett-Packard Network Analyzer HP 8510 Using the Matrix Pencil Method," *IEEE Transactions on Microwave Theory and Techniques*, Vol. 39, No. 3, pp. 538–547, 1991.

23 B. Fourestie, Z. Altman, and M. Kanda, "Anechoic Chamber Evaluation Using the Matrix Pencil Method," *IEEE Transactions on Electromagnetic Compatibility*, Vol. 41, No. 3, pp. 169–174, 1999.

24 B. Foureste, Z. Altman, J. Wiart, and A. Azoulay, "On the Use of the Matrix-Pencil Method to Correlate Measurements at Different Test Sites," *IEEE Transactions on Antennas and Propagation*, Vol. 47, pp. 1569–1573, 1999.

25 S. Loredo, M. R. Pino, F. Las-Heras, and T. K. Sarkar, "Echo Identification and Cancellation Techniques for Antenna Measurement in Non-Anechoic Test Sites," *IEEE Antennas and Propagation Magazine*, Vol. 46, No. 1, pp. 100–107, 2004.

26 Z. Du, J. I. Moon, S.-S. Oh, J. Koh, and T. K. Sarkar, "Generation of Free Space Radiation Patterns from Non-Anechoic Measurements Using Chebyshev Polynomials," *IEEE Transactions on Antennas and Propagation*, Vol. 58, No. 8, pp. 2785–2790, 2010.

27 D. N. Black and E. B. Joy, "Test Zone Field Compensation," *IEEE Transactions on Antennas and Propagation*, Vol. AP-43, No. 4, pp. 362–368, 1995.

28 D. A. Leatherwood and E. B. Joy, "Plane Wave, Pattern Subtraction, Range Compensation," *IEEE Transactions on Antennas and Propagation*, Vol. AP-49, No. 12, pp. 1843–1851, 2001.

29 P. S. H. Leather and D. Parsons, "Equalization for Antenna-Pattern Measurements: Established Technique – New Application," *IEEE Antennas and Propagation Magazine*, Vol. 45, No. 2, pp. 154–161, 2003.

30 F. J. Harris, "On the Use of Windows for Harmonic Analysis with the Discrete Fourier Transform," *Proceedings of the IEEE*, Vol. 66, No. 1, pp. 51–84, 1978.

31 C. E. Baum, "The Singularity Expansion Method," in *Transient Electromagnetic Fields*, L. B. Felsen, editor, Springer-Verlag, New York, 1976.

32 E. K. Miller, "Time Domain Modeling in Electromagnetics," *Journal of Electromagnetic Waves and Applications*, Vol. 8, No. 9–10, pp. 1125–1172, 1994.

33 R. S. Adve, T. K. Sarkar, O. M. Pereira-Filho, and S. M. Rao, "Extrapolation of Time-Domain Responses from Three-Dimensional Conducting Objects Utilizing the Matrix Pencil Technique," *IEEE Transactions on Antennas and Propagation*, Vol. 45, No. 1, pp. 147–156, 1997.

34 J. Ritter and F. Arndt, "Efficient FDTD/Matrix-Pencil Method for the Full-Wave Scattering Parameter Analysis of Waveguiding Structures," *IEEE Transactions on Microwave Theory and Techniques*, Vol. 44, No. 12, pp. 2450–2456, 1996.

35 S. Ciochina, R. Cacoveanu, G. Lojewski, F. Ndagijimana, and P. Saguet, "Resonator Analysis in TLM Using the Matrix Pencil Method," *Electronics Letters*, Vol. 32, No. 3, pp. 226–228, 1996.

36 R. Rezaiesarlak and M. Manteghi, "Accurate Extraction of Early-/Late-Time Responses Using Short-Time Matrix Pencil Method for Transient Analysis of Scatterers," *IEEE Transactions on Antennas and Propagation*, Vol. 63, No. 11, pp. 4995–5002, 2015.

37 R. Rezaiesarlak and M. Manteghi, "Short-Time Matrix Pencil Method for Chipless RFID Detection Applications," *IEEE Transactions on Antennas and Propagation*, Vol. 61, No. 5, pp. 2801–2806, 2013.

38 N. Yilmazer, R. Fernandez-Recio, and T. K. Sarkar, "Matrix Pencil Method for Simultaneously Estimating Azimuth and Elevation Angles of Arrival Along with the Frequency of the Incoming Signals," *Digital Signal Processing*, Vol. 16, No. 6, pp. 796–816, 2006.

39 Y. Zhengand Y. Yongzhi, "Joint Estimation of DOA and TDOA of Multiple Reflections by Matrix Pencil in Mobile Communications," *IEEE Access*, Vol. 7, pp. 15469–15477, 2019.

40 A. Gaber and A. Omar, "A Study of Wireless Indoor Positioning Based on Joint TDOA and DOA Estimation Using 2-D Matrix Pencil Algorithms and IEEE 802.11AC," *IEEE Transactions on Wireless Communications*, Vol. 14, No. 5, pp. 2440–2454, 2015.

41 S. Hwang, S. Burintramart, T. K. Sarkar, and S. R. Best, "Direction of Arrival (DOA) Estimation Using Electrically Small Tuned Dipole Antennas," *IEEE*

Transactions on Antennas and Propagation, Vol. 54, No. 11, pp. 3292–3301, 2006.

42 N. Yilmazer, J. Koh, and T. K. Sarkar, "Utilization of a Unitary Transform for Efficient Computation in the Matrix Pencil Method to find the Direction of Arrival," *IEEE Transactions on Antennas and Propagation*, Vol. 54, No. 1, pp. 175–181, 2006.

43 K. Kim, T. K. Sarkar, H. Wang, and M. Salazar-Palma, "Direction of Arrival Estimation Based on Temporal and Spatial Processing Using a Direct Data Domain (D^3) Approach," *IEEE Transactions on Antennas and Propagation*, Vol. 52, No. 2, pp. 533–541, 2004.

44 J. E. F. Del Rio and M. F. Catedra-Perez, "The Matrix Pencil Method for Two-Dimensional Direction of Arrival Estimation Employing an L-Shaped Array," *IEEE Transactions on Antennas and Propagation*, Vol. 45, No. 11, pp. 1693 – 1694, 1997.

45 Y. Li, K. J. Ray Liu, and J. Razavilar, "A Parameter Estimation Scheme for Damped Sinusoidal Signals Based on Low-Rank Hankel Approximation," *IEEE Transactions on Signal Processing*, Vol. 45, No. 2, pp. 481–486, 1997.

46 Y. Hua and H. Yang, "High SNR Perturbation in Single 2D Frequency Estimate Using Matrix Pencil," *IEEE Transactions on Signal Processing*, Vol. 43, No. 5, pp. 1291–1292, 1995.

47 C. K. E. Lau, R. S. Adve, and T. K. Sarkar, "Minimum Norm Mutual Coupling Compensation with Applications in Direction of Arrival Estimation," *IEEE Transactions on Antennas and Propagation*, Vol. 52, No. 8, pp. 2034–2041, 2004.

48 Y. Hua, "Estimating Two-Dimensional Frequencies by Matrix Enhancement and Matrix Pencil," *IEEE Transactions on Signal Processing*, Vol. 40, No. 9, pp. 2267–2280, 1992.

49 F.-J. Chen, C. C. Fung, C.-W. Kok, and S. Kwong, "Estimation of Two-Dimensional Frequencies Using Modified Matrix Pencil Method," *IEEE Transactions on Signal Processing*, Vol. 55, No. 2, pp. 718–724, 2007.

50 Y. Liu, Q. H. Liu, and Z. Nie, "Reducing the Number of Elements in the Synthesis of Shaped-Beam Patterns by the Forward-Backward Matrix Pencil Method," *IEEE Transactions on Antennas and Propagation*, Vol. 58, No. 2, pp. 604–608, 2010.

51 K. A. Aliouche and D. Benazzouz, "Split Array of Antenna Sensors and Matrix Pencil Method for Azimuth and Elevation Angles Estimation," *IET Signal Processing*, Vol. 11, No. 6, pp. 687–694, 2017.

52 M. Yuan and T. K. Sarkar, "Computation of the Sommerfeld Integral Tails Using the Matrix Pencil Method," *IEEE Transactions on Antennas and Propagation*, Vol. 54, No. 4, pp. 1358–1362, 2006.

53 M. Yuan, T. K. Sarkar, and M. Salazar-Palma, "A Direct Discrete Complex Image Method from the Closed-Form Green's Functions in Multilayered Media," *IEEE Transactions on Microwave Theory and Techniques*, Vol. 54, No. 3, pp. 1025–1032, 2006.

54 W. Tan and Z. Shen, "Efficient Analysis of Open-Ended Coaxial Line Using Sommerfeld Identity and Matrix Pencil Method," *IEEE Microwave and Wireless Components Letters*, Vol. 18, No. 1, pp. 7–9, 2008.

55 Y.-C. Chen, C.-K. C. Tzuang, T. Itoh, and T. K. Sarkar, "Modal Characteristics of Planar Transmission Lines with Periodical Perturbations: Their Behaviors in Bound, Stopband, and Radiation Regions," *IEEE Transactions on Antennas and Propagation*, Vol. 53, No. 1, pp. 47–58, 2005.

56 E. Drake, R. R. Boix, M. Horno, and T. K. Sarkar, "Effect of Substrate Dielectric Anisotropy on the Frequency Behavior of Microstrip Circuits," *IEEE Transactions on Microwave Theory and Techniques*, Vol. 48, No. 8, pp. 1394–1403, 2000.

57 S.-G. Hsu and R.-B. Wu, "Full Wave Characterization of a Through Hole Via Using the Matrix-Penciled Moment Method," *IEEE Transactions on Microwave Theory and Techniques*, Vol. 42, No. 8, pp. 1540–1547, 1994.

58 W. Lee, T. K. Sarkar, H. Moon, and M. Salazar-Palma, "Identification of Multiple Objects Using Their Natural Resonant Frequencies," *IEEE Antennas and Wireless Propagation Letters*, Vol. 12, pp. 54–57, 2013.

59 K. Li, W. Zhang, H. Wang, Y. Li, and M. Liu, "Parameter Identification of Subsynchronous Oscillation Based on Fastica and MP Algorithms," *The Journal of Engineering*, Vol. 2019, No.16, pp. 2454–2457, 2019.

60 F. Sarrazin and E. Richalot, "Accurate Characterization of Reverberation Chamber Resonant Modes from Scattering Parameters Measurement," *IEEE Transactions on Electromagnetic Compatibility*, (Early Access), pp. 1–12, 2019.

61 N. Chantasen, A. Boonpoonga, S. Burintramart, K. Athikulwongse, and P. Akkaraekthalin, "Automatic Detection and Classification of Buried Objects Using Ground-Penetrating Radar for Counter-Improvised Explosive Devices," *Radio Science*, Vol. 53, No. 2, pp. 210–227, 2018.

62 A. Boonpoonga, P. Chomdee, S. Burintramart, and P. Akkaraekthalin, "Simple Estimation of Late-Time Response for Radar Target Identification," *Radio Science*, Vol. 52, No. 6, pp. 743–756, 2017.

63 F. Sarrazin, P. Pouliguen, A. Sharaiha, J. Chauveau, and P. Potier, "Antenna Physical Poles Extracted from Measured Backscattered Fields," *IEEE Transactions on Antennas and Propagation*, Vol. 63, No. 9, pp. 3963–3972, 2015.

64 J. A. Garzon-Guerrero, D. P. Ruiz, and M. C. Carrion, "Classification of Geometrical Targets Using Natural Resonances and Principal Components Analysis," *IEEE Transactions on Antennas and Propagation*, Vol. 61, No. 9, pp. 4881–4884, 2013.

65 Y. Wang, I. D. Longstaff, C. J. Leat, and N. V. Shuley, "Complex Natural Resonances of Conducting Planar Objects Buried in a Dielectric Half-Space," *IEEE Transactions on Geoscience and Remote Sensing*, Vol. 39, No. 6, pp. 1183–1189, 2001.

66 K.-T. Kim, D.-K. Seo, and H.-T. Kim, "Radar Target Identification Using One-Dimensional Scattering Centres," *IEE Proceedings – Radar, Sonar and Navigation*, Vol. 148, No. 5, pp. 285–296, 2001.

67 A. Gallego, D. P. Ruiz, and A. Medouri, "High-Resolution Frequency Method to Recover Natural Resonances of a Conducting Target from Its Transient Response," *Electronics Letters*, Vol. 34, No. 17, pp. 1693–1695, 1998.

68 M. Bhuiyan, E. V. Malyarenko, M. A. Pantea, F. M. Seviaryn, and R. G. Maev, "Advantages and Limitations of Using Matrix Pencil Method for the Modal Analysis of Medical Percussion Signals," *IEEE Transactions on Biomedical Engineering*, Vol. 60, No. 2, pp. 417–426, 2013.

69 F. Boussaid, F. Olivie, M. Benzohra, and A. Martinez, "On the Use of the Matrix Pencil Method for Deep Level Transient Spectroscopy: MP-DLTS," *IEEE Transactions on Instrumentation and Measurement*, Vol. 47, No. 3, pp. 692–697, 1998.

70 S. Jang, W. Choi, T. K. Sarkar, and E. L. Mokole, "Quantitative Comparison Between Matrix Pencil Method and State-Space-Based Methods for Radar Object Identification," *URSI Radio Science Bulletin*, Vol. 2005, No. 313, pp. 27–38, 2005.

71 K. Sheshyekani, H. R. Karami, P. Dehkhoda, M. Paolone, and F. Rachidi, "Application of the Matrix Pencil Method to Rational Fitting of Frequency-Domain Responses," *IEEE Transactions on Power Delivery*, Vol. 27, No. 4, pp. 2399–2408, 2012.

72 R. Zhang, W. Xia, F. Yan, and L. Shen, "A Single-Site Positioning Method Based on TOA and DOA Estimation Using Virtual Stations in NLOS Environment," *China Communications*, Vol. 16, No. 2, pp. 146–159, 2019.

73 C. Chang, Z. Ding, S. F. Yau, and F. H. Y. Chan, "A Matrix Pencil Approach to Blind Separation of Colored Nonstationary Signals," *IEEE Transactions on Signal Processing*, Vol. 48, No. 3, pp. 900–907, 2000.

74 K. Chahine, V. Baltazart, and Y. Wang, "Parameter Estimation of Dispersive Media Using the Matrix Pencil Method with Interpolated Mode Vectors," *IET Signal Processing*, Vol. 5, No. 4, pp. 397–406, 2011.

75 F. Declercq, H. Rogier, and C. Hertleer, "Permittivity and Loss Tangent Characterization for Garment Antennas Based on a New Matrix Pencil Two-Line Method," *IEEE Transactions on Antennas and Propagation*, Vol. 56, No. 8, pp. 2548–2554, 2008.

76 F. Sagnard and G. E. Zein, "In Situ Characterization of Building Materials for Propagation Modeling: Frequency and Time Responses," *IEEE Transactions on Antennas and Propagation*, Vol. 53, No. 10, pp. 3166–3173, 2005.

77 S. J. Howard and K. Palahvan, "Measurement and Analysis of the Indoor Radio Channel in the Frequency Domain," *IEEE Transactions on Instrumentation and Aeasurement*, Vol. 39, No. 5, pp. 751–754, 1990.

78 H. Zaghoul, G. Morrison, and M. Fattouche, "Frequency Response and Path Loss Measurements of Indoor Channel," *Electronics Letters*, Vol. 27, No. 12, pp. 1021–1022, 1991.

79 L. Yang, Z. Jiao, X. Kang, G. Song, and J. Suonan, "Fast Algorithm for Estimating Power Frequency Phasors Under Power System Transients," *IET Generation, Transmission & Distribution*, Vol. 9, No. 4, pp. 395–403, 2015.

80 M. L. Crow and A. Singh, "The Matrix Pencil for Power System Modal Extraction," *IEEE Transactions on Power Systems*, Vol. 20, No. 1, pp. 501–502, 2005.

81 Y. Terriche, S. Golestan, J. M. Guerrero, D. Kerdoune, and J. C. Vasquez, "Matrix Pencil Method-Based Reference Current Generation for Shunt Active Power Filters," *IET Power Electronics*, Vol. 11, No. 4, pp. 772–780, 2018.

82 J. Guo, Y.-Z. Xie, and F. Rachidi, "A Semi-Analytical Method to Evaluate Lightning-Induced Overvoltages on Overhead Lines Using the Matrix Pencil Method," *IEEE Transactions on Power Delivery*, Vol. 33, No. 6, pp. 2837–2848, 2018.

83 R. J. Hamidi, H. Livani, and R. Rezaiesarlak, "Traveling-Wave Detection Technique Using Short-Time Matrix Pencil Method," *IEEE Transactions on Power Delivery*, Vol. 32, No. 6, pp. 2565–2574, 2017.

84 Y. Hua and T. K. Sarkar, "Parameter Estimation of Multiple Transient Signals," *Signal Processing*, Vol. 28, pp. 109–115, 1992.

85 J. Wang, P. Aubry, and A. Yarovoy, "Wavenumber-Domain Multiband Signal Fusion with Matrix Pencil Approach for High-Resolution Imaging," *IEEE Transactions on Geoscience and Remote Sensing*, Vol. 56, No. 7, pp. 4037–4049, 2018.

86 O. Tulgar and A. A. Ergin, "Improved Pencil Back-Projection Method with Image Segmentation for Far-Field/Near-Field SAR Imaging and RCS Extraction," *IEEE Transactions on Antennas and Propagation*, Vol. 63, No. 6, pp. 2572–2584, 2015.

87 H. Shen and B. Wang, "An Effective Method for Synthesizing Multiple-Pattern Linear Arrays with a Reduced Number of Antenna Elements," *IEEE Transactions on Antennas and Propagation*, Vol. 65, No. 5, pp. 2358–2366, 2017.

88 Y. Liu, L. Zhang, C. Zhu, and Q. H. Liu, "Synthesis of Nonuniformly Spaced Linear Arrays with Frequency-Invariant Patterns by the Generalized Matrix Pencil Methods," *IEEE Transactions on Antennas and Propagation*, Vol. 63, No. 4, pp. 1614–1625, 2015.

89 A. Orlandi and C. R. Paul, "An Efficient Characterization of Interconnected Multiconductor-Transmisslon-Line Networks," *IEEE Transactions on Microwave Theory and Techniques*, Vol. 48, No. 3, pp. 466–470, 2000.

3

The Cauchy Method

Summary

In the mathematical field of numerical analysis, interpolation is a method of estimating an unknown data within the range of known data from the available information. Extrapolation is also the process of approximating unknown data outside the range of known data from the available data. In this chapter, the concept of the Cauchy method is presented for the interpolation `and extrapolation of data. The Cauchy method is to interpolate and extrapolate the data from calculating coefficients simultaneously of the numerator and denominator polynomials which are used to approximate the data by a ratio of two polynomials. First, the Cauchy method is illustrated starting from the basic theory, its procedures, and then present some applications. The applications deal with extending the efficiency of the moment method through frequency extrapolation, Interpolating results for optical computations, generation of pass band using stop band data and vice versa, efficient broadband device characterization, effect of noise on the performance of the Cauchy method, applications to extrapolating amplitude-only data for the far-field or RCS extrapolation, Using it to generate the non-minimum phase response from amplitude-only data, and adaptive interpolation for sparsely sampled data. In addition, it has been applied to characterization of filters extracting resonant frequencies of objects using frequency domain data. Other applications include non-destructive evaluation of fruit and quality of fruit juices, RCS applications and to multidimensional extrapolation followed by come conclusions. A computer program implementing the Cauchy method has been provided in the Appendix followed by a list of references.

3.1 Introduction

The origin of the Cauchy method starts from interpolating data with a ratio of two polynomials [1]. The Cauchy method is a generalization of the Taylor's theorem. In Taylor's theorem, the value of the function and its derivatives at one

Modern Characterization of Electromagnetic Systems and Its Associated Metrology, First Edition.
Tapan K. Sarkar, Magdalena Salazar-Palma, Ming Da Zhu, and Heng Chen.
© 2021 John Wiley & Sons, Inc. Published 2021 by John Wiley & Sons, Inc.

point is generally used to extrapolate the function analytically to the nearest singularity which can be located at infinity. However, it is difficult to implement as one can seldom compute/measure the higher order derivatives beyond say four or five depending on the number of significant digits available. This has been illustrated in [2–8]. The uniqueness of the Cauchy method is that it is a generalization of the Taylor's theorem where instead of evaluating the function and its derivatives at one point the values of the function and its derivatives are computed at a number of points so that the derivative information does not become inaccurate.

This concept is expanded to interpolate or extrapolate the wide-band response of Electromagnetic (EM) systems using the narrow-band data [9–12]. It means that the Cauchy method can be used to speed up the numerical computations of parameters including the input impedance, currents, and the scattering data of any linear time-invariant (LTI) EM system. The Cauchy method starts by assuming that the parameter of interest that is to be extrapolated and/or interpolated, as a function of frequency, can be performed using a ratio of two polynomials. This procedure holds for an LTI system [9] as the ratio of two rational polynomials are related to the eigenfunctions of the system in the transform domain. Let us assume that the system response is an LTI system. The transfer function $H(f)$ for an LTI system, as a rational function of frequency, can be characterized by

$$H(f_i) \approx \frac{A(f_i)}{B(f_i)} = \frac{\sum_{k=0}^{P} a_k f_i^k}{\sum_{k=0}^{Q} b_k f_i^k}, \quad i = 1, 2, \cdots N, \tag{3.1}$$

where the numerator and denominator polynomials are given by $A(f)$ and $B(f)$, respectively. For convenience and computational simplicity, it is assumed that

$$Q = P + 1, \tag{3.2}$$

where P is the order of the numerator polynomial and Q is the order of the denominator polynomial. As seen from (3.1), the unknown coefficients a_k and b_k can be put into the following form:

$$[A]_{N \times (P+1)} a_{(P+1) \times 1} = [B]_{N \times (Q+1)} b_{(Q+1) \times 1}, \tag{3.3}$$

or equivalently they can be rewritten as

$$[A \quad -B]_{N \times (P+Q+2)} \begin{bmatrix} a \\ b \end{bmatrix}_{(P+Q+2) \times 1} = 0 \quad \Rightarrow$$

$$[C]_{N \times (P+Q+2)} \begin{bmatrix} a \\ b \end{bmatrix}_{(P+Q+2) \times 1} = 0 \tag{3.4}$$

where

$$[a] = [a_0, a_1, a_2, \cdots, a_P]^T,$$ (3.5)

$$[b] = [b_0, b_1, b_2, \cdots, b_Q]^T,$$ (3.6)

and

$$[C] = \underbrace{\begin{bmatrix} 1 & f_1 & \cdots & f_1^P & -H(f_1) \\ 1 & f_2 & \cdots & f_2^P & -H(f_2) \\ \vdots & \vdots & \ddots & \vdots & \vdots \\ 1 & f_N & \cdots & f_N^P & -H(f_N) \end{bmatrix}}_{[A]} \underbrace{\begin{bmatrix} -H(f_1)f_1 & \cdots & -H(f_1)f_1^Q \\ -H(f_2)f_2 & \cdots & -H(f_2)f_2^Q \\ \vdots & \ddots & \vdots \\ -H(f_N)f_N & & -H(f_N)f_N^Q \end{bmatrix}}_{[B]}.$$ (3.7)

Here, the superscript T denotes the transpose of a matrix. The size of the matrix $[C]$ is $N \times (P + Q + 2)$, so the solution of $[a]$ and $[b]$ are unique only if the total number of given frequency sample points are greater than or equal to the number of unknown coefficients $P+Q+2$.

$$N \geq P + Q + 2.$$ (3.8)

The singular value decomposition (SVD) [9, 13–15] of the matrix $[C]$ will provide an estimate for the values of P and Q. An SVD decomposition of the matrix $[C]$ results in

$$[U][\Sigma][V]^H \begin{bmatrix} a \\ b \end{bmatrix} = 0,$$ (3.9)

where the matrices $[U]$ and $[V]$ are unitary matrices, i.e.,

$$[U]^H[U] = [I] \quad \text{and} \quad [V]^H[V] = [I].$$ (3.10)

The superscript H denotes the conjugate transpose of a matrix. $[\Sigma]$ is a diagonal matrix with the singular values of the matrix $[C]$ in descending order as its entries and is of the form

$$\Sigma = \begin{bmatrix} \sigma_1 & & & & & \\ & \sigma_2 & & O & & \\ & & \ddots & & & \\ & & & \sigma_R & & \\ & O & & & 0 & \\ & & & & & \ddots \end{bmatrix} \quad \sigma_1 \geq \sigma_2 \geq \cdots \sigma_R > 0.$$ (3.11)

The columns of the matrix $[U]$ are the left singular vectors of the matrix $[C]$ or the eigenvectors of the matrix $[C][C]^H$. The columns of the matrix $[V]$ are the right singular vectors of matrix $[C]$ or the eigenvectors of the matrix $[C]^H[C]$. The singular values are the square roots of the eigenvalues of the matrix $[C]^H[C]$. Therefore, the singular values of any matrix are real and positive. The number of nonzero singular values of the matrix $[C]$ is the rank (R) of the matrix $[C]$ and they contain the information content or the number of independent degrees of freedom of the system transfer function $H(f)$. Therefore, the knowledge of the number of nonzero singular values does provide useful information for the rank of the system.

For the validation of this approximation given by (3.4), the smallest singular value should be less than or equal to the number of accurate significant decimal digits of the data. It means that if the data is corrupted by additive noise including numerical noise, the parameters P and Q are estimated by observing the ratio of the various singular values to the largest one as defined by [9]

$$\frac{\sigma_R}{\sigma_{\max}} \approx 10^{-w}, \tag{3.12}$$

where w is the number of accurate significant decimal digits of the given data. Based on w, one can choose the required parameters P and Q to interpolate or extrapolate the data. The computed number of nonzero singular values from the selected parameter P and Q is the rank of the matrix in (3.9), so it provides an idea about the information in this system of simultaneous equations. Since the rank R is the number of nonzero singular values, the dimension of the right null space of $[C]$ is $(P + Q + 2 - R)$. The solution vector belongs to this null space. Therefore, to make this solution unique, one needs to make the dimension of this null space approximately 1 so that only one vector defines this space. Hence, P and Q must satisfy the relation

$$R + 1 = P + Q + 2. \tag{3.13}$$

Using (3.13), better estimates for the parameter P and Q are obtained. Letting P and Q represent these new estimates of the polynomial orders, one can regenerate the matrix $[C]$ using (3.7) resulting in

$$[C]_{N \times (P+Q+2)} \begin{bmatrix} a \\ b \end{bmatrix}_{(P+Q+2) \times 1} = [A \quad -B]_{N \times (P+Q+2)} \begin{bmatrix} a \\ b \end{bmatrix}_{(P+Q+2) \times 1} = [0] \tag{3.14}$$

where the matrix $[C]$ is a rectangular matrix with more rows than columns. The above equation can be solved by using the total least squares (TLS) method [13–15]. In the matrix equation of (3.14), the submatrix $[A]$ is related to the sample values of the frequency only, and does not depend on the data being observed or measured as illustrated in (3.7). Hence, this matrix is not affected by measurement errors and noise. However, the submatrix $[B]$ is affected by the measurement and computational errors in the evaluation of the transfer function. To take this non-uniformity of noise in the data into account, one first needs to perform a QR decomposition [13–15] of the composite matrix $[A -B]$ up to its first $P+1$ columns. First, one performs a QR decomposition of the submatrix $[A]$ to correct for the measurement and computational errors because submatrix $[A]$ is the samples of the frequency variable only and does not include any noise. A QR decomposition of the matrix results in

$$[A] = [Q][R], \tag{3.15}$$

$$[Q^T][A \quad -B]_{N \times (P+Q+2)} \begin{bmatrix} a \\ b \end{bmatrix}_{(P+Q+2) \times 1} = [R \quad -Q^T B] \begin{bmatrix} a \\ b \end{bmatrix}$$

$$= \begin{bmatrix} R_{11} & R_{12} \\ 0 & R_{22} \end{bmatrix} \begin{bmatrix} a \\ b \end{bmatrix} = 0 \tag{3.16}$$

where $[Q]$ is a $N \times N$ orthogonal matrix and the matrix $[R]$ is an upper triangular matrix. $[R_{11}]$ is also an upper triangular matrix whose elements are not affected by noise. Elements of both $[R_{12}]$ and $[R_{22}]$ are affected by the noise in the data. From (3.16), one can now equate

$$[R_{22}][b] = 0, \tag{3.17}$$

$$[R_{11}][a] = -[R_{12}][b] \quad \Rightarrow \quad [a] = -[R_{11}]^{-1}[R_{12}][b]. \tag{3.18}$$

A SVD of $[R_{22}]$ results in

$$[U'][\Sigma'][V']^H[b] = 0. \tag{3.19}$$

By the theory of the total least squares solution (TLS), the solution vector $[b]$ is proportional to the last column of the matrix $[V']$ as

$$[b] = [V']_{Q+1}. \tag{3.20}$$

This is the optimal solution for the coefficients of the denominator polynomial under the given conditions. Using (3.20) and (3.18), the coefficients of the numerator polynomial can be computed and one can interpolate or extrapolate the system response by utilizing the numerator and denominator polynomials. Finally, the transfer function $H(f)$ can be rewritten as

$$H(f) \approx \frac{A(f)}{B(f)} = \frac{\sum_{k=0}^{P} a_k f^k}{\sum_{k=0}^{Q} b_k f^k} \approx \sum_{m=1}^{M} \left(\frac{R_m}{f - \left(\dfrac{\sigma_m}{j2\pi} + f_m\right)} + \frac{R_m^*}{f - \left(\dfrac{\sigma_m}{j2\pi} - f_m\right)} \right) \tag{3.21}$$

where R_m is the residue (R_m^* is the complex conjugate of R_m), σ_m is the damping factor, and f_m is the natural frequency associated with the mth pole.

3.2 Procedure for Interpolating or Extrapolating the System Response Using the Cauchy Method

There are four steps involved to estimate the system response using the Cauchy method from a given data set.

Step 1. Choose the number of frequency sample points (N) among the total data.

Step 2. Select the optimal orders of the denominator and numerator polynomials (P and Q) using the SVD and which satisfies the following conditions:

$$P + Q \leq N - 2, \quad \frac{\sigma_R}{\sigma_{max}} \approx 10^{-w}, \quad R + 1 = P + Q + 2, \quad Q = P + 1.$$

$$(3.22)$$

Step 3. Compute the coefficients $[a]$ and $[b]$ of the numerator and denominator polynomial using the QR decomposition and the TLS method.

Step 4. Generate the system response as a ratio of two polynomials using the computed coefficients of the numerator and denominator polynomials.

3.3 Examples to Estimate the System Response Using the Cauchy Method

3.3.1 Example 1

Consider a system function consisting of a ratio of two polynomials

$$H(f) = \frac{2f^2 + 11f + 50}{2f^3 - f^2 - 25f - 12}, \quad \text{for} \quad f = 0.10, 0.11, 0.12, \cdots, 3 \ (GHz),$$

$$(3.23)$$

where the unit of f is GHz. The number of data samples is 291. Next, it is illustrated on how to estimate the parameters which will help to interpolate or extrapolate the data using a few number of samples from the given data.

Step 1. First, choose 9 samples among the 300 data samples ($N=9$) as follows:

$$f = 0.1000, 0.2100, 0.3200, 0.4300, 0.5400, 0.6500, 0.7600, 0.8700, 0.9800$$

$$H(f) = -3.5236, -3.0331, -2.6813, -2.4192, -2.2186, -2.0622,$$
$$-1.9386, -1.8404, -1.7623$$

Step 2. Compute the maximum orders of the denominator and numerator polynomials from the number of selected samples. Using (3.22), initial values of P and Q obtained are 3 and 4, respectively. Now one can obtain the Matrix $[C]$ from (3.7) as

		Matrix $[A]$					Matrix $[-B]$		

$$[C] = \begin{bmatrix} 1.0000 & 0.1000 & 0.0100 & 0.0010 & 3.5236 & 0.3524 & 0.0352 & 0.0035 & 0.0004 \\ 1.0000 & 0.2100 & 0.0441 & 0.0093 & 3.0331 & 0.6369 & 0.1338 & 0.0281 & 0.0059 \\ 1.0000 & 0.3200 & 0.1024 & 0.0328 & 2.6813 & 0.8580 & 0.2746 & 0.0879 & 0.0281 \\ 1.0000 & 0.4300 & 0.1849 & 0.0795 & 2.4192 & 1.0403 & 0.4473 & 0.1923 & 0.0827 \\ 1.0000 & 0.5400 & 0.2916 & 0.1575 & 2.2186 & 1.1981 & 0.6470 & 0.3494 & 0.1887 \\ 1.0000 & 0.6500 & 0.4225 & 0.2746 & 2.0622 & 1.3404 & 0.8713 & 0.5663 & 0.3681 \\ 1.0000 & 0.7600 & 0.5776 & 0.4390 & 1.9386 & 1.4734 & 1.1198 & 0.8510 & 0.6468 \\ 1.0000 & 0.8700 & 0.7569 & 0.6585 & 1.8404 & 1.6012 & 1.3930 & 1.2119 & 1.0544 \\ 1.0000 & 0.9800 & 0.9604 & 0.9412 & 1.7623 & 1.7271 & 1.6926 & 1.6587 & 1.6255 \end{bmatrix}_{(9 \times 9)}$$

The first part of this matrix is $[A]$ supplemented by the matrix $[-B]$. The size of the matrix $[C]$ is $N \times (P + Q + 2) = 9 \times 9$. The SVD of matrix $[C]$ will provide the required values of P and Q. Figure 3.1 displays the normalized singular values (dB scale) from the SVD of matrix $[C]$. From (3.7) and (3.11), one can obtain the optimal orders P and Q. If the number of accurate significant decimal digits (w) is 5, the number of singular values (R) is 6 as

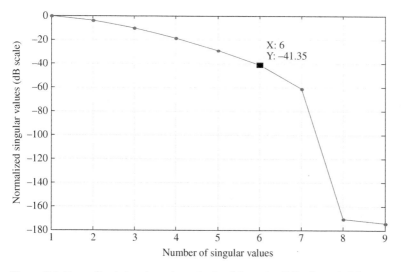

Figure 3.1 Normalized singular values obtained from the SVD of matrix $[C]$.

shown in Figure 3.1. From (3.22), one can select the optimal orders $P = 2$ and $Q = 3$. They are the same orders of the numerator and denominator polynomials chosen for original system function.

Step 3. Now one can remake matrix $[C]$ by now using the optimal orders for P and Q.

$$
[C] =
\begin{array}{cc}
\text{Matrix } [A] & \text{Matrix } [\text{-}B]
\end{array}
$$

	Matrix [A]			Matrix [-B]		
1.0000	0.1000	0.0100	3.5236	0.3524	0.0352	0.0035
1.0000	0.2100	0.0441	3.0331	0.6369	0.1338	0.0281
1.0000	0.3200	0.1024	2.6813	0.8580	0.2746	0.0879
1.0000	0.4300	0.1849	2.4192	1.0403	0.4473	0.1923
1.0000	0.5400	0.2916	2.2186	1.1981	0.6470	0.3494
1.0000	0.6500	0.4225	2.0622	1.3404	0.8713	0.5663
1.0000	0.7600	0.5776	1.9386	1.4734	1.1198	0.8510
1.0000	0.8700	0.7569	1.8404	1.6012	1.3930	1.2119
1.0000	0.9800	0.9604	1.7623	1.7271	1.6926	1.6587

9×7

First, a QR decomposition is performed of submatrix $[A]$ to consider the measurement and computational errors because the submatrix $[A]$ contains only the numerical values associated with the frequency only and is not contaminated by noise barring the truncation errors involved in making a number with a fixed number of digits. Using equations (3.15)–(3.17), the matrix $[A \,\text{-}B]$ can be reformulated as

	$[R_{11}]$			$[R_{12}]$		
-3	-1.6200	-1.1168	-7.1598	-3.4092	-2.2048	-1.6497
0	0.8521	0.9202	-1.6093	1.2810	1.6165	1.5586
0	0	0.2124	0.4673	-0.1746	0.2385	0.5285
0	0	0	-0.0178	0.0095	-0.0012	0.0112
0	0	0	0.0132	-0.0066	0.0028	-0.0022
0	0	0	0.0298	-0.0155	0.0036	-0.0137
0	0	0	0.0208	-0.0113	0.0006	-0.0155
0	0	0	-0.0213	0.0105	-0.0054	0.0008
0	0	0	-0.1017	0.0529	-0.0130	0.0450

9×7

| | $[R_{21}]$ | | | $[R_{22}]$ | | |

Finally, the solution vector $[b]$ can be obtained from (3.18)-(3.21).

$$[V'] = \begin{bmatrix} 0.8208 & 0.2995 & 0.2247 & \boxed{0.4313} \\ -0.4273 & -0.0876 & -0.0471 & 0.8986 \\ 0.1045 & 0.3575 & -0.9273 & 0.0359 \\ -0.3643 & 0.8802 & 0.2955 & -0.0719 \end{bmatrix}$$

Therefore, the estimated coefficients of the denominator polynomial are computed as

$$b_0 = 0.4313, b_1 = 0.8986, b_2 = 0.0359, b_3 = -0.0719. \tag{3.24}$$

Using (3.18), one can also calculate the estimated coefficients of the numerator polynomial as

$$a_0 = -1.7972, a_1 = -0.3954, a_2 = -0.0719. \tag{3.25}$$

Step 4. Based on the estimated coefficients of the denominator and numerator polynomials, one can estimate the transfer function as shown in Figure 3.2. To evaluate the performance of the interpolation and extrapolation, it is useful to compute the *estimated error* following the normalized mean square errors (MSEs) in the frequency domain as

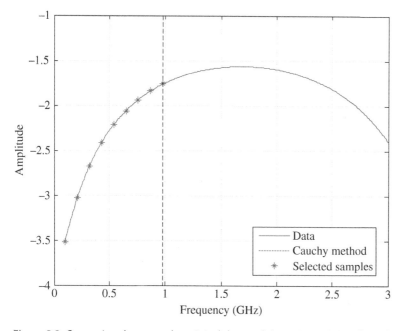

Figure 3.2 Comparison between the original data and the estimated data from the Cauchy method using 9 selected samples.

$$E_{est} = \frac{\|\hat{H} - H\|_2}{\|H\|_2},$$
(3.26)

where $\|\cdot\|_2$ is the L^2-norm of a vector. $\hat{H}$ are the estimated frequency domain data. In this case, the estimated mean squares error (MSE) is 1.4438×10^{-13}.

a_k	b_k	$R's$	$P's$
$a_0 = -1.7972$	$b_0 = 0.4313$	$R_1 = 2.0000$	$P_1 = 4.0000$
$a_1 = -0.3954$	$b_1 = 0.8986$	$R_2 = 1.0000$	$P_2 = -3.0000$
$a_2 = -0.0719$	$b_2 = 0.0359$	$R_3 = -2.0000$	$P_3 = -0.5000$
	$b_3 = -0.0719$		

Also, the transfer function $H(f)$ can be now changed to a new expression using scalars with parameters $R's$ (residues) and $P's$ (poles). Therefore, (3.24) can be rewritten as

$$H(f) = \frac{1}{f+3} + \frac{2}{f-4} - \frac{2}{f+0.5}, \quad f = 0.10, 0.11, 0.12, \cdots, 3 \ (GHz).$$
(3.27)

Here the values of $R's$ are 1, 2, –2, and their corresponding $P's$ are –3, 4, –0.5, respectively. From the estimated coefficients of the two polynomials, one can also obtain residues and poles. From (3.24) and (3.25), one can calculate the exactly same $R's$ and $P's$ as in (3.27).

3.3.2 Example 2

Next, consider the system function with complex coefficients of the denominator and numerator polynomials. The system function $H(f)$ is given by

$$H(f) = \frac{2f^2 + (11 - 20i)f + 36 - 11i}{2f^3 - (1 + 2i)f^2 + (-25 + 2i)f - 12 + 24i}, \quad f = 0.10, 0.11, 0.12, \cdots, 3 \ (GHz).$$
(3.28)

All other conditions are the same as in example 1 of 3.3.1 except for the complex coefficients of the denominator and numerator polynomials. The same computational procedure to estimate the parameters to interpolate or extrapolate the data as illustrated in example 1 are invoked.

Step 1. Choose 9 samples among 291 data samples ($N = 9$) as

$$f = 0.1000, 0.2100, 0.3200, 0.4300, 0.5400, 0.6500, 0.7600, 0.8700, 0.9800$$
$$H(f) = -1.0726 - j0.8916, \ -1.1603 - j0.7543, \ -1.2229 - j0.6229,$$
$$-1.2661 - j0.5008, \ -1.2948 - j0.3893, \ -1.3134 - j0.2887,$$
$$-1.3252 - j0.1983, \ -1.3330 - j0.1173, \ -1.3388 - j0.0445$$

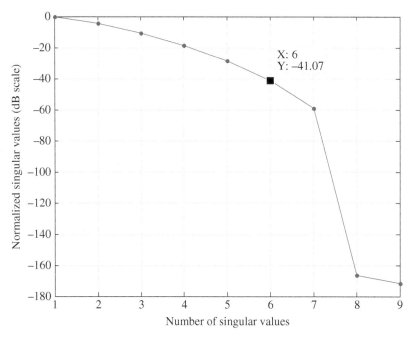

Figure 3.3 Normalized singular values from the SVD of matrix [C].

Step 2. One can select the optimal orders of the denominator and numerator polynomials (P and Q) using the SVD satisfying (3.22). Here, it is briefly explained on how to get the optimal orders. Figure 3.3 shows the normalized singular values (dB scale) from the SVD of matrix [C]. The parameter w is set to be 5. Thus the number of singular values (R) is 6 and the optimal orders for P and Q are 2 and 3, respectively.

Step 3. Compute the coefficients [a] and [b] of the numerator and denominator polynomials by using the TLS method. Using equations (3.15)-(3.20), one can obtain the coefficients [a] and [b] as

$$b_0 = 0.7281, b_1 = 0.3519 + j\,0.5825, b_2 = -0.0364 + j0.0485,$$
$$b_3 = -0.0243 - j\,0.0485,$$

(3.29)

$$a_0 = -0.7039 - j\,0.7403, a_1 = -0.6189 - j\,0.0243, a_2 = -0.0243 - j\,0.0485.$$

(3.30)

Step 4. From the estimated coefficients of the denominator and numerator polynomials, one can estimate the transfer function as shown in Figure 3.4. The estimated MSE is 1.1845×10^{-12}.

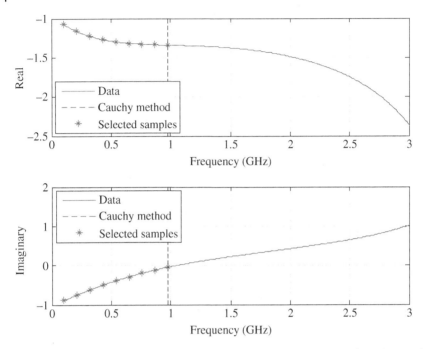

Figure 3.4 Comparison between the original data and the estimated data from the Cauchy method using 9 selected samples.

Also, from (3.28), the transfer function $H(f)$ can be reformulated as

$$H(f) = \frac{1+j}{f+3} + \frac{2-j}{f-4} - \frac{2}{f+0.5+j}, \text{ for } f = 0.01, 0.02, \cdots, 3 \ (GHz).$$

$$(3.31)$$

where $j = \sqrt{-1}$. From the estimated coefficients of the denominator and numerator polynomials, one can obtain the same residues and poles as in (3.31). The errors in the computed values for the residues and the poles are close to zero.

$R's$	$P's$
$R_1 = 2.0000 - j1.0000$	$P_1 = 4.0000 - j\,3.4849 \times 10^{-12}$
$R_2 = 1.0000 + j1.0000$	$P_2 = -3.0000 + j\,1.1846 \times 10^{-11}$
$R_3 = -2.0000 - j5.0797 \times 10^{-12}$	$P_3 = -0.5000 - j\,1.0000$

3.3.3 Example 3

Let the data of example 2 be now contaminated with additive white Gaussian noise to the data as shown in Figure 3.5. All conditions are the same as in example 2 except for added noise to the data resulting in 40 dB SNR. This implies

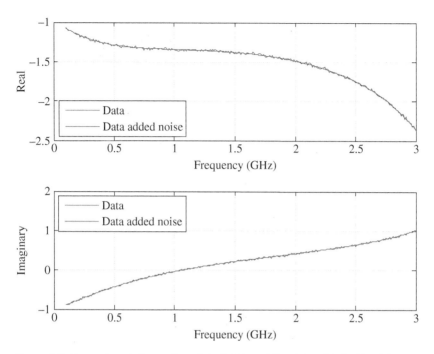

Figure 3.5 Comparison between the original data and the data added with white Gaussian noise.

there are approximately 7 effective bits of the data as the addition of one extra bit provides a gain of 6 dB. One can also apply the same steps to estimate the parameters and then interpolate the results as described in example 2.

Step 1. First, choose the entire 291 data samples (N=291).

Step 2. Select the optimal orders of the denominator and numerator polynomials (P and Q) using the SVD satisfying the constraints of (3.22). Figure 3.6 shows the normalized singular values (dB scale) determined from the SVD of matrix $[C]$. The value of w is set as 8. Thus the number of singular values (R) is 6 and the optimal orders P and Q are 2 and 3, respectively.

Step 3. One can compute the coefficients $[a]$ and $[b]$ of the numerator and denominator polynomials using the TLS method. Using (3.15)-(3.20), one can obtain the coefficients $[a]$ and $[b]$ as

$$b_0 = -0.5786, b_1 = 0.2261 \ -j\,0.6134, b_2 = 0.1108 + j\,0.4674,$$
$$b_3 = -0.0315 - j\,0.0784, \tag{3.32}$$

$$a_0 = 0.5450 + j\,0.5753, a_1 = -0.0519 - j\,0.2465, a_2 = -0.1360 - j\,0.0195. \tag{3.33}$$

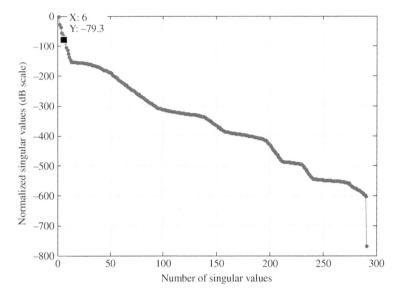

Figure 3.6 Normalized singular values from the SVD of matrix [C].

Step 4. From the estimated coefficients of the denominator and numerator polynomials, one can estimate the system response as shown in Figure 3.7. The estimated MSE is 4.66×10^{-3}. From the estimated coefficients of the denominator and numerator, one can obtain the value for the residues and the poles as

R's	P's
$R_1 = 2.0583 - j\,1.3532$	$P_1 = 4.0689 + j\,0.08176$
$R_2 = 0.00114 - j\,0.00134$	$P_2 = 1.8688 - j\,0.0833$
$R_3 = -1.2453 - j\,0.05302$	$P_3 = -0.3148 - j\,0.8427$

Observe that one does not have the numerically exact values for the residues and the poles that one started with, due to the effect of the additive white Gaussian noise. Next some examples are presented to illustrate the application of this methodology.

3.4 Illustration of Extrapolation by the Cauchy Method

3.4.1 Extending the Efficiency of the Moment Method Through Extrapolation by the Cauchy Method

The usefulness of the Cauchy method is now demonstrated for the ease in which it can be incorporated into a Method of Moments (MoM) program to speed up the computation process over a broad band. The MoM converts an

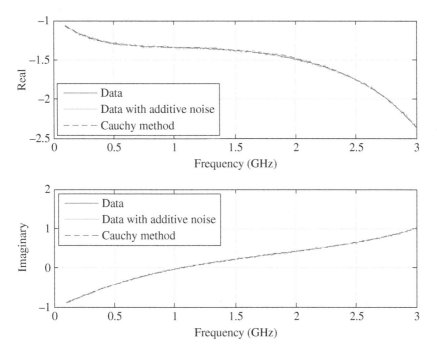

Figure 3.7 Comparison between the original data and the estimated data from the Cauchy method using 291 selected samples.

integro-differential operator equation to a matrix equation of the form relating the induced voltage, the impedance matrix and the current distribution on the structure by

$$[V] = [Z][I] \tag{3.34a}$$

Differentiating the above equation with respect to frequency results in a binomial expansion

$$[V]' = [Z]'[I] + [I]'[Z] \text{ or equivalently } [I]' = [Z]^{-1}\left\{[V] - [Z]'[I]\right\} \tag{3.34b}$$

where the superscript V' represents the first derivative with frequency. In general

$$[V]^{(n)} = \sum_{i=1}^{n} {}^{n}C_i [Z]^{(n-i)} [I]^{(i)} \text{ and therefore}$$

$$[I]^{(n)} = [Z]^{-1}\left\{[V]^{(n)} - \sum_{i=1}^{n-1} {}^{n}C_i [Z]^{(n-i)} [I]^{(i)}\right\} \tag{3.35}$$

In the above equations $[V]^{(n)}$ is the vector with each element of $[V]$ differentiated with respect to frequency n times. Similarly $[Z]^{(n)}$ is the matrix generated by differentiating each element of the matrix $[Z]$ with respect to frequency n times. This is easy to perform as the explicit form of the frequency exists in the expression of the wavenumber $k = \dfrac{2\pi}{\lambda} = \dfrac{2\pi f}{c}$, where λ is the wavelength and c is the velocity of light in free space. Hence using a MOM program, one can generate all the information needed to apply the Cauchy method by differentiating the free space Green's function only with respect to frequency. Each element to the solution $[I]$ matrix is treated as the function $H(s)$ described earlier. Given the functional values and their derivatives at some frequency points, one can evaluate the function at many more interpolated/extrapolated points.

To test the accuracy of the Cauchy method, the radar cross section (RCS) of a sphere is plotted over a wide frequency band. A program to calculate the RCS of an arbitrarily shaped closed or open body using an electromagnetic simulator using triangular patches [11, 12, 16, 17] was used. It was modified to also calculate the derivatives of the currents with respect to frequency. This information was used as input to the Cauchy methodology. The same MOM program was used to calculate the RCS without the Cauchy method. The RCS of a sphere was plotted as a function of a/λ, where a is he radius of the sphere and λ is the wavelength of the incident field. The Cauchy method saves execution time while retaining the accuracy of the original program. To get a measure of how much time could be saved by using the Cauchy method, the same program was run for a test case. Here the points chosen for the MoM program were between $\lambda = 0.3$ m and $\lambda = 0.84$ m at intervals of 0.135 m. Using the above frequencies, the RCS of a sphere of radius 0.3 m was evaluated using 180 nodes resulting in 540 unknowns. Using the Cauchy method, the RCS was calculated at 51 frequency points. The program was executed on an IBM RS6000 platform running AIX. The time taken for the above evaluation is compared to the time taken by the MoM program to evaluate the RCS at these five frequency points.

Using the MoM to compute the solution at 5 points: 47 min, 55 s.

The Cauchy Method then to extrapolate the solution to 51 points: 57 min, 56 s.

In contrast, to generate the same information for those 51 points the MOM program would take approximately 8 h 8 min.

The Cauchy method thus can be used to generate information about three dimensional objects over a very broad frequency range in an efficient fashion. This method was tested over a decade bandwidth. Figure 3.8 shows the results using the Cauchy method and the MOM program. As can be seen from the figure, the approximation is very accurate over this broad frequency range using 51 computed points.

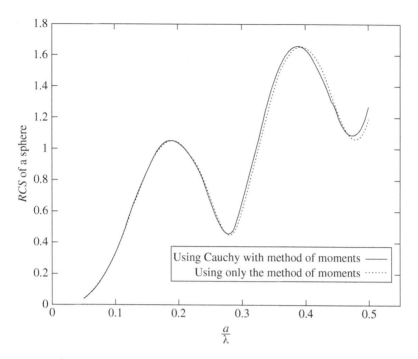

Figure 3.8 RCS of a sphere as a function of its radius and generated over a wideband using narrow band data.

3.4.2 Interpolating Results for Optical Computations

The calculation of either the scattering efficiency or the intensity in optical propagation is highly computationally intensive. If these parameters are desired over a broad range and at finely spaced points of the size parameter, the time required for the calculations may be prohibitive. The Cauchy method would solve this problem by needing the calculations to be done at a much coarser spacing and then interpolate the parameter of interest in between the sampled data.

This application was tested on the scattering efficiency of a sphere as a function of the size parameter [18]. The sphere had an index of refraction of 2.0. The original data calculated the scattering efficiency at a 0.002 in spacing as a function of the size parameter. The range of the size parameter was varied from 7.0 to 8.0. Hence, the original data had 501 points. The Cauchy method needed a spacing of 0.01 in the size parameter, without any derivative information, to accurately calculate the scattering efficiency of the sphere at the original 501 points. This cuts down the program execution time by a factor of 5. The input to the Cauchy program is shown in Figure 3.9(a).

(a)

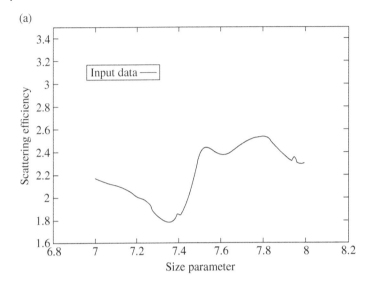

(b)

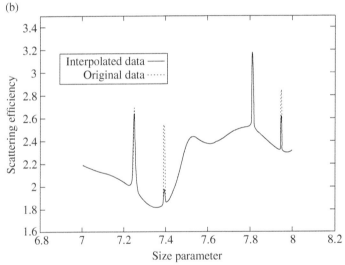

Figure 3.9 Scattering efficiency as a function of size parameter. (a) Input to the Cauchy program. (b) Results of application of the Cauchy method to optical computations.

Because all computer calculations suffer from round-off error, most of the singular values returned from the SVD subroutine are not exactly zero. The choice of the threshold was such that a singular value was considered to be zero if it was 18 orders of magnitude lower than the largest singular value. This is because the data was available in double precision. Only 72 singular values are above the chosen threshold, i.e., $R = 72$. Using this estimate for R and (3.8), the choice for the polynomial orders was reduced to 35 for the numerator and 36 for the denominator. Using these polynomials, the scattering efficiency was calculated at the original 501 points.

Figure 3.9(b) shows the results of the application of the Cauchy method to optical computations. The dotted line represents the original data while the unbroken line represents the interpolated data. As can be seen, the two plots are nearly visually indistinguishable. Also, even though the input data to the program did not have the peaks of the scattering parameter, the Cauchy method was able to reproduce them.

3.4.3   Application to Filter Analysis

The Cauchy method can also be used in the analysis of filters over a broad frequency range. A filter response is a ratio of two polynomials and, hence, lends itself easily to the use of the Cauchy method to analyze the data. This has practical application to the problem of generating the stopband response given the passband response or the reverse, i.e., generating the passband response given some data from the stopband.

A filter transfer function was measured using a network analyzer HP8510B at various sampled frequency points both in and outside the passband of the filter. The filter had its 3-dB points at 4.98 and 6.61 GHz, respectively. Hence, the filter had a passband of 1.63 GHz with a center frequency of 5.80 GHz. The filter response was measured at 415 equally spaced points in the frequency range of 4.31–7.42 GHz.

In the first application, the response over the entire band of measurement was recovered using mostly passband information at 51 equally spaced points. The data in the frequency range of 4.79–6.96 GHz, were chosen as input to the Cauchy method. As this is measured data, it is difficult to generate the derivative information of the transfer function with respect to frequency as it is contaminated with noise.

The threshold for the analysis was chosen such that a singular value was considered to be zero if it was 14 orders of magnitude lower than the largest singular value. After the method checked for the number of nonzero singular values, the estimate for the order was 16. The order of the numerator polynomial was chosen to be 7 while that of the denominator polynomial was chosen to be 8.

In Figure 3.10(a) and 3.10(b) the results from the Cauchy program are described. Figure 3.10(a) shows the magnitude response while Figure 3.10(b) shows the phase response of the filter. As is often the case with filters, the magnitude response was considered more important. Hence, the phase response was allowed to show a poor agreement so as to maximize the agreement of the magnitude response. If a 10% error in the magnitude were acceptable, the extrapolation is valid over 0.39 GHz. This is 6.7% of the center frequency

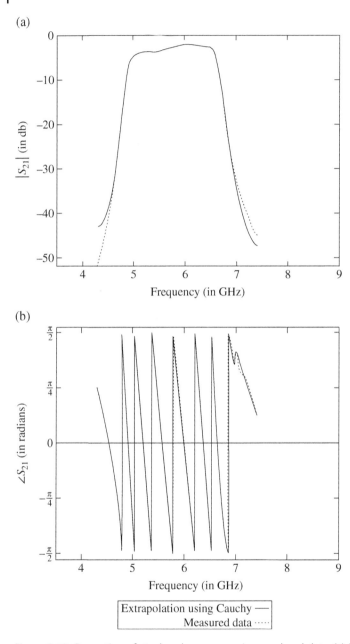

Figure 3.10 Generation of stopband response using passband data. (a) Reconstructed Magnitude response. (b) Reconstructed Phase response.

and 23.9% of the bandwidth. For frequencies beyond the passband, the extrapolation is accurate within 10% up to 7.42 GHz, the frequency until which data was available. Hence, an accurate data over a bandwidth of 3.32 GHz starting with data over a bandwidth of 2.16 GHz is generated.

In the second application, data from the stopband and a little data from the passband was used to interpolate the results into the passband. Here too, the choice of the threshold is very important. Using the same threshold as in the first application, the estimates remained of similar order. Hence, in this case too, the numerator polynomial had order 7 while the denominator had order 8. In this case, 23 equally spaced points from 4.31 up to 5.35 GHz and 28 equally spaced points from 6.20 to 7.42 GHz were used to interpolate into the passband. This represents an interpolation of 0.85 GHz, which is 14.6% of the center frequency or 52.7% of the bandwidth. Figure 3.11(a) and 3.11(b) show the results of this application. Figure 3.11(a) is the reconstructed magnitude response and Figure 3.11(b) is the reconstructed phase response. Again, since more attention is paid to the magnitude response, the phase response shows poorer agreement with the true response. In both figures the dotted line represents the measured data while the continuous line represents the results of the Cauchy method.

3.4.4 Broadband Device Characterization Using Few Parameters

An application of the Cauchy method is in creating a data-base of many devices working in varying operating conditions. The Cauchy method would require the value of a parameter at a few frequency points and use this information to evaluate the parameter of interest over a wide frequency band. Over many devices, and their operating conditions, this would yield significant savings in memory requirements. To test this application, the parameters of a pseudo-morphic high electron mobility transistor (PHEMT) was measured over the range of 1.0 – 40.0 GHz. Just five of these points were used as input to the Cauchy method. The points chosen were at the frequency points 1.0, 10.0, 20.0, 30.0, and 40.0 GHz, respectively. This resulted in a numerator polynomial of order 1 and a denominator polynomial of order 2. Just five of these points were used as the input to the Cauchy program. Figure 3.12(a) shows the magnitude of $|Y_{11}|$ reconstructed over the broad frequency range. Figure 3.12(b) shows the phase $\measuredangle Y_{11}$ over the same frequency range. As can be seen, the agreement between the measured values and the interpolated values are excellent.

In summary, the Cauchy method as presented has many practical applications. It starts with assuming that the parameter of interest as a function of frequency is a rational function. Using this form, the parameter is evaluated at many frequency points. It has been shown that the technique has applications for many practical problems. In this section the technique is applied to the MoM, filter analysis and device characterization. In all applications, the Cauchy

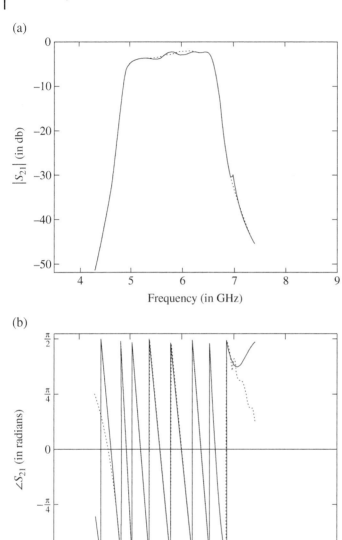

Figure 3.11 Generation of passband response using stopband data. (a) Reconstructed magnitude response. (b) Reconstructed phase response.

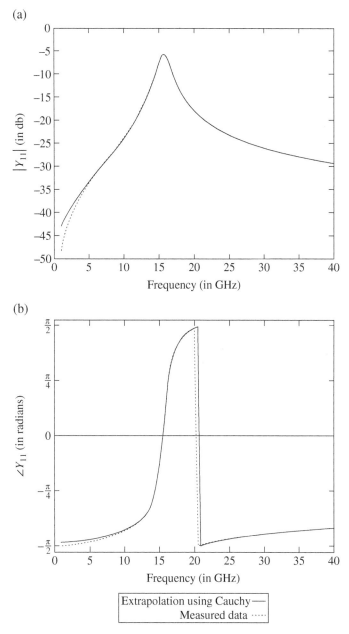

Figure 3.12 Broadband device characterization. Solid lines, extrapolation using Cauchy method; dotted lines, measured data. (a) Magnitude response: $|Y_{11}|$ (b) Phase response $\angle Y_{11}$.

method saved time and memory for the computations of the parameters of interest.

It must be pointed out that the Cauchy method is completely general and can be used to interpolate or extrapolate both theoretical and experimental data as a function of any variable. However, in many applications in electromagnetics, frequency is the variable of interest.

3.5 Effect of Noise Contaminating the Data and Its Impact on the Performance of the Cauchy Method

The solution vector of (3.20) belongs to the invariant subspace that is spanned by the right singular vector $[V]_{P+Q+2}$ This singular vector is associated with the smallest singular value. However, because of the noise in the data, the entries of the matrix $[C]$ are perturbed from their true values. Hence, the solution vector is also perturbed. We need to quantify the perturbation of this subspace.

Notation. In this article a perturbed parameter or matrix will be represented by a tilde (.......,) above the corresponding unperturbed parameter or matrix.

3.5.1 Perturbation of Invariant Subspaces

Let $\Re$ denote the set of real numbers, $\Re^n$ the set of real vectors of length N, and $\Re^{N \times P}$ the set of real matrices of order $N \times P$. Consider an arbitrary matrix $A \in \Re^{N \times P}$ with $P \leq N$. Let $\tilde{A} = A + E$, where E is the permutation to matrix A, and

$$
[U]^T [A] [V] = \begin{pmatrix} \sigma_1 & 0 \\ 0 & \Sigma_2 \\ 0 & 0 \\ 1 & P-1 \end{pmatrix} \quad \begin{bmatrix} 1 \\ P-1 \\ N-P \end{bmatrix} \tag{3.36}
$$

Here figures below the matrix indicate the number of columns in each submatrix, whereas the figures to the side of the matrix represent the number of columns in each submatrix. Also

$$
U = (u_1 \mid U_2 \mid U_3) \text{ and } V = (v_1 \mid V_2) \tag{3.37}
$$

Here $u_1 \in \Re^N$, $U_2 \in \Re^{N \times (P-1)}$, $U_3 \in \Re^{N \times (N-P)}$, $v_1 \in \Re^P$ and $v_2 \in \Re^{P \times (P-1)}$. σ_1 is the singular value corresponding to the left singular vector u_1 and the right singular vector v_1. This singular value can be the one of interest and not just the largest singular value. In the Cauchy method, the singular value of interest is the smallest or the zero singular value. Σ is the diagonal matrix with the rest of the singular values of C as its entries. These singular values can be ordered arbitrarily as long as the columns of U and V are permuted appropriately so as to maintain the equality of equation (3.36). If

$$[U]^T \left[\sum \right] [V] = \begin{pmatrix} \gamma_{11} & g'_{12} \\ g_{21} & G_{22} \\ g_{31} & G_{32} \end{pmatrix} \tag{3.38}$$

where $\gamma_{11} \in \mathfrak{R}$. $g_{12}, g_{21}, g_{31} \in \mathfrak{R}^{P-1}$, $G_{22} \in \mathfrak{R}^{(P-1) \times (P-1)}$, and $G_{32} \in \mathfrak{R}^{(P-1) \times (P-1)}$, and if σ_1 is not repeated as a singular value then [19]

$$\tilde{v}_1 = v_1 + V_2 \left(\sigma_1^2 I - \Sigma_2^2 \right)^{-1} h + O \left(\|E\|_2^2 \right), \text{ here } h = \sigma_1 g_{12} + \Sigma_2 g_{21} \tag{3.39}$$

3.5.2 Perturbation of the Solution of the Cauchy Method Due to Additive Noise

In all the measurements the true value of the measured parameter (here H_i) is perturbed by an additive noise component. Hence $\tilde{H}_i = H_i + e_i$ where $\tilde{H}_i$ is the value of H_i after it has been perturbed by noise e_i. In the ensuing discussions one is assuming the following:

1) The noise is only in the measurement of the parameter $[H(s)]$, and not in the measurement of the frequency (s).
2) $C x = 0$ has a solution which is unique to within a constant.
   (a) $x = v_1$, with $\sigma_1 = 0$
   (b) $\sigma_1 = 0$ is a simple singular value. This assumption is valid because in the solution procedure it is constrained that the rank of the null space of C is one.
3) $\tilde{H}(s_i) = H(s_i) + e_i$, $\{e_i\}_{i=1}^N$ are zero mean, Gaussian, uncorrelated, and have equal variance σ^2.

Using the forgoing notation for a perturbed matrix and equation (3.14), one obtains

$$[\tilde{C}] \begin{bmatrix} \tilde{a} \\ \tilde{b} \end{bmatrix} = 0 \text{ where } [\tilde{C}] = \begin{bmatrix} 1 & s_1 & \cdots & s_N^P & -\tilde{H}_1 & -\tilde{H}_1 s_1 & \cdots & -\tilde{H}_1 s_1^Q \\ 1 & s_2 & \cdots & s_N^P & -\tilde{H}_2 & -\tilde{H}_2 s_2 & \cdots & -\tilde{H}_2 s_2^Q \\ \vdots & \vdots & \cdots & \vdots & \vdots & \vdots & & \vdots \\ 1 & s_N & \cdots & s_N^P & -\tilde{H}_N & -\tilde{H}_N s_N & \cdots & -\tilde{H}_N s_N^Q \end{bmatrix}$$

$$\tag{3.40}$$

where P = estimate of the order of the numerator, Q = estimate of the order of the denominator, N = number of sample points. Hence

$$[\tilde{C}] = [C] + [E] \tag{3.41}$$

where E is the additive error to the matrix C due to noise in the data. Hence $[E] = [\,0\,|E_1]$ where $[0]$ is a zero matrix of order $N \times (P-1)$ and

$$[E_1] = \begin{bmatrix} e_1 & e_1 s_1 & e_1 s_1^2 & \cdots & e_1 s_1^Q \\ e_2 & e_2 s_2 & e_2 s_2^2 & \cdots & e_2 s_2^Q \\ \vdots & \vdots & \cdots & \vdots & \vdots \\ e_N & e_N s_N & e_N s_N^2 & \cdots & e_N s_N^Q \end{bmatrix}_{N \times (Q+1)} \tag{3.42}$$

and so

$$[E_1] = \begin{bmatrix} e_1 & 0 & 0 & \cdots & 0 \\ 0 & e_2 & 0 & \cdots & 0 \\ \vdots & \vdots & & \cdots & \vdots \\ 0 & 0 & 0 & \cdots & e_N \end{bmatrix}_{N \times N} \begin{bmatrix} 1 & s_1 & s_1^2 & \cdots & s_1^Q \\ 1 & s_2 & s_2^2 & \cdots & s_2^Q \\ \vdots & \vdots & \vdots & \cdots & \vdots & \vdots \\ 1 & s_N & s_N^2 & \cdots & s_N^Q \end{bmatrix}_{N \times (Q+1)} \tag{3.43}$$

$$[U]^T [E] [''V] = [U]^T [0\,|E_1] [V] \begin{pmatrix} v_1' & | & V_2' \\ v_1'' & | & V_2'' \end{pmatrix} \begin{bmatrix} P+1 \\ \\ Q+1 \end{bmatrix} = [U^T E_1 v_1'' \,|\, U^T E_1 V_2'']$$

$$1 \quad P+Q+2 \tag{3.44}$$

and therefore

$$[U]^T [E] [V] = [U^T E_1 v_1'' \,|\, U^T E_1 V_2''] = \begin{pmatrix} \gamma_{11} & g_{12}^T \\ g_{21} & G_{22} \\ g_{31} & G_{32} \end{pmatrix} \tag{3.45}$$

Because v_1 is the solution of the unperturbed Cauchy problem $[C] \begin{bmatrix} a \\ b \end{bmatrix} = 0$ and v_1'' is the vector of the last $Q + 1$ entries of v_1 and forms the vector of the denominator coefficients. Also, the singular value of interest (σ_1) is zero. Hence, $h = \sigma_1 g_{12} + \Sigma_2 g_{21} = \Sigma_2 g_{21}$. Also using (3.39)

$$\tilde{v}_1 = v_1 + V_2 \Sigma_2^{-1} g_{21} + O(\|E\|_2^2) \tag{3.46}$$

Hence, g_{12} is of no consequence. From (3.45) $[U]^T [E_1] [v_1''] = \begin{pmatrix} \gamma_{11} \\ g_{21} \\ g_{31} \end{pmatrix}$. Using (3.43) and the fact that v_1'' is the vector of the denominator coefficients

$$E_1 v_1'' = - \begin{bmatrix} e_1 & 0 & 0 & \cdots & 0 \\ 0 & e_2 & 0 & \cdots & 0 \\ \vdots & \vdots & & \cdots & \vdots \\ 0 & 0 & 0 & \cdots & e_N \end{bmatrix} \begin{bmatrix} de\,(s_1) \\ de\,(s_2) \\ \vdots \\ de\,(s_N) \end{bmatrix} \tag{3.47}$$

where $de(s_i) = \sum_{k=0}^{Q} b_k s_i^k$ is the value off the unperturbed polynomial evaluated at s_i. For convenience one can define a new vector $\tilde{e}$ as

$$\tilde{e} = E_1 v_1'' = - \begin{bmatrix} e_1\,de\,(s_1) \\ e_2\,de\,(s_2) \\ \vdots \\ e_N\,de\,(s_N) \end{bmatrix} \tag{3.48}$$

Using this and the fact that $U = (u_1 \mid U_2 \mid U_3)$ and equation (3.45), one obtains $g_{21} = U_2^T \tilde{e}$. Therefore, using (3.46) one obtains

$$\tilde{v}_1 = v_1 + V_2 \Sigma_2^{-1} U_2^T \tilde{e} + O\big(\|E\|_2^2\big). \tag{3.49}$$

Because of the elements of $\tilde{e}$ are Gaussian random variables, $\tilde{v}_1$ is, to the *first order of approximation*, a Gaussian random vector. Now using (3.36) and the fact that $\sigma_1 = 0$, $C = U_2 \Sigma_2 V_2^T$ or equivalently $C^\dagger = V_2 \Sigma_2^{-1} U_2^T$ where $C^\dagger$ is the pseudo inverse of C. Therefore to the first order of approximation $\tilde{v}_1 = v_1 + C^\dagger \tilde{e}$. Using the fact that $C^\dagger$ is unperturbed and the noise is zero mean, the expectation value of the solution vector (v_1) is given by

$$\mathbf{E}(v_1) = v_1 + C^\dagger\,\mathbf{E}(\tilde{e}) = v_1 \tag{3.50}$$

where $\mathbf{E}$ is the expectation operator and not the error matrix. Therefore, to the *first order of approximation*, the estimator is unbiased. The covariance matrix of v_1 is given by

$$\mathrm{cov}\,(v_1) = \mathbf{E}\Big[(\tilde{v}_1 - v_1)(\tilde{v}_1 - v_1)^T\Big] \tag{3.51}$$

Using (3.50) results in

$$\mathbf{E}\Big[(\tilde{v}_1 - v_1)(\tilde{v}_1 - v_1)^T\Big] = \mathbf{E}\Big[C^\dagger \tilde{e}\,\tilde{e}^T C^{\dagger T}\Big] = C^\dagger \mathbf{E}\Big[\tilde{e}\,\tilde{e}^T\Big]C^{\dagger T} \tag{3.52}$$

$$\text{Now } \tilde{e}\,\tilde{e}^T \;=\; - \begin{bmatrix} e_1\, de\,(s_1) \\ e_2\, de\,(s_2) \\ \vdots \\ e_N\, de\,(s_N) \end{bmatrix} \{e_1\, de\,(s_1) \quad e_2\, de\,(s_2) \quad \cdots \quad e_N\, de\,(s_N)\}.$$

(3.53)

Therefore the *ij*th entry of the matrix is given by $\left[\tilde{e}\,\tilde{e}^T\right]_{ij} = e_i e_j\, de\,(s_i)\, de\,(s_j)$.
Because e_i and e_j are assumed to be zero mean independent and identically distributed with variance σ^2, i.e.,

$$\mathbf{E}\!\left[e_i\, e_j\right] \;=\; \sigma^2 \delta_{ij} \text{ where } \delta_{ij} = \begin{cases} 1 & \text{if } i = j \\ 0 & \text{otherwise} \end{cases}$$

$$\mathbf{E}\!\left[\tilde{e}\,\tilde{e}^T\right] \;=\; \sigma^2 \begin{bmatrix} de^2(s_1) & 0 & 0 & \cdots & 0 \\ 0 & de^2(s_2) & 0 & \cdots & 0 \\ \vdots & \vdots & & \ddots & \vdots \\ 0 & 0 & 0 & \cdots & de^2(s_N) \end{bmatrix}$$

(3.54)

$$\mathbf{E}\!\left[(\tilde{v}_1 - v_1)(\tilde{v}_1 - v_1)^T\right] \;=\; \sigma^2 C^\dagger \begin{bmatrix} de^2(s_1) & 0 & 0 & \cdots & 0 \\ 0 & de^2(s_2) & 0 & \cdots & 0 \\ \vdots & \vdots & & \ddots & \vdots \\ 0 & 0 & 0 & \cdots & de^2(s_N) \end{bmatrix} C^{\dagger T}$$

(3.55)

If $C_{ij}^\dagger = c_{ij}$, the autocorrelation of the *i*th entry of $\tilde{v}_1$ is given by

$$\mathbf{E}\!\left[C^\dagger\, \tilde{e}\,\tilde{e}^T\, C^{\dagger T}\right]_{ii} \;=\; \sigma^2 \sum_{j=1}^{N} c_{ij}^2\, de^2\,(s_j)$$

(3.56)

This is the variance of the *i*th entry in the vector of coefficients. Hence, if $i \le P + 1$, one is dealing with a numerator coefficient, else one is dealing with a denominator coefficient.

Because one has solved a matrix equation in which the elements of the matrix are Gaussian random variables, each element of the solution vector is a Gaussian random variable. Also, the numerator and denominators are linear combinations of the coefficients. Hence, the numerator and denominator are Gaussian random variables as functions of frequency. Hence, to completely characterize the numerator and denominator random variables, one only need their expectation values and variances.

To make this problem of the ratio of two Gaussians solvable, one has to assume that any two coefficients are independent of each other. Hence, the cross-covariance matrix of v_1 is assumed to be diagonal. Now,

$$\tilde{A}(s) = \sum_{k=0}^{P} \tilde{a}_k\, s^k \text{ and}$$

$$\tilde{B}(s) = \sum_{k=0}^{Q} \tilde{b}_k\, s^k \text{ and therefore}$$

$$\mathbf{E}[\tilde{A}(s)] = \sum_{k=0}^{P} \mathbf{E}[\tilde{a}_k]\, s^k \text{ and}$$

$$\mathbf{E}[\tilde{B}(s)] = \sum_{k=0}^{Q} \mathbf{E}[\tilde{b}_k]\, s^k.$$

However, because to the first order of approximation the coefficients are unbiased, then

$$\mathbf{E}[\tilde{A}(s)] = \sum_{k=0}^{P} \tilde{a}_k\, s^k \text{ and } \mathbf{E}[\tilde{B}(s)] = \sum_{k=0}^{Q} \tilde{b}_k\, s^k \qquad (3.57)$$

Therefore, the estimators for the numerator and denominator as a function of frequency are unbiased. Because the ratio of two variables is not a linear function, this does not imply that the final estimator is unbiased.

To calculate the variances of the numerator and denominator as a function of frequency,

$$\operatorname{var}[A(s)] = \operatorname{var}\left[\sum_{k=0}^{P} \tilde{a}_k\, s^k\right].$$

Using the assumption that each coefficient is independent of the others,

$$\operatorname{var}[A(s)] = \sum_{k=0}^{P} \operatorname{var}(\tilde{a}_k)\, s^{2k}.$$

Therefore

$$\operatorname{var}[A(s)] = \sigma^2 \sum_{i=1}^{P+1} s^{2i} \sum_{j=1}^{N} c_{ij}^2\, de^2(s_j) \text{ and } \operatorname{var}[B(s)] = \sigma^2 \sum_{i=P+2}^{P+Q+2} s^{2i} \sum_{j=1}^{N} c_{ij}^2\, de^2(s_j).$$

$$(3.58)$$

Let $\overline{N} = \mathbf{E}[A(s)]$, $\overline{D} = \mathbf{E}[B(s)]$, $a^2 = \operatorname{var}[A(s)]$, and $b^2 = \operatorname{var}[B(s)]$. Therefore, the problem is reduced to: Given the means and variances of two independent Gaussian random variables, then what is the probability density function (PDF) of their ratio? This problem has been solved in [20]. In the notation of [19, 20], if

N and D are independent Gaussian random variables with means $\overline{N}$ and $\overline{D}$, respectively, and variance a_2 and b_2, respectively, and if $R = N / D$, then the probability density function of R is given by

$$f_R(r) = \sqrt{\frac{(ab)^3}{\pi} \frac{1}{b^2 r^2 + a^2}} \exp\left(\frac{N^2}{2a^2} - \frac{D^2}{2b^2}\right) \left[Z \, erf\,(Z) \exp\,(Z^2) + \frac{1}{\sqrt{\pi}}\right]$$

(3.59)

where

$$Z = \frac{1}{\sqrt{2\,b^2 a^2}} \left(\frac{b^2 \overline{N}\, r + a^2 \overline{D}}{\sqrt{b^2 r^2 + a^2}}\right)$$

(3.60)

and the error function (*erf*) is defined by $erf\,(Z) = \dfrac{2}{\pi} \displaystyle\int_0^Z \exp\,(-t^2)\, dt.$

Hence, one has the theoretical PDF of the ratio of two random variables. However, this density function is an approximation of the true density function. To obtain the true density function one would need to take into account the cross correlation between the coefficients. This leads to a problem that is highly difficult to solve.

3.5.3 Numerical Example

To test the presented theory, the Cauchy method was tested with a simple example. As an example, the function chosen to be the testing function was

$$H(s) = \frac{\displaystyle\sum_{k=0}^{4} k\, s^k}{\displaystyle\sum_{k=0}^{5} (k+1)\, s^k}.$$

This ratio of two polynomials was evaluated at 31 points in the range $s = 2.0$ and $s = 4.0$. Two tests were performed on these data. In the first test, Gaussian noise was added to the data directly. A numerical Gaussian random number generator was used. The power in the noise was chosen such that the signal-to-noise ratio (SNR) was 30 dB. These perturbed data were used to evaluate the parameter at $s = 3.0$. This was considered to be one sample of the random variable at $s = 3.0$. Samples taken numbered 1001. A PDF estimator was used to estimate the PDF at $s = 3.0$. Figure 3.13 shows the PDF found using this method. This is the plot marked "Adding Noise to the data".

In the second test, the original unperturbed data between $s = 2.0$ and $s = 4.0$ were used as inputs to the Cauchy program. The unperturbed numerator and

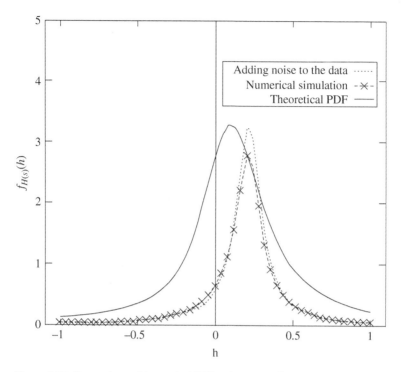

Figure 3.13 Comparison of theoretical PDF and numerically simulated PDF's. SNR = 30 dB (≈ 5 effective bits as 1 bit of data adds 6 dB of Signal).

the denominator coefficients were evaluated. The means of the numerator and the denominator coefficients were evaluated using (3.57). Also the variances off the numerator and the denominator were evaluated using (3.58). Using these values of mean and variances, a Gaussian random variable, with the numerator mean and variance, was divided with another Gaussian random variable with the denominator mean and variance. This was repeated 1001 times. The Gaussian random numbers were generated using the same random number generator as in the first test. The 1001 samples obtained from this test were used as input to the same PDF estimator. The result from this estimation of the PDF is shown in Figure 3.13. This plot is labeled "Numerical Simulation." Finally, these two PDFs are compared with the theoretical PDF in (3.59). The choices $\overline{N}, \overline{D}, a^2$ and b^2 are obtained from the theoretical means and variances used in the second test. At $s = 3.0$, using the above function, one can obtain

Actual Value: 0.2126
Mean (adding noise to the data): 0.2124
Mean (dividing two Gaussians with the theoretical means and variances): 0.1804

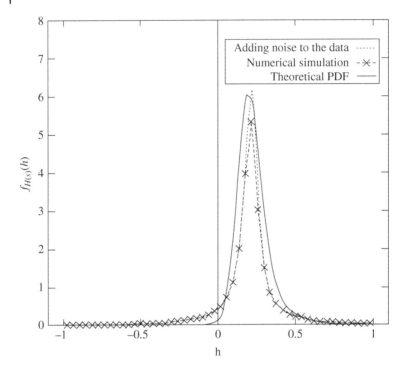

Figure 3.14 Comparison of theoretical PDF and numerically simulated PDFs. SNR = 40 dB ($\approx$ 6 effective bits).

Figure 3.14 plots the same three PDF's for a signal-to-noise ratio of 40 dB. Here the agreement is better than from the earlier case. This is to be expected, because the assumptions come closer to being satisfied as the amount of noise is reduced.

3.6 Generating High Resolution Wideband Response from Sparse and Incomplete Amplitude-Only Data

The first application of the Cauchy method is to generate an accurate rational model for the scattering analysis using amplitude-only data. The Cauchy method is applied to interpolate/extrapolate the radar cross section (RCS) amplitude-only data over a frequency band.

In electromagnetic analysis, field quantities are usually assumed to be time harmonic. This suggests that the solution lies in the frequency domain. The conventional approach is to carry out a frequency-domain analysis which is to sample the response at a set of discrete frequency points and utilize the MoM to calculate the frequency response at those sample points [21, 16]. This has been implemented in many commercial software packages such as

HOBBIES [22, 23]. However, for wideband analysis, if the electromagnetic response contains very sharp peaks or nulls, one needs a fine frequency step in order to use the sample points to generate a broadband solution. That means one will have a very large number of sample points and the MoM program needs to be executed at all the frequency points of interest. Therefore, the conventional approach becomes computationally very intensive and the size of the MoM matrix to be solved for can be prohibitively large. Another separate situation where this type of problem occurs is when one needs to measure experimentally the RCS of a structure. It may be difficult in some cases to have the measured data with a high resolution in frequency. So, it is very desirable if one can reduce the number of sample points and apply interpolation techniques to reconstruct the high resolution frequency response accurately. The Cauchy method is employed in this section to interpolate the high resolution frequency response from sparse and incomplete amplitude-only data.

The main difference between what has been presented so far and the materials to be presented in this section is that, here the Cauchy method is applied for amplitude-only data, since most of the time it is very hard to measure the phase information of the frequency response particularly at high frequencies. And the Cauchy method is implemented for the non-uniform sampling case so that it can deal with the incomplete data in which some frequency band information is missing. Also, the Cauchy method is very sensitive to the selected orders of the polynomials, which is conventionally decided by the ratio of diagonal components of SVD. It is possible to get a good estimate generally for determining the orders of the polynomials from the given data however it does not guarantee to provide the best result. In this presentation, an exhaustive search is made to get the optimum polynomial order. Another improvement is after the coefficients of the denominator polynomials are estimated by the TLS technique, the coefficients of the numerator polynomials are calculated using a matrix inversion. However, that matrix is very ill-conditioned. Here the calculation is carried out automatically by using the conjugate gradient method in order to achieve a stable result. It is also important to point out that the Cauchy method becomes unstable when the dataset is very large. Here the input data is automatically partitioned into a number of small frequency bands and the Cauchy method is applied separately to each segment.

3.6.1 Development of the Interpolatory Cauchy Method for Amplitude-Only Data

Define the far-field power spectrum density $|H(f)|^2$ as a rational function of frequency f. Then one can approximate $|H(f)|^2$ by a ratio of two real polynomial functions $A(f)$ and $B(f)$ as follows:

$$|H(f)|^2 = H(f)H^*(f) \approx \frac{A(f)}{B(f)} = \frac{\sum_{k=0}^{P} a_k f^k}{\sum_{k=0}^{Q} b_k f^k} \tag{3.61}$$

Please note that these a_k and b_k are completely different from the ones described in (3.5) and (3.6). Hence the system function $|H(f)|^2$ is represented by a few parameters a_k and b_k which are all real numbers. The given information is the value of $|H(f)|^2$ at some sampled nonequally spaced frequency points. If the parameters a_k and b_k can be estimated correctly from the given values of $|H(f)|^2$, then the amplitude-only data can be reproduced and the amplitude response at any frequency band can be interpolated/extrapolated accurately. Assume $|H(f)|^2$ is known at the sampled frequency points f_j, $j = 1, 2, ...N$. In that case the Cauchy problem becomes:

Given $|H(f_j)|^2$, $j = 1, 2, ...N$, find the proper P, Q and estimate $\{a_k, k = 0, 1, ...P\}$ and $\{b_k, k = 0, 1, ...Q\}$ to interpolate/extrapolate the sampled data.

It is natural to enforce the equality of (3.61) at the measured/observed frequency points f_j and it can be rewritten as:

$$A(f_j) - |H(f_j)|^2 B(f_j) = 0 \tag{3.62}$$

Next, expand $A(f)$ and $B(f)$ by polynomials and use the notation $|H|^2_j$ to represent $H(f_j)$, Equation (3.61) becomes:

$$a_0 + a_1 f_j + a_2 f_j^2 + \cdots + a_P f_j^P - |H|^2_j b_0 - |H|^2_j b_1 f_j - |H|^2_j b_2 f_j^2 - \cdots - |H|^2_j b_Q f_j^Q = 0 \tag{3.63}$$

Now, convert Equation (3.63) into the following matrix form as

$$[C] \begin{bmatrix} \mathbf{a} \\ \mathbf{b} \end{bmatrix} = 0 \tag{3.64}$$

where

$$[\mathbf{a}] = [a_0 \quad a_1 \quad a_2 \quad \cdots \quad a_P]^T \tag{3.65}$$

$$[\mathbf{b}] = [b_0 \quad b_1 \quad b_2 \quad \cdots \quad b_Q]^T \tag{3.66}$$

and

$$[C] = \begin{bmatrix} 1 & f_1 & \cdots & f_1^P & -|H|^2_1 & -|H|^2_1 f_1 & \cdots & -|H|^2_1 f_1^Q \\ 1 & f_2 & \cdots & f_2^P & -|H|^2_2 & -|H|^2_2 f_2 & \cdots & -|H|^2_2 f_2^Q \\ \vdots & \vdots & \cdots & \vdots & \vdots & \vdots & \cdots & \vdots \\ 1 & f_N & \cdots & f_N^P & -|H|^2_N & -|H|^2_N f_N & \cdots & -|H|^2_N f_N^Q \end{bmatrix} \tag{3.67}$$

The size of matrix $[C]$ is $N \times P + Q + 2$, so the solution of $\mathbf{a}$ and $\mathbf{b}$ are unique only if the total number of sample points are greater than or equal to the number of the unknown coefficients $P + Q + 2$. The singular value Decomposition (SVD) of matrix [**C**] will provide an estimate for the required values of P and Q. A SVD of [**C**] results in:

$$[\mathbf{U}][\mathbf{\Sigma}][\mathbf{V}]^H \begin{bmatrix} \mathbf{a} \\ \mathbf{b} \end{bmatrix} = 0 \tag{3.68}$$

The matrices **U** and **V** are unitary matrices and $\mathbf{\Sigma}$ is a diagonal matrix with the singular values of **C** in descending order as its entries. The columns of **U** are the left singular vectors of **C** or the eigenvectors of $\mathbf{CC}^H$. The columns of **V** are the right singular vectors of **C** or the eigenvectors of $\mathbf{C}^H\mathbf{C}$. The singular values are the square roots of the eigenvalues of the matrix $\mathbf{C}^H\mathbf{C}$. Therefore, the singular values of any matrix are real and positive. The number of nonzero singular values is the rank of the matrix in (3.67) and so this provides one an idea of the amount of unique information in this system of simultaneous equations. If R is the number of nonzero singular values, the dimension of the right null space of **C** is $P + Q + 2 - R$. The solution vector then belongs to this null space. Therefore, to make this solution unique, one needs to make the dimension of this null space 1 so that only one vector defines this space. Hence, P and Q must satisfy the relation:

$$R + 1 = P + Q + 2 \tag{3.69}$$

The solution algorithm must include a method to estimate R. This is done by starting out with the choices of P and Q that are higher than can be expected for the system at hand. Then, one gets an estimate for R from the number of non-zero singular values of the matrix **C**. Now, using (3.69) better estimates for P and Q are obtained. Letting P and Q stand for these new estimates of the polynomial orders, one can regenerate the matrix **C**. Therefore, one essentially comes back to Equation (3.64):

$$[\mathbf{C}] \begin{bmatrix} \mathbf{a} \\ \mathbf{b} \end{bmatrix} = [\mathbf{A} \quad -\mathbf{B}] \begin{bmatrix} \mathbf{a} \\ \mathbf{b} \end{bmatrix} = 0 \tag{3.70}$$

where **C** is a rectangular matrix with more rows than columns. The above equation can be solved by the total least square method. In the matrix of (3.70), the submatrix **A** is a function of the frequencies only and does not depend on the parameter measured. Hence, this matrix is not affected by measurement errors and noise. However, the submatrix **B** is affected by the errors. To take this non-uniformity into account, one needs to perform a QR decomposition of the matrix [**A** $-$**B**] up to its first $P+1$ columns. A QR decomposition of a matrix is a decomposition of a matrix **A** into a product $\mathbf{A} = \mathbf{QR}$ of an orthogonal matrix **Q** and an upper triangular matrix **R**. A QR decomposition of the matrix **C** results in:

$$\begin{bmatrix} \mathbf{R}_{11} & \mathbf{R}_{12} \\ \mathbf{0} & \mathbf{R}_{22} \end{bmatrix} \begin{bmatrix} \mathbf{a} \\ \mathbf{b} \end{bmatrix} = 0 \tag{3.71}$$

where R_{11} is the upper triangular and has exact entries for the elements where the elements of R_{22} is completely affected by the noise. This results in,

$$R_{22}b = 0 \tag{3.72}$$

and

$$R_{11}a = -R_{12}b \tag{3.73}$$

A SVD of R_{22} provides

$$[U][\Sigma][V]^{H}b = 0 \tag{3.74}$$

By the theory of the TLS, the solution of the vector b is proportional to the last column of the matrix V. Hence, one can choose

$$b = [V]_{Q+1} \tag{3.75}$$

This is the optimal solution even in the case that the matrix does not have a null space. Using this solution for the denominator coefficients and using (3.73), the numerator coefficients using the conventional least squares (LS) solution can be solved. The above TLS approach removes some of the errors involved with the associated processing of the conventional LS approach.

3.6.2 Interpolating High Resolution Amplitude Response

In this section, a numerical example is presented to illustrate the applicability of the Cauchy method in extracting the nulls of a system response from the sparse and the incomplete power spectrum density data. To illustrate this point, consider a cubic conducting hollow box composed of perfect electric conductors (PEC) with dimension 1 m by 1 m by 1 m. There is a rectangular hole on the top surface perpendicular to x axis. The size of the hole in the PEC box is 0.1 m by 0.1 m. This conducting box, which is shown in Figure 3.15, is excited

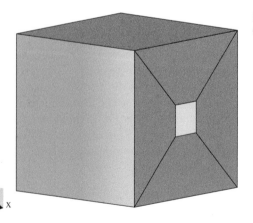

Figure 3.15 A Conducting Cube with a Square Hole.

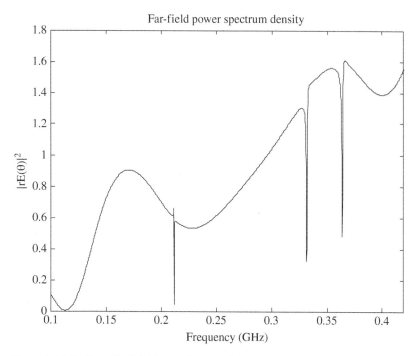

Figure 3.16 Radiated far-field Power density of the PEC Box due to an incident field.

by a plane wave incident from the x-direction and then the scattered far-field along the x-axis is measured from the frequency of 0.1 GHz to 0.424 GHz. The actual data with 1800 sample points are generated by using the software HOBBIES [22]. The far-field power spectrum density is plotted in Figure 3.16. From Figure 3.16 one can observe that there are three narrow nulls at around 0.2114, 0.3315 and 0.3638 GHz. These nulls are extremely narrow so that very fine frequency step is required in order to capture them. In this example, the frequency interval of interest is 0.18 MHz.

The original data will now be down sampled by a certain rate at first to get a sparse data set, which is next used as the input data to the present Cauchy method. Then the interpolated data will be used to compare with the original data. Define the percentage error rate through:

$$Error\ Percentage\ Rate = \left| \frac{Original - Interpolated}{Original} \right| \times 100\% \qquad (3.76)$$

Figures 3.17, 3.19 and 3.21 show the comparison between original and the interpolated far-field spectrum density when downsampling rates of 10, 20 and 30,

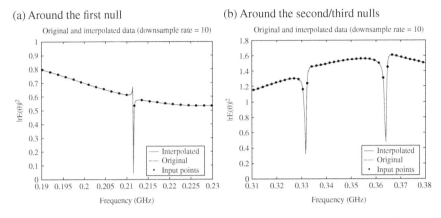

Figure 3.17 Comparison of Original and Interpolated Data (Down sample Rate = 10).

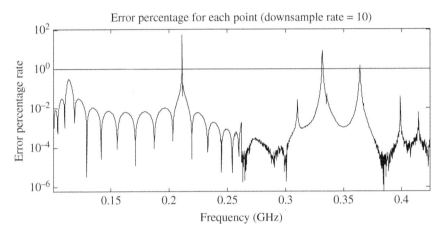

Figure 3.18 Error Percentage Rate over the entire band (Down sample Rate = 10).

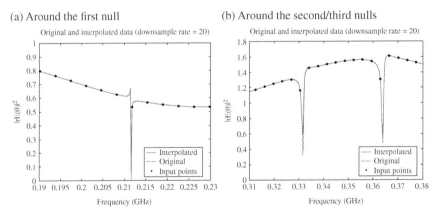

Figure 3.19 Comparison of the Original and Interpolated Data (Down sample Rate = 20).

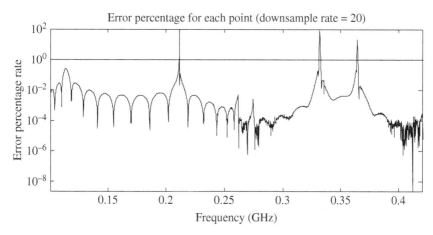

Figure 3.20 Error Percentage Rate (Down sample Rate = 20).

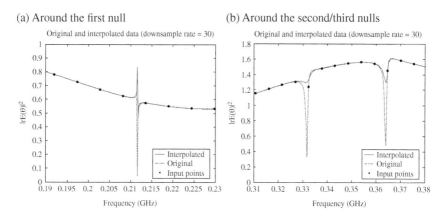

Figure 3.21 Comparison of Original and Interpolated Data (Down sample Rate = 30).

respectively, are applied to the originally sampled data. While Figures 3.18, 3.20 and 3.22 provides the corresponding error percentage rate for the extrapolation of the downsampled data by the Cauchy method. The plots are related to the error between the original data and the interpolated data.

From the above figures it is seen that even when the down sampling rate is below 20, and even though the nulls in the responses are not very clear in the down sampled data, the Cauchy method can still not only detect the right location of the nulls, but also reconstruct the position of the nulls in the

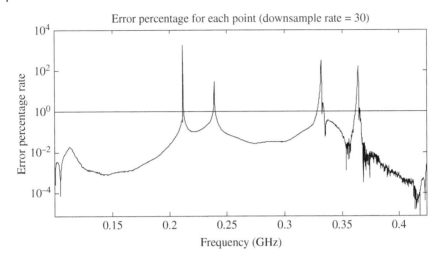

Figure 3.22 Error Percentage Rate (Down sample Rate = 30).

response quite accurately. The error is relatively large just at one or two points around the extreme points of the null. When the down sampling rate is very large, i.e., the input data is sparser, the error gets larger, as expected, and one may not be able to recover the exact shape of the nulls. However, the Cauchy method still provides a good indication as to where the nulls exist in the band.

Next, the Cauchy method is applied to retrieve the nulls in the response, when they completely disappear from the given sampled data. One can now apply the non-uniform sampling which removes the sample points around the nulls. The base sample rate is set to be 10 and the data is incomplete since part of the data is missing in some frequency bands. Figures 3.23(a)-(d) show the comparison of the original and interpolated data when 3, 7, 11 and 15 sample points around the first null are removed respectively, and Figures 3.24(a)-(d) show the comparison of the original and interpolated data when 3, 5, 7 and 9 sample points around the second and third nulls are removed respectively.

The above figures clearly illustrate that although the nulls of the far-field power spectrum density are completely excluded from the input data, the modified Cauchy method is still able to reconstruct the nulls. One may thus conclude that the Cauchy method can restore the nulls of the in the missing data. Even when the missing frequency band is quite large (7 points missing for the single null and 5 points missing for the dual nulls), the interpolated data is still very close to the original data. It is also reasonable that the larger

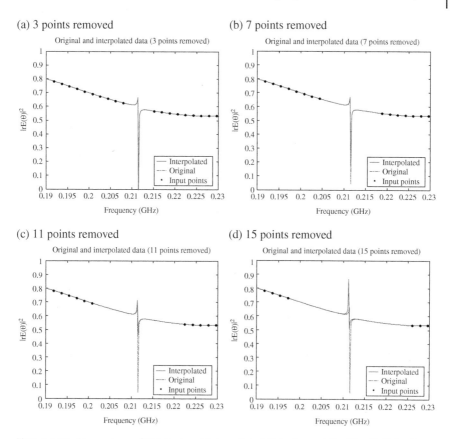

Figure 3.23 Comparison of Original and Interpolated Data Around the First Null for the sampled data.

missing band results in a worse reproduction of the response around the immediate vicinity of the nulls. The error around the nulls becomes larger when more points are removed. However, the location of the null is correctly detected.

Here the focus has been on extrapolation and interpolation of amplitude-only data without the phase information. The question that is now addressed is: can one regenerate the phase from the given amplitude only data? The next section illustrates that since one is performing a model based parameter modelling using a ratio of two polynomials, it is possible to use the model obtained from the analysis of amplitude-only data to generate the non-minimum phase response from the extracted model which is discussed next.

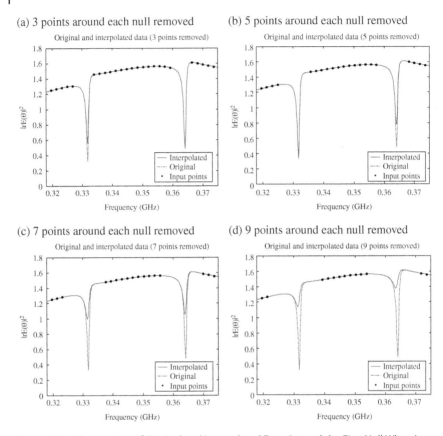

Figure 3.24 Comparison of Original and Interpolated Data Around the First Null When Input Data are Incomplete.

3.7 Generation of the Non-minimum Phase Response from Amplitude-Only Data Using the Cauchy Method

In electromagnetic analysis, for high frequencies, the phase response is hard to obtain but on the contrary the magnitude response can easily be measured. Therefore, in reality it is very useful to reconstruct the phase response from amplitude-only data. If the far-field power pattern of a causal system is obtained as a function of space, the associated phase function can be retrieved by utilizing the fact that the Discrete Fourier transform of the far-field power pattern is equal to the autocorrelation of the equivalent spatial current distribution on the electromagnetic structure [24, 25]. In reference [24] the all-pass representation is also used to reduce the computational load. However, for the problem where the far-field system response is measured as a function of frequency where one encounters a non-minimum phase response, the approach introduced in [25] is not applicable.

In this section, a new approach is presented which focuses on retrieving the poles and zeros from a non-minimum phase response when the amplitude-only data is the only available information. To regenerate the non-minimum phase over the given frequency band, the whole frequency band of interest is sub-divided into several sub-bands with certain overlaps. By applying the Cauchy method presented in the previous section, to the square of the magnitude response, i.e., the power spectrum density and solve for the two polynomial functions in both the numerator and the denominator, one obtains a set of conjugate symmetric pairs of poles and zeros for each sub-band. One pole or zero for each pair is selected based on the fact that the most appropriate reconstructed phase should have the linear phase difference with the actual phase. The addition of a linear phase difference to the actual phase response is equivalent to a pure delay in the time domain, and that information is not contained in the power spectrum. Also a delay in the response does not change the amplitude spectrum. Since one is dealing with a LTI (linear time-invariant) system, changing the impulse response of the system by a time shift does not change the transfer function of the original system, except that the phase spectrum is modified by a linear phase function, and the amplitude spectrum is unchanged. The slope of the linear phase difference corresponds to the time delay.

3.7.1 Generation of the Non-minimum Phase

In general, knowledge about the amplitude spectrum provides no information about the phase, and vice versa. There are typically two different methodologies to generate the non-minimum phase function from amplitude-only data. One such method will be discussed in this section while the other technique based on the Hilbert transform approach will be described in the next chapter 4. For systems described by linear constant-coefficient difference equations, i.e., rational system functions, there is some constraint between the magnitude and the phase response. Such system transfer function can be expressed by a set of poles and zeros in s-plane as follows:

$$H(f) = H\left(\frac{s}{j2\pi}\right) = \frac{c_0 \prod\limits_{k=1}^{M} (s - jc_k)}{d_0 \prod\limits_{k=1}^{N} (s - jd_k)} \tag{3.77}$$

Then we have:

$$\left|H^2(f)\right| = H\left(\frac{s}{j2\pi}\right)H^*\left(\frac{s^*}{-j2\pi}\right) = \frac{c_0 \prod\limits_{k=1}^{M}(s-jc_k)\ c_0^* \prod\limits_{k=1}^{M}\left(s^*+jc_k\right)^*}{d_0 \prod\limits_{k=1}^{N}(s-jd_k)\ d_0^* \prod\limits_{k=1}^{N}\left(s^*+jd_k\right)^*}$$

$$= \frac{\left|c_0^2\right| \prod\limits_{k=1}^{M}(s-jc_k)(s-jc_k^*)}{\left|d_0^2\right| \prod\limits_{k=1}^{N}(s-jd_k)(s-jd_k^*)}$$

(3.78)

Equation (3.78) is exactly the same as Equation (3.61) when the numerator and denominator polynomials are expanded. So the coefficients in Equation (3.78) can be estimated by the Cauchy method, and then the poles and zeros in the s-plane can be solved for by searching for the roots of the polynomial functions. Note that the poles and zeros consist of symmetric pairs which are symmetric along the imaginary axis in the s-plane. One pole or zero in each pair is associated with the system function $H(f)$ and the other pole or zero is associated with $H^*(f)$. Now the problem is simplified to choosing the right poles and zeros to model $H(f)$. If $H(f)$ is assumed to correspond to a causal and stable system, which is the case for most electromagnetic systems, all the poles must lie in the left hand side of the s-plane as first stated by James Clerk Maxwell [24, 25]. With this constraint, it is straight forward to identify the poles of $H(f)$ from the poles of $|H^2(f)|$. Only half of the poles will satisfy this constraint and will be used to construct the system response $H(f)$. However, the zeros of $H(f)$ cannot be uniquely identified from the zeros of $|H^2(f)|$ with this constraint alone. If there are n pairs of zeros, the total number of possible system response $H(f)$ is 2^n.

Another key constraint is that assuming the right zeros are selected, there should be a linear phase difference between the reconstructed and the actual phase functions. Because for the LTI (linear time-invariant) system, changing the impulse response of the system by a time shift does not change the amplitude response of the transfer function of the original system, except that the phase spectrum is modified by a linear phase function. So given a particular amplitude response, the solution of the phase response is not unique. Any phase function with a linear phase difference to the actual phase should fall into the solution set, and the slope of the linear phase difference corresponds to the time delay. To apply this constraint, at first one needs to split the entire frequency band into several small sub-bands with certain overlaps. Secondly one can apply the Cauchy method on every sub-band and get all the possible phase functions for each sub-band. There should be a linear phase difference between the actual phase (which is unknown) and are both the right candidates for sub-bands 1 and 2. Since these two sub-bands are overlapping, the phase difference in the overlapping part between the reconstructed phase functions in sub-bands 1 and 2 should also be linear. So one can just list all the combinations for the phase functions for sub-bands 1 and 2 and find the most linear phase difference, and then select the corresponding phase functions as the desired ones. Similarly, one can choose the phase function for sub-bands 3, 4… by looking at the following overlaps, and at last the phase functions for all the sub-bands are assembled together and the phase response for the whole frequency band is generated. To compare the linearity in the overlapping bands, one can apply a linear regression to the sample points in the overlapping frequency range.

However, there are several important issues that one needs to be careful about in this approach. At first the length of each sub-band cannot be very large. Because the Cauchy method does not work very well for large sample size, the optimal range for the band is between 50 to 100 sample points. Also, the amplitude change in each sub-band should not be complex, so that a small number of computed pole/zero pairs are adequate to reproduce the amplitude response accurately. Because the number of possible candidates increases exponentially with the number of zero pairs, it is necessary to control the number of zero pairs in each sub-band under 4, normally 2 or 3. Therefore the number of possible phase function combinations at each overlap is normally less than 100. Secondly the overlapping band cannot be very small otherwise the right function cannot be distinguished from the other candidates. Normally the overlap is around 1/4 to 1/3 of the sub-band. Sometimes one may get pure imaginary poles and zeros which simply scale the system response. They will not affect the phase functions and thus can be ignored.

3.7.2 Illustration Through Numerical Examples

In this subsection a numerical example is presented to illustrate the applicability of this approach in reconstructing the phase response from the magnitude-only data. For the first example, consider a horn antenna as shown in Figure 3.25. The probe at the end of the horn is excited with 1 volt and is oriented

Figure 3.25 A Horn Antenna.

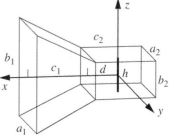

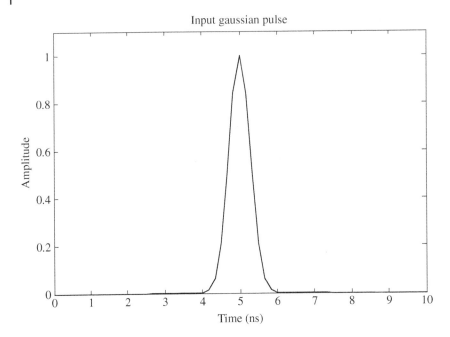

Figure 3.26 An input Gaussian Pulse.

along the z-axis. Its length is 40 mm. The dimensions of the horn antenna are: $a_1 = 72$ mm, $b_1 = 80$ mm, $c_1 = 50$ mm, $a_2 = 60$ mm, $b_2 = 30$ mm, $c_2 = 50$ mm and $d = 25$ mm. The input exciting the horn antenna is a Gaussian pulse as shown in Figure 3.26.

The far-field power spectrum from 1.4 GHz to 2.8 GHz is given in Figure 3.27. There are in total 1401 sample points, which is down sampled by 10 before processing by the Cauchy method. The 1.4 GHz frequency band is divided into three sub-bands of 0.6 GHz duration, and the overlap between the adjacent sub-bands is 0.2 GHz. Figure 3.27 shows the actual computed phase by HOB-BIES [22] (solid line), the reconstructed phase (dashed line) and the phase difference (dotted line) responses for the horn antenna. From Figure 3.28 it is seen that the difference between the actual phase and the estimated phase is very linear. This is because the reconstruction of phase from amplitude-only data does not provide a unique solution. Any true phase response related to the amplitude-only data should be related to the reconstructed phase by a linear phase function which in the time domain is a pure time delay of the true temporal function. Therefore, from Figure 3.28 it is clear that the actual phase has been recovered!

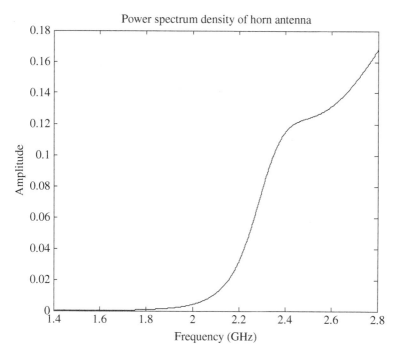

Figure 3.27 Power Spectrum Density of Horn Antenna.

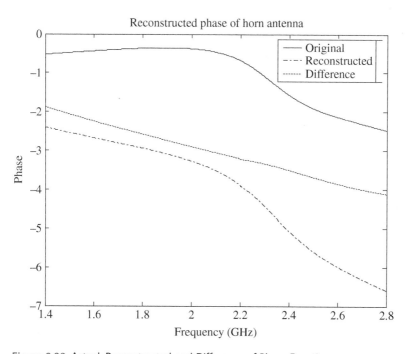

Figure 3.28 Actual, Reconstructed and Difference of Phase Functions.

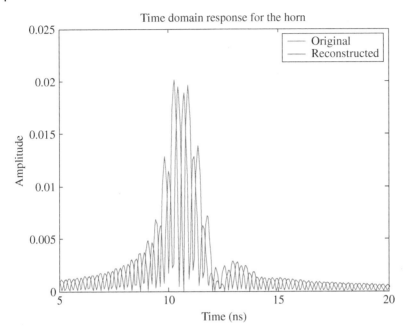

Figure 3.29 Time Domain Responses.

Figures 3.29 and 3.30 illustrate the time domain response of the horn antenna. The difference between Figures 3.29 and 3.30 is that in Figure 3.30 the reconstructed time domain response is shifted to match the original time domain response to compensate for the linear phase representing the time delay. From these two figures one can observe that the distortion in time domain of the computed response is very small.

For the next example, consider a microstrip patch antenna as shown in Figure 3.31. The probe exciting the patch is attached to a long rectangular metallic strip and it connects to a square patch which are printed over a dielectric substrate with $\varepsilon_r = 2.6$. The dimension of the patch and the substrate are: $a_1 = 76.56$ mm, $b_1 = 48.72$ mm, $a_2 = b_2 = 34.8$ mm, $a_3 = 34.8$ mm, $b_3 = 7.36$ mm, $h_1 = 1.575$ mm, $h_2 = 0.4725$ mm. The input excitation is the same Gaussian pulse as in the previous example shown in Figure 3.26. Figure 3.32 shows the power spectrum density of the radiated fields from the microstrip patch antenna from 2 GHz to 3 GHz.

There are in total 2001 sample points, which is down sampled by 10 before being processed by the Cauchy method. The 1 GHz frequency band is divided into three sub-bands of 0.4 GHz, and the overlap between the adjacent sub-bands is 0.1 GHz. The overlap is to align the phase between the overlapping

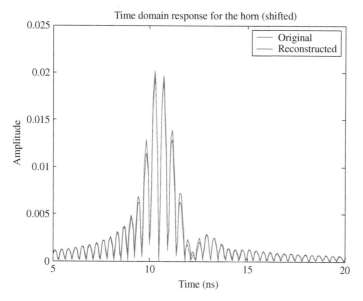

Figure 3.30 Sifted Time Domain Responses.

Figure 3.31 A Microstrip Patch Antenna.

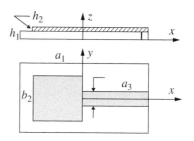

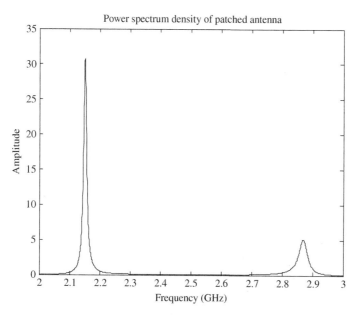

Figure 3.32 Power Spectrum Density of the Microstrip Patch Antenna.

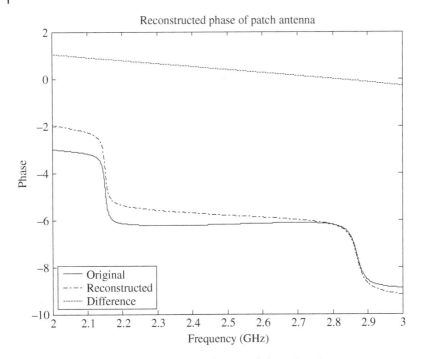

Figure 3.33 Actual, Reconstructed and Difference of Phase Functions.

bands. Figure 3.33 shows the actual computed phase (solid line), the reconstructed phase (dashed line) and the phase difference (dotted line) responses for the microstrip patch antenna. From Figure 3.33 one can see that the phase difference is a straight line. The original and the shifted reconstructed time domain responses are plotted in Figure 3.34. These two-time domain responses are almost identical.

For the last example, consider the example of the PEC box of Figure 3.15. There are totally 1801 sample points, which is down sampled by 10 before being processed by the Cauchy method. The 0.324 GHz frequency band is divided into three sub-bands of 0.081 GHz, and the overlap between the adjacent sub-bands is 0.027 GHz. From Figure 3.35 one can see that the phase difference is very linear over the whole frequency band. The phase function is reconstructed accurately even if the nulls in the magnitude response does not exist. The original and the shifted reconstructed time domain responses are plotted in Figure 3.36. It is seen that the shapes in time domain are very similar except that there are some amplitude distortions around the peaks.

In summary, the Cauchy Method in this section has been used to retrieve the phase information from the magnitude response of the far-field electromagnetic system. The whole frequency band of interest is divided into several suitable

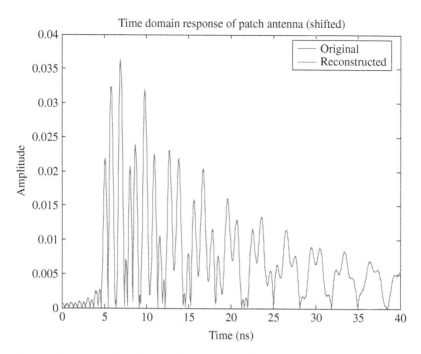

Figure 3.34 Aligned Time Domain Response from the Patch Antenna.

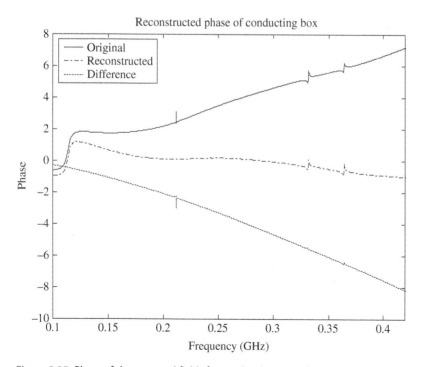

Figure 3.35 Phase of the scattered fields from a Conducting Cubic Box with a Square Hole.

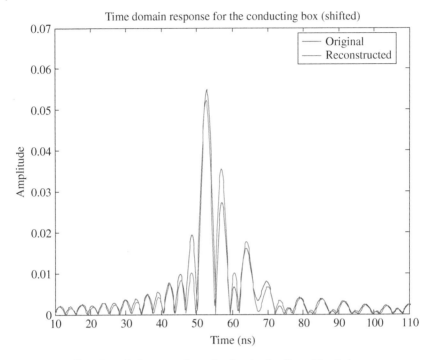

Figure 3.36 Time Domain Response from the Conducting Box with a hole.

overlapping sub-bands and the Cauchy method is applied in each sub-band to obtain the pole and zeros pairs of the power spectrum density which consists of symmetric pairs. The poles of the frequency response are determined based on the constraint of stable and causal systems; the zeros of the frequency response are selected based on the fact that the phase difference of the reconstructed phase functions should have linear phase differences over the overlapping frequency ranges. Three numerical examples, including far-field radiation and the reflection power spectrum of different antennas indicate that this approach can be widely used for phase reconstruction purposes.

3.8 Development of an Adaptive Cauchy Method

3.8.1 Introduction

A fast and accurate interpolation algorithm is proposed to reconstruct the high resolution amplitude-only frequency domain response such as radar cross section and antenna radiation patterns from sparse and non-uniform samples based on the Cauchy method. Non-uniform sampling is implemented in the

Cauchy method such that any additional samples and any a priori information can be added to the existing sample set to continuously improve the interpolation result. In this proposed algorithm, the sample set is automatically adjusted according to the nature of the given data, so that the computational load is minimized by taking the least number of sample points while still maintaining high interpolation accuracy. Even though the Cauchy method can accurately interpolate the data containing both amplitude and phase, in this section only the case for the amplitude-only data is presented where the function may not be differentiable at all points.

In this section, the Cauchy method is implemented to allow non-uniform sampling. Furthermore, a novel algorithm is developed to predict the variation of the interpolated frequency response. Once an expected peak/null structure is obtained, additional samples are adaptively added to the existing sample pool to improve the result of the interpolation. Alternately an adapting sampling procedure can also be applied [26]. As such, very accurate frequency response can be generated by the presented algorithm with minimum computational load. The literature on this general topic is quite vast and illustrative references are provided where additional references can be obtained. However, this section is concerned not with the general problem of interpolation of complex data in the frequency domain but using only amplitude-only data.

Therefore, attention is focused on the extrapolation of amplitude-only data where the derivative information may not be available. This eliminates most of the conventional model based parameter estimation techniques. However, the Cauchy method can still be implemented using samples of the waveform which may be nonuniformly spaced.

3.8.2 Adaptive Interpolation Algorithm

The number of sampling points required for accurately interpolating a curve generally depends on the smoothness of the curve. In this section a novel adaptive algorithm is introduced for accurate interpolation of data using a minimum number of sample points. The basic concept of this adaptive algorithm is an iterative process that starts with a sparse uniformly sampled set, and then depending on the interpolation result, continuously adds new samples to the original set to improve the accuracy of the simulation result. To implement this adaptive algorithm, two questions need to be addressed: 1) what is the proper base sample rate, and 2) how to determine the number and locations of the new samples.

Intuitively the number of samples required to interpolate a curve accurately depends on the complexity of the curve. A curve with more features would need more samples to interpolate than a simple and smooth curve. However, the problem is that one does not know how complex the actual response is beforehand. To determine the base sample rate, one can always start with very sparse samples. Since in the Cauchy method the parameters are estimated based on the TLS approach, the interpolated curve may not pass through all the given sample points. It is in fact a good indication on whether or not the sample rate is adequate. When the interpolated curve doesn't pass through the given sample points, the base sample rate is too low and has to be increased. Then, keep increasing the sample rate till the interpolated curve passes through all the given sample points, which indicates that the proper base sample rate has been reached. However, passing through all the given sample points does not mean that the interpolation is accurate. In the interpolated curve, error may occur at the locations between any two sample points. The error most likely happens when the curve has a sudden change in shape, which can be represented by the curvature of the curve. The curvature of a curve $y = f(x)$ is given by:

$$k(x) = \frac{|y''|}{\left[1 + (y')^2\right]^{3/2}} \tag{3.79}$$

Once the high resolution interpolated data is obtained, the first and second derivatives of the function can be evaluated and the curvature at any point can be calculated using (3.79). [It is important to point out that an estimate of the derivatives from the given sample data points is quite different than the derivative information required at every given sample points by the usual Cauchy method and other model based parameter estimation techniques]. If any curvature value is beyond a certain threshold, this point will be sampled and used as a validation point to check the accuracy of the interpolated data. As long as the error is not satisfactory, this validation sample will be added to the original sample set, and the interpolation will be iterated till all high curvature points on the given curve are completely validated. The flowchart of this iterative process is summarized in Figure 3.37. Several examples will be given in the next section to illustrate the procedure of the proposed adaptive interpolation algorithm.

3.8.3 Illustration Using Numerical Examples

In this section, three numerical examples using different scenarios are presented to illustrate the applicability of the proposed interpolation algorithm in reconstruction of high resolution frequency domain response and antenna power pattern from sparse samples. In these examples, the original high resolution data is obtained using the EM simulation package HOBBIES [22]. Then the original data is down-sampled by a certain rate at first to generate a sparsely sampled data, which will serve as the input data for the interpolation. Once the adaptive interpolation is completed, the interpolated data with the same high resolution will be used to compare with the original data to verify if the interpolation is accurate. One can also look at the relative error which is defined by

$$Error = \frac{|\,Interpolated - Original\,|}{Original} \times 100\% \tag{3.80}$$

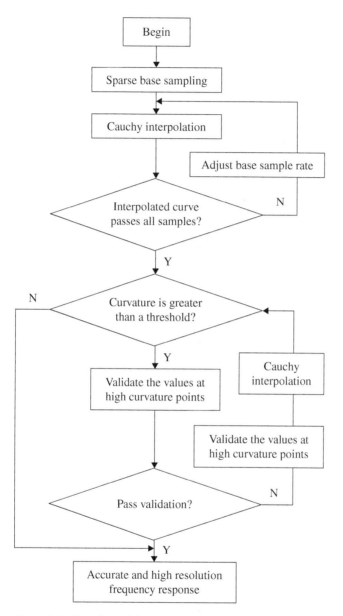

Figure 3.37 Flowchart of the proposed interpolation algorithm for adaptive interpolation algorithm based on the Cauchy method.

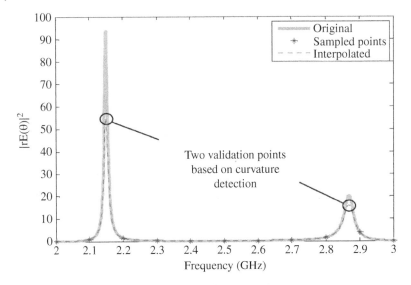

Figure 3.38 The initial interpolation result for the patch antenna.

In all these examples, the threshold value for the error is set to be 1%.

For the first example, consider a microstrip patch antenna as illustrated in Figure 3.31. The antenna is excited by a Gaussian pulse and the resulting amplitude-only response in the frequency range from 2 GHz to 3 GHz is depicted in Figure 3.32. The simulation data contains 2001 sample points with a frequency increment of 1 MHz. A very narrow and high peak at 2.15 GHz and a relatively lower peak at 2.87 GHz can be observed in the power spectrum of the patch antenna.

At first, the simulation data is down-sampled by 100, resulting in 11 sample points located at 100 MHz frequency intervals. The Cauchy interpolation approach described in the previous section for amplitude-only data is then applied to these sample points. Figure 3.38 shows the initial interpolation result for the patch antenna. It is surprising to see that although the two peaks seem completely missing in the sparse samples, both of them have been successfully detected and recovered using the Cauchy interpolation, even with the right shapes. Curvature detection of the interpolated data suggests two high curvature points at 2.153 GHz and 2.8695 GHz. Comparison between the interpolation result and the simulation data reveals relatively large errors at these two frequency points, which means a new iteration is required and the process is continued till the error is minimized.

Next two additional frequency validation points is included in the sample pool and the Cauchy interpolation is applied on the updated sample set. The second

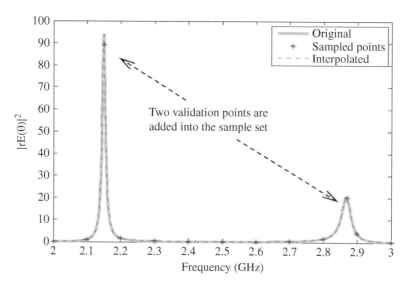

Figure 3.39 The second interpolation result for the patch antenna, the interpolated and the original curves are completely overlapping.

interpolation result of the patch antenna is displayed in Figure 3.39. To verify the result, the relative error is calculated and plotted in Figure 3.40. From the result one can conclude that extremely accurate extrapolated frequency response is obtained once the two new sample points are added. The error is less than 0.01% over the entire frequency band, including at the high peak value that is very challenging to interpolate. In this example, only 13 sample points are used to obtain a comparable frequency domain response originally generated from simulation at 1001 frequency points.

The second example considered is a hollow conducting box with a dimension of 1 m by 1m by 1 m. There is a rectangular hole on the top surface perpendicular to *x* axis. The size of the hole is 0.1 m by 0.1 m. This conducting box, which is shown in Figure 3.15, is excited by a plane wave incident from *x* direction and the far-field along the *x* axis is measured from 0.3 GHz to 0.516 GHz. In this frequency segment, the far-field power spectrum density with 1201 sample is shown in Figure 3.16. It can be seen that the frequency response of this hollow box is very complex, which contains two extremely narrow nulls at around 0.332 GHz and 0.364 GHz, and other less deep nulls at 0.443 GHz and 0.470 GHz, respectively. These nulls (especially the first two) are extremely narrow so that a very fine frequency step is required in order to capture them. In this example, the frequency interval of the original simulation data is 0.18 MHz. Due to the complex features of this curve, it is very difficult to reproduce the curve accurately using traditional interpolation methods.

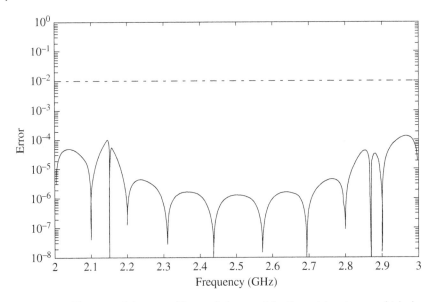

Figure 3.40 The error of the second interpolation result for the patch antenna, which shows the maximum error is less than 1%.

In the initial iteration, 21 uniformly distributed samples are used in the interpolation process, i.e., the frequency interval of the sample data is 10.8 MHz. The interpolated frequency response is shown in Figure 3.41. Form the figure one can observe that the interpolated curve passes through the three beginning and four ending sampling points, as the actual frequency response is relatively smooth in these regions. However, for other frequency components, the interpolated curve moves away from the sample points, indicating that a smaller frequency step is required. It is also interesting to see that even though the samples are sparse, the first and third nulls are still detected.

In the second iteration, 31 samples are taken from this frequency band, resulting in the base frequency interval of 7.2 MHz. The simulation result and the errors are shown in Figure 3.42. This time the interpolated curve passes through all the sample points, thus this base sample rate is appropriate. It is also verified from the simulation result, in which all four nulls are correctly detected. However, apparently there are glitches at the first null. Based on curvature detection, five validation points need to be checked, as marked in Figure 3.42 The computed errors at these frequency points are not satisfactory.

Next, the five validation points from the previous iteration are included into the sample set such that the third iteration for the interpolation is performed. The updated sample set, the new interpolation result from the updated sample

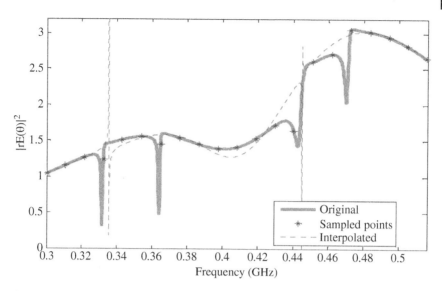

Figure 3.41 The initial interpolation result for the hollow PEC box while the base frequency interval is 10.8 MHz, which shows the base sample rate is too low.

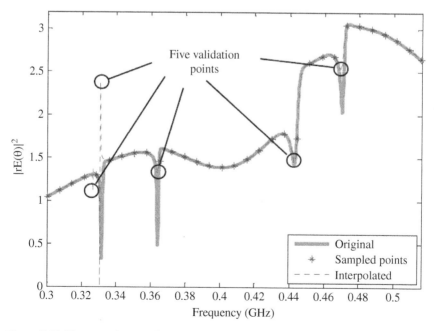

Figure 3.42 The second interpolation result of the hollow box while the base frequency interval is 7.2 MHz, with 5 validation points marked.

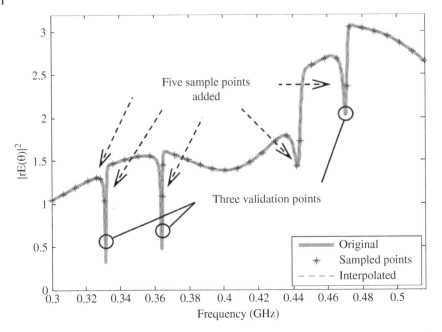

Figure 3.43 The third interpolation result of the hollow box from updated sample set, with three validation points marked.

set, and the original data are compared in Figure 3.43. Three new validation points are also determined from the curvature detection of the new interpolation curve, as marked in Figure 3.43. In this iteration, the interpolation result is much improved. The last two nulls are accurately reproduced with less than 1% error, therefore no more new samples are required around these regions. The glitches at the first null are successfully removed in the new interpolation result, and the errors at the first two nulls are also significantly reduced. Compared to the original simulation data, the interpolation result is considerably more accurate over the entire frequency band, except slight offsets right at the tip of the first two nulls.

To see if the result can be further improved, the first two validation points from the third iteration is absorbed into the sample set and the fourth iteration is performed. The interpolation result and the error are depicted in Figure 3.44 and Figure 3.45, respectively. As expected, the errors around the first two nulls are further reduced, as the maximum error over this entire frequency band is now less than 1%. Therefore, the frequency response of this hollow conducting box with such complex response is accurately reconstructed by using only 38 sample points.

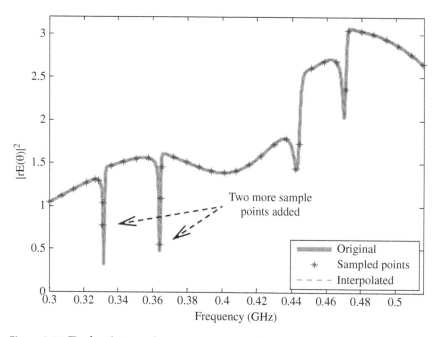

Figure 3.44 The fourth interpolation result for the hollow PEC box with two more sample points added, the interpolated and the original curves are completely overlapped.

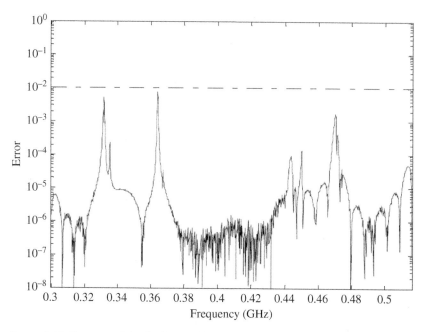

Figure 3.45 The error of the final interpolation result for the hollow PEC box, which shows the maximum error is less than 1%.

Table 3.1 Computational Load Calculation.

Approach	Sample Points	Computational Load %
MoM without interpolation	1201	100%
Traditional Cauchy method	241	20.1%
Cauchy with adaptive sampling	38	3.2%

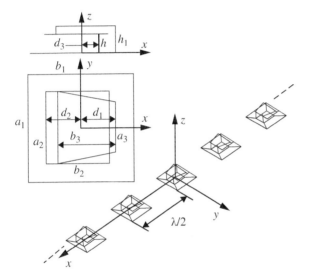

Figure 3.46 A SDPA array with 11 elements.

Given less than 1% error, the computational loads for the original simulation, the traditional Cauchy interpolation and the Cauchy interpolation based on adaptive sampling are summarized in Table 3.1. Since the electromagnetic field computation is much more computationally intensive than the interpolation process, the computational load for interpolation can be neglected. Table 3.1 indicates that significant advantage can be gained from the proposed adaptive Cauchy interpolation.

Besides processing the power spectrum, one can also use the adaptive Cauchy interpolation method in processing antenna radiation patterns. Figure 3.46 shows a Short Dual Patch Antenna (SDPA) array containing 11 elements placed along the x-axis, with a half-wavelength distance in between. The dimensions of each SPDA element, designed for operating at 2.4153 GHz, are: $a_1 = b_1 = 27.5$ mm, $a_2 = 17.5$ mm, $b_2 = 14$ mm, $a_3 = 10$ mm, $b_3 = 15$ mm, $h = 6$ mm,

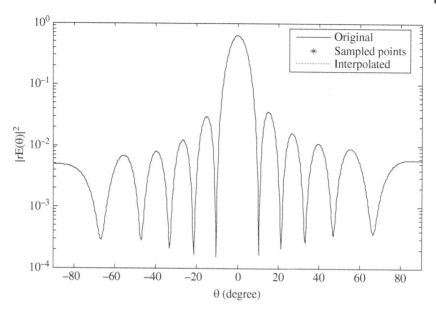

Figure 3.47 Far-field radiation pattern of the SDPA array.

$h_1 = 8$ mm, $d_1 = d_2 = 8.75$ mm, and $d_3 = 4.75$ mm. All antennas in this array are excited simultaneously with an excitation of 1 volt. The far-field power pattern of this antenna array is calculated at every quarter degree from $-90°$ to $90°$, and the result is depicted in Figure 3.47. It is a typical antenna pattern with a main beam of approximately 20 degrees with a set of sidelobes. The main obstacle of reconstructing the power pattern from the sparse sample is the nulls between the adjacent lobes. Since the radiation pattern is symmetric around $0°$, only half of the data from $0°$ to $90°$ is processed.

In the first iteration, 16 data sampled at every $6°$ is taken as input, and the interpolated output using the Cauchy method is shown in Figure 3.48. The result clearly indicates that the sample rate of the input data is too sparse.

In the second iteration, the sample rate is doubled, resulting in 31 samples with $3°$ interval. Figure 3.49 shows the result of the second interpolation, which is close to the original data. Curvature detection of the result suggests five validation points at $10.5°$, $21.5°$, $33.25°$, $47°$ and $66.25°$, as marked in Figure 3.49. The error at the last three validation points are all below 1%, so only two new samples at first two nulls are required in the next iteration.

In the third iteration, the validation points at $10.5°$ and $21.5°$ are added into the sample set. The comparison between the interpolated radiation pattern and

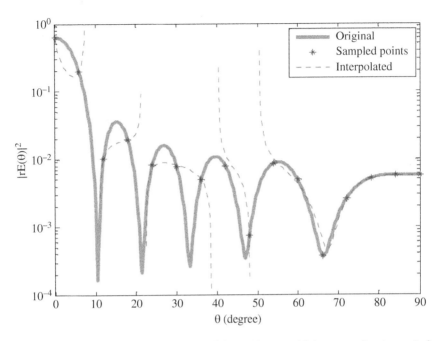

Figure 3.48 The initial interpolation result of the SDPA array with base sampling interval of 6°, which shows the base samples are too sparse.

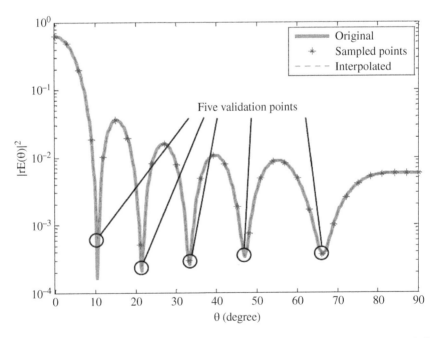

Figure 3.49 The second interpolation result of the SDPA array with base sampling interval of 3°, indicating the appropriate base sample rate is achieved.

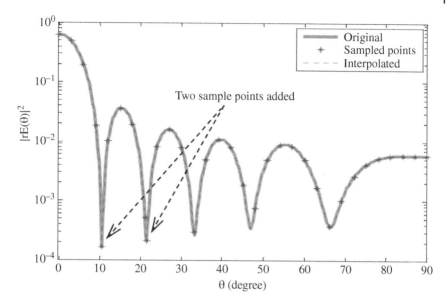

Figure 3.50 The third and final interpolation result of the SDPA array with two sample points added.

the original data, and the error of the interpolation are shown in Figure 3.50 and Figure 3.51, respectively. As expected, by including the two additional sample points, the maximum error is reduced to far below 1%, and very accurate radiation pattern is reconstructed from only 33 samples, which is an order of magnitude lower than the original data size.

3.8.4 Summary

In this sub-section, a novel adaptive interpolation algorithm capable of accurately reproducing high resolution frequency domain response and radiation pattern with complex patterns using very sparse samples is proposed. The interpolation is based on the non-uniform sampling implementation of the Cauchy method, assuming that the parameter of interest, as a function of frequency, can be approximated by a ratio of two rational polynomial functions. Numerical example indicates that, by iteratively involving sampling points at specific locations into the sample pool, the interpolation result is progressively improved and very accurate and high resolution frequency domain response and radiation pattern can be reproduced even at the tip of the peak/null, the most challenging task for an interpolation methodology. The proposed algorithm is able to reliably generate high resolution wideband frequency response and antenna radiation pattern at a very low computational load. The advantage of this algorithm

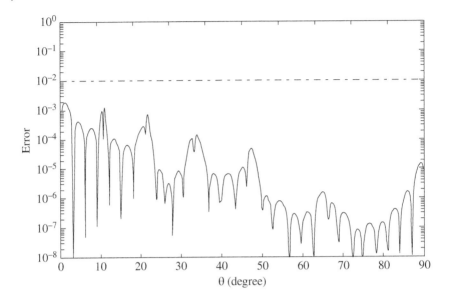

Figure 3.51 The error of the last interpolation result for the SDPA array showing the error at all the data sample points are less than 1%.

is that it can reduce not only the computation time and storage requirement dramatically in the simulation, but also the cost and effort in performing measurements in a real experiment.

3.9 Efficient Characterization of a Filter

The next application of the Cauchy method is to generate an accurate rational model of lossy systems from either measurements or an electromagnetic analysis. The Cauchy method can be used to extract the coupling structure of the filter [27–30]. Two examples are presented. One deals with measured data and the other one uses numerical simulation data from an electromagnetic analysis.

The first example of the filter is an E-plane Waveguide Filter. The Cauchy method has been applied first to a set of measured data samples of a Ku-band bandpass filter response. The device is a sixth-order E-plane filter built in a WR28 rectangular waveguide. The original design presents a Chebyshev-type response with a reflection parameter better than −20 dB in the pass band. However, the measured response is far from this ideal response due to low mechanical tolerances.

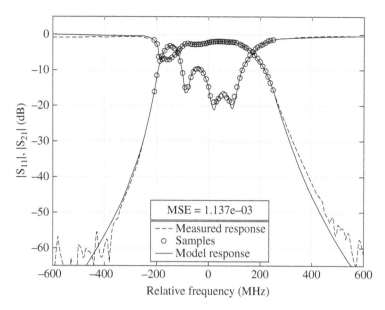

Figure 3.52 Waveguide filter. Model of S_{11} and S_{22} with common denominator.

Table 3.2 E-Plane Filter: Low-Pass Polynomial Coefficients.

k	$a_k^{(1)}$	$a_k^{(2)}$	b_k
0	$-0.0244-0.0110_j$	$0.1133+0.0545_j$	$0.1123-0.1158_j$
1	$-0.0453-0.2532_j$		$0.6002-0.5472_j$
2	$-0.0745-0.1796_j$		$1.3800-1.3009_j$
3	$-0.1619-1.5168_j$		$2.4933-1.9912_j$
4	$0.6856-0.2242_j$		$2.4082-1.5359_j$
5	$-0.0139-1.3188_j$		$1.9356-1.3007_j$
6	1.0000		1.0000

Figure 3.52 shows the measured data and the response of a sixth-order model with no finite transmission zeros, including the mean square error (MSE) between both responses. Larger order models cannot obtain further improvements. Table 3.2 contains the polynomial coefficients of the model. It should be noted out that as this is a low-pass equivalent of a bandpass filter, the coefficients can be complex.

As the second example related to the filter, two different models corresponding to a microstrip cross-coupled bandpass filter have been generated. The structure is a four-pole filter formed by four identical folded hairpin resonators tuned to $f_0 = 2.46$ GHz with significant magnetic and electric coupling between each contiguous pair. The quadruplet structure with a cross coupling generates a pair of symmetrical finite transmission zeros, allowing the synthesis of an elliptic transfer function. The device has been analyzed using the MoM and both narrow-band and wide-band models have been generated using the Cauchy method.

Figure 3.53 shows the narrow-band system response obtained using MoM and the corresponding rational model. As only data samples around the pass band have been used the model can be considered a narrow-band one. The polynomials of the low-pass equivalent, whose coefficients are shown in Table 3.3,

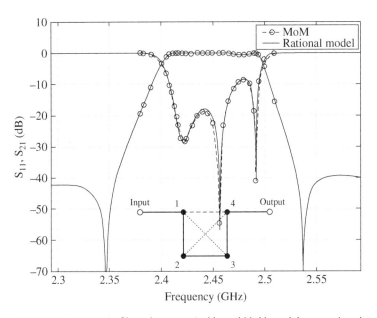

Figure 3.53 Microstrip filter, the numerical based MoM model generating the narrow-band response, and coupling topology (solid: main coupling; dashed: cross coupling; dotted: spurious coupling).

Table 3.3 Microstrip Filter: Narrow-Band Model Polynomial Coefficients.

k	$a_k^{(1)}$	$a_k^{(2)}$	b_k
0	−0.0479+0.0285j	0.2156+0.3849j	0.4187−0.1452j
1	−0.0290−0.0255j	0.0042−0.0027j	1.1584−0.3911j
2	−0.5488+0.3179j	0.0568+0.1067j	1.7892−0.5378j
3	0.1395+0.3070j		1.5244−0.3386j
4	−0.8745+0.4851j		1.0000

correspond to a fourth-order filter with two finite transmission zeros. As can be seen, the location of the reflection zeros and the in-band return loss have been accurately modeled.

From this computed rational model a coupling matrix can be synthesized with the topology shown in Figure 3.53. This coupling matrix reveals some features of the physical filter: first, although the resonators have been designed to be identical some frequency shifts are necessary to model the response. Second, the model is more accurate if diagonal couplings (1-3, 2-4) are allowed (their value is low).

A wide-band rational model has also been generated using data samples from both the main pass band and the first upper band (Figure 3.54). In this case, two resonant modes are considered for each physical hairpin resonator, therefore the order of the filter is doubled. The resulting eight-pole structure is roughly equivalent to two four-order networks in parallel with an additional direct coupling between the source and the load. Figure 3.54 shows that some second-order effects have also been modeled, i.e., the reduction of the maximum attenuation of the filter to approximately 40 dB and the shape of the upper band, including an isolated transmission zero at $f \approx 5.8$ GHz.

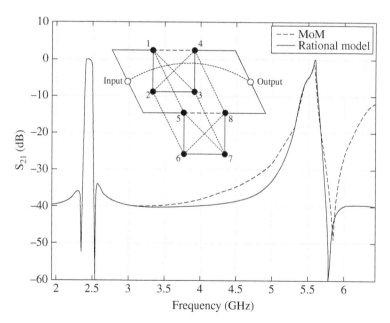

Figure 3.54 Microstrip filter, result from the MoM model and its wide-band response, and coupling topology (solid: main coupling; dashed: cross coupling; dotted: spurious coupling).

3.10 Extraction of Resonant Frequencies of an Object from Frequency Domain Data

The last application of the Cauchy method is to obtain the external resonant frequencies of an object using wideband frequency domain data which may be used for target identification [31–34]. The EM scattering response from a 0.15-m-diameter PEC sphere in free space is computed as shown in Figure 3.55(a) using the HOBBIES (Higher Order Basis Based Integral Equation Solver) simulation program. HOBBIES can be used for solution of various types of electromagnetic field analysis involving electrically large objects of arbitrary shapes composed of complex metallic and dielectric structures in the frequency domain. In this case, a plane wave is applied as an incident wave ($\phi = 0°$, $\theta = 0°$, $E_\phi = 0$, $E_\theta = 1$) and for three different observation angles ($\phi = 0°$, $\theta = 0°$; $\phi = 0°$, $\theta = 30°$; and $\phi = 0°$, $\theta = 60°$). The observed responses are shown in Figure 3.55(b). In Figure 3.55(a), the horizontal green arrow is the propagation vector corresponding to the incident wave and the vertical red arrow is the orientation for the incident E-field vector. The result is generated for the frequency range from 0.01 GHz to 5 GHz ($\Delta f = 0.01 GHz$), and the

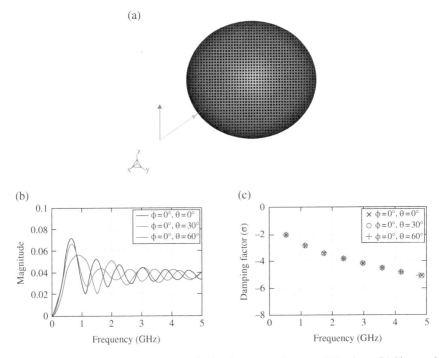

Figure 3.55 (a) HOBBIES simulation model for the 0.15-m-diameter PEC sphere. (b) Observed response of three observation angles. (c) Natural poles.

number of samples selected is 500. Not only the results from the positive frequency data generated by HOBBIES, but also it is complemented with the response for the negative frequencies (conjugate) to reflect its mirror image in the frequency domain.

If the Cauchy method is utilized to get the natural poles of the 0.15-m-diameter PEC sphere without transforming the frequency domain data into time domain, one can obtain eight natural poles corresponding to the physical resonances. Figure 3.55(c) shows the natural poles of the 0.15-m-diameter PEC sphere for three different observation angles using the Cauchy method. One can notice that the poles from these three different realizations corresponding to three different observation angles almost overlap each other because the natural poles are directly related to the late time response (creeping wave) of the object.

Figure 3.56 shows the natural poles of the 0.15 m diameter sphere using the singularity expansion method (SEM) methodology in the time domain and the Cauchy method is applied in the frequency domain. The natural poles from the SEM and the Cauchy method match very well. Therefore, it is proved that one can generate the natural poles of the 0.15 m diameter PEC sphere using high fidelity frequency domain data using the Cauchy method as a library of poles to identify objects, which does not require any late time characterization of the data.

Different simulation examples are presented to illustrate the application of this method to make a library of poles for different objects. The EM scattering data from five additional PEC objects, such as a wire, disk, ellipsoid, sphere, and a cone are also generated using the HOBBIES simulation program. Each object has the same characteristic size (0.1 m diameter, or length, or height) but they are of different shape. The simulation setup is the same as outlined before.

Figures 3.57-3.61 show the five HOBBIES simulation models and their natural poles for the three observation angles. From the Figures 3.57(b), 3.58(b), 3.59(b),

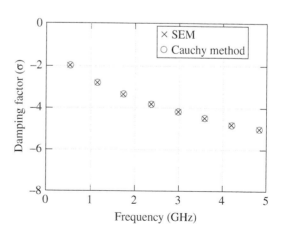

Figure 3.56 Natural poles of the 0.15-m-diameter PEC sphere from the SEM and the Cauchy method.

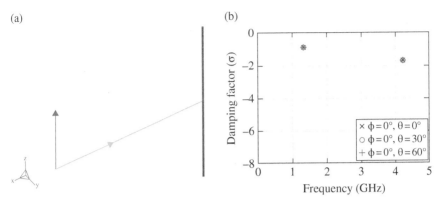

Figure 3.57 (a) HOBBIES simulation model for the PEC wire with 0.1 m length and 1 mm radius. (b) Natural poles of the observed response using frequency domain data.

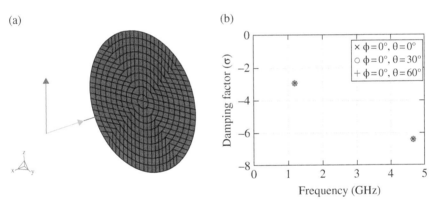

Figure 3.58 (a) HOBBIES simulation model for the PEC disk with 0.1 m diameter. (b) Natural poles of observed response using frequency domain data.

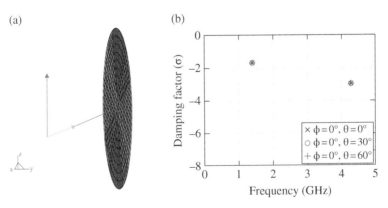

Figure 3.59 (a) HOBBIES simulation model for the PEC ellipsoid with 0.02 m diameter and 0.1 m length. (b) Natural poles of observed response using frequency domain data.

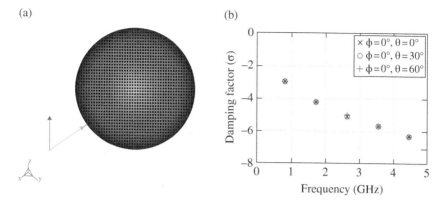

Figure 3.60 (a) HOBBIES simulation model for the PEC sphere with 0.1 m diameter. (b) Natural poles extracted from the frequency domain data.

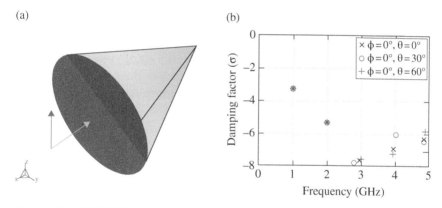

Figure 3.61 (a) HOBBIES simulation model for the PEC cone with 0.1 m diameter and 0.1 m height. (b) Natural poles of extracted from frequency domain data.

and 3.60(b), it is seen that the poles overlap each other even though the observation angles are different. However, one can see that the higher order poles (third, fourth, and fifth) are slightly different when using the frequency domain data than when using the time domain in the real part, the damping factor as shown in Figure 3.61(b). Perhaps, the damping factor of the higher order poles can be fluctuated due to the illuminated surface of the object for the different observation angle.

In Table 3.4 a comparison is provided between the lower frequency poles obtained using the classical late time response and the presented Cauchy method applied to the frequency domain response. Figure 3.57 presents the results for a PEC wire, whereas Figure 3.58 provides the data for a PEC disk.

Table 3.4 Pole Library of the Six PEC Objects.

Object	First Order Pole (σ_1; f_1)	Higher Order Frequency (GHz)
Wire	(−0.8811, 1.3339)	4.2265
Disk	(−2.9610, 1.1775)	4.6855
Ellipsoid	(−1.5828, 1.3967)	4.2956
Cone	(−3.3024, 0.9700)	1.9722, 2.9350, 3.9375, 4.8416
Sphere (D=0.1 m)	(−2.9988, 0.8267)	1.7243, 2.6322, 3.5454, 4.4602
Sphere (D=0.1 5m)	(−2.0011, 0.5504)	1.1490, 1.7534, 2.3638, 2.9761 3.5885, 4.2081, 4.8299

The poles for the PEC ellipsoid are presented in Figure 3.61. The poles for a PEC sphere are presented in Figure 3.60 and for a PEC cone in Figure 3.61. The classical SEM poles and the resonant frequencies are presented in Table 3.4 for all of these objects. Since the values for the damping factors are not reliable for the higher order poles we present the values only for the imaginary part for the natural poles. Finally, one can identify the unknown object by comparing the pole library with the extracted poles measured using the data in the time domain.

This frequency domain methodology for computing the poles has also been applied to the nondestructive evaluation of the maturity of fruits and determining the quality of the fruit juice [34–38] and other RCS applications [39, 40]. This methodology has also been extended to multidimensional extrapolation in [41, 42].

3.11 Conclusion

In this chapter, the Cauchy method is presented starting with the initial concept and its application has been illustrated using some examples. Since the Cauchy method is assuming a data model which consists of a ratio of two polynomials, the problem then is how to estimate accurately its coefficients. This amounts to approximating a response by the natural eigen functions of a LTI system in the frequency domain which forms a ratio of two polynomials. In contrast, in the time domain the eigen functions are a sum of complex exponentials which was discussed in Chapter 2. To solve the approximation problem in the frequency domain, the four steps that are required to generate an accurate approximation are presented from the beginning with choosing the number of samples to reconstruct the system response with computed coefficients of two polynomials. For the noiseless case, one can calculate coefficients of two polynomials and also interpolate/extrapolate the data with high accuracy using as little as 9 samples. Furthermore, one can extract the same residues and poles as the original system functions. When there is additive noise in the data, one can still interpolate the data. As a result, the Cauchy method is very good also in inter-polating/extrapolating the given data with accuracy and efficiency. But the Cau-chy method is sensitive in calculating the residues at the poles for the additive noise case.

Some applications of the Cauchy method are presented to illustrate this novel and accurate methodology. Therefore, one can perform the scattering analysis, the filter model generation, the computation of a library of poles of objects for the purpose of target identification, and so on using the Cauchy method. This method can also be used in the nondestructive evaluation of fruits, fruit juices in addition.

Appendix 3A   MATLAB Codes for the Cauchy Method

Example 1.1

%% Cauchy method: Example 1

```
format short                % Format for displaying numbers
f=0.1:0.01:3; f=f.';        % Frequency (unit: GHz)
data=(2*f.^2+11*f+50)./(2*f.^3-f.^2-25*f-12);   % System
response

%% STEP 1. Choose the number of frequency sample points
Nf=f(1:11:floor(end/3));    % Choose 9 samples (Frequency)
N=length(Nf);               % Number of samples
Hf_s=data(1:11:floor(end/3));  % Choose 9 samples
(System response)

%% STEP 2. Select optimal orders of P and Q
P=floor((N-3)/2);   % P+Q<=N-2, P=(N-3)/2, Order of the
numerator polynomial
Q=P+1;              % Order of the denominator polynomial

C=Matrix_c_draft(P,Q,N,Nf,Hf_s);        % Matrix C
[U,S,V]=svd(C);                         % Singular value
decomposition
N_SV=10*log10(diag(S)./diag(S(1,1)));   % Normalized
singular values

figure(1)                               % Plot normalized
singular values
plot(1:length(N_SV),N_SV,'b-',1:length(N_SV),
N_SV,'b.','MarkerSize',11);
```

```
grid on;
xlabel('Number of Singular values', 'fontsize',11);
ylabel('Normalized singular values (dB scale)',
'fontsize',11);

NR=input('Please input the number of singular values
(R): ');
NP=ceil((NR-2)/2), NQ=NP+1      % Optimal P and Q
NC=Matrix_c_draft(NP,NQ,N,Nf,Hf_s);

%% STEP 3. Compute coefficients ak_e and bk_e
[SV A B R22 ak_e,bk_e]=cauchyy(Nf,Hf_s,NP,NQ);
% Finding residues and poles
[r,p,k]=residue(ak_e(NP+1:-1:1), bk_e(NQ+1:-1:1));

%% STEP 4. Generate the system response
Hf_e=polyval(ak_e(length(ak_e):-1:1),f)./polyval(bk_e
(length(bk_e):-1:1),f);
r                       % Residues
damp=-imag(p);          % Damping factor
poles=real(p)+j*damp    % Poles
MSE=norm(data-Hf_e)/norm(data)    % Mean square errors

% Plot for the comparison between original data and the
estimated data
figure(2);
plot(f,(data),'b-','LineWidth',1.2); hold on; grid on;
plot(f,(Hf_e),'r--','LineWidth',1.2);
plot(Nf,Hf_s,'r*');
axis([0 3 -4 -1]);
legend('Data','Cauchy Method','Selected samples');
plot([Nf(end) Nf(end)],[-1 -4],'k--');
xlabel('Frequency(GHz)','fontsize',11);
ylabel('Amplitude','fontsize',11);
hold off

function [C]=Matrix_C(P,Q,N,Nf,Hf_s)

% Creating Matrix C
% C:      Matrix C
% P:      Order of numerator polynomial
% Q:      Order of denominator polynomial
% N:      Number of samples
```

```
% Nf:     Selected sampling data (Frequency)
% Hf_s:   Selected sampling data (System response)

for n=1:N
for ak=1:P+1
    C(n,ak)=Nf(n)^(ak-1);
end
for hbk=1:Q+1
    C(n,hbk+P+1)=-Hf_s(n)*Nf(n)^(hbk-1);
end
end


function [SV A B R22 ak_e,bk_e]=cauchyy(Nf,Hf_s,NP,NQ)

% Function of the Cauchy Method
% SV:     Normalized singular values
% A:      Matrix A
% B:      Matrix B
% R22:    Matrix R22
% ak_e:   Estimated coefficients of numerator polynomial
% bk_e:   Estimated coefficients of denominator polynomial
% Nf:     Selected sampling data (Frequency)
% Hf_s:   Selected sampling data (System response)
% NP:     Optimal order of numerator polynomial
% NQ:     Optimal order of denominator polynomial

N=length(Hf_s);
A=ones(N,1);     % Matrix A
for kk=1:NP
    A=[A Nf.^kk];
end

B=Hf_s;          % Matrix B
for kk=1:NQ
    B=[B Hf_s.*Nf.^kk];
end

%% Total Least Square Method
[Qa,Ra]=qr(A);               % QR decomposition
Total=[Ra -Qa'*B];
```

```
[q,r]=qr(-B);
R11=Total(1:NP+1,1:NP+1);
R12=Total(1:NP+1,NP+2:NP+NQ+2);
R22=Total(NP+2:N,NP+2:NP+NQ+2);

[U,S,V]=svd(R22);              % Singular value decompositon
bk_e=V(:,NQ+1)
ak_e=-inv(R11)*(R12*bk_e)
SV=diag(S)./diag(S(1,1));
```

Code for Example 2

%% Cauchy method: Example 2
```
format short                   % Format for displaying numbers
f=0.1:0.01:3; f=f.';           % Frequency (unit: GHz)
% System response
data=(2*f.^2+(11-20*j)*f+36-11*j)./(2*f.^3-(1+2*j)*f.
^2+(-25+2*j)*f-12+24*j);


%% STEP 1. Choose the number of frequency sample points
Nf=f(1:11:floor(end/3));       % Selected sampling data
(Frequency)
N=length(Nf);                  % Number of samples
Hf_s=data(1:11:floor(end/3));  % Selected sampling data
(System response)


%% STEP 2. Select optimal orders of P and Q
P=floor((N-3)/2);  % P+Q<=N-2, P=(N-3)/2, Order of the
numerator polynomial
Q=P+1;             % Order of the denominator polynomial

C=Matrix_c_draft(P,Q,N,Nf,Hf_s);        % Matrix C
[U,S,V]=svd(C);                         % Singular value
decomposition
N_SV=10*log10(diag(S)./diag(S(1,1)));  % Normalized
singular values

figure(1)                % Plot normalized singular values
plot(1:length(N_SV),N_SV,'b-',1:length(N_SV),
N_SV,'b.','MarkerSize',11);
grid on;
xlabel('Number of Singular values', 'fontsize',11);
ylabel('Normalized singular values (dB scale)',
'fontsize',11);
```

```
NR=input('Please input the number of singular values (R) : ');
NP=ceil((NR-2)/2), NQ=NP+1        % Optimal P and Q
NC=Matrix_c_draft(NP,NQ,N,Nf,Hf_s);

%% STEP 3. Compute coefficients ak_e and bk_e
[SV A B R22 ak_e,bk_e]=cauchyy(Nf,Hf_s,NP,NQ);
% Finding residues and poles
[r,p,k]=residue(ak_e(NP+1:-1:1), bk_e(NQ+1:-1:1));

%% STEP 4. Generate the system response
Hf_e=polyval(ak_e(length(ak_e):-1:1),f)./polyval(bk_e
(length(bk_e):-1:1),f);
r                          % Residues
damp=-imag(p);             % Damping factor
poles=real(p)+j*damp       % Poles
MSE=norm(data-Hf_e)/norm(data)   % Mean square errors

% Plot for the comparison between original data and the
estimated data
figure(2); subplot(211)
plot(f,real(data),'b-','LineWidth',1.2); hold on; grid on;
plot(f,real(Hf_e),'r--','LineWidth',1.2);
plot(Nf,real(Hf_s),'r*');
legend('Data','Cauchy Method','Selected samples');
plot([Nf(end) Nf(end)],[-2.5 -1],'k--');
xlabel('Frequency(GHz)','fontsize',11); ylabel
('Real','fontsize',11); hold off
subplot(212)
plot(f,imag(data),'b-','LineWidth',1.2); hold on; grid on;
plot(f,imag(Hf_e),'r--','LineWidth',1.2);
plot(Nf,imag(Hf_s),'r*');
legend('Data','Cauchy Method','Selected samples');
plot([Nf(end) Nf(end)],[2 -1],'k--');
xlabel('Frequency(GHz)','fontsize',11); ylabel
('Imaginary','fontsize',11); hold off
```

Code for Example 3

%% Cauchy method: Example 3
```
format short                       % Format for displaying numbers
f=0.1:0.01:3; f=f.';               % Frequency (unit: GHz)
% System response
data=(2*f.^2+(11-20*j)*f+36-11*j)./(2*f.^3-(1+2*j)*f.
^2+(-25+2*j)*f-12+24*j);
```

```
data_noise=awgn(data,40,'measured');    % Added white
Gaussian noise

%% STEP 1. Choose the number of frequency sample points
Nf=f;                          % Selected sampling data
(Frequency)
N=length(Nf);                  % Number of samples
Hf_s=data_noise;               % Selected sampling data
(System response)

%% STEP 2. Select optimal orders of P and Q
P=floor((N-3)/2);  % P+Q<=N-2, P=(N-3)/2, Order of the
numerator polynomial
Q=P+1;             % Order of the denominator polynomial

C=Matrix_c_draft(P,Q,N,Nf,Hf_s);        % Matrix C
[U,S,V]=svd(C);         % Singular value decomposition
N_SV=10*log10(diag(S)./diag(S(1,1)));  % Normalized
singular values

figure(1)               % Plot normalized singular values
plot(1:length(N_SV),N_SV,'b-',1:length(N_SV),
N_SV,'b.','MarkerSize',11);
grid on;
xlabel('Number of Singular values', 'fontsize',11);
ylabel('Normalized singular values (dB scale)',
'fontsize',11);

NR=input('Please input the number of singular values (R): ');
NP=ceil((NR-2)/2), NQ=NP+1     % Optimal P and Q
NC=Matrix_c_draft(NP,NQ,N,Nf,Hf_s);

%% STEP 3. Compute coefficients ak_e and bk_e
[SV A B R22 ak_e,bk_e]=cauchyy(Nf,Hf_s,NP,NQ);
% Finding residues and poles
[r,p,k]=residue(ak_e(NP+1:-1:1), bk_e(NQ+1:-1:1));

%% STEP 4. Generate the system response
Hf_e=polyval(ak_e(length(ak_e):-1:1),f)./polyval(bk_e
(length(bk_e):-1:1),f);
r                       % Residues
damp=-imag(p);          % Damping factor
```

```
poles=real(p)+j*damp   % Poles
MSE=norm(data-Hf_e)/norm(data)   % Mean square errors

% Plot for the comparison between original data and the
estimated data
figure(2); subplot(211)
plot(f,real(data),'b-','LineWidth',1.2); hold on; grid on;
plot(f,real(data_noise),'r--','LineWidth',1.2);
legend('Data','Data added noise')
xlabel('Frequency(GHz)','fontsize',11); ylabel
('Real','fontsize',11); hold off
subplot(212); plot(f,imag(data),'b-','LineWidth',1.2);
hold on; grid on;
plot(f,imag(data_noise),'r-','LineWidth',1.2); legend
('Data','Data added noise')
xlabel('Frequency(GHz)','fontsize',11); ylabel
('Imaginary','fontsize',11); hold off

figure(3); subplot(211)
plot(f,real(data),'b-','LineWidth',1.2); hold on; grid on;
plot(f,real(data_noise),'g-','LineWidth',1.2);
plot(f,real(Hf_e),'r--','LineWidth',1.2);
legend('Data','Data with additive noise','Cauchy
Method');
plot([Nf(end) Nf(end)],[-2.5 -1],'k--');
xlabel('Frequency(GHz)','fontsize',11); ylabel
('Real','fontsize',11); hold off
subplot(212); plot(f,imag(data),'b-','LineWidth',1.2);
hold on; grid on;
plot(f,imag(data_noise),'g-','LineWidth',1.2);
plot(f,imag(Hf_e),'r--','LineWidth',1.2);
legend('Data','Data with additive noise','Cauchy
Method');
plot([Nf(end) Nf(end)],[2 -1],'k--');
xlabel('Frequency(GHz)','fontsize',11); ylabel
('Imaginary','fontsize',11); hold off
```

References

1 A. L. Cauchy, *"Sur la formule de Lagrange relative a l'interpolation,"* Analyse Albebrique, Paris, 1821.
2 K. Kottapalli, T. K. Sarkar, Y. Hua, E. K. Miller, and G. J. Burke, "Accurate Computation of Wide-Band Response of Electromagnetic Systems Utilizing

Narrow-Band Information," *IEEE Transactions on Microwave Theory and Techniques*, Vol. 39, No. 4, pp. 682–687, 1991.

3 K. Kottpalli, T. K. Sarkar, X. Yang, E. K. Miller, and G. J. Burke, "Use of Frequency-Derivative Information to Reconstruct the Scattered Electric Field of a Conducting Cylinder Over a Wide Frequency Band," *Journal of Electromagnetic Waves and Applications*, Vol. 5, No. 6, pp. 653–663, 1991.

4 E. K. Miller and T. K. Sarkar, "An Introduction to the Use of Model-Based Parameter Estimation Electromagnetics," in Review of Radio Science, 1996-99, R. Stone, editor, Oxford University Press, Oxford, UK, pp. 139–174, 1999.

5 E. K. Miller and T. K. Sarkar, "Model-Order Reduction in Electromagnetics Using Model-Based Parameter Estimation," Chapter 9 in Frontiers in Electromagnetics, D. H. Werner and R. Mittra, editors, IEEE Press, Piscataway, NJ, pp. 371–436, 1999.

6 E. K. Miller, R. S. Roberts, and S. Chakrabarti, "Using Adaptive Frequency Sampling for more Efficient Determination of Broad Band Transfer Functions," in Boundary Element Technology VII, C. A. Brebbia and M. S. Ingber, editors, Computational Mechanics Publications, Boston, MA, pp. 745–756, 1992.

7 C. Brezinski, Pade-type Approximation and General Orthogonal Polynomials, Birkhauser Verlag, Basel, Switzerland, 1980.

8 B. Gustavsen and A. Smlyen, "Rational Approximation of Frequency Domain Responses by Vector Fitting," *IEEE Transactions on Power Delivery*, Vol. 14, pp. 1052–1061, 1999.

9 T. K. Sarkar, M. Salazar-Palma, and E. L. Mokole, "Application of the Principle of Analytic Continuation to Interpolate/Extrapolate System Responses Resulting in Reduced Computations – Part A: Parametric Methods," *IEEE Journal on Multiscale and Multiphysics Computational Techniques*, Vol. 1, pp. 48–59, 2016.

10 R. S. Adve and T. K. Sarkar, "Generation of Accurate Broadband Information from Narrowband Data Using the Cauchy Method," *Microwave and Optical Technology Letters*, Vol. 6, No. 10, pp. 569–573, 1993.

11 R. S. Adve, T. K. Sarkar, S. M. Rao, E. K. Miller, and D. R. Pflug, "Application of the Cauchy Method for Extrapolating/Interpolating Narrow-Band System Responses," *IEEE Transactions on Microwave Theory and Techniques*, Vol. 45, No. 5, pp. 837–845, 1997.

12 K. Kottapalli, T. K. Sarkar, R. Adve, Y. Hua, E. K. Miller, and G. J. Burke, "Accurate computation of wideband response of electromagnetic systems utilizing narrowband information," *Computer Physics Communications*, Vol. 68, pp. 126–144, 1991.

13 S. Van Huffel, *Analysis of the Total Least Squares Problem and Its Use in Parameter Estimations*, Ph.D. Dissertation, Department Electrotechnick. Kattrolicke Universiterit Leuven, 1987.

14 G. H. Golub and C. F. Van Loan, Matrix Computations, Johns Hopkins University Press, Baltimore, MD, Second Edition, 1989.

15 G. W. Stewart and J. Sun. Matrix Perturbation Theory, Academic Press, Orlando, FL, 1990.

16 R. F. Harrington, Field Computation by Moment Methods, Robert E. Krieger Publishing Co., Piscataway, NJ, 1982.

17 E. H. Newman, "Generation of Wide-Band Data from the Method of Moments by Interpolating the Impedance Matrix," *IEEE Transactions on Antennas and Propagation*, Vol. 36, pp 1820–1824, 1993.

18 P. W. Barber and S. C. Hill, Light Scattering by Particles – Computational Methods, World Scientific, Singapore, 1990.

19 R. S. Adve and T. K. Sarkar, "The Effect of Noise in the Data on the Cauchy Method," *Microwave and Optical Technology Letters*, Vol. 7, No. 5, pp. 242–247, 1994.

20 I. Kanter, "The Ratios of Functions of Random Variables," *IEEE Transactions on Aerospace and Electronic Systems*, Vol. AES-13, pp. 624–630, 1977.

21 J. Yang and T. K. Sarkar, "Interpolation/Extrapolation of Radar Cross Section (RCS) Data in the Frequency Domain Using the Cauchy Method," *IEEE Transactions on Antennas and Propagation*, Vol. 55, No. 10, pp. 2844–2851, 2007.

22 Y. Zhang, T. K. Sarkar, X. Zhao, D. Garcia-Donoro, W. Zhao, M. Salazar-Palma, and S. Ting, Higher Order Basis Based Integral Equation Solver (HOBBIES), John Wiley & Sons, Malabar, FL, 2012.

23 Y. Zhang and T. K. Sarkar, Parallel Solution of Integral Equation-Based EM Problems in the Frequency Domain, John Wiley & Sons, Hoboken, NJ, 2009.

24 J. Yang, J. Koh, and T. K. Sarkar, "Reconstructing a Non-Minimum Phase Response from the Far-Field Power Pattern of an Electromagnetic System," *IEEE Transactions on Antennas and Propagation*, Vol. AP-53, No. 2, pp. 833–842, 2005.

25 A. V. Oppenheim and R. W. Schafer, Discrete-time Signal Processing, Prentice Hall, Englewood Cliffs, NJ, Second Edition, 1989.

26 J. Yang and T. K. Sarkar, "Accurate Interpolation of Amplitude-Only Frequency Domain Response Based on an Adaptive Cauchy Method," *IEEE Transactions on Antennas and Propagation*, Vol. 64, No. 3, pp. 1005–1013, 2016.

27 A. G. Lamperez, T. K. Sarkar, and M. Salazar-Palma, "Generation of Accurate Rational Models of Lossy Systems Using the Cauchy Method," *IEEE Microwave and Wireless Components Letters*, Vol. 14, No. 10, pp. 490–492, 2004.

28 D. Traina, G. Macchiarella, and T. K. Sarkar, "Robust Formulations of the Cauchy Method Suitable for Microwave Diplexer Modeling," *IEEE Transactions on Microwave Theory and Techniques*, Vol. 55, No. 5, pp. 974–982, 2007.

29 G. Macchiarella and D. Traina, "A Formulation of The Cauchy Method Suitable for the Synthesis of Lossless Circuit Models of Microwave Filters from Lossy Measurements," *IEEE Microwave and Wireless Components Letters*, Vol. 16, No. 5, pp. 243–245, 2006.

30 A. Lamecki, P. Kozakowski, and M. Mrozowski, "Efficient Implementation of the Cauchy Method for Automated CAD-Model Construction," *IEEE Microwave and Wireless Components Letters*, Vol. 13, No. 7, pp. 268–270, 2003.

31 W. Lee, T. K. Sarkar, H. Moon, A. G. Lamperez, and M. Salazar-Palma, "Effect of Material Parameters on the Resonant Frequencies of a Dielectric Object," *IEEE Antennas and Wireless Propagation Letters*, Vol. 12, pp. 1311–1314, 2013.

32 W. Lee, T. K. Sarkar, H. Moon, and M. Salazar-Palma, "Identification of Multiple Objects Using Their Natural Resonant Frequencies," *IEEE Antennas and Wireless Propagation Letters*, Vol. 12, pp. 54–57, 2013.

33 W. Lee, T. K. Sarkar, H. Moon, and M. Salazar-Palma, "Computation of the Natural Poles of an Object in the Frequency Domain Using the Cauchy Method," *IEEE Antennas and Wireless Propagation Letters*, Vol. 11, pp. 1137–1140, 2012.

34 T. Tantisopharak, H. Moon, P. Youryon, K. Bunyaathichart, M. Krairiksh, and T. K. Sarkar, "Nondestructive Determination of the Maturity of the Durian Fruit in the Frequency Domain Using the Change in the Natural Frequency," *IEEE Transactions on Antennas and Propagation*, Vol. 64, No. 5, pp. 1779–1787, 2016.

35 T. Tantisopharak, M. Krairiksh, and T. K. Sarkar, "Experimental Study Using Natural Frequencies for Fruit Identification," *IET International Radar Conference*, Xian, China, pp. 1–4, 2013.

36 T. Tantisopharak, M. Krairiksh, H. Moon, W. Lee, and T. K. Sarkar, "Identification of Maturity of Fruit in the Frequency Domain Using Its Natural Frequencies," *2012 IEEE Asia-Pacific Conference on Antennas and Propagation*, Singapore, pp. 1–2, 2012.

37 T. Tantisopharak, P. Keowsawat, R. Kanahna, M. Krairiksh, and T. K. Sarkar, "Investigation of the Natural Resonant Frequencies of Palmyrah Palm Juice for Quality Control," *2015 IEEE Conference on Antenna Measurements & Applications (CAMA)*, Chiang Mai, Thailand, pp. 1–5, 2015.

38 P. Leekul, M. Krairiksh, and T. K. Sarkar, "Application of the Natural Frequency Estimation Technique for Mangosteen Classification," *Proceedings of 2014 3rd Asia-Pacific Conference on Antennas and Propagation*, Harbin, China, pp. 928–930, 2014.

39 F. Sarrazin, J. Chauveau, P. Pouliguen, P. Potier, and A. Sharaiha, "Accuracy of Singularity Expansion Method in Time and Frequency Domains to Characterize Antennas in Presence of Noise," *IEEE Transactions on Antennas and Propagation*, Vol. 62, No. 3, pp. 1261–1269, 2014.

40 F. Roussafi, N. Fortino, and J. Y. Dauvignac, "Compact Modeling of UWB Antenna 3-D Near-Field Radiation Using Spherical Vector Wave Expansion and Cauchy Methods," *IEEE Antennas and Wireless Propagation Letters*, Vol. 16, pp. 173–176, 2017.

41 S. F. Peik, R. R. Mansour, and Y. L. Chow, "Multidimensional Cauchy Method and Adaptive Sampling for an Accurate Microwave Circuit Modeling," *IEEE Transactions on Microwave Theory and Techniques*, Vol. 46, No. 12, pp. 2364–2371, 1998.

42 P. A. W. Basl, R. H. Gohary, M. H. Bakr, and R. Mansour, "Modelling of Electromagnetic Responses Using a Robust Multi-Dimensional Cauchy Interpolation Technique," *IET Microwaves, Antennas & Propagation*, Vol. 4, No. 11, pp. 1955–1964, 2010.

4

Applications of the Hilbert Transform – A Nonparametric Method for Interpolation/Extrapolation of Data

Summary

The last two chapters discussed the parametric methods in the context of the principle of analytic continuation and provided its relationship to reduced rank modelling using the total least squares based singular value decomposition methodology. The problem with a parametric method is that the quality of the solution is determined by the choice of the basis functions and use of bad basis functions can generate bad solutions. A priori it is quite difficult to recognize what are good basis functions and what are bad basis functions even though methodologies exist in theory on how to choose good ones. The advantage of the nonparametric methods are that no such choices of the basis functions need to be made as the solution procedure by itself develops the nature of the solution and no a priori information is necessary. This is accomplished through the use of the Hilbert transform which exploits one of the fundamental properties of nature which is causality. The Hilbert transform illustrates that the real and imaginary parts of any nonminimum phase transfer function from a causal system satisfy this relationship. In addition, some parametrization can also be made of this procedure which can enable one to generate a nonminimum phase function from its amplitude response and from that generate the phase response and thereby one can compute the time domain data when amplitude-only data is available barring a delay in the response. This uncertainty is removed in holography as in such a procedure an amplitude and phase information is measured for a specific look angle thus eliminating the phase ambiguity. An overview of the technique along with examples are presented to illustrate this methodology.

The Hilbert transform can also be used to speed up the spectral analysis of nonuniformly spaced data samples. Therefore in this section a novel least squares methodology is presented so that it can be applied to a finite data set for spectral estimation. This can be applied for the analysis of the far field pattern from unevenly spaced antennas. The advantage of using a non-uniformly sampled data is that it is not necessary to satisfy the Nyquist sampling criterion as long as the average value of the sampling rate is less than the Nyquist rate.

Modern Characterization of Electromagnetic Systems and Its Associated Metrology, First Edition.
Tapan K. Sarkar, Magdalena Salazar-Palma, Ming Da Zhu, and Heng Chen.
© 2021 John Wiley & Sons, Inc. Published 2021 by John Wiley & Sons, Inc.

Accurate and efficient computation of the spectrum using a least squares method applied to a finite unevenly spaced data is studied. The well-known Lomb periodogram which also computes a least squares solution for the spectrum for a finite data set and has lots of benefits but cannot discern between the positive and the negative frequencies of the spectrum. Using a modified scheme, the positive and negative frequencies of the spectrum can now be discerned without losing any of the benefits of the Lomb periodogram approach. One of the properties of the periodogram approach is that the there is a Hilbert transform relationship between the coefficients of the parameters used to evaluate the spectrum. By utilizing this property, the processing time can be reduced in half providing a fast accurate least square way to compute the spectrum from unevenly spaced finite sized non-periodic samples of the data.

4.1 Introduction

A function $x(t)$ is said to be "causal" if

$$x(t) = 0 \text{ whenever } t < 0. \tag{4.1}$$

These type of functions arise in the study of causal systems and are of obvious importance in describing all natural phenomena that have well-defined *starting points* [1–3]. Let $x(t)$ be a real causal function with Fourier transform $X(\omega)$ and let $R(\omega)$ and $I(\omega)$ be the real and the imaginary parts of $X(\omega)$. Then, [1–5]

$$X(\omega) = R(\omega) + jI(\omega) = |X(\omega)| e^{j\Phi(\omega)} = |X(\omega)| \angle[X(\omega)] \tag{4.2}$$

where $j = \sqrt{(-1)}$ and $\angle$ represents the phase angle of the transfer function. If $x(t)$ is real, $R(\omega)$ is even and $I(\omega)$ is an odd function of ω. A general question of whether a specified amplitude characteristic can be realized as a causal system response is answered by the Paley–Wiener criterion [1, 2]. Consider a specific magnitude $|X(\omega)|$ of a transfer function $X(\omega)$. It can be realized by means of a causal system if and only if the integral

$$\int_{-\infty}^{\infty} \frac{ln\,|X(\omega)|}{1 + \omega^2} \, d\omega < \infty \tag{4.3}$$

is bounded. Then, a phase function associated with $X(\omega)$ {represented by $\angle[X(\omega)]$} exists such that the impulse response $x(t)$ is causal. The Paley–Wiener criterion is satisfied only if the support of $|X(\omega)|$ is unbounded. Otherwise $|X(\omega)|$ would be zero over finite intervals of frequency and this would result in infinite values for the numerator in (4.3) as $ln\,|X(\omega)| = \infty$.

Since $x(t)$ has a causal representation one can write [1–5]:

$$x(t) = \frac{2}{\pi} \int_0^{\infty} R(\omega) \cos(\omega t) \, d\omega \text{ for } t > 0; \text{ or}$$

$$x(t) = -\frac{2}{\pi} \int_0^{\infty} I(\omega) \sin(\omega t) \, d\omega \text{ for } t > 0 \tag{4.4}$$

and also

$$\int_{-\infty}^{\infty} |x(t)|^2 \, dt = \frac{1}{\pi} \int_{-\infty}^{\infty} |R(\omega)|^2 d\omega = \frac{1}{\pi} \int_{-\infty}^{\infty} |I(\omega)|^2 d\omega \tag{4.5}$$

In addition if $x(t)$ is bounded at the origin, then [6–10]

$$R(\omega) = -\frac{1}{\pi} \int_{-\infty}^{\infty} \frac{I(s)}{\omega - s} \, ds = \mathcal{H}[I(\omega)] \tag{4.6}$$

$$I(\omega) = \frac{1}{\pi} \int_{-\infty}^{\infty} \frac{R(s)}{\omega - s} \, ds = \mathcal{H}[R(\omega)] \tag{4.7}$$

where $\mathcal{H}[\bullet]$ defines the Hilbert Transform. The Hilbert transform constitutes a convolution operation with the function $1/s$, which is not defined at the origin. For a real valued function $x(t)$, in the interval $-\infty < t < \infty$, its Hilbert transform, denoted by $\hat{x}(t)$, is defined by [1–5]

$$\hat{x}(t) = \frac{1}{\pi} \mathcal{PV} \int_{-\infty}^{\infty} \frac{x(\tau)}{t - \tau} \, d\tau \tag{4.8}$$

where $\mathcal{PV}$ denotes the principal value of the integral. For notational simplicity, the symbol $\mathcal{PV}$ will be omitted from the integrals. Hence, both (4.6) and (4.7) have to be interpreted in the principal value sense. So, they both are defined in terms of the Hilbert transforms and has been defined using Cauchy principal values. Note that (4.6) and (4.7) hold for nonminimum phase systems. The only restriction is that the temporal response be causal. This restriction holds for most practical systems.

As an example, consider the reconstruction of phase from amplitude-only data which is a relatively straight forward problem for a certain restricted class of problems defined as minimum phase systems as the phase response is given by the Hilbert transform of the log of the magnitude of the amplitude data [1–13]. For a minimum phase system both the poles and the zeros of the transfer function are located in the left half of the complex frequency s plane with $s = \sigma + j\omega$, with $\sigma < 0$. For stability, as Maxwell was the first to point out that the poles need to be in the left half plane but zeros can be located either on the right or in the left half plane. Unfortunately most electromagnetic systems which exhibits delays in their response are nonminimum phase as some of their zeros can be located in the right half plane. Another characteristic of a minimum phase function is that the maximum of the energy is concentrated at $t = 0$, whereas for a nonminimum phase system it occurs later in time. For general class of problems the relationship between the amplitude and phase is a little involved as we will see later on. In summary, for minimum phase systems, the reconstruction of phase from amplitude-only data is relatively straightforward as the phase response is given by the Hilbert transform of the log of the magnitude $ln\,|X(j\omega)|$ of the amplitude data [14, 15]. The minimum phase response of a transfer function in terms of the amplitude only response as a function of frequency is given by $arg\ |X(j\omega)|$ and is given by

$$arg\ |X(\omega)| = \frac{1}{\pi} \mathcal{PV} \int_{-\infty}^{\infty} \frac{ln\ |X(\omega)|}{\lambda - \omega} d\lambda \qquad (4.9)$$

since it is a principal value integral, as the integrand has a singularity that is not integrable. Therefore the integral in (4.9) only exists in a principal-value sense. However, this property given by (4.9) of a linear-time invariant system (LTI) does not hold if the system is not minimum phase. If the system is not minimum phase (i.e., when some of the zeros of the transfer function may be on the right half-plane), then (4.9) does not hold. Hence, (4.9) has very little use for the practical problems in electromagnetics even though they are useful for minimum phase acoustic systems and are available in most signal processing text books.

However, there is a more general result (which is not very well known) for the Hilbert transform, which is based on the principle of causality and is valid for nonminimum phase systems. The principle of causality implies that the function $x(t) = 0$ for $t < 0$ and is nonzero otherwise. It is important at the onset to point out that the phase realization (be it minimum or nonminimum phase) is a problem that does not have a unique solution given amplitude-only data. A linear-phase term may be added to any phase function without altering its amplitude spectrum. This is because the addition of a linear phase to the phase of the transfer function with a uniform amplitude is equivalent to a pure delay in the time domain. Since one is dealing with linear-shift invariant systems (as the response of the system is the same independent of the time origin), changing the starting point of the impulse response of the system by a time shift does not alter the magnitude of the transfer function of the original system, except that the phase spectrum is modified by a linear-phase function. The slope of this linear-phase function is equivalent to the time delay. Also, the amplitude spectrum of the transfer function is unaltered by providing a delay to the impulse response of the system at hand.

4.2 Consequence of Causality and Its Relationship to the Hilbert Transform

All time domain responses of physical systems are causal, in that the signal is nonzero only after a certain interval of initial time. In other words the system output is observed only after a finite delay and cannot be observed before the application of the input. However, since bandlimited complex frequency domain data does not guarantee causality in the time domain or a real time domain response, computations or measurements carried out in the frequency domain do not truly represent the transient response of the system. Even so, one can establish that it is possible to extract a causal response by interpolating the complex frequency domain data under the premise that the time domain signal must be causal. The principle of causality is used to extrapolate/interpolate the frequency domain response [13, 14].

In general, the real and imaginary parts of the complex frequency domain data are independent of each other. However, the causality of the time domain signal, denoted as $h(t)$, assures one that the real and imaginary components of the frequency domain response are related through the Hilbert transform. The physical principle of causality imposes some constraints on the real and imaginary parts of the transfer function. The relationship was originally developed by Kramers and Kronig [14, 15]. This is equivalent to equations (4.6) and (4.7) of this presentation.

James and Andrasic [16] used this approach to minimize the effects of noise on experimental data. Arabi et al. [17] used the Hilbert transform technique to generate causal time domain responses of multiconductor transmission lines by enforcing the Kramers–Kronig [14, 15] relationship between the dielectric constant and the loss tangent of any dielectric material. Bruck and Sodin applied them for image reconstruction [18]. Tesche and Pyati used this technique [19, 20] to generate causal time domain response from band-limited frequency domain data. If we denote $H_R(\omega)$ as the real part and $H_I(\omega)$ as the imaginary part of the transfer function $H(\omega)$ obtained from $h(t)$, then, invoking the principle of causality, indicates that they have to be related by the Hilbert transform [21]. The property that the real and imaginary parts of the frequency domain data correspond to the even and odd parts of $h(t)$ is exploited in extracting a causal response from complex band limited frequency domain data.

The various properties of the Hilbert transform are summarized next for completeness as it is not well known in the electromagnetic literature.

4.3 Properties of the Hilbert Transform

By a change of variable, (4.8) transforms to

$$\hat{x}(t) = -\frac{1}{\pi} \int_{-\infty}^{\infty} \frac{x(t+\tau)}{\tau} \, d\tau = \frac{1}{\pi} \int_{-\infty}^{\infty} \frac{x(t-\tau)}{\tau} \, d\tau \tag{4.10}$$

The various pertinent properties are:

(A) The Relation between the Fourier transform of the Hilbert transform In this case,

$$\hat{X}(\omega) = \mathcal{F}[\hat{x}(t)] = -j X(\omega) \, sgn(\omega) \tag{4.11}$$

where $\mathcal{F}$ denotes the Fourier transform, $X(\omega)$ is the Fourier transform of $x(t)$, and $sgn(\omega)$ is the signum function defined by

$$sgn\,(\omega) = \begin{bmatrix} 1 & \text{for } f > 0 \\ 0 & \text{for } f = 0 \\ -1 & \text{for } f < 0 \end{bmatrix} \tag{4.12}$$

and $\omega = 2\pi f$. To establish this result, it is seen that

$$\hat{X}(\omega) = \mathcal{F}[\hat{x}(t)] = \int_{-\infty}^{\infty} \hat{x}(t)\, e^{-j\omega t}\, dt = \frac{1}{\pi}\int_{-\infty}^{\infty} e^{-j\omega t}\, dt \int_{-\infty}^{\infty} \frac{x(t-\tau)}{\tau}$$

$$d\tau = \frac{1}{\pi}\int_{-\infty}^{\infty} e^{-j\omega y}\, dy\, x(y) \int_{-\infty}^{\infty} \frac{e^{-j\omega \tau}}{\tau}\, d\tau$$

with $y = t - \tau$, results in

$$= \frac{X(\omega)}{\pi}\left[\int_{0}^{\infty} \frac{e^{-j\omega \tau}}{\tau}\, d\tau + \int_{-\infty}^{0} \frac{e^{j\omega p}}{p}\, dp\right]$$

$$= \frac{2j\,X(\omega)}{\pi}\int_{0}^{\infty} \frac{\sin\,(\omega\,\tau)}{\omega\,\tau}\,(\omega\, d\tau) = -j\, sgn\,(\omega)\, X(\omega) \tag{4.13}$$

(B) The inner product defined by $<x;\hat{x}>$ is zero, i.e. the function and its Hilbert transform are orthogonal. To prove this

$$<x;\hat{x}> \; = \int_{-\infty}^{\infty} x(t)\,\hat{x}(t)\, dt = \int_{-\infty}^{\infty} X^*(\omega)\,\hat{X}(\omega)d\omega$$

$$= j\int_{-\infty}^{\infty} |X(\omega)|^2\, sgn\,(-\omega)\, d\omega = 0 \tag{4.14}$$

where $*$ denotes a complex conjugate.

(C) The energies in $x(t)$ and $\hat{x}(t)$ are the same, i.e.,

$$<x;x> \; = \int_{-\infty}^{\infty} x^2(t)\, dt = \int_{-\infty}^{\infty} X^*(\omega)X(\omega)\, d\omega = \int_{-\infty}^{\infty} |X(\omega)|^2\, d\omega$$

$$<\hat{x};\hat{x}> \; = \int_{-\infty}^{\infty} \hat{x}^2(t)\, dt = \int_{-\infty}^{\infty} \hat{X}^*(\omega)\hat{X}(\omega)\, d\omega = \int_{-\infty}^{\infty} |\hat{X}(\omega)|^2\, d\omega = \int_{-\infty}^{\infty} |X(\omega)|^2\, d\omega \tag{4.15}$$

Hence the Fourier transforms of a signal and its Hilbert transform have identical amplitude spectra. However, the phase spectra differ from one another only by a constant

$$\measuredangle\hat{X}(\omega) = \measuredangle X(\omega) - \frac{\pi}{2}\, sgn\,(\omega) \tag{4.16}$$

where $\measuredangle$ denotes the phase angle of the complex function. This relationship can be used to compute the Hilbert Transform from the Fourier transform.

(D) Convolution product:
If $z(t) = x(t) \otimes y(t)$ then $\hat{z}(t) = \hat{x}(t) \otimes y(t) = x(t) \otimes \hat{y}(t)$. This arises from

$$\hat{Z}(\omega) = -j\,Z(\omega)\,\operatorname{sgn}(\omega) = -j\,\operatorname{sgn}(\omega)\,X(\omega)\,Y(\omega) = \hat{X}(\omega)\,Y(\omega) = X(\omega)\,\hat{Y}(\omega)$$
(4.17)

(E) An extension of the above property is

$$x(t) \otimes y(t) = -\hat{x}(t) \otimes \hat{y}(t)$$
(4.18)

(F) The double Hilbert transform of a function is the negative of the function itself, i.e.,

$$\hat{\hat{x}}(t) = -x(t)$$
(4.19)

Or equivalently, the inverse of the Hilbert transform is the negative of the function itself! Since

$$x(t) \otimes y(t) = -\hat{x}(t) \otimes \hat{y}(t)$$
(4.20)

then if one takes the Hilbert transform of both sides and utilizing property (4.20) one obtains $\hat{x}(t) \otimes y(t) = \hat{x}(t) \otimes \hat{y}(t) = -\hat{\hat{x}}(t) \otimes \hat{y}(t)$ and this is equivalent to $\hat{\hat{x}}(t) = -x(t)$. This self-mapping property of a Hilbert transform can be used very effectively to extrapolate computational data, or fill up missing data or can be used to reduce noise in the measurements without having any structural knowledge of the waveform that one is dealing with! This is what is implied by extrapolation of the transfer function of a LTI system without any parametric model. Here a nonparametric modeling scheme based on the physical principles are utilized to extrapolate limited data without any a priori knowledge about the structure of the waveform as illustrated in [8, 9, 11].

(G) Properties of an analytic signal
For an analytic signal which is a complex function of time, the Fourier transform, exists only for positive frequency components and not for negative frequencies. So an analytic signal $\underline{x}(t)$ is given by

$$\underline{x}(t) = x(t) + j\,\hat{x}(t)$$
(4.21)

with $X(\omega) = X(\omega) + X(\omega)\,\operatorname{sgn}(\omega) = 2\,\Theta\,X(\omega)$ where

$$\Theta(\omega) = \begin{cases} 0.5 & \text{for } \omega = 0 \\ 1.0 & \text{for } \omega > 0 \end{cases}$$
(4.22)

(H) Expansion of the real and imaginary parts of an analytic function
As Hilderbrand has shown [3] that if a function $x(\theta)$ is defined by the Fourier cosine series, then

$$x(\theta) = a_0 + \sum_{n=1}^{\infty} a_n \cos(n\theta) \quad (\text{for } 0 < \theta < \pi)$$
(4.23)

and it can be shown that its Hilbert transform $\hat{x}(\theta)$ is of the form

$$\hat{x}(\theta) = -\sum_{n=1}^{\infty} a_n \sin(n\theta) \quad (\text{for } 0 < \theta < \pi) \tag{4.24}$$

as the Hilbert transform of a constant is zero. The principles of modulation and demodulation techniques are based on this principle and so on ALL instruments implemented in hardware. This arises from the fact that the inphase and the quadrature phase channels are related by the Hilbert transform. It can be illustrated as follows.

When a low frequency signal f_m is modulated by a carrier frequency signal f_0 then one essentially generates two sidebands located at either side of the carrier frequency. The upper sideband is located above the value of the carrier frequency at $f_0 + f_m$ and the lower sideband is located at $f_0 - f_m$ below the carrier frequency. The modulated signal is translated up in frequency so as to reduce dispersion as the effective bandwidth of the total waveform with respect to the carrier frequency now becomes very small as opposed to the percentage bandwidth of the signal of interest f_m. So once the signal traverses to its desired destination with little distortion, it is now necessary to demodulate the composite signals and remove the carrier frequency and get back the baseband signal. However, the problem is that one cannot simply demodulate the high frequency signal by beating it with a local oscillator as the two sidebands will simultaneously get translated into the baseband and will interfere with each other. Hence the first step in the demodulation process is to generate an analytic signal through I (in-phase) and the Q (quadrature phase) components of the modulated signal so that the signal is defined only for positive frequencies as the negative frequencies will be eliminated through (4.21). Now the analytic signal beats with the local oscillator generating a single sideband generating a translated version of the baseband signal of interest. Besides this important point there is another economic issue related to every equipment having two separate independent channels for the in phase and the quadrature phase channels and thereby almost doubling the cost of any instrument. A more innovative way will be to generate the I channel only, digitize it and use a Hilbert transform of the I channel to generate the Q-channel. The latter can now be done in software rather than in hardware reducing the cost. This will cut the cost of any instrument almost by a factor of half as two independent channels are not required in this process! Some modern instrumentations including some ground probing radars have embraced such a methodology. This will be a very interesting application of a nonparametric processing algorithm in many equipment manufacturing besides having some advantages in computational methodology.

(I) *Inverse of a transform*

If one considers the Hilbert transform over a finite region defined by the singular integral equation [2–5], i.e.,

$$G(q) = \frac{1}{\pi} \, \mathcal{PV} \int_{-1}^{1} \frac{w\,(y)}{q-y} \, dy \; \text{for} \; -1 < q < 1 \tag{4.25}$$

then if $w(-1)$ is finite, and the solution for w can be given by

$$w(y) = \frac{1}{\pi} \sqrt{\frac{1+y}{1-y}} \, \mathcal{PV} \int_{-1}^{1} \sqrt{\frac{1+q}{1-q}} \, G(q) \, \frac{dq}{y-q} \tag{4.26}$$

Many of these properties have been utilized in phase reconstruction [22], material parameter characterization [23–25] and on causality for the S-parameters [26]. Causality for transmission line responses have been carried out in [24–28]. Since one extensively computes the Hilbert transform using Fourier transforms some important related computational properties is discussed next. Often one deals with digitized signals resulting in discrete frequency domain data, and hence it is necessary to process frequency and time domain signals in the form of sequences.

4.4 Relationship Between the Hilbert and the Fourier Transforms for the Analog and the Discrete Cases

For an analog causal signal given by (4.1), its Fourier and the Hilbert transforms are related by (4.6) and (4.7). For the discrete case, some of the properties of sequences and their Fourier transforms are discussed. The relevant point here is that for the discrete case, one needs to develop an alternate form of the Hilbert transform where the singular kernel periodically repeats itself and not just exists as a single monotonically decaying function. Since one is going to perform computations, it is necessary to consider the digital sequence $h[n]$. The details of the computations are available in Chapter 10 of Reference [6].

Any complex sequence $h[n]$ can be expressed as the sum of a symmetric sequence $h_e[n]$ and an antisymmetric sequence $h_o[n]$. In the case of real sequences, these are called even and odd sequences [6] represented by a subscript e and o. Therefore,

$$h[n] = h_e[n] + h_o[n]$$
$$h_e[n] = h_e[-n] \tag{4.27}$$
$$h_o[n] = -h_o[-n]$$

where the Fourier transform of any complex sequence $h[n]$ is represented by $H(e^{j\omega})$, where

$$H\left(e^{j\omega}\right) = \sum_{n=-\infty}^{\infty} h\left[n\right] e^{-j\omega n} \tag{4.28}$$

Therefore,

$$H\left(e^{-j\omega}\right) = \sum_{n=-\infty}^{\infty} h\left[n\right] e^{j\omega n}. \tag{4.29}$$

This implies that

$$H_R\left(e^{j\omega}\right) = \mathcal{F}\{h_e[n]\} \text{ and } j H_I\left(e^{j\omega}\right) = \mathcal{F}\{h_o[n]\} \tag{4.30}$$

Also, for real $h[n]$, $H_R\left(e^{j\omega}\right) = H_R\left(e^{-j\omega}\right)$, which is an even function and $H_I\left(e^{j\omega}\right) = -H_I\left(e^{-j\omega}\right)$, which is an odd function.

The procedure outlined in section 10.2 of [6] forms the basis of the computation for the Hilbert transform technique which is used for the extraction of a real causal time domain response from bandlimited complex valued frequency domain data. The theoretical development assures one that by computing the discrete Fourier transform (DFT)s and inverse discrete Fourier transform (IDFT)s the original real time sequence will not lose its causal nature.

4.5 Methodology to Extrapolate/Interpolate Data in the Frequency Domain Using a Nonparametric Methodology

A technique to extrapolate/interpolate data in the frequency domain in a nonparametric fashion utilizing the Hilbert transform is described next. Before the algorithm is described, it is useful to know something about the properties of a frequency domain data samples. Assume that one has a complex frequency domain data between frequencies f_1 and f_4. Consider a missing band of data between the frequencies f_2 and f_3. The frequency domain data is sampled at $(n_2 - n_1)$ frequency points between f_2 and f_1, and at $(n_4 - n_3)$ points between f_4 and f_3. The sampled data at these frequencies is expressed as a vector

$$H[n_1 : n_4] = [H_{n_1} \cdots H_{n_2}, 0 \cdots 0, H_{n_3} \cdots H_{n_4}] \tag{4.31}$$

The objective now is to interpolate the missing data between n_2 and n_3.

As a first step:

1) The available bandlimited frequency domain data is padded with zeros to ensure a length of n points where n is given by $N/2 + 1$, and N ls [2 4 8 ... 1024 2048 · · ·], providing a sequence of data as

$$[H_{n_3} \cdots H_{n_4}, 0, 0, \cdots 0,] \tag{4.32}$$

2) This complex sequence is altered to obtain a modified sequence of length N. This is done by appending the complex conjugate of the sequence to the original data as

$$H[1:N] = \left[H[1:N/2 + 1], \overline{H}[N/2:2] \right] \qquad (4.33)$$

where the overbar denotes the complex conjugate of a complex valued sequence.

3) The complex sequence is now split into its real and imaginary parts

$$H_R = \text{Real}[H] \; ; \; H_I = \text{Imag}[H] \qquad (4.34)$$

4) An Inverse Discrete Fourier transform of H_R results in an even sequence $h_e[n]$ so that

$$h_e[1:N] = Real \left[IFFT(H_R) \right]; \text{and } h_e[n] = h_e[-n] \qquad (4.35)$$

This in fact is the even part of the time domain sequence. The numerical implementation of the Hilbert transform may be found elsewhere [6].

5) Before proceeding further, it is important to know that there are sharp discontinuities in the frequency domain data as portions of the data is missing. In order to deal with this situation, it is necessary to multiply the time domain sequence with a window function. A Hanning window [6] of length N is multiplied with the time domain sequence. The resulting frequency domain sequence is now filtered or "smoothed" [11]. The Hanning window is given by

$$W(n) = \left[\begin{array}{ll} 0.5 - \dfrac{0.5 \cos (2\pi n)}{N} & 0 \leq n < N \\ 0 & otherwise \end{array} \right] \qquad (4.36)$$

Hence

$$h_e[1:N] = h_e[1:N] \otimes W[1:N] \qquad (4.37)$$

where the $\otimes$ denotes the convolution.

6) The odd sequence is obtained from the even sequence by making use of the available relationships [5, 7, 8]. This results in

$$h_o[1:N] = [0, h_e(2:N/2), 0, -h_e(N/2 + 2:N)] \qquad (4.38)$$
$$h_e[n] = h_e[-n] \qquad (4.39)$$
$$h_o[n] = -h_o[-n] \qquad (4.40)$$

7) The discrete Fourier transform of this odd sequence will provide the imaginary part of the spectrum as stated earlier

$$H_I^{new} = Imag \left[FFT(h_o) \right] \qquad (4.41)$$

8) A substitution for the missing points is made in the imaginary part of the original sequence using the sequence obtained in Step 7) as

$$H_I^{sub} = \left[H_I^{new}(1:n_1-1), H_I(n_1:n_2), H_I^{new}(n_2+1:n_3-1), H_I(n_3:n_4),\right.$$
$$\left.H_I^{new}(n_4+1:N/2+1)\right]$$

(4.42)

9) This sequence is copied to obtain a sequence of length N which is an improved version of the original sequence

$$H_I^{sub} = \left[H_I^{sub}(1:N/2+1), -H_I^{sub}(N/2:2)\right]$$

(4.43)

10) The inverse discrete Fourier transform of this sequence will yield the odd sequence as

$$h_o^{new} = IFFT\left[j\, H_I^{subs}\right] \text{ since } h[n] = h_e[n] + h_o[n]$$

(4.44)

11) A modified version of h_e is generated from

$$h_e^{new} = \left[h_e(1), h_o^{new}(2:N/2), h_e(N_2+1), -h_o^{new}(N/2:N)\right]$$

(4.45)

12) The discrete Fourier transform of this sequence obtained in the previous step will yield the real part of the spectrum as stated earlier

$$H_R^{new} = \text{Real}\left[FFT\left(h_e^{new}2\right)\right]$$

(4.46)

13) A substitution for the missing points is made in the Real part of the original sequence using the sequence obtained in Step 12), as

$$H_R^{sub} = \left[H_R^{new}(1:n_1-1), H_R(n_1:n_2), H_R^{new}(n_2+1:n_3-1), H_I(n_3:n_4),\right.$$
$$\left.H_R^{new}(n_4+1:N/2+1)\right]$$

(4.47)

14) This sequence is copied to obtain another sequence of length N

$$H_R^{subs} = \left[H_R^{sub}(1:N/2+1), H_R^{sub}(N/2:2)\right]$$

(4.48)

which is an improved version of the original sequence.

15) The resulting sequence is subjected to an inverse discrete Fourier transform to obtain the even sequence

$$h_e[1:N] = \text{Real}\left[IFFT\left(H_R^{subs}\right)\right]$$

(4.49)

16) As in Step 5), this time domain sequence is multiplied with the Hanning window to make the sequence rather smooth.

17) Subsequent processing involves iterations of Steps (6-16).

The above set of procedures will interpolate the missing band of frequencies. The reconstructed sequence will now be the complex sequence given by

$$H^{rec}[1:n_4] = H_R^{subs}[1:n_4] + j\, H_I^{subs}[1:n_4]$$

(4.50)

And by comparing it with (4.45) illustrates that the missing data is iteratively being reconstructed as

$$H^{rec}[1:n_4] = \left[H_{n_1}, \cdots H_{n_2}, H^{rec}_{n_2+1}, \cdots H^{rec}_{n_3-1}, H_{n_3} \cdots H_{n_4} \right] \qquad (4.51)$$

It is worthwhile to note that by making use of the Hanning window, although one has overcome the difficulties due to discontinuities at the ends of the missing band, one might suffer a loss of resolution. This is not a serious problem and its effects can be minimized as shown in the numerical examples.

4.6 Interpolating Missing Data

Consider the frequency domain data of a microstrip filter measured using the HP 8510B Network Analyzer. The device is a bandpass filter and its characteristics are measured (it can also be computed) at 415 points from 4.31 to 7.415 GHz. The objective here is to compare the performance of the two methods (the Cauchy method described in Chapter 3 which is a parametric method and the iterative nonparametric technique based on the Hilbert transform as the number of missing points are gradually increased. These missing points are created by deleting portions of the measured data. However, the data samples could as well have been taken from some numerical computations in an electromagnetic simulation or from other systems.

Figure 4.1(a) shows the real and imaginary parts of the original data. We now discard 40 points (which is about 10% of the data) from 200 – 240. Figure 4.1(b) shows this deleted band of data. These modified data points are now given as input to the parametric interpolation/extrapolation based on the Cauchy method. The entire data set is not required for this. Only a few points before and after the missing band is sufficient. The program returns the interpolated data. Next, these same missing data is given as input to the iterative nonparametric methodology described in the earlier section based on the Hilbert transform presented in Section 4.5. The missing points are first zero padded. Figure 4.1(c) compares the output of both methods (the parametric Cauchy method and the nonparametric iterative method based on the Hilbert transform) with the original real part, while Figure 4.1(d) compares the corresponding imaginary part. Clearly the reconstruction is quite accurate using either technique.

Next, the number of deleted points were increased to 60; i.e., points 200–260 were discarded. The same procedure was repeated. Figure 4.2(a) displays the truncated data. Figure 4.2(b) shows the reconstructed real part and Figure 4.2(c) shows the reconstructed imaginary part of the response utilizing both the techniques. It is clear from these figures that the reconstruction obtained using the iterative method based on the Hilbert transform has a

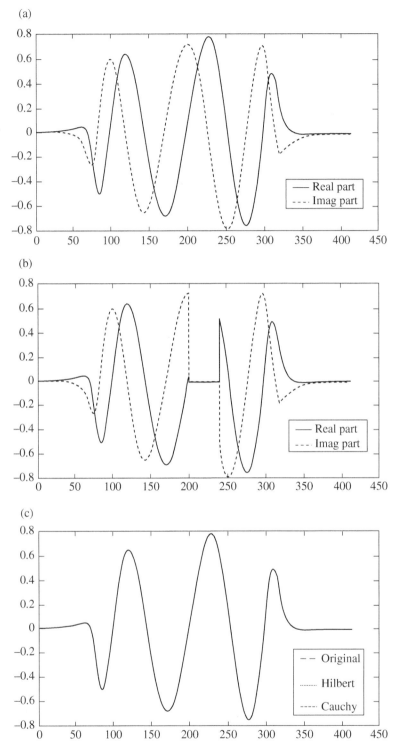

Figure 4.1 (a) Original data – 415 points, (b) Truncated Data -40 points missing
(c) Comparison of the reconstructed Data using the two methods – Cauchy and the Iterative
Hilbert transform -real part (d) Comparison of the reconstructed Data using the two methods
– Cauchy and the Iterative Hilbert transform -Imaginary part.

(d)

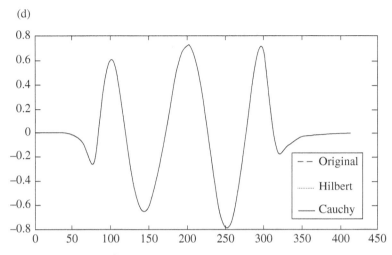

Figure 4.1 (Continued)

slightly better performance than that obtained using the direct method. Note that the amplitude of the reconstructed part using the Cauchy method came out slightly higher than the actual amplitude.

When the number of deleted points was increased to 80 [about 20% of the data has been deleted from the middle of the band]; i.e., from 200 – 280, the iterative method again proved to be better than the direct method. But this time, a slight modification was made in the initial guess for the missing points in the iterative method. A straight line extrapolation between the ends of the missing band was made in the initial guess instead of zero padding it. Figure 4.3(a) shows the original data with the initial guess. Figures 4.3(b) and 4.3(c) show the reconstructed real and imaginary parts using both the methods. Clearly the iterative method gave better results. But it should be noted that a better interpolation can be obtained by the Cauchy method if the cutoff for the singular values (explained earlier in the theory in [1]) is chosen appropriately. Figures 4.3(d) and 4.3(e) show the improved result. The cutoff for the singular values was changed to 10^{-21} from the previous value of 10^{-16}. However determining the cutoff for the singular values in practical situations may not be possible!

As the second example consider the frequency domain measured data for a microstrip filter performed using the HP 8510B Network Analyzer. The device is a band-pass filter and its characteristics are measured at 415 points from 4.2069-8.5 GHz as shown in Figure 4.4(a). Since in this example, the final result of interest is extrapolation/interpolation of the data in the frequency domain. translating the frequency axis from 4.2069 to 0 GHz does not really affect the results. In this example, the data points from 161-219 is discarded, which

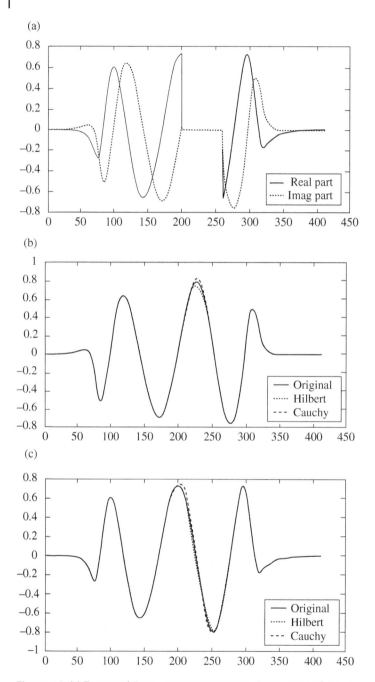

Figure 4.2 (a) Truncated Data – 60 points missing. Comparison of the reconstructed Data using the two methods – Cauchy and the Iterative Hilbert transform -(b) real part and (c) -Imaginary part.

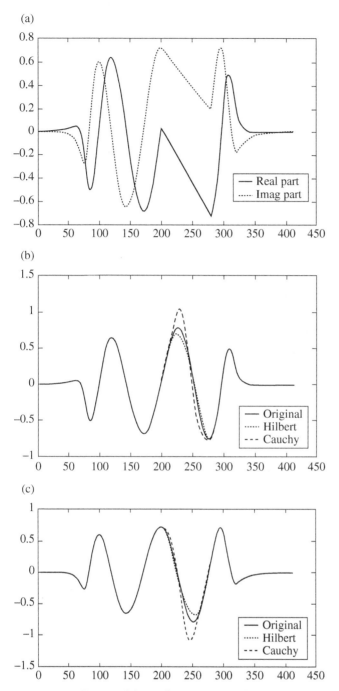

Figure 4.3 (a) Truncated data with a nonzero initial guess (80 missing points) (b) Comparison of the reconstructed Data using the two methods – Cauchy and the Iterative Hilbert transform -real part and a singular value cutoff of 10^{-16} for the Cauchy method (c) Comparison of the reconstructed Data using the two methods – Cauchy and the Iterative Hilbert transform -imaginary part and a singular value cutoff of 10^{-16} for the Cauchy method (d) Comparison of the reconstructed Data using the two methods – Cauchy and the Iterative Hilbert transform -real part and a singular value cutoff of 10^{-21} for the Cauchy method (e) Comparison of the reconstructed Data using the two methods – Cauchy and the Iterative Hilbert transform -imaginary part and a singular value cutoff of 10^{-21} for the Cauchy method.

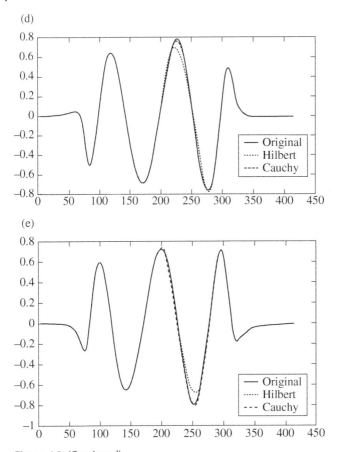

Figure 4.3 (Continued)

correspond to the frequency points between 5.4875-6.2375 GHz, as shown in Figure 4.4(b). The missing data points are replaced by zeros, and the data is padded by zeros from 416-1025 sample points. The objective is to interpolate the missing data values by utilizing the principles outlined in the previous section. Figure 4.4(c) describes the interpolated data points utilizing the iterative principles to interpolate samples 161-219. In Figures 4.4(c) and 4.4(d) the reconstructed data is compared to that of the original data, both in the real and in the imaginary parts, respectively. Figure 4.4(e) plots the log-magnitude response of the bandpass filter with both the real data and the reconstructed data superimposed. So for this example, the objective has been to interpolate part of the

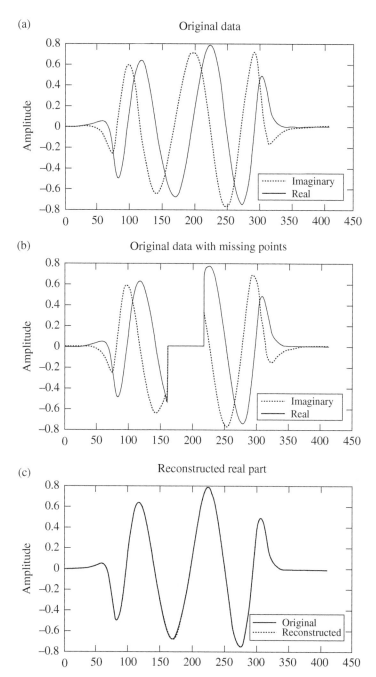

Figure 4.4 These arc plots of the frequency domain data of a microstrip band-pass filter. [Interpolation Results]. (a) Plot of the original data, (b) Plot of the real and imaginary parts of the original data showing the missing band, (c) Plot of reconstructed real part of the original real part of the data (d) Plot of reconstructed imaginary part of the original imaginary part of the data. (e) Plot of the log-magnitude of both the original and the reconstructed data.

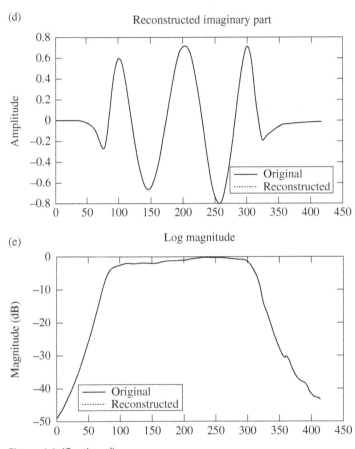

Figure 4.4 (Continued)

pass band response from the stop band data. The extrapolation of the missing data for the bandpass filter is shown in Figure 4.5(a)-(c).

For the fourth example consider the measured data of a microstrip filter between 4.2069-8.0013 GHz using 468 points. The data is shown in Figure 4.6(a). We now remove a large number of data points in the pass band from 201-270. The Hilbert transform technique was used to fill in the missing data points producing interpolated responses for the real and imaginary parts of the data as shown in Figures 4.6(b) and 4.6(c). The interpolated data agrees well with the original data shown in Figure 4.6(a).

For the fifth example consider the interpolation of the input impedance of a dipole antenna. The antenna is considered to be 2 m (= L) long and of radius 0.1 mm (= R). The input impedance of the center fed dipole was computed at every 1 MHz interval till 800 MHz and 801 data samples are considered.

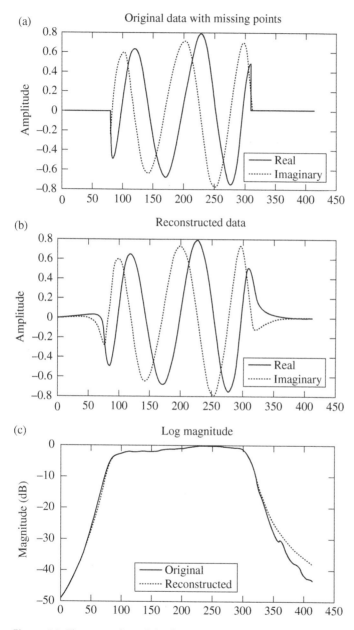

Figure 4.5 These arc plots of the frequency domain data of a microstrip band-pass tilter. [Extrapolation results]. (a) Plot of the real and imaginary parts of the original data showing the missing band. (b) Plot of real and imaginary parts of the reconstructed data. (c) Plot of log-magnitude of both the original and the reconstructed data.

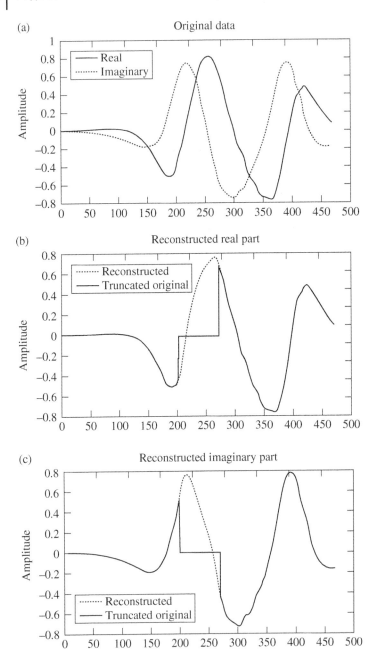

Figure 4.6 These are plots of the frequency domain data of another microstrip filter. [Interpolation of a considerably large number of missing points]. (a) Plot of the real and imaginary parts of the original data. (b) Plot of reconstructed real part and the original real part showing the missing points. (c) Plot of reconstructed imaginary part of the original imaginary part showing the missing points.

The data measured was generated utilizing the commercially available code AWAS 1121. The original data is shown in Figure 4.7(a). Next we delete the data from 401-470 MHz which is equivalent to removing a peak in the real part and a fraction of the peak of the input reactance of the imaginary part. Next the Hilbert transform relationship is utilized to interpolate the input impedance of the dipole antenna in the missing hand. The interpolated data are shown in Figures 4.7(b) and 4.7(c). Even though the peak is positioned correctly, the amplitudes are underestimated. It has been observed that for thick dipole antennas (where the L/R ratio is small) the peak is reproduced more accurately than for the thin dipole antennas. The interpolated results more accurately match the actual data, since for an antenna with small L/R, the peaks in the impedances are wider and the sampled values of the computed FFT becomes much more well behaved.

As the final example, consider the measured data of a microstrip notch filter between the frequencies 2.0-6.0 GHz. Figure 4.8(a) shows the original data with the real and imaginary parts. In this case, most of the first peak is removed, i.e., data points from 35-41. Figure 4.8(b) and Figure 4.8(c) shows the reconstructed real and imaginary parts respectively, while Figure 4.8(d) shows the plot of the log-magnitude response. The reconstructed data generated from this methodology closely matches with the original data.

Applications in antenna measurements [29–33] have been implemented using this technique, causal responses have been computed in [34], causal parameter characterization [35–38] and in electromagnetic time reversal [39]. The final application of generating the temporal response from amplitude only data is discussed next. Other related applications can be seen in [40–43].

Finally, the application of the Hilbert transform for computing the spectrum of a waveform in a fast and efficient way from nonuniformly spaced data is illustrated.

4.7 Application of the Hilbert Transform for Efficient Computation of the Spectrum for Nonuniformly Spaced Data

The concept of spectral analysis using a least squares methodology for a finite non-periodic data set was first proposed by Vanicek in 1970 [44]. Lomb [45] developed this method and showed that there is a correlation between the heights of the spectrum at any two frequencies which is equal to the mean height of the spectrum due to a sinusoidal signal of frequency f_1, at the frequency f_2. These correlations reduce the distortion in the spectrum of a signal affected by noise which is an additional benefit of using unevenly spaced data [45]. Further studies have been done by Scargle [46] and he provided a

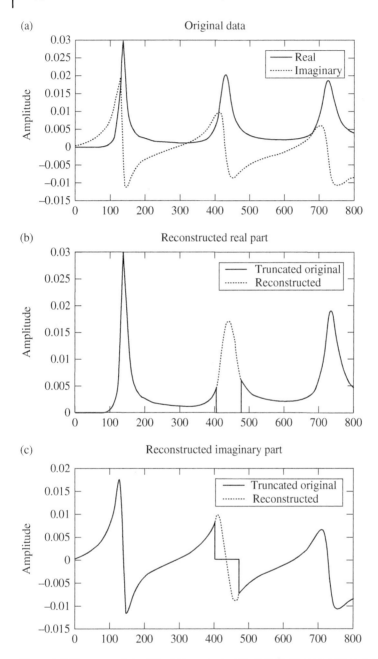

Figure 4.7 These arc plots of the frequency domain data for the input impedance of a dipole antenna. (a) Plot of the real and imaginary parts of the original data. (b) Plot of the reconstructed real part and the original real part of the data with the missing points, (c) Plot of the reconstructed imaginary part and the imaginary part of the original data with the missing points.

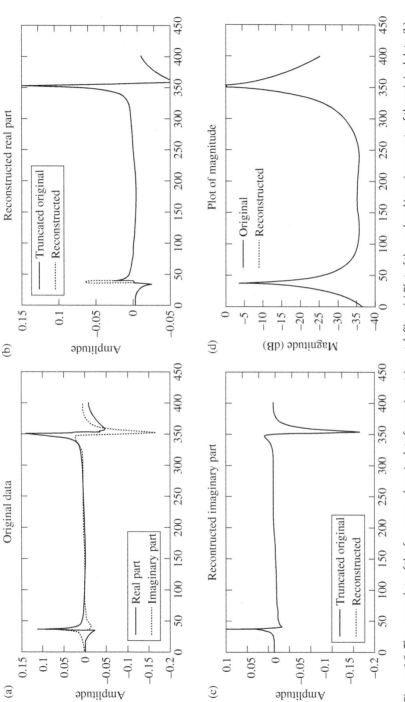

Figure 4.8 These are plots of the frequency domain data for a microstrip notch filter. (a) Plot of the real and imaginary parts of the original data. (b) Plot of the reconstructed real part and the original real part showing the missing points, (c) Plot of reconstructed imaginary part and the original imaginary part showing the missing points, (d) Plot of log-magnitude of both the original and the reconstructed data.

simple estimate of the significance of the height of a peak in the power spectrum through the false alarm probability. Spectral analysis has also been for uniformly spaced data [46]. Feraz-Mello [47] also tried to solve the same problem by applying nonorthogonality of the basis functions when the sampling is uneven using the Gram-Schmidt orthogonalization procedure which was basically the same as a periodogram method.

The periodogram approach to compute the spectrum from an unevenly sampled data provides a scan of a given frequency range obtained by fitting sines and cosines in a least squares fashion to the data and plotting the correlation of the data at each frequency. The frequency increment can be determined with any precision and that is a benefit of using this method. More benefits derived from the uneven or random sampling of the data results in absence of aliasing if the sampling were to be a completely random, Poisson process [48–50]. It was proved that the spectrum would be completely alias free in such a situation [51]. This can be simply shown from an extension of the Fourier transform of a uniformly spaced discrete sequence. Consider a uniformly sampled signal $x(nT)$ and Fourier transformation of the signal. The Fourier transform is periodic since

$$
X\left(\omega + \frac{2\pi}{T}\right) = \sum_{n=-\infty}^{\infty} x(nT)e^{-j\left(\omega + \frac{2\pi}{T}\right)nT} = \sum_{n=-\infty}^{\infty} x(nT)e^{-j\omega nT - j2\pi n}
$$

$$
= \sum_{n=-\infty}^{\infty} x(nT)e^{-j\omega nT} = X(\omega)
$$

(4.52)

For a nonuniformly sampled case, the Fourier transform will be

$$
X(\omega) = \int_{-\infty}^{\infty} x(t_n)e^{-j\omega t_n}dt = \sum_{n=-\infty}^{\infty} x(nT)e^{-j\omega nT}
$$

(4.53)

$X(\omega)$ is periodic only when the exponential part of $X(\omega)$ is periodic as shown for the uniformly sampled case. That is, $e^{j\omega t_{ni}} = e^{j(\omega t_{nj} + 2\pi)}$ should hold for all t_n. If the time steps are not deterministic quantities then the probability of t_{ni} and t_{nj} having the same time increment, $\frac{2\pi}{\omega}$, as in a uniformly sampled case for all time duration will be zero. Therefore $X(\omega)$ is not periodic in ω and $x(t)$ does not need to be a band-limited sequence and that leads to an alias free condition [52–54].

The least squares spectrum provides the best measure of the power contributed by the different frequencies to the overall variance of the data. This can be regarded as the natural extension of the Fourier methods to nonuniformly spaced data. And Lomb periodogram analysis is exactly equivalent to the least squares fitting of a curve to the data.

Even though the periodogram analysis has all these benefits, it has some drawbacks. For example it cannot evaluate the spectrum for negative frequencies since it is the power spectrum of a real sequence. Moreover the estimated peak does not correspond to the true magnitude of the signal. The error in the peak is mainly from the nonuniform spacing which is not the same source of error as in the fast Fourier transform (FFT) where the main contribution to the error comes from finiteness of the sequence or noise. A formulation that can resolve positive and negative frequencies without losing any of the benefits of the periodogram approach has been illustrated in this section. Additional property of the least square approach has been investigated in the sub sections 4.7.1 and 4.7.2 and utilized to reduce the computation time. Section 4.7.3 deals with the estimation of the magnitude of the spectrum using a least square method.

4.7.1 Formulation of the Least Square Method

Let a continuous complex signal $x(t)$ be sampled at the time instants, $t = t_k$, $k = 0$, 1, 2,..., $N-1$. The goal is to look for some harmonic components, i.e.,

$$h(t) = (a + jb) \cos [\omega(t - \tau)] + \left(c + j \lim_{x \to \infty} d \right) \sin [\omega(t - \tau)] \tag{4.54}$$

of frequency ω, where a, b, c, d and τ are real constants and $j = \sqrt{-1}$. The delay parameter τ enables us to elect any arbitrary origin of time. To estimate a and b the mean square difference is minimized,

$$F = \sum_{k=0}^{N-1} \left| x(t_k) - (a + jb) \cos [\omega(t_k - \tau)] - (c + jd) \sin [\omega(t_k - \tau)] \right|^2 \tag{4.55}$$

with respect to the unknowns. Taking derivative of F with respect to the unknowns a, b, c and d will produce the normal equations which are:

$$\frac{dF}{da} = 0$$

or

$$\sum_{k=0}^{N-1} \overline{x}(t_k) \cos [\omega(t_k - \tau)] + \sum_{k=0}^{N-1} x(t_k) \cos [\omega(t_k - \tau)]$$

$$= 2a \sum_{k=0}^{N-1} \cos^2 [\omega(t_k - \tau)] + 2c \sum_{k=0}^{N-1} \cos [\omega(t_k - \tau)] \sin [\omega(t_k - \tau)] \tag{4.56}$$

where $\overline{x}(t_k)$ is the complex conjugate of $x(t_k)$. Since τ is a free parameter, one can select it so as to simplify the normal equations; that is,

$$\sum_{k=0}^{N-1} \cos\left[\omega(t_k - \tau)\right] \sin\left[\omega(t_k - \tau)\right] = 0 \tag{4.57}$$

Solving for τ will yield

$$\tan(2\omega\tau) = \frac{\displaystyle\sum_{k=0}^{N-1} \sin(2\omega\, t_k)}{\displaystyle\sum_{k=0}^{N-1} \cos(2\omega\, t_k)} \tag{4.58}$$

For the parameters b, c and d, one needs to enforce

$$\frac{dF}{db} = \frac{dF}{dc} = \frac{dF}{dd} = 0, \tag{4.59}$$

Use of (4.59) will yield the following equations

$$\sum_{k=0}^{N-1} j\,\overline{x}(t_k) \cos\left[\omega(t_k - \tau)\right] - \sum_{k=0}^{N-1} j\,x(t_k) \cos\left[\omega(t_k - \tau)\right] = 2b \sum_{k=0}^{N-1} \cos^2\left[\omega(t_k - \tau)\right] \tag{4.60}$$

$$\sum_{k=0}^{N-1} \overline{x}(t_k) \sin\left[\omega(t_k - \tau)\right] + \sum_{k=0}^{N-1} x(t_k) \sin\left[\omega(t_k - \tau)\right] = 2c \sum_{k=0}^{N-1} \sin^2\left[\omega(t_k - \tau)\right] \tag{4.61}$$

$$\sum_{k=0}^{N-1} j\,\overline{x}(t_k) \sin\left[\omega(t_k - \tau)\right] - \sum_{k=0}^{N-1} j\,x(t_k) \sin\left[\omega(t_k - \tau)\right] = 2d \sum_{k=0}^{N-1} \sin^2\left[\omega(t_k - \tau)\right] \tag{4.62}$$

The resulting values are:

$$a = \frac{\displaystyle\sum_{k=0}^{N-1} \text{Re}\left[x(t_k)\right] \cos\left[\omega(t_k - \tau)\right]}{\displaystyle\sum_{k=0}^{N-1} \cos^2\left[\omega(t_k - \tau)\right]}, \tag{4.63}$$

$$b = \frac{\displaystyle\sum_{k=0}^{N-1} \text{Im}\left[x(t_k)\right] \cos\left[\omega(t_k - \tau)\right]}{\displaystyle\sum_{k=0}^{N-1} \cos^2\left[\omega(t_k - \tau)\right]} \tag{4.64}$$

$$c = \frac{\displaystyle\sum_{k=0}^{N-1} \text{Re}\left[x(t_k)\right] \sin\left[\omega(t_k - \tau)\right]}{\displaystyle\sum_{k=0}^{N-1} \sin^2\left[\omega(t_k - \tau)\right]}, \tag{4.65}$$

$$d = \frac{\sum\limits_{k=0}^{N-1} Im[x(t_k)] \sin[\omega(t_k - \tau)]}{\sum\limits_{k=0}^{N-1} \sin^2[\omega(t_k - \tau)]} \qquad (4.66)$$

and $Re(\cdot)$ and $Im(\cdot)$, stands for the real and the imaginary parts of a complex function. Therefore,

$$a + jb = \frac{\sum\limits_{k=0}^{N-1} x(t_k) \cos[\omega(t_k - \tau)]}{\sum\limits_{k=0}^{N-1} \cos^2[\omega(t_k - \tau)]}, \qquad (4.67)$$

$$c + jd = \frac{\sum\limits_{k=0}^{N-1} x(t_k) \sin[\omega(t_k - \tau)]}{\sum\limits_{k=0}^{N-1} \sin^2[\omega(t_k - \tau)]} \qquad (4.68)$$

Substituting (4.67) and (4.68) into (4.54) results in

$$h(t) = \frac{\sum\limits_{k=0}^{N-1} x(t_k) \cos[\omega(t_k - \tau)]}{\sum\limits_{k=0}^{N-1} \cos^2[\omega(t_k - \tau)]} \cos[\omega(t - \tau)] + \frac{\sum\limits_{k=0}^{N-1} x(t_k) \sin[\omega(t_k - \tau)]}{\sum\limits_{k=0}^{N-1} \sin^2[\omega(t_k - \tau)]} \sin[\omega(t - \tau)]$$

$$(4.69)$$

The power in the harmonic component at frequency ω is given by

$$P(\omega) = \sum_{k=0}^{N-1} |h(t_k)|^2 = |a + jb|^2 \sum_{k=0}^{N-1} \cos^2[\omega(t_k - \tau)] + |c + jd|^2 \sum_{k=0}^{N-1} \sin^2[\omega(t_k - \tau)]$$

$$= \frac{\left\{\sum\limits_{k=0}^{N-1} x(t_k) \cos[\omega(t_k - \tau)]\right\}^2}{\sum\limits_{k=0}^{N-1} \cos^2[\omega(t_k - \tau)]} + \frac{\left\{\sum\limits_{k=0}^{N-1} x(t_k) \sin[\omega(t_k - \tau)]\right\}^2}{\sum\limits_{k=0}^{N-1} \sin^2[\omega(t_k - \tau)]} \qquad (4.70)$$

Equation (4.70) is a complex version of the Lomb periodogram [45]. Observe that (4.70) yields the same value for ω and $-\omega$ since it occurs inside an expression which is squared. Obviously this expression is not suitable for computing the spectrum for negative frequencies. To obtain the phase component of the spectrum from the power representation (4.70) one can assume the frequency response $F(\omega)$ to be,

$$F(\omega) = \sum_{k=0}^{N-1} \alpha \cos[\omega(t_k - \tau)] + j \sum_{k=0}^{N-1} \beta \sin[\omega(t_k - \tau)] \qquad (4.71)$$

where τ is a free parameter same as defined in (4.54) and (4.58). The corresponding power spectrum can be written as

$$P(\omega) = |F(\omega)|^2 = \left| \sum_{k=0}^{N-1} \alpha \cos\left[\omega(t_k - \tau)\right] \right|^2 + \left| \sum_{k=0}^{N-1} \beta \sin\left[\omega(t_k - \tau)\right] \right|^2$$

(4.72)

Matching (4.70), (4.71) and (4.72) for all $x(t_k)$ will define the unknown coefficients α and β. That is,

$$\sum_{k=0}^{N-1} \alpha \cos\left[\omega(t_k - \tau)\right] = \pm \frac{\sum_{k=0}^{N-1} x(t_k) \cos\left[\omega(t_k - \tau)\right]}{\sqrt{\sum_{k=0}^{N-1} \cos^2\left[\omega(t_k - \tau)\right]}},$$

(4.73)

$$\sum_{k=0}^{N-1} \beta \sin\left[\omega(t_k - \tau)\right] = \pm \frac{\sum_{k=0}^{N-1} x(t_k) \sin\left[\omega(t_k - \tau)\right]}{\sqrt{\sum_{k=0}^{N-1} \sin^2\left[\omega(t_k - \tau)\right]}}$$

(4.74)

Thus,

$$F(\omega) = \pm \frac{\sum_{k=0}^{N-1} x(t_k) \cos\left[\omega(t_k - \tau)\right]}{\sqrt{\sum_{k=0}^{N-1} \cos^2\left[\omega(t_k - \tau)\right]}} \pm j \frac{\sum_{k=0}^{N-1} x(t_k) \sin\left[\omega(t_k - \tau)\right]}{\sqrt{\sum_{k=0}^{N-1} \sin^2\left[\omega(t_k - \tau)\right]}}$$

(4.75)

There can be four possible choices for the sign and three of them are their images. The expression for the correct frequency can be easily obtained from the analogy of the conventional Fourier transformation or by applying a test signal $x(t_k) = e^{j\,\omega_1 t_k}$ and observing (4.74) at the frequency ω_1 and $-\omega_1$. The resulting expression turns out to take positive signs for both the terms.

$$F(\omega) = \frac{\sum_{k=0}^{N-1} x(t_k) \cos\left[\omega(t_k - \tau)\right]}{\sqrt{\sum_{k=0}^{N-1} \cos^2\left[\omega(t_k - \tau)\right]}} + j \frac{\sum_{k=0}^{N-1} x(t_k) \sin\left[\omega(t_k - \tau)\right]}{\sqrt{\sum_{k=0}^{N-1} \sin^2\left[\omega(t_k - \tau)\right]}}$$

(4.76)

As an example, consider a signal

$$x(t_k) = e^{-2\pi \cdot 20 \cdot j\, t_k} + e^{2\pi \cdot 30 \cdot j\, t_k} + e^{2\pi \cdot 40 \cdot j\, t_k}$$

where t_k is a random number uniformly distributed between $[0, 10]$ for $k = 1, 2, \ldots, 100$. Now (4.71) and (4.76) are used to estimate the spectrum of the nonuniformly sampled data and the result is shown in the two figures of Figure 4.9. The signal has three frequency components at (−20, 30 and 40 Hz) with magnitude unity. The average sampling frequency is 10 Hz which is much lower than the uniformly spaced Nyquist sampling rate of 40 Hz. As seen in Figure 4.9(b) the response at the negative frequency (−20 Hz) is distinguished by the equation (4,76) but not in figure 1(a) which is using the Lomb periodogram. Note that the absolute value was taken and squared in figure 4.9(b) for suitable comparison with Figure 4.9(a). This example illustrates the advantage for computing the spectrum using a non-uniformly sampled data using a lower frequency than the Nyquist sampling of the data.

4.7.2 Hilbert Transform Relationship

We can reduce the processing time of computing the spectrum using non-uniformly spaced data by using a Hilbert transform relationship between the real and the imaginary part of (4.76). This means that the time domain response of (4.76) is causal as will be shown in the next case. For an antenna problem this implies that the array is finite. Since the Hilbert transformation can be carried out by performing two Fast Fourier transformations and one multiplication. If the number of frequency step is M, the operation count for computing a Hilbert transform will be $2M \log M + M$. The operation count for equation (4.76) is $(4N+4)M$ if N is the number of time domain data. Therefore, the processing time will be reduced in our computations from $(4N+4)M$ to $(2N+2)M + 2M \log M + M$. If M and N are large numbers, we will have maximum of 50% of reduction of computation time by utilizing the Hilbert transform relationship.

Assume that one has a causal time domain signal $x(t)$ that exist only on $[0, \alpha]$ where α is not an infinite number, then

$$F(\omega) = \frac{\sum\limits_{k=0}^{N-1} x(t_k)\cos(\omega t_k)}{\sqrt{\sum\limits_{k=0}^{N-1} \cos^2(\omega t_k)}} + j\,\frac{\sum\limits_{k=0}^{N-1} x(t_k)\sin(\omega t_k)}{\sqrt{\sum\limits_{k=0}^{N-1} \sin^2(\omega t_k)}} \qquad (4.77)$$

Here the term τ is ignored since α is assumed to be a large number compared to the period of the signal and assume $x(t)$ as a time invariant sequence. When the time step is small, one can approximate the summation of (4.77) by an integration resulting in

$$F(\omega) = \frac{\int\limits_0^\alpha x(t)\cos(\omega t)dt}{\sqrt{\int\limits_0^\alpha \cos^2(\omega t)dt}} + j\,\frac{\int\limits_0^\alpha x(t)\sin(\omega t)dt}{\sqrt{\int\limits_0^\alpha \sin^2(\omega t)dt}} \qquad (4.78)$$

(a)

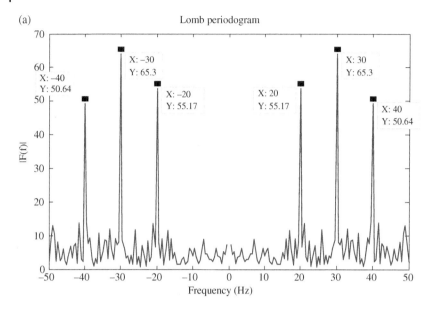

(b)

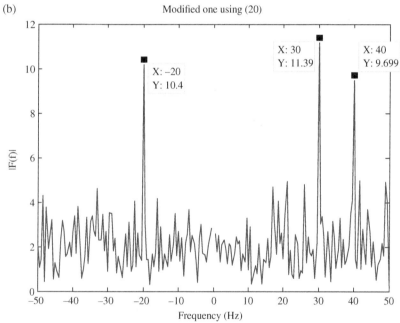

Figure 4.9 Comparison between the Lomb periodogram and the modified method presented here. (a) Lomb Periodogram using Eq (4.71). (b) Modified Spectrogram using Eq (4.77).

Since $\int_0^\alpha \cos^2(\omega t)dt = \dfrac{\alpha}{2} + \dfrac{\sin(2\omega\alpha)}{4\omega}$ and, $\int_0^\alpha \sin^2(\omega t)dt = \dfrac{\alpha}{2} - \dfrac{\sin(2\omega\alpha)}{4\omega}$, one

can transform (4.78) into

$$F(\omega) = \frac{\displaystyle\int_0^\alpha x(t)\cos(\omega t)dt}{\sqrt{\dfrac{\alpha}{2} + \dfrac{\sin(2\omega\alpha)}{4\omega}}} + j\,\frac{\displaystyle\int_0^\alpha x(t)\sin(\omega t)dt}{\sqrt{\dfrac{\alpha}{2} - \dfrac{\sin(2\omega\alpha)}{4\omega}}} \tag{4.79}$$

Usually we take $\omega\alpha \gg 1$, thus $|\sin(2\omega\alpha)| \ll |2\omega\alpha|$ and this transforms (4.79) to

$$F(\omega) \approx \sqrt{\frac{2}{\alpha}} \int_0^\alpha x(t)\cos(\omega t)dt + j\sqrt{\frac{2}{\alpha}} \int_0^\alpha x(t)\sin(\omega t)dt \tag{4.80}$$

Therefore the real and imaginary part of $F(\omega)$ are a Hilbert transformation pair when $\omega\alpha \gg 1$ and $x(t)$ is causal.

Figure 4.10 compares between the values of $F_2(\omega)$ and Hilbert transform of $F_1(\omega)$ where $F(\omega) = F_1(\omega) + iF_2(\omega)$ and $F_1(\omega)$, $F_2(\omega)$ are real. Figure 4.10(a) corresponds to the case, when $\omega\alpha$ is a relatively large number on the frequency axis and the two results coincide with each other. Figure 4.10(b) results when $\omega\alpha$ is small and there are some differences between the two curves at the middle where ω has a small value. In this example, the function chosen is $x(t_k) = e^{2\pi \cdot 10 \cdot j\, t_k}$; $0 < t_k < \alpha$ and the Matlab function HILBERT was used to compute the Hilbert transform of $F_2(\omega)$.

Processing time to get $F(\omega)$ was measured by changing the number of frequency steps and the number of time domain data and the result is shown in Figure 4.11. In both the cases, the processing time was reduced by 46% and 47 % respectively by utilizing the Hilbert transformation.

4.7.3 Magnitude Estimation

As seen in Figure 4.9, none of the Lomb periodogram or the modified one provide the exact magnitude of the signal. The error obviously comes from the uneven spacing and the aliasing between different frequencies.

Since the estimates of the frequencies giving rise to the peaks are not much different than the originals, we can estimate the amplitude of the desired signals from these frequencies by using a least square method. If the signal is a sum of exponentials,

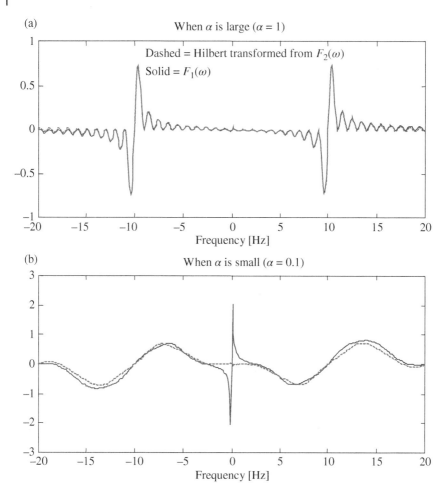

Figure 4.10 $F_1(\omega)$ and Hilbert transform of $F_2(\omega)$ for different values of $\omega \alpha$. If $\omega \alpha \gg 1$ they coincide with each other as shown in (a), while for α small there are differences between the two curves for ω also small as shown in (b).

$$x(t_k) = \sum_{l=1}^{L} A_l e^{j \omega_l t_k}; k = 1, 2, \ldots N, \tag{4.81}$$

where

ω_l= frequencies which provide the highest peak values
$x(t_k)$ = given data with respect to unevenly spaced point t_k
L = Number of frequency components
A_l = Unknown amplitudes

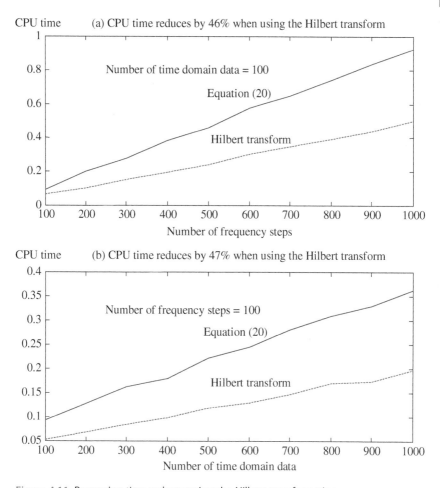

Figure 4.11 Processing time reduces using the Hilbert transformation.

By rewriting the above equation as

$$
\begin{bmatrix} x(t_1) \\ x(t_2) \\ \vdots \\ x(t_N) \end{bmatrix} = \begin{bmatrix} e^{j\omega_1 t_1} & e^{j\omega_2 t_1} & \cdots & e^{j\omega_L t_1} \\ e^{j\omega_1 t_2} & e^{j\omega_2 t_2} & \cdots & e^{j\omega_L t_2} \\ \vdots & \vdots & \ddots & \vdots \\ e^{j\omega_1 t_N} & e^{j\omega_2 t_N} & \cdots & e^{j\omega_L t_N} \end{bmatrix} \begin{bmatrix} A_1 \\ A_2 \\ \vdots \\ A_L \end{bmatrix}
\tag{4.82}
$$

and using the pseudo inverse, a magnitude vector A can be obtained using the unevenly sampled points t_k, signal $x(t_k)$ and the estimated frequency ω_k

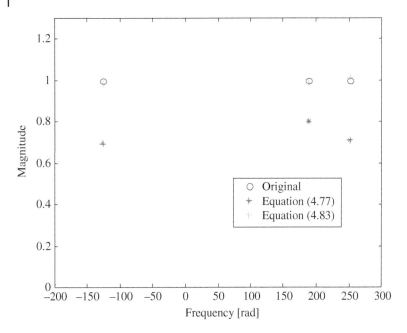

Figure 4.12 Accurate Estimate for the amplitude is given by equation (4.83), as the periodogram does not provide accurate results for the magnitudes.

$$A = (E^*E)^{-1}E^*f \tag{4.83}$$

$$\text{where } E = \begin{bmatrix} e^{j\omega_1 t_1} & e^{j\omega_2 t_1} & \cdots & e^{j\omega_L t_1} \\ e^{j\omega_1 t_2} & e^{j\omega_2 t_2} & \cdots & e^{j\omega_L t_2} \\ \vdots & \vdots & \ddots & \vdots \\ e^{j\omega_1 t_N} & e^{j\omega_2 t_N} & \cdots & e^{j\omega_L t_N} \end{bmatrix}_{N \times L} \tag{4.84}$$

and $E*$ is the conjugate transpose of E.

The same signal for the first example has been used to verify (4.83) and the result is shown in Figure 4.12. Three signal components with unit magnitudes can be obtained precisely by utilizing (4.83) while the magnitudes obtained from (4.77) have some differences from the true values.

As the final example, consider the phasor voltage X_n induced at the nth antenna element at a particular instance of time by a sum of P complex exponentials ignoring interferers and thermal noise as

$$X_n = \sum_{p=1}^{P} A_p e^{\frac{j 2 \pi n \Delta \cos\phi_p}{\lambda}} \tag{4.85}$$

Now assume that three signals impinge on this array from $60°, 70°$, and $120°$. All of their amplitudes are considered to be 1. Therefore, the received signal can be defined as

$$x(s_k) = e^{j2\pi s_k \cos 60°} + e^{j2\pi s_k \cos 70°} + e^{j2\pi s_k \cos 120°} \tag{4.86}$$

First, consider a 15 element uniformly spaced antenna array so that all the elements are spaced half wavelength apart. Using the voltages received at the antenna elements one can compute the estimated direction of arrival using the concept of the spectrum from a uniformly spaced array and it is shown in Figure 4.13(a). The estimated amplitudes of the signal are now obtained as [0.9835; 0.9463; 1.0049].

Next the elements are redistributed so that the 15 elements are now non-uniformly spaced as $\left[0, \dfrac{1}{4}, \dfrac{2}{4}, \dfrac{3}{4}, \dfrac{7}{4}, \dfrac{11}{4}, \dfrac{14}{4}, \dfrac{17}{4}, \dfrac{18}{4}, \dfrac{19}{4}, \dfrac{20}{4}, \dfrac{22}{4}, \dfrac{24}{4}, \dfrac{26}{4}, \dfrac{28}{4}\right]$

which occupies the same aperture size of 7λ, but there are some elements which are more closely spaced and some that are quite widely spaced violating the Nyquist sampling criteria. The Direction of arrival of the signal is now obtained as shown in Figure 4.13(b). Their respective amplitudes are estimated as [1.0055; 0.9298; 0.9872].

The following features can be observed from Figure 4.13(a) and Figure 4.13(b):

1) For a 7λ antenna array, the resolution in DOA using the Fourier techniques is about $15°$ and yet the least the squares solution can resolve signals that are spaced closer than this.
2) The antenna elements need not be placed in a way that satisfies the Nyquist sampling criteria. This is an advantage for a non-uniformly spaced antenna array as the Nyquist sampling theory needs to be enforced for the average inter element spacing and not for the spacing of each element individually.
3) The disadvantage of this new procedure is that the spectrum does not yield the correct amplitudes and they need to be estimated solving a separate least squares problem.

In summary, unevenly spaced spectrum using the least square method has been described. The well-known Lomb periodogram approach has lots of benefits but cannot discern positive and negative frequencies. Using a modified scheme, positive and negative frequencies can be discerned without losing any of the benefits of the Lomb periodogram. A method to accurately estimate the amplitudes of the signal is also described. One of the properties of the periodogram approach is the relationship that exists for the coefficients as they form a Hilbert transformation pair. By utilizing this property, the processing time can be reduced in half.

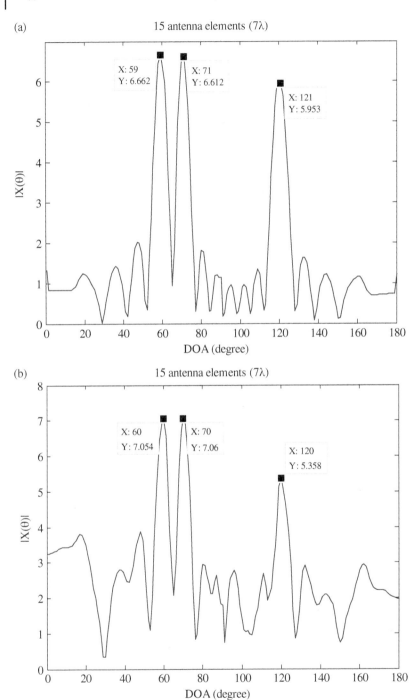

Figure 4.13 (a) DOA Estimation using a half wavelength spaced 15 element array. (b) DOA Estimation using a 15 element non-uniformly spaced array having the same aperture.

4.8 Conclusion

In this chapter, a nonparametric method has been presented for the application of the principle of analytic continuation to interpolate/extrapolate system responses resulting in reduced computations. The advantage of the nonparametric methods are that no specific choices of the basis functions need to be made as the solution procedure by itself develops the nature of the solution and no a priori information is necessary. This is accomplished through the use of the Hilbert transform which exploits one of the fundamental properties of nature and that is causality. The Hilbert transform illustrates that the real and imaginary parts of any nonminimum phase transfer function from a causal system satisfy this relationship. In addition, some parametrization can also be made of this procedure which can enable one to generate a nonminimum phase function from its amplitude response and from that generate the phase response. In following such a procedure, one can compute the time domain data for amplitude-only data except a delay in the response. This uncertainty is removed in holography as in such a procedure an amplitude and phase information is measured for a specific look angle thus eliminating the phase ambiguity. Examples have been presented to illustrate this methodology. In addition, a methodology has been presented for the case of unevenly spaced data and to generate its spectrum using the least square method has been described. The advantage of using nonuniformly sampled data is that the spacing between the samples need not satisfy the Nyquist sampling rate but it is necessary for the average sampling period taken over all the samples to satisfy the Nyquist rate.

References

1 T. K. Sarkar, M. Salazar-Palma, and E. L. Mokole, "Application of the Principle of Analytic Continuation to Interpolate/Extrapolate System Responses Resulting in Reduced Computations: Part B: Nonparametric Methods," *IEEE Journal on Multiscale and Multiphysics Computational Techniques*, Vol. 1, pp. 60–72, 2016.

2 A. D. Poularikas, *The Transforms and Applications Handbook*, IEEE Press, Piscataway, NJ, 1996.

3 F. B. Hilderbrand, *Method of Applied Mathematics*, Prentice-Hall, Englewood Cliffs, NJ, 1952.

4 L. D. Landau and E. M. Lifshitz, *Electrodynamics of Continuous Media*, Addison Wesley, New York, 1960.

5 V. Cizek, "Discrete Hilbert Transform," *IEEE Transactions on Audio and Electroacoustics*, Vol. AU-18, No. 4, pp. 340–343, 1970.

6 A. V. Oppenheim and R. W. Schafer, *Discrete-Time Signal Processing*, Prentice-Hall, Englewood Cliffs, NJ, Second Edition, 1989.

7 J. R. James and F. Andrasic, "Assessing the Accuracy of Wideband Electrical Data Using Hilbert Transforms," *IEE Proceedings H Microwaves, Antennas and Propagation*, Vol. 177, No. 3, pp. 184–188, 1990.

8 S. Narayana, G. Rao, R. Adve, T. K. Sarkar, V. C. Vannicola, M. Wicks, and S. Scott, "Interpolation/Extrapolation of Frequency Responses Using the Hilbert Transform," *IEEE Transactions on Microwave Theory and Techniques*, Vol. 44, pp. 1621–1627, 1996.

9 S. Narayana, T. K. Sarkar, R. S. Adve, M. Wicks, and V. Vannicola, "A Comparison of Two Techniques for the Interpolation/Extrapolation of Frequency Domain Responses," *Digital Signal Processing*, Vol. 6, No. 2, pp. 51–67, 1996.

10 T. K. Sarkar and B. Hu, "Generation of Nonminimum Phase from Amplitude-Only Data," *IEEE Transactions on Microwave Theory and Techniques*, Vol. 46, No. 8, pp. 1079–1084, 1998.

11 R. S. Adve and T. K. Sarkar, "Simultaneous Time- and Frequency-Domain Extrapolation," *IEEE Transactions on Antennas and Propagation*, Vol. 46, No. 4, pp. 484–493, 1998.

12 T. Quatieri and A. Oppenheim, "Iterative Techniques for Minimum Phase Signal Reconstruction from Phase or Magnitude," *IEEE Transactions on Acoustics, Speech, and Signal Processing*, Vol. 29, No. 6, pp. 1187–1193, 1981.

13 M. H. Hayes, "The Reconstruction of a Multidimensional Sequence from the Phase or Magnitude of Its Fourier Transform," *IEEE Transactions on Acoustics, Speech, and Signal Processing*, Vol. ASSP-30, pp. 140–154, 1982.

14 R. de L. Kronig, "On the Theory of Dispersion of X-Rays," *Journal of the Optical Society of America*, Vol. 12, pp. 547–551, 1926.

15 H. A. Kramers, "Die Dispersion und Absorption von Rontgenstralen," *Physikalishce Zeitschrift*, Vol. 30, pp. 522–523, 1929.

16 J. R. James and G. Andresic, "On the Use of the Hilbert Transform for Processing Measured CW Data," *IEEE Transactions on Electromagnetic Compatibility*, Vol. 35, No. 3, pp. 408, 1993.

17 T. R. Arabi, A. T. Murphy, and T. K. Sarkar, "An Efficient Technique for the Time Domain Analysis of Multi-Conductor Transmission Lines Using the Hilbert Transform," *Microwave Symposium Digest, IEEE MTT-S International Symposium*, Atlanta, GA, Vol. 1; pp. 185–188, 1991.

18 Y. M. Bruck and L. G. Sodin, "On the Ambiguity of the Image Reconstruction Problem," *Optics Communication*, Vol. 30, No. 3, pp. 304–308, 1979.

19 F. M. Tesche, "On the Use of the Hilbert Transform for Processing Measured CW Data," *IEEE Transactions on Electromagnetic Compatibility*, Vol. 34, No. 3, pp. 259–266, 1992.

20 V. Pyati, "On the Use of Hilbert Transform for Processing Measured CW Data," *IEEE Transactions on Electromagnetic Compatibility*, Vol. 35, p. 485, 1993.

21 G. C. James, "Phase Retrieval and Near Field Antenna Measurements," in *Proceedings of the 1992 URSI International Symposium on Electromagnetic Theory*, Sydney, Australia, pp. 510–512, 1992.

22 T. R. Arabi and R. Suarez-Gartner, "Time Domain Aanalysis of Lossy Multi-Conductor Transmission Lines Using the Hilbert Transform," *Microwave Symposium Digest, IEEE MTT-S International*, Atlanta, GA, Vol. 2, pp. 987–990, 1993.

23 A. Carballar and M. A. Muriel, "Phase Reconstruction from Reflectivity in Fiber Bragg Gratings," *Journal of Lightwave Technology*, Vol. 15, No. 8, pp. 1314–1322, 1997.

24 R. E. Diaz and N. G. Alexopoulos, "An Analytic Continuation Method for the Analysis and Design of Dispersive Materials," *IEEE Transactions on Antennas and Propagation*, Vol. 45, No. 11, pp. 1602–1610, 1997.

25 A. R. Djordjevic, R. M. Biljie, V. D. Likar-Smiljanic, and T. K. Sarkar, "Wideband Frequency-Domain Characterization of FR-4 and Time-Domain Causality," *IEEE Transactions on Electromagnetic Compatibility*, Vol. 43, No. 4, pp. 662–667, 2001.

26 S. Grivet-Talocia; H. Huang; A. E. Ruehli; F. Canavero, and I. M. Elfadel, "Transient Analysis of Lossy Transmission Lines: An Efficient Approach Based on the Method of Characteristics," *IEEE Transactions on Advanced Packaging*, Vol. 27, No. 1, pp. 45–56, 2004.

27 P. A. Perry and T. J. Brazil, "Forcing Causality on S-Parameter Data Using the Hilbert Transform," *IEEE Microwave and Guided Wave Letters*, Vol. 8, No. 11, pp. 378–380, 1998.

28 M. Condon, R. Ivanov, and C. Brennan, "A Causal Model for Linear RF Systems Developed from Frequency-Domain Measured Data," *IEEE Transactions on Circuits and Systems II: Express Briefs*, Vol. 52, No. 8, pp. 457–460, 2005.

29 J. Yang, J. Koh, and T. K. Sarkar, "Reconstructing a Nonminimum Phase Response from the Far-Field Power Pattern of an Electromagnetic System," *IEEE Transactions on Antennas and Propagation*, Vol. 53, No. 2, pp. 833–841, 2005.

30 Z. Du, J. Moon, S. s. Oh, J. Koh, and T. Sarkar, "Generation of Free Space Radiation Patterns From Non-Anechoic Measurements Using Chebyshev Polynomials," *IEEE Transactions on Antennas and Propagation*, Vol. 58, No. 8, pp. 2785–2790, 2010.

31 M. Yuan, J. Koh, T. K. Sarkar, W. Lee, and M. Salazar-Palma, "A Comparison of Performance of Three Orthogonal Polynomials in Extraction of Wide-Band Response Using Early Time and Low Frequency Data," *IEEE Transactions on Antennas and Propagation*, Vol. 53, No. 2, pp. 785–792, 2005.

32 J. Koh, A. De, T. K. Sarkar, H. Moon, W. Zhao, and M. Salazar-Palma, "Free Space Radiation Pattern Reconstruction from Non-Anechoic Measurements Using an Impulse Response of the Environment," *IEEE Transactions on Antennas and Propagation*, Vol. 60, No. 2, pp. 821–831, 2012.

33 J. Koh, W. Lee, T. K. Sarkar, and M. Salazar-Palma, "Calculation of Far-Field Radiation Pattern Using Nonuniformly Spaced Antennas by a Least Square Method," *IEEE Transactions on Antennas and Propagation*, Vol. 62, No. 4, pp. 1572–1578, 2014.

34 F. Fan, T. K. Sarkar, C. Park, and J. Koh, "Far-field pattern reconstruction using an iterative Hilbert transform," *IEICE Transactions on Communications*, Vol. E98B, No. 6, pp. 1032–1039, 2015.

35 S. Luo and Z. Chen, "Iterative Methods for Extracting Causal Time-Domain Parameters," *IEEE Transactions on Microwave Theory and Techniques*, Vol. 53, No. 3, pp. 969–976, 2005.

36 P. Triverio and S. Grivet-Talocia, "Robust Causality Characterization via Generalized Dispersion Relations," *IEEE Transactions on Advanced Packaging*, Vol. 31, No. 3, pp. 579–593, 2008.

37 J. Zhang, J. L. Drewniak, D. J. Pommerenke, M. Y. Koledintseva, R. E. DuBroff, W. Cheng, Z. Yang, Q. B. Chen, and A. Orlandi, "Causal RLGC(f) Models for Transmission Lines from Measured S-Parameters," *IEEE Transactions on Electromagnetic Compatibility*, Vol. 52, No. 1, pp. 189–198, 2010.

38 S. Lalgudi, "On Checking Causality of Tabulated S-Parameters," *IEEE Transactions on Components, Packaging and Manufacturing Technology*, Vol. 3, No. 7, pp. 1204–1217, 2013.

39 W. M. Galal Dyab, T. Kumar Sarkar, A. Garcia-Lamperez, M. Salazar-Palma, and M. A. Lagunas, "A Critical Look at the Principles of Electromagnetic Time Reversal and Its Consequences," *IEEE Antennas and Propagation Magazine*, Vol. 55, No. 5, pp. 28–62, 2013.

40 Y. Wang, S. Liu, and T. J. Brazil, "Discrete-Time Representation of Band-Pass S-Parameter Data," *IEEE Microwave and Wireless Components Letters*, Vol. 25, No. 11, pp. 697–699, 2015.

41 L. L. Barannyk, H. A. Aboutaleb, A. Elshabini, and F. D. Barlow, "Spectrally Accurate Causality Enforcement Using SVD-Based Fourier Continuations for High-Speed Digital Interconnects," *IEEE Transactions on Components, Packaging and Manufacturing Technology*, Vol. 5, No. 7, pp. 991–1005, 2015.

42 T. R. Arabi, A. T. Murphy, T. K. Sarkar, R. F. Harrington, and A. R. Djordjevic, "On the Modeling of Conductor and Substrate Losses in Multiconductor, Multidielectric Transmission Line Systems," *IEEE Transactions on Microwave Theory and Techniques*, Vol. 39, No. 7, pp. 1090–1097, 1991.

43 T. R. Arabi, A. T. Murphy, and T. K. Sarkar, "An Efficient Technique for the Time Domain Analysis of Multi-Conductor Transmission Lines Using the Hilbert Transform," *1991 IEEE MTT-S International Microwave Symposium Digest*, Boston, MA, pp. 185–188, 1991.

44 P. Vanicek, "Further Development and Properties of the Spectral Analysis by Least Squares," *Astrophysics and Space Science*, Vol. 12, pp. 10–33, 1971.

45 N. R. Lomb, "Least Squares Frequency Analysis Unequally Spaced Data," *Astrophysics and Space Science*, Vol. 39, pp. 447–462, 1975.

46 J. D. Scargle, "Studies in Astronomical Time Series Analysis. II. Statistical Aspects of Spectral Analysis of Unevenly Spaced Data," *The Astrophysical Journal*, Vol. 263, pp. 835–853, 1982.

47 J. H. Horne and S. L. Baliunas, "A Prescription for Period Analysis of Unevenly Sampled Time Series," *The Astrophysical Journal*, Vol. 302, pp. 757–763, 1986.

48 S. Ferraz-Mello, "Estimation of Periods from Unequally Spaced Observations," *Astronomy Journal*, Vol. 86, No. 4, pp. 619–624, 1981.

49 D. D. Meisel, "Fourier Transforms of Data Sampled at Unequal Observational Intervals," *Astronomy Journal*, Vol. 83, No. 5, pp. 538–545, 1978.

50 P. A. Gorry, "General Least Squares Smoothing and Differentiation of Nonuniformly Spaced Data by the Convolution (Savitzky-Golay) Method," *Analytical Chemistry*, Vol. 63, pp. 534–536, 1991.

51 W. H. Press, S. A. Teukolsky, W. T. Vetterling, and B. P. Flannery, *Numerical Recipes in C*, University Press, Cambridge, UK, 1992.

52 I. Bilinskis and A. Mikelsons, *Randomized Signal Processing*, Prentice Hall, New York, 1992.

53 H. S. Black, *Modulation Theory*, Van Nostrand, New York, 1953.

54 F. J. Beutler, "Error-Free Recovery of Signals from Irregularly Spaced Samples," *SIAM Review*, Vol. 8, No. 3, pp 328–335, 1966.

5

The Source Reconstruction Method

Summary

The source reconstruction method (SRM) is a recent technique developed for antenna diagnostics and for carrying out near-field (NF) to far-field (FF) transformation. The SRM is based on the application of the electromagnetic Equivalence Principle, in which one establishes an equivalent current distribution that radiates the same fields as the actual currents induced in the antenna under test (AUT). The knowledge of the equivalent currents allows the determination of the antenna radiating elements, as well as the prediction of the AUT-radiated fields outside the equivalent currents domain. The unique feature of the novel methodology presented in this section is that it can resolve equivalent currents that are smaller than half a wavelength in size, thus providing super-resolution. Furthermore, the measurement field samples can be taken at spacings greater than half a wavelength, thus going beyond the classical sampling criteria. These two distinctive features are possible due to the choice of a model-based parameter estimation methodology where the unknown sources are approximated by a basis in the computational Method of Moment (MoM) context and, secondly, through the use of the analytic free space Green's function. The latter condition also guarantees the invertibility of the electric field operator and provides a stable solution for the currents even when evanescent waves are present in the measurements. In addition, the use of the singular value decomposition in the solution of the matrix equations provides the user with a quantitative tool to assess the quality and the quantity of the measured data. Alternatively, the use of the iterative conjugate gradient (CG) method in solving the ill-conditioned matrix equations for the equivalent currents can also be implemented. Two different methods are presented in this section. One that deals with the equivalent magnetic current and the second that deals with the equivalent electric current. If the formulation is sound then either of the methodologies will provide the same far field using the same near field data. Examples are presented to illustrate the applicability and accuracy of the proposed methodology using either of the equivalent currents and applied to experimental data. This methodology is then

Modern Characterization of Electromagnetic Systems and Its Associated Metrology, First Edition.
Tapan K. Sarkar, Magdalena Salazar-Palma, Ming Da Zhu, and Heng Chen.
© 2021 John Wiley & Sons, Inc. Published 2021 by John Wiley & Sons, Inc.

used for near-field to near/far-field transformations for arbitrary near-field geometry to evaluate the safe distance for commercial antennas followed by some conclusions and a list of references.

5.1 Introduction

Antenna diagnostic methods based on the analysis of the near fields close to the antenna surface are becoming of great interest for antenna manufacturers. These methods allow an accurate determination of the antenna anomalies from the measurement of the fields emanating from the AUT, without requiring any interaction with the antenna [1–11].

The basic idea in this methodology is that, given the measured (near or far) field, it is possible to determine a set of equivalent sources (both electric and magnetic) that generate the same fields as that of the AUT. Thus, if the sources are known, the field at any point outside the source domain can be calculated. The determination of the equivalent currents from the emanating fields is based on the application of the electromagnetic Equivalence Principle and the solution of the integral equations relating fields and sources through the analytical free space Green's function methodology. This last condition guarantees the stability of this computational methodology. In addition, the solution methodology does not blow up even when evanescent waves are present in the measured data. In contrast, the classical fast Fourier trans-form (FFT)-based method becomes unstable in the presence of evanescent waves. In addition, the use of the analytic Green's function is not influenced by the sampling criteria for the fields.

With respect to other antenna diagnostics and near-field (NF) to far-field (FF) transformation methods, the SRM has the following advantages.

1) The SRM makes use of the integral equations relating an arbitrary electromagnetic currents distribution with its radiated fields. Thus, the SRM can work with both arbitrary-geometry field acquisition (measurement) domain as well as arbitrary-geometry sources domain (an example is shown in Figure 5.1). Other antenna diagnostics methods based on the expansion of the field in wave modes [1–3, 12] are limited to canonical geometries (i.e., planar, cylindrical, and spherical).

2) In this SRM, it is not necessary to have the measured field points located at half a wavelength apart.

3) It is possible to have a resolution smaller than half a wavelength for the reconstructed currents.

4) Use of the singular value decomposition (SVD) technique can be used to assess the quality and the quantity of the measured data in this methodology.

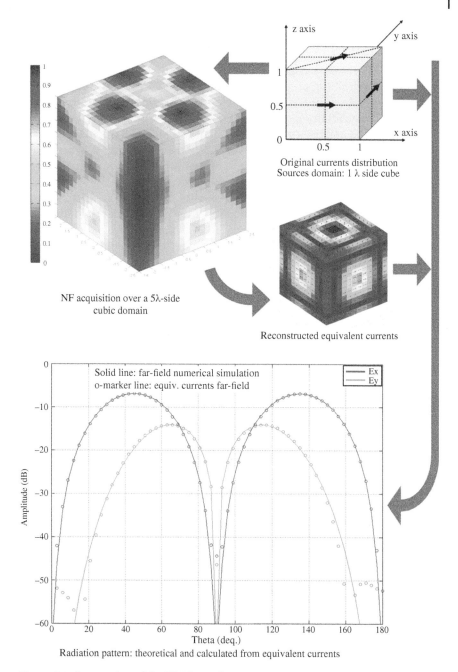

Figure 5.1 An overview of the NF-FF transformation using the SRM.

A drawback of the SRM technique is its computational cost. The calculation of the equivalent currents requires solution of large systems of equations. Earlier, the application of the SRM was limited to the analysis of medium-sized problems (in terms of wavelength) or complex geometry antennas which requires a dense discretization of their surface. Nowadays, with the development powerful parallel computers it is possible to extend this methodology to fault diagnosis of large arbitrary shaped antenna structures. In addition, the implementation of the iterative conjugate Gradient (CG) method along with the fast Fourier Transform (FFT) can solve very large matrix equations in a fast and accurate way for certain canonical structures on which the equivalent currents are placed and the field points measured.

Section 5.2 reviews the general methodology of the SRM [10], focusing on the Equivalence Principle and the computation of the currents given the measured fields and the general mathematical formulation is given succinctly in section 5.3. The SRM formulation specialized for the magnetic currents is presented in detail in Section 5.4, while the application of the electric current is described in section 5.5. Section 5.6 presents a unique application of this methodology where near field radiation patterns of commercial antennas can be evaluated. This is useful to find the safety distance from transmitting antennas. The methodology is applied to some experimental data followed by some conclusions in 5.7, followed by related references.

5.2   An Overview of the Source Reconstruction Method (SRM)

This section summarizes the main theoretical bases of the SRM, which have been presented in the works of [4–13]. The SRM is based on the electromagnetic Equivalence Principle [14, 15], which states that: given a set of sources bounded by the volume v', operating in free space, it is possible to characterize the original source distribution by equivalent currents distributed on a surface S' that encloses the original sources, so that the generated fields outside the domain containing the sources are the same in both the original and the equivalent problem (see Figure 5.1).

The next step is to establish a relationship between the equivalent currents and the generated fields (near field or far field), which is done through the integral equations [12, 13]. These equations can be seen as two linear systems (one for the electric fields, and the other for the magnetic fields) where the inputs are the equivalent currents and the outputs are the fields. Thus, due to the relationship between the electric and magnetic fields, it will be possible to determine the equivalent current distribution by inverting one of the two system matrix, depending on which kind of field is available. The knowledge of just one field

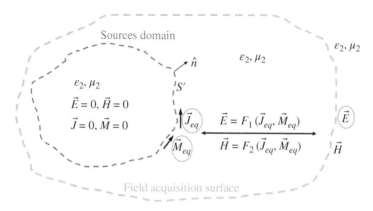

Figure 5.2 Relationship between the field acquisition domain and the equivalent sources on their respective surfaces.

(electric or magnetic) is enough for the calculation of both the equivalent currents (see Figure 5.2) as explained in [4, 12]. The utilization of the measured electric or the magnetic fields depends on the particular application. However, most of the SRM applications for antenna diagnostics and NF-FF transformations are based on the electric field measurements. It must be noted that the formulation of the SRM based on the magnetic field acquisition is dual to the formulation that uses the acquired electric field as described in [4].

Another question related to the radiated field measurement is the number of field components to be considered for the source reconstruction. According to the Equivalence Principle, the knowledge of the tangential fields (i.e. the equivalent currents) on a closed acquisition domain allows one to determine the total field in all points of space. Thus, only the two tangential field components in the acquisition domain are required for the determination of the equivalent currents, and the knowledge of the normal component is not necessary. This is one of the reasons why most of the antenna measurement ranges usually provides the tangential field components: E_x, E_y for planar acquisition systems; E_ϕ, E_z for cylindrical; and E_ϕ, E_θ for spherical [5–8].

5.3 Mathematical Formulation for the Integral Equations

The total electric field generated in a homogeneous medium, by an arbitrary distribution of electric and magnetic currents is given by the superposition of the responses due to the electric and magnetic currents as follows:

$$\vec{E}\left(\vec{r}\right) = \vec{E}_J\left(\vec{r}\right) + \vec{E}_M\left(\vec{r}\right) \tag{5.1}$$

$$\vec{E}_J\left(\vec{r}\right) = -\frac{j\eta}{4\pi k_0} \int\limits_{S'} \left\{ \left(k_0{}^2 + \nabla^2\right) \left(\vec{J}_{eq}\left(\vec{r}'\right) \frac{e^{-jk_0 R\left(\vec{r}; \vec{r}'\right)}}{R\left(\vec{r}; \vec{r}'\right)}\right) \right\} dS' \tag{5.2}$$

$$\vec{E}_M\left(\vec{r}\right) = -\frac{1}{4\pi} \nabla \times \int\limits_{S'} \left\{ \vec{M}_{eq}\left(\vec{r}'\right) \frac{e^{-jk_0 R\left(\vec{r}; \vec{r}'\right)}}{R\left(\vec{r}; \vec{r}'\right)} \right\} dS' \tag{5.3}$$

η is the intrinsic impedance of the medium, k_0 the wavenumber, $R\left(\vec{r}; \vec{r}'\right) = \left|\vec{r} - \vec{r}'\right|$ where $\vec{r}$ is the observation point vector and $\vec{r}'$ is the current source position vector. $\vec{J}_{eqs}\left(\vec{r}'\right)$ and $\vec{M}_{eq}\left(\vec{r}'\right)$ are the equivalent electric and magnetic current distribution defined on the surface S' that encloses the original sources. This integro-differential equation is solved by the method of moments, where we need to expand the unknown currents in terms of some known basis functions multiplied by unknown constants [16].

One can further simplify the problem by considering the bounding surface encompassing the source domain to be either a perfectly electrically conducting surface in which case only the equivalent magnetic current resides on the surface or a perfectly magnetically conducting surface in which case the electric currents reside on that surface. Both are equivalent and either one can be used to address our problem as we shall see. In the next chapter when dealing with electrically conducting surface enclosing the structure in which case the equivalent magnetic current resides on the structure. The equations decouple and very efficient computational techniques can be used to solve that problem. However, in this section we present both the methodologies and compare their predictions using experimental data to illustrate the concepts of this novel methodology.

5.4 Near-Field to Far-Field Transformation Using an Equivalent Magnetic Current Approach

A method is presented for computing near- and far-field patterns of an antenna from its near-field measurements taken over an arbitrary geometry. This method utilizes near-field data to determine an equivalent magnetic current source over a fictitious surface which encompasses the antenna. This magnetic current, once determined, can be used to ascertain the near and the far fields. This method demonstrates that once the values of the electromagnetic field are known over an arbitrary geometry, its values for any other region can be obtained. An electric field integral equation is developed to relate the near fields to the equivalent magnetic current. A moment method procedure [15] is employed to solve the integral equation by transforming it into a matrix equation. A least squares solution via singular value decomposition [17, 18] is used to solve the matrix equation. Computations with both synthetic and experimental data, where the near field of several antenna configurations are measured over various geometric surfaces, illustrate the accuracy of this method.

Presented here is a method for near-field to near/far field transformation which requires no specific geometry for near-field measurements. In this approach, by using the equivalence principle [14, 15], an equivalent magnetic current replaces the radiating antenna. Furthermore, it is assumed that the near field is produced by the equivalent magnetic current and therefore, via Maxwell's equation relate to the measured near-field data, the current source can be determined. Once this is accomplished, the near field and the far field of the radiating antenna in all regions in space in front or all around the radiating antenna can be determined directly from the equivalent magnetic current. An electric field integral equation is developed which pertains to the measured near fields and the equivalent magnetic current. This integral equation has been solved for the unknown magnetic current source through a moment method procedure [16] with point matching, where the equivalent current is expanded as linear combinations of two-dimensional (2-D) pulse basis functions and, through this construct, the integral equation is then transformed into a matrix equation. In general, the related matrix is rectangular whose dimensions depend on the number of field and source points chosen. The matrix equation is solved by the method of least squares via the singular value decomposition [17, 18] and also as explained in Chapter 1. By this, the moment matrix is decomposed into a set of orthogonal matrices which can be easily inverted.

Another aspect of this approach is that the numerical integrations in the process of creating the moment matrix elements have been avoided by taking a limiting case. Since the field points and the source points are to never coincide, and if their distances are much larger than the sizes of the current patches, then the pulse basis functions expanding the current source can be approximated by Hertzian dipoles. The formulation and theoretical basis for the equivalent magnetic current approach along with the formation of the corresponding matrix equation using the method of moments and its solution using the subdomain basis functions is described next.

5.4.1 Description of the Proposed Methodology

Consider an arbitrarily shaped antenna radiating into free space. The aperture of the antenna is assumed to be in a plane for illustration purposes. The plane surface is separating all space into left half and right half spaces as shown in Figure 5.3. The aperture of the antenna is placed in the xy-plane and is facing

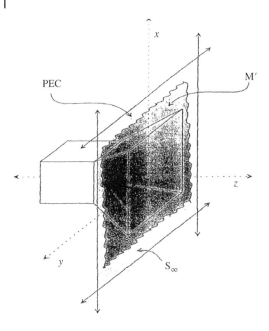

Figure 5.3 Original and the Equivalent problem with a magnetic current sheet over an electric conductor.

the positive z-axis. Since we are interested only in the electromagnetic field in front of the radiating antenna (i.e., the space where $z > 0$), we place a perfect electric conductor, extending to infinity in the x- and y-directions, on the xy-plane in front of the aperture of the antenna (Figure 5.3). By the equivalence principle [14, 15], a surface magnetic current M' may be placed on this perfect electric conductor whose value is equal to the tangential value of the electric field on the xy-plane

$$M' = \widehat{E} \times \widehat{n} \ \text{on} \ S_\infty \tag{5.4}$$

where E is the electric field on the xy-plane, S_∞, which is the entire xy-plane at $z = 0$, and $\widehat{n}$ is the unit outward normal to the xy-plane pointing in the direction of the positive z-axis.

Using image theory [14, 15], an equivalent magnetic current M can be introduced as seen in Figure 5.4 as

$$M = 2 M' \tag{5.5}$$

which is now radiating in free space. For computational purposes the surface S_∞ is truncated to a finite sized surface S_0 as seen in Figure 5.4. Therefore

$$\widehat{M} = 2 \widehat{E} \times \widehat{n} \ \text{on} \ S_\infty \tag{5.6}$$

where M now replaces the source antenna and radiates into free space (Figure 5.4), producing exactly the same field as the original antenna in the

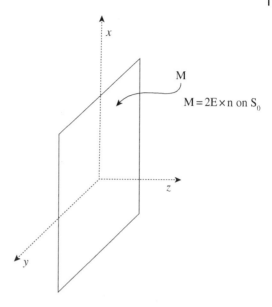

Figure 5.4 Equivalent magnetic current covering the aperture of the antenna and radiating in free space.

$M = 2E \times n$ on S_0

region $z > 0$. The measured near field of the antenna may now be used to determine M from

$$E_{meas} = E(M) \tag{5.7}$$

where E_{meas} is the electric near field measured over a geometry at a distance away from the aperture of the radiating antenna. We now solve for M using the measured Electric fields as an input to the Maxwell's equations, where

$$E(M) = -\nabla \times F(M) \tag{5.8}$$

where $F(M)$ is the electric vector potential and is given by

$$F(M) = -\frac{1}{4\pi} \int_{S_\infty} M(r') \, G(r, r') \, ds' \tag{5.9}$$

where r is the field coordinate and r' is the source coordinate. ∇ is the gradient operator and in the Cartesian coordinates is given by

$$\nabla = \frac{\partial}{\partial x} \widehat{x} + \frac{\partial}{\partial y} \widehat{y} + \frac{\partial}{\partial z} \widehat{z} \tag{5.10}$$

where $\widehat{x}, \widehat{y}$ and $\widehat{z}$ are the unit vectors along the x-, y-, and z-axes, respectively. The free space Green's function $G(r, r')$ is given by

$$G(r, r') = \frac{e^{-j k_0 |r - r'|}}{|r - r'|} \tag{5.11}$$

and $k_0 = \frac{2\pi}{\lambda}$ is the free space wave-number and λ is the wavelength of the radiated fields.

Combining equations (5.8) and (5.9) one obtains

$$E_{meas} = E(M) = -\frac{1}{4\pi} \nabla \times \int_{S_\infty} M(r')\, G(r,r')\, ds'$$

$$= \frac{1}{4\pi} \int_{S_\infty} [M(r') \times \nabla G(r,r')]\, ds' \qquad (5.12)$$

where E_{meas} is the electric near field measured over a geometry at a distance away from the aperture of the radiating antenna.

In a matrix form (5.12) can be written as

$$\begin{bmatrix} E_{meas,\,x}(r) \\ E_{meas,\,y}(r) \\ E_{meas,\,z}(r) \end{bmatrix} = \frac{1}{4\pi} \int_{S_\infty} \begin{bmatrix} 0 & \dfrac{\partial G}{\partial z} \\[2mm] -\dfrac{\partial G}{\partial z} & 0 \\[2mm] \dfrac{\partial G}{\partial y} & \dfrac{\partial G}{\partial x} \end{bmatrix} \times \begin{bmatrix} M_x(r') \\ M_y(r') \end{bmatrix} ds' \qquad (5.13)$$

where

$$\frac{\partial G}{\partial x} = \frac{e^{-j k_0 R}}{R^2}(x-x')\left\{ j k_0 + \frac{1}{R} \right\}$$

$$\frac{\partial G}{\partial y} = \frac{e^{-j k_0 R}}{R^2}(y-y')\left\{ j k_0 + \frac{1}{R} \right\} \qquad (5.14)$$

$$\frac{\partial G}{\partial z} = \frac{e^{-j k_0 R}}{R^2}(z-z')\left\{ j k_0 + \frac{1}{R} \right\}$$

and $R = |r - r'|$. Now the relationship between the various components of the measured near field in rectangular and spherical coordinates are given by

$$\begin{bmatrix} E_{meas,\,\theta} \\ E_{meas,\,\phi} \end{bmatrix} = \begin{bmatrix} \cos\theta\cos\phi & \cos\theta\sin\phi & -\sin\theta \\ -\sin\phi & \cos\phi & 0 \end{bmatrix} \begin{bmatrix} E_{meas,\,x}(r) \\ E_{meas,\,y}(r) \\ E_{meas,\,z}(r) \end{bmatrix} \qquad (5.15)$$

By combining (5.14) and (5.15) one obtains

$$\begin{bmatrix} E_{meas,\,\theta}(r) \\ E_{meas,\,\phi}(r) \end{bmatrix} = \frac{1}{4\pi} \int_{S_\infty} \begin{bmatrix} H_{11}(r,r') & H_{12}(r,r') \\ H_{21}(r,r') & H_{22}(r,r') \end{bmatrix} \begin{bmatrix} M_x(r') \\ M_y(r') \end{bmatrix} ds' \qquad (5.16)$$

where

$$H_{11}(r,r') = -\cos\theta\,\sin\phi\,\frac{\partial G}{\partial z} - \sin\theta\,\frac{\partial G}{\partial y} \tag{5.17}$$

$$H_{12}(r,r') = -\cos\theta\,\cos\phi\,\frac{\partial G}{\partial z} + \sin\theta\,\frac{\partial G}{\partial x} \tag{5.18}$$

$$H_{21}(r,r') = -\cos\phi\,\frac{\partial G}{\partial z} \tag{5.19}$$

$$H_{22}(r,r') = -\sin\phi\,\frac{\partial G}{\partial z} \tag{5.20}$$

The procedure for solving this coupled integral equation using the Method of moments is described next.

5.4.2 Solution of the Integral Equation for the Magnetic Current

As stated earlier, the equivalent Magnetic current is equal to twice the value of the incident tangential electric field on the xy-plane. In the plane on which the current is located the tangential part of the electric field decays very rapidly as the observation point moves away from the aperture. For computational purposes, we need to truncate the surface S_∞ to S_0 as depicted in Figure 5.4. Therefore we assume $(E \times n)$ to be zero everywhere on the xy-plane except in the vicinity of the aperture of the antenna as shown in Figure 5.4. Hence

$$M = \begin{cases} 2\,E \times n & \text{on} \quad S_0 \\ 0 & \text{otherwise} \end{cases} \tag{5.21}$$

where S_0 is the part of the xy-plane in the vicinity of the aperture of the original antenna. The region S_0 where the equivalent magnetic currents reside is taken to be the rectangle in the xy-plane for which $-\frac{w_x}{2} \leq x \leq \frac{w_x}{2}$ and $-\frac{w_y}{2} \leq y \leq \frac{w_y}{2}$ as shown in Figure 5.5. The region S_0 is now divided into $N_x N_y$ rectangular patches, each with the same dimensions Δx and Δy which are given by

$$\Delta x = \frac{w_x}{N_x}$$
$$\Delta y = \frac{w_y}{N_y} \tag{5.22}$$

In Figure 5.5 x_i and y_j are the x- and y-coordinates of the center of the ijth patch and are given by

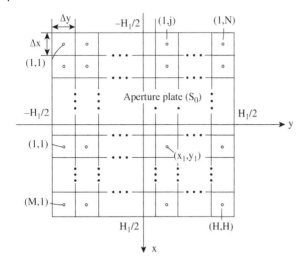

Figure 5.5 Plate S_o on the xy-plane where the equivalent magnetic current resides.

$$x_i = -\frac{w_x}{2} - \frac{\Delta x}{2} + i\,\Delta x$$

$$y_j = -\frac{w_y}{2} - \frac{\Delta y}{2} + j\,\Delta y \qquad (5.23)$$

The components of the equivalent magnetic currents M_x and M_y are approximated by equally spaced two-dimensional pulse basis functions as

$$M_x(x',y') = \sum_{i=1}^{N_x} \sum_{j=1}^{N_y} \alpha_{ij}\,\Pi_{ij}(x',y') \qquad (5.24)$$

$$M_y(x',y') = \sum_{i=1}^{N_x} \sum_{j=1}^{N_y} \beta_{ij}\,\Pi_{ij}(x',y') \qquad (5.25)$$

where α_{ij} and β_{ij} are the unknown coefficients of the x- and y-directed magnetic currents, respectively on the ijth patch. $\Pi_{ij}(x', y')$ is the two dimensional pulse basis function pertaining to the ijth patch and is defined by

$$\Pi_{ij}(x',y') = \begin{cases} 1 & \text{if} \begin{cases} x_i - \dfrac{\Delta x}{2} \le x' \le x_i + \dfrac{\Delta x}{2} \\[2mm] y_j - \dfrac{\Delta y}{2} \le y' \le y_j + \dfrac{\Delta y}{2} \end{cases} \\[6mm] 0, & \text{otherwise} \end{cases} \qquad (5.26)$$

Since it is assumed that the measured field points are somewhat away from the current carrying region S_0, the two dimensional pulse basis function may be further approximated in the integrals of (5.16) by Hertzian dipoles located at the center of the *ij*th patch. This is equivalent to utilizing a delta function as the expansion function for the source over each rectangular patch as shown in Figure 5.5. Therefore,

$$\Pi_{ij}(x', y') \approx \Delta x \, \Delta y \, \delta\left(x' - x_i; y' - y_j\right) \tag{5.27}$$

where δ is the Dirac delta function. Utilization of (5.27) in (5.24) and (5.25) leads to

$$M_x(x', y') = \sum_{i=1}^{N_x} \sum_{j=1}^{N_y} \alpha_{ij} \, \Delta x \, \Delta y \, \delta(x' - x; y' - y) \tag{5.28}$$

$$M_y(x', y') = \sum_{i=1}^{N_x} \sum_{j=1}^{N_y} \beta_{ij} \, \Delta x \, \Delta y \, \delta(x' - x; y' - y) \tag{5.29}$$

The effect of the delta function in (5.28) and (5.29) results in replacing the integrals in (5.16) by their integrals evaluated at the location of the δ-function. Since the electric near field is known at discrete points on the geometry over which it has been measured, a point matching procedure [5, 16] is chosen.

Use of (5.28) and (5.29) in (5.16) leads to

$$\begin{bmatrix} E_{meas,\theta} \\ E_{meas,\phi} \end{bmatrix} = \begin{bmatrix} H_{11} & H_{21} \\ H_{12} & H_{22} \end{bmatrix} \begin{bmatrix} M_x \\ M_y \end{bmatrix} \tag{5.30}$$

where $E_{meas,\,\theta}$ and $E_{meas,\,\phi}$ are column vectors whose elements are the θ- and ϕ-components of the electric near field, respectively, measured at discrete points. M_x and M_y are column vectors whose elements are the unknown coefficients α_{ij} and β_{ij}, respectively. H_{11}, H_{12}, H_{21}, and H_{22} are the submatrices of the entire moment matrix and are given by

$$\{H_{11}\}_{k,m} = \left\{ \cos\theta_k \, \sin\phi_k \, \frac{e^{-j\,k_0\,R_{k,m}}}{4\,\pi\,R_{k,m}^2} z_k^f \left(j\,k_0 + \frac{1}{R_{k,m}} \right) \right.$$
$$\left. + \sin\theta_k \, \frac{e^{-j\,k_0\,R_{k,m}}}{4\,\pi\,R_{k,m}^2} \left(y_k^f - y_m^s \right) \left(j\,k_0 + \frac{1}{R_{k,m}} \right) \right\} \Delta x \, \Delta y$$

$$\{H_{12}\}_{k,m} = \left\{ \cos\theta_k \, \cos\phi_k \, \frac{e^{-j\,k_0\,R_{k,m}}}{4\,\pi\,R_{k,m}^2} z_k^f \left(j\,k_0 + \frac{1}{R_{k,m}} \right) \right.$$
$$\left. + \sin\theta_k \, \frac{e^{-j\,k_0\,R_{k,m}}}{4\,\pi\,R_{k,m}^2} \left(x_k^f - x_m^s \right) \left(j\,k_0 + \frac{1}{R_{k,m}} \right) \right\} \Delta x \, \Delta y$$

$$\{H_{21}\}_{k,m} = \left\{ \cos\phi_k \, \frac{e^{-j\,k_0\,R_{k,m}}}{4\,\pi\,R_{k,m}^2} z_k^f \left(j\,k_0 + \frac{1}{R_{k,m}} \right) \right\} \Delta x \, \Delta y$$

$$\{H_{22}\}_{k,m} = \left\{ \sin\phi_k \, \frac{e^{-j\,k_0\,R_{k,m}}}{4\,\pi\,R_{k,m}^2} z_k^f \left(j\,k_0 + \frac{1}{R_{k,m}} \right) \right\} \Delta x \, \Delta y \tag{5.31}$$

where θ_k and ϕ_k are the θ and ϕ coordinates, respectively, of the kth field measuring point x_m^s and y_m^s are the x- and y-coordinates, respectively, of the mth source point. $R_{k,m}$ is the distance between the kth field point (r_k) and the mth source point (r_m^s) and is defined by

$$R_{k,m} = \sqrt{\left(x_k^f - x_m^s\right)^2 + \left(y_k^f - y_m^s\right)^2 + \left(z_k^f\right)^2} \tag{5.32}$$

The superscript f denotes the field measuring point and the superscript s denotes the source point. Note that in (5.31) the two subscripts i and j have been replaced by the single subscript m. That is, $\left(x_m^s, y_m^s\right)$ is (x_i, y_i) where i and j are determined by m. Δx and Δy represent the size of the patch on S_0 carrying the dipoles. The resulting matrix (5.30) along with (5.31), when solved, determines the elements of M_x and M_y. This matrix equation is solved using the method of total least squares with singular value decomposition [5].

Rewriting (5.30) in the following form

$$H X = Y \tag{5.33}$$

where H is the resulting impedance matrix arising from the application of the Method of Moment and in general is rectangular with dimensions $M \times N$, where M is the number of θ- and ϕ-components of the measured electric field (i.e., M is the number of equations) and N is twice the number of patches in Figure 5.5 (i.e., N is the number of unknowns). X is the $N \times 1$ unknown column vector of the elements M_x and M_y, and Y is the right hand side of the known $M \times 1$ column vector containing the measured values for the electric field.

Using the singular value decomposition, the matrix H can be decomposed into

$$H = U \Sigma V^* \tag{5.34}$$

where the matrix U is $P \times P$ and contains the right singular vectors. The superscript $*$ denotes the complex conjugate transpose of a matrix. Σ is of dimensions $P \times Q$ whose diagonal elements are the singular values as shown

$$\Sigma = \begin{bmatrix} \sigma_1 & & & & & & & \\ & \sigma_2 & & & & & & \\ & & \ddots & & & & & \\ & & & \sigma_{p'} & 0 & & & \\ & & & & 0 & & & \\ & & & & & 0 & & \\ & & O & & & & \ddots & \\ & & & & & & & 0 \end{bmatrix}_{p \times p} \tag{5.35}$$

Since U and V are unitary matrices, one obtains

$$U^{-1} = U^* \text{ and } V^{-1} = V^* \tag{5.36}$$

Hence the inverse of H can be written as

$$H^{-1} = V \Sigma^{-1} U^* \tag{5.37}$$

Therefore the solution for X is given by

$$X = H^{-1}Y = V \Sigma^{-1} U^* Y \tag{5.38}$$

Next this theory is applied to some experimental data to illustrate the performance of this novel methodology.

5.4.3   Numerical Results Utilizing the Magnetic Current

The methodology is now validated using experimentally measured data. Consider a microstrip array consisting of 32×32 uniformly distributed patch antennas located on a 1.5 m $\times$ 1.5 m surface. The near fields are measured at discrete points on a spherical surface at a distance of 1.23 m away from the antenna and at a frequency of 3.3 GHz. The data is taken every $4°$ in $< \varphi$ for $0° \leq \varphi \leq 360°$ and every $2°$ in θ for $0° \leq \theta \leq 90°$. Measurements have been performed using an open ended cylindrical WR284 wave guide fed by TE_{11} mode. The measured data was provided by Dr. Carl Stubenrauch of NIST [19].

To solve now for the unknown equivalent currents, fictitious planar surfaces on the xy-plane of several different dimensions were used to form magnetic current sheets that would radiate the appropriate far field. Also, the number of unknowns used in each experiment was varied to determine its numerical effects on the results. Figures 5.6 to 5.53 depict these results.

In Figure 5.6 a fictitious planar surface on the xy-plane of dimensions 1.5 m $\times$ 1.5 m is used to form a magnetic current sheet. This magnetic current sheet is divided into 49×49 equally spaced magnetic current patches, 0.33λ apart. Figure 5.6 compares the co-polarization characteristic of the electric far field pattern, E_φ, obtained by the equivalent current method with the result obtained analytically using the formulation described in Chapter 7 and in [20]. These analytical results are computed from a near field to far field transformation using spherical wave expansions where the fields are expanded in terms of TM and TE to r modes. The methodology will be described in details in Chapter 7. Figure 5.6 describes $20 \log_{10} |E_\varphi|$ for $\varphi = 0°$ and $-89° < \theta < 89°$. Near field data was measured on 8100 discrete points located on the surface of a hemisphere of radius 1.23 m from the center of the antenna's aperture. The cut-off value (described in Chapter 1) chosen for the normalized singular values of the moment matrix is 10^{-2}, resulting in a rank of 1628 for the moment matrix. As seen in Figure 5.6, the agreement between the two methods is good for $55° < \theta < 55°$.

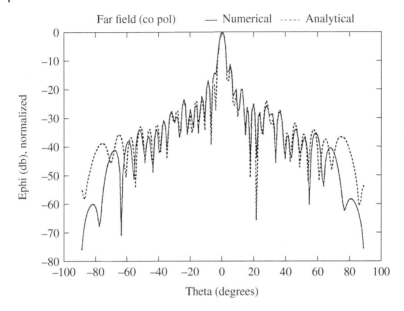

Figure 5.6 Co-polarization characteristic for $\varphi = 0°$ cut for a 32 × 32 patch microstrip array using numerical and analytical results. Number of equally spaced source patches are 49 × 49 covering 1.5 m × 1.5 m. Number of measured field points are 8100. Number of unknowns is 4802 and each patch is 0.33 λ apart. The rank of the Moment Matrix is 1628 with a cut off for the singular values to be 10^{-2}.

Figure 5.7 depicts the same co-polarization characteristics as expressed in Figure 5.6. Here, however, in order to investigate the effects that increasing the dimensions of the fictitious aperture would have on the far field patterns obtained, the dimensions of the size of the source current patch have been increased to 1.6 m × 1.6 m and all other variables such as the number of unknowns and cut-off values have remained the same. Figure 5.7 confirms that the agreement between the two methods has increased to − 59° < θ < 59°. In Figure 5.8 the dimension of the aperture has increased to 1.8 m × 1.8 m, and everything else remains the same. In Figure 5.9 the aperture's dimensions increase to 1.9 m × 1.9 m and again all other parameters are kept constant, and the agreement improves to − 65° < θ < 65°. This pattern is expected, since the electric field does not sharply jump to zero outside the aperture. Therefore, by increasing the size of the fictitious aperture, we have compensated for the lack of accuracy in the original assumption.

Figures 5.10 to 5.13 investigate the effects of increasing the size of the source current sheet on the θ-component of the co-polarization characteristic of the electric far-field for $\varphi = 90°$, and the same effects as before are observed.

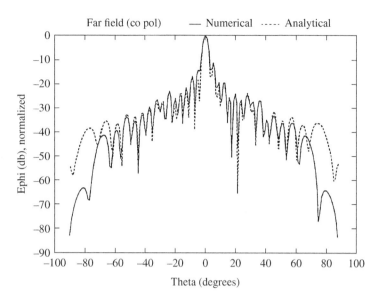

Figure 5.7 Co-polarization characteristic for $\varphi = 0°$ cut for a 32×32 patch microstrip array using numerical and analytical results. Number of equally spaced source patches are 49×49 covering 1.6 m × 1.6 m. Number of measured field points are 8100. Number of unknowns is 4802 and each patch is 0.36 λ apart. The rank of the Moment Matrix is 1821 with a cut off for the singular values to be 10^{-2}.

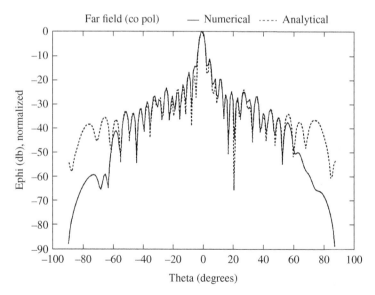

Figure 5.8 Co-polarization characteristic for $\varphi = 0°$ cut for a 32×32 patch microstrip array using numerical and analytical results. Number of equally spaced source patches are 49×49 covering 1.8 m × 1.8 m. Number of measured field points are 8100. Number of unknowns is 4802 and each patch is 0.40 λ apart. The rank of the Moment Matrix is 2198 with a cut off for the singular values to be 10^{-2}.

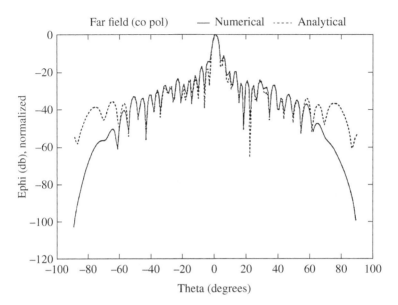

Figure 5.9 Co-polarization characteristic for $\varphi = 0°$ cut for a 32 × 32 patch microstrip array using numerical and analytical results. Number of equally spaced source patches are 49 × 49 covering 1.9 m × 1.9 m. Number of measured field points are 8100. Number of unknowns is 4802 and each patch is 0.43 λ apart. The rank of the Moment Matrix is 2375 with a cut off for the singular values to be 10^{-2}.

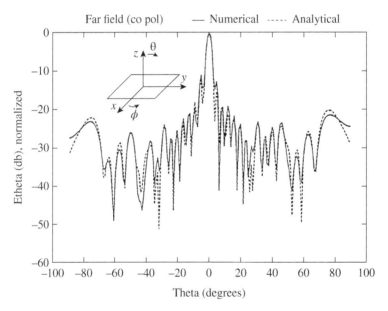

Figure 5.10 Co-polarization characteristic for $\varphi = 90°$ cut for a 32 × 32 patch microstrip array using numerical and analytical results. Number of equally spaced source patches are 49 × 49 covering 1.5 m × 1.5 m. Number of unknowns is 4802 and each patch is 0.33 λ apart. The rank of the Moment Matrix is 1628 with a cut off for the singular values to be 10^{-2}.

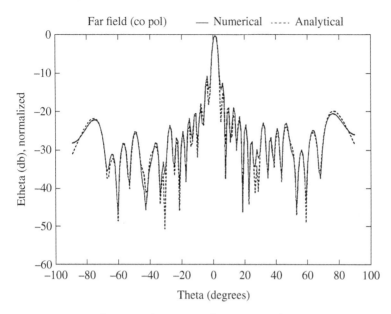

Figure 5.11 Co-polarization characteristic for $\varphi = 90°$ cut for a 32×32 patch microstrip array using numerical and analytical results. Number of equally spaced source patches are 49×49 covering 1.6 m × 1.6 m. Number of unknowns is 4802 and each patch is 0.36 λ apart. The rank of the Moment Matrix is 1821 with a cut off for the singular values to be 10^{-2}.

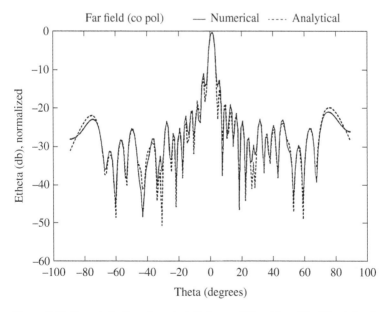

Figure 5.12 Co-polarization characteristic for $\varphi = 90°$ cut for a 32×32 patch microstrip array using numerical and analytical results. Number of equally spaced source patches are 49×49 covering 1.8 m × 1.8 m. Number of unknowns is 4802 and each patch is 0.40 λ apart. The rank of the Moment Matrix is 2198 with a cut off for the singular values to be 10^{-2}.

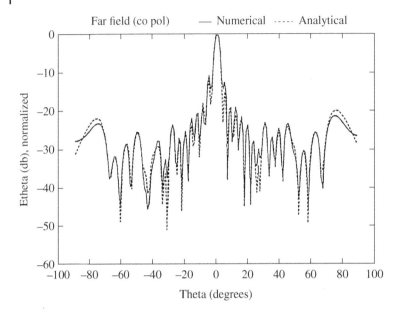

Figure 5.13 Co-polarization characteristic for $\varphi = 90°$ cut for a 32 × 32 patch microstrip array using numerical and analytical results. Number of equally spaced source patches are 49 × 49 covering 1.9 m × 1.9 m. Number of unknowns is 4802 and each patch is 0.42 λ apart. The rank of the Moment Matrix is 2375 with a cut off for the singular values to be 10^{-2}.

Figure 5.14 depicts the cross-polarization characteristic of the far field patterns for the same experiment as described in Figure 5.6. Figure 5.14 describes $20 \log_{10} |E_\theta|$ for $\varphi = 0°$ and $-89° < \theta < 89°$. Figures 5.15 through 5.17 describe the effects of increasing the dimensions of the antenna's aperture and keeping all other variables constant on the far field patterns. Figures 5.18 through 5.21 describe $20 \log_{10} |E_\varphi|$ for $\varphi = 90°$ of the cross-polarization characteristic of the far field pattern. Again it is observed that enlarging the size of the source plane improves the agreement between the numerical and the analytical approaches.

We now demonstrate how changing the cut-off values of the normalized singular values in the moment matrix affects the far field patterns obtained. The dimensions of the aperture and the number of unknowns will remain the same and the cut-off values will vary. In Figure 5.22 the source planar surface on the xy-plane of dimensions 1.9 m × 1.9 m is used. A magnetic current sheet is located on it. This magnetic current sheet is divided into 49 × 49 equally spaced current patches, 0.42λ, apart containing two components of the magnetic current. This figure compares the co-polarization characteristic of the electric far field pattern, E_φ, obtained by the present method with the result obtained analytically in Chapter 7 [20]. Figure 5.22 describes $20 \log_{10} |E_\varphi|$ for $\varphi = 0°$ and $-89° < \theta < 89°$. Near-field data was measured at 8100 discrete points located

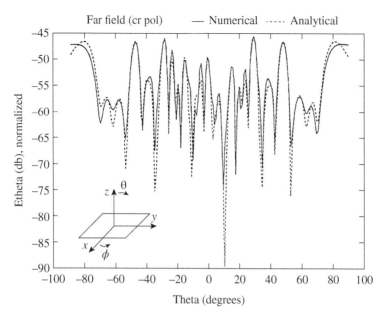

Figure 5.14 Cross-polarization characteristic for $\varphi = 0°$ cut for a 32 × 32 patch microstrip array using numerical and analytical results. Number of equally spaced source patches are 49 × 49 covering 1.5 m × 1.5 m. Number of unknowns is 4802 and each patch is 0.33 λ apart. The rank of the Moment Matrix is 1628 with a cut off for the singular values to be 10^{-2}.

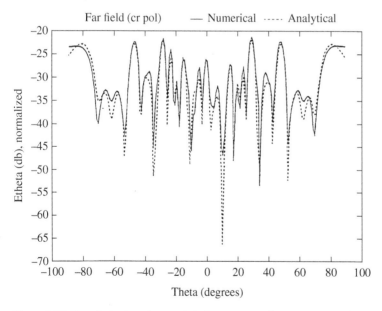

Figure 5.15 Co-polarization characteristic for $\varphi = 0°$ cut for a 32 × 32 patch microstrip array using numerical and analytical results. Number of equally spaced source patches are 49 × 49 covering 1.6 m × 1.6 m. Number of unknowns is 4802 and each patch is 0.36 λ apart. The rank of the Moment Matrix is 1821 with a cut off for the singular values to be 10^{-2}.

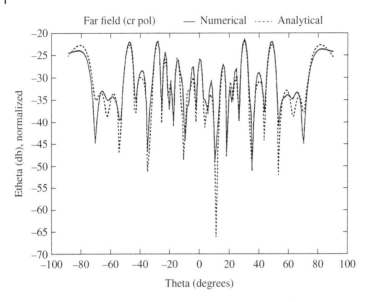

Figure 5.16 Cross-polarization characteristic for $\varphi = 0°$ cut for a 32 × 32 patch microstrip array using numerical and analytical results. Number of equally spaced source patches are 49 × 49 covering 1.8 m × 1.8 m. Number of unknowns is 4802 and each patch is 0.40 λ apart. The rank of the Moment Matrix is 2198 with a cut off for the singular values to be 10^{-2}.

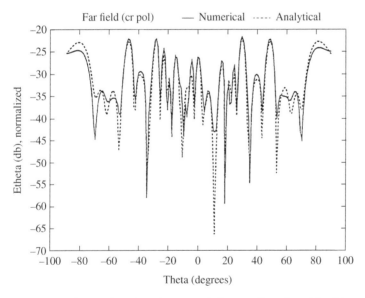

Figure 5.17 Cross-polarization characteristic for $\varphi = 0°$ cut for a 32 × 32 patch microstrip array using numerical and analytical results. Number of equally spaced source patches are 49 × 49 covering 1.9 m × 1.9 m. Number of unknowns is 4802 and each patch is 0.43 λ apart. The rank of the Moment Matrix is 2375 with a cut off for the singular values to be 10^{-2}.

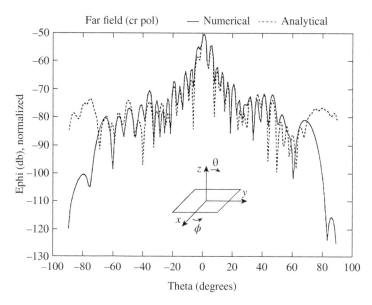

Figure 5.18 Cross-polarization characteristic for $\varphi = 90°$ cut for a 32×32 patch microstrip array using numerical and analytical results. Number of equally spaced source patches are 49×49 covering 1.5 m $\times$ 1.5 m. Number of unknowns is 4802 and each patch is 0.33 λ apart. The rank of the Moment Matrix is 1628 with a cut off for the singular values to be 10^{-2}.

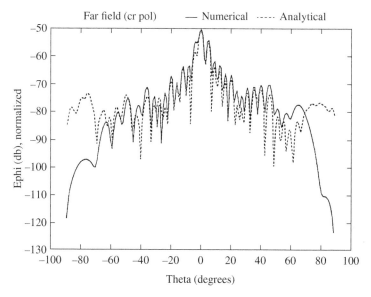

Figure 5.19 Cross-polarization characteristic for $\varphi = 90°$ cut for a 32×32 patch microstrip array using numerical and analytical results. Number of equally spaced source patches are 49×49 covering 1.6 m $\times$ 1.6 m. Number of unknowns is 4802 and each patch is 0.36 λ apart. The rank of the Moment Matrix is 1821 with a cut off for the singular values to be 10^{-2}.

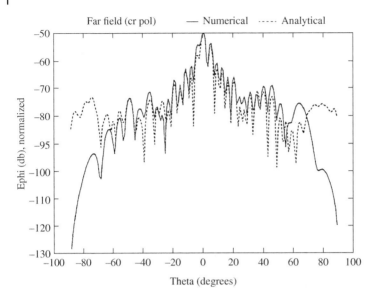

Figure 5.20 Cross-polarization characteristic for $\varphi = 90°$ cut for a 32 × 32 patch microstrip array using numerical and analytical results. Number of equally spaced source patches are 49 × 49 covering 1.8 m × 1.8 m. Number of unknowns is 4802 and each patch is 0.40 λ apart. The rank of the Moment Matrix is 2198 with a cut off for the singular values to be 10^{-2}.

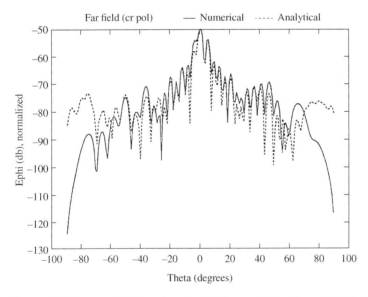

Figure 5.21 Cross-polarization characteristic for $\varphi = 90°$ cut for a 32 × 32 patch microstrip array using numerical and analytical results. Number of equally spaced source patches are 49 × 49 covering 1.9 m × 1.9 m. Number of unknowns is 4802 and each patch is 0.43 λ apart. The rank of the Moment Matrix is 2375 with a cut off for the singular values to be 10^{-2}.

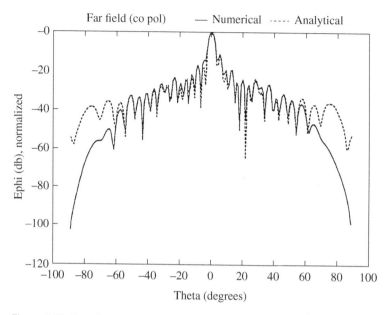

Figure 5.22 Co-polarization characteristic for $\varphi = 0°$ cut for a 32 × 32 patch microstrip array using numerical and analytical results. Number of equally spaced source patches are 49 × 49 covering 1.9 m × 1.9 m. Number of unknowns is 4802 and each patch is 0.40 λ apart. The rank of the Moment Matrix is 2375 with a cut off for the singular values to be 10^{-2}.

on the surface of a hemisphere of radius 1.23 m measured from the center of the antenna array. The cut-off value chosen for the normalized singular values of the moment matrix was 10^{-2} resulting in a rank of 1628 for the moment matrix. In Figure 5.23 the result of the same experiment is depicted except that the cut-off value is chosen to be 10^{-3}, and in Figure 5.24 the cut-off value of 10^{-4} is chosen. It is observed that best agreement between our numerical and analytical methods are achieved when the cut-off value is 10^{-3}. In Figures 5.25 through 5.27 the co-polarization characteristics of the electric far field patterns, E_θ for $\varphi = 90°$ are depicted. In these figures again the effects that reducing the cut-off of the normalized singular values would have on the E_θ component of the far field pattern is investigated. It is again verified that the cut-off value of 10^{-3} results in closest agreement between the numerical and analytical methods. Figures 5.28 through 5.33 depict experiments investigating effects that reducing the cut-off values have on the cross-polarization characteristics of the far field patterns. It is observed that for the cut-off value of 10^{-3} the agreement between numerical and analytical approaches is good for far field values as low as – 70 db.

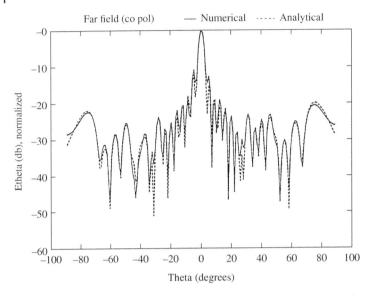

Figure 5.23 Co-polarization characteristic for $\varphi = 0°$ cut for a 32 × 32 patch microstrip array using numerical and analytical results. Number of equally spaced source patches are 49 × 49 covering 1.9 m × 1.9 m. Number of unknowns is 4802 and each patch is 0.4 λ apart. The rank of the Moment Matrix is 2663 with a cut off for the singular values to be 10^{-3}.

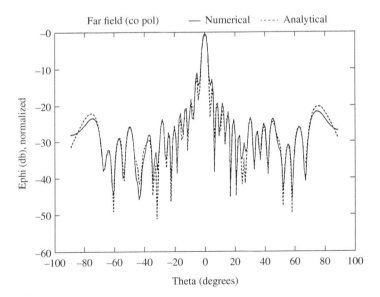

Figure 5.24 Co-polarization characteristic for $\varphi = 0°$ cut for a 32 × 32 patch microstrip array using numerical and analytical results. Number of equally spaced source patches are 49 × 49 covering 1.9 m × 1.9 m. Number of unknowns is 4802 and each patch is 0.43 λ apart. The rank of the Moment Matrix is 2790 with a cut off for the singular values to be 10^{-4}.

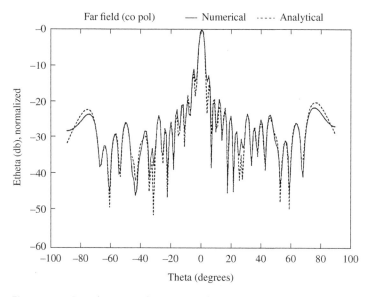

Figure 5.25 Co-polarization characteristic for $\varphi = 90°$ cut for a 32 × 32 patch microstrip array using numerical and analytical results. Number of equally spaced source patches are 49 × 49 covering 1.9 m × 1.9 m. Number of unknowns is 4802 and each patch is 0.4 λ apart. The rank of the Moment Matrix is 2375 with a cut off for the singular values to be 10^{-2}.

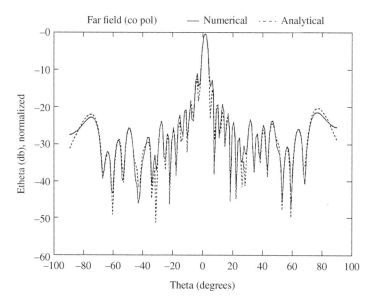

Figure 5.26 Co-polarization characteristic for $\varphi = 90°$ cut for a 32 × 32 patch microstrip array using numerical and analytical results. Number of equally spaced source patches are 49 × 49 covering 1.9 m × 1.9 m. Number of unknowns is 4802 and each patch is 0.4 λ apart. The rank of the Moment Matrix is 2663 with a cut off for the singular values to be 10^{-3}.

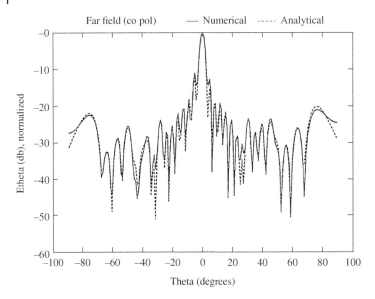

Figure 5.27 Co-polarization characteristic for $\varphi = 90°$ cut for a 32×32 patch microstrip array using numerical and analytical results. Number of equally spaced source patches are 49×49 covering 1.9 m × 1.9 m. Number of unknowns is 4802 and each patch is 0.4 λ apart. The rank of the Moment Matrix is 2790 with a cut off for the singular values to be 10^{-4}.

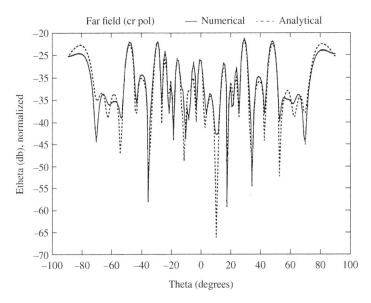

Figure 5.28 Cross-polarization characteristic for $\varphi = 0°$ cut for a 32×32 patch microstrip array using numerical and analytical results. Number of equally spaced source patches are 49×49 covering 1.9 m × 1.9 m. Number of unknowns is 4802 and each patch is 0.40 λ apart. The rank of the Moment Matrix is 2375 with a cut off for the singular values to be 10^{-2}.

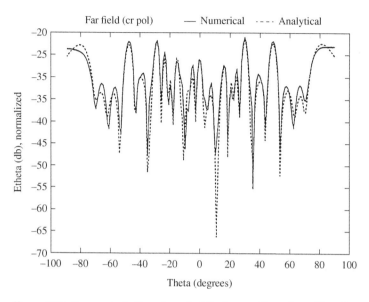

Figure 5.29 Cross-polarization characteristic for $\varphi = 0°$ cut for a 32 × 32 patch microstrip array using numerical and analytical results. Number of equally spaced source patches are 49 × 49 covering 1.9 m × 1.9 m. Number of unknowns is 4802 and each patch is 0.4 λ apart. The rank of the Moment Matrix is 2663 with a cut off for the singular values to be 10^{-3}.

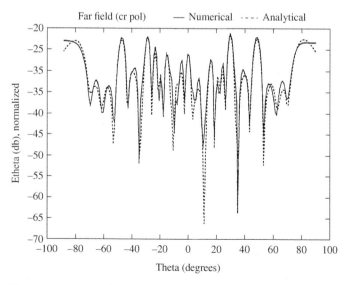

Figure 5.30 Cross-polarization characteristic for $\varphi = 0°$ cut for a 32 × 32 patch microstrip array using numerical and analytical results. Number of equally spaced source patches are 49 × 49 covering 1.9 m × 1.9 m. Number of unknowns is 4802 and each patch is 0.4 λ apart. The rank of the Moment Matrix is 2790 with a cut off for the singular values to be 10^{-4}.

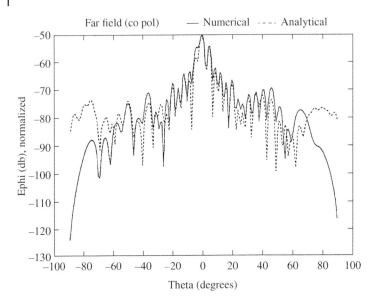

Figure 5.31 Cross-polarization characteristic for $\varphi = 90°$ cut for a 32 × 32 patch microstrip array using numerical and analytical results. Number of equally spaced source patches are 49 × 49 covering 1.9 m × 1.9 m. Number of unknowns is 4802 and each patch is 0.4 λ apart. The rank of the Moment Matrix is 2375 with a cut off for the singular values to be 10^{-2}.

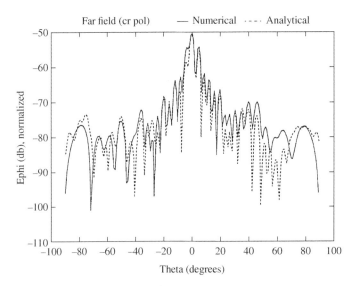

Figure 5.32 Cross-polarization characteristic for $\varphi = 90°$ cut for a 32 × 32 patch microstrip array using numerical and analytical results. Number of equally spaced source patches are 49 × 49 covering 1.9 m × 1.9 m. Number of unknowns is 4802 and each patch is 0.40 λ apart. The rank of the Moment Matrix is 2663 with a cut off for the singular values to be 10^{-3}.

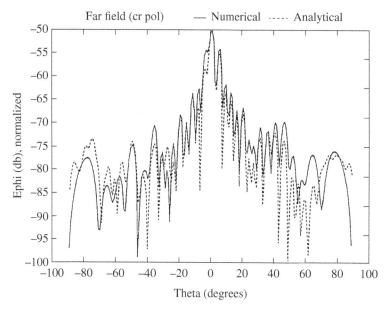

Figure 5.33 Cross-polarization characteristic for $\varphi = 90°$ cut for a 32×32 patch microstrip array using numerical and analytical results. Number of equally spaced source patches are 49×49 covering 1.9 m $\times$ 1.9 m. Number of unknowns is 4802 and each patch is 0.4 λ apart. The rank of the Moment Matrix is 2790 with a cut off for the singular values to be 10^{-4}.

Next the effects of increasing the number of unknowns on the far field patterns are investigated. These experiments are depicted in Figures 5.34 through 5.53. In Figure 5.34 a fictitious planar surface on the xy-plane of dimensions 1.9 m $\times$ 1.9 m is used to form a magnetic current sheet. This magnetic current sheet is divided into 3 8 $\times$ 3 8 equally spaced magnetic current patches, 0.55λ apart. This figure compares the co-polarization characteristic of the electric far field pattern, E_φ, obtained by the present method, with the result obtained analytically in Chapter 7. Figure 5.34 describes $20 \log_{10} |E_\varphi|$ for $\varphi = 0°$ and $-89° < \theta < 89°$. Near-field data was measured at 8100 discrete points located on the surface of a hemisphere at a radius of 1.23 m from the center of the antenna's aperture. The cut-off value chosen for the normalized singular values of the moment matrix was 10^{-3} resulting in a rank of 1187 for the moment matrix. In Figure 5.35, results for the same experiment are depicted, except that the fictitious planar surface simulating the aperture of the original antenna is divided into 44 $\times$ 44 equally spaced patches. In Figure 5.36 the number of unknowns is increased to 45 $\times$ 45 patches and in Figure 5.37 to 49 $\times$ 49. Best result was obtained in Figure 5.38 where the number of unknowns chosen were 55 $\times$ 55 equally spaced patches. We expect the result to improve in accuracy by

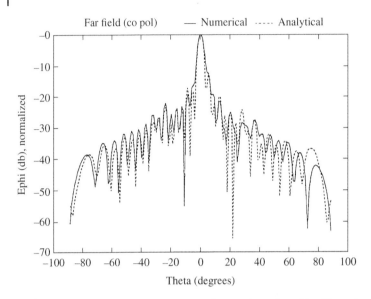

Figure 5.34 Co-polarization characteristic for $\varphi = 0°$ cut for a 32 × 32 patch microstrip array using numerical and analytical results. Number of equally spaced source patches are 38 × 38 covering 1.9 m × 1.9 m. Number of unknowns is 4802 and each patch is 0.55 λ apart. The rank of the Moment Matrix is 1187 with a cut off for the singular values to be 10^{-3}.

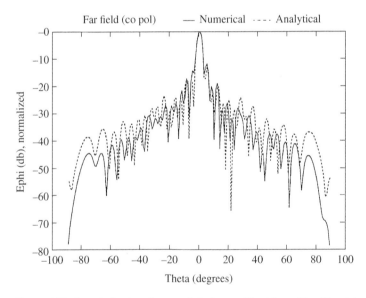

Figure 5.35 Co-polarization characteristic for $\varphi = 0°$ cut for a 32 × 32 patch microstrip array using numerical and analytical results. Number of equally spaced source patches are 44 × 44 covering 1.9 m × 1.9 m. Number of unknowns is 4802 and each patch is 0.475 λ apart. The rank of the Moment Matrix is 2017 with a cut off for the singular values to be 10^{-3}.

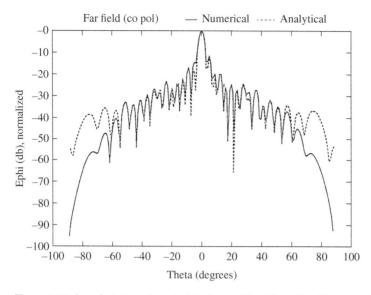

Figure 5.36 Co-polarization characteristic for $\varphi = 0°$ cut for a 32 × 32 patch microstrip array using numerical and analytical results. Number of equally spaced source patches are 45 × 45 covering 1.9 m × 1.9 m. Number of unknowns is 4802 and each patch is 0.46 λ apart. The rank of the Moment Matrix is 2968 with a cut off for the singular values to be 10^{-3}.

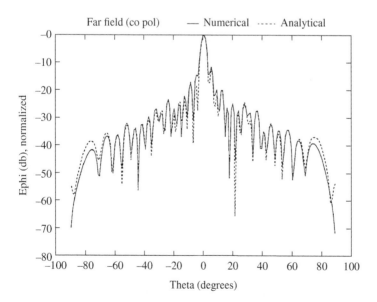

Figure 5.37 Co-polarization characteristic for $\varphi = 0°$ cut for a 32 × 32 patch microstrip array using numerical and analytical results. Number of equally spaced source patches are 49 × 49 covering 1.9 m × 1.9 m. Number of unknowns is 4802 and each patch is 0.43 λ apart. The rank of the Moment Matrix is 2663 with a cut off for the singular values to be 10^{-3}.

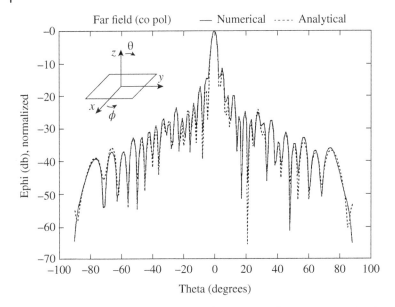

Figure 5.38 Co-polarization characteristic for $\varphi = 0°$ cut for a 32 × 32 patch microstrip array using numerical and analytical results. Number of equally spaced source patches are 55 × 55 covering 1.9 m × 1.9 m. Number of unknowns is 4802 and each patch is 0.38 λ apart. The rank of the Moment Matrix is 3033 with a cut off for the singular values to be 10^{-3}.

increasing the number of unknowns, since the more unknowns used, the more accurate simulation of the original antenna is possible. Figure 5.39 through 5.53 depict other co-polarization and cross-polarization characteristics of the far field patterns. Again it is observed that increasing the number of unknowns improves the final result.

5.4.4 Summary

The method presented here determines the fields for $z > 0$ in front of the radiating antenna simply from the knowledge of the near field on any arbitrary oriented geometry in space. Using various antenna configurations and near-field geometries, an investigation of the accuracy of this method has been performed. For cases where actual experimental sources have been used, the far fields were compared with a semi-analytical approach and the results look extremely promising.

Next we study the use of an equivalent electric current.

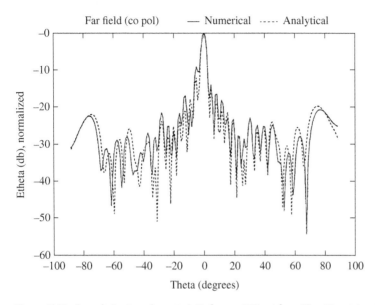

Figure 5.39 Co-polarization characteristic for $\varphi = 90°$ cut for a 32×32 patch microstrip array using numerical and analytical results. Number of equally spaced source patches are 38×38 covering 1.9 m $\times$ 1.9 m. Number of unknowns is 4802 and each patch is 0.55 λ apart. The rank of the Moment Matrix is 1187 with a cut off for the singular values to be 10^{-3}.

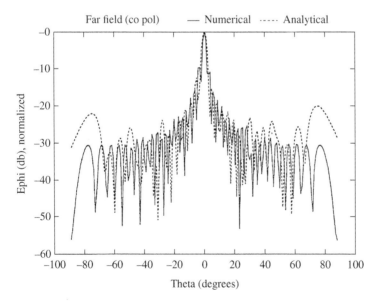

Figure 5.40 Co-polarization characteristic for $\varphi = 90°$ cut for a 32×32 patch microstrip array using numerical and analytical results. Number of equally spaced source patches are 44×44 covering 1.9 m $\times$ 1.9 m. Number of unknowns is 4802 and each patch is 0.475 λ apart. The rank of the Moment Matrix is 2017 with a cut off for the singular values to be 10^{-3}.

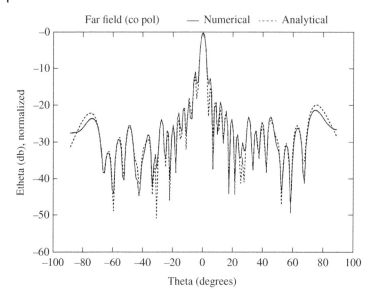

Figure 5.41 Co-polarization characteristic for $\varphi = 90°$ cut for a 32 × 32 patch microstrip array using numerical and analytical results. Number of equally spaced source patches are 45 × 45 covering 1.9 m × 1.9 m. Number of unknowns is 4802 and each patch is 0.46 λ apart. The rank of the Moment Matrix is 2968 with a cut off for the singular values to be 10^{-3}.

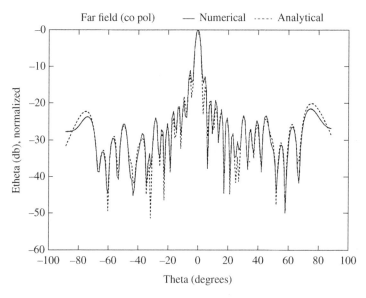

Figure 5.42 Co-polarization characteristic for $\varphi = 90°$ cut for a 32 × 32 patch microstrip array using numerical and analytical results. Number of equally spaced source patches are 49 × 49 covering 1.9 m × 1.9 m. Number of unknowns is 4802 and each patch is 0.43 λ apart. The rank of the Moment Matrix is 2663 with a cut off for the singular values to be 10^{-3}.

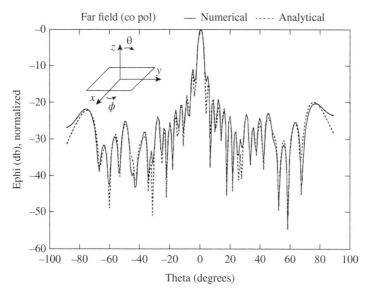

Figure 5.43 Co-polarization characteristic for $\varphi = 90°$ cut for a 32×32 patch microstrip array using numerical and analytical results. Number of equally spaced source patches are 55×55 covering 1.9 m $\times$ 1.9 m. Number of unknowns is 4802 and each patch is 0.38 λ apart. The rank of the Moment Matrix is 3033 with a cut off for the singular values to be 10^{-3}.

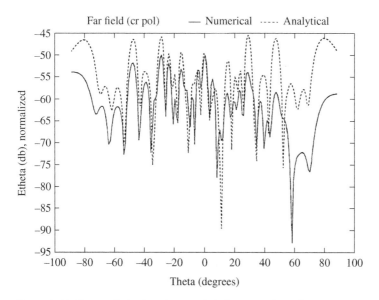

Figure 5.44 Cross-polarization characteristic for $\varphi = 90°$ cut for a 32×32 patch microstrip array using numerical and analytical results. Number of equally spaced source patches are 38×38 covering 1.9 m $\times$ 1.9 m. Number of unknowns is 4802 and each patch is 0.55 λ apart. The rank of the Moment Matrix is 1187 with a cut off for the singular values to be 10^{-3}.

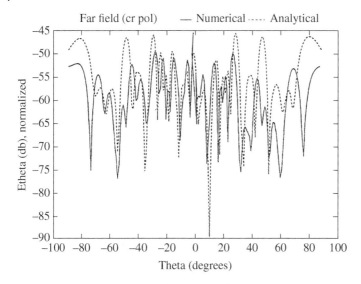

Figure 5.45 Cross-polarization characteristic for $\varphi = 90°$ cut for a 32×32 patch microstrip array using numerical and analytical results. Number of equally spaced source patches are 44×44 covering 1.9 m $\times 1.9$ m. Number of unknowns is 4802 and each patch is 0.475 λ apart. The rank of the Moment Matrix is 2017 with a cut off for the singular values to be 10^{-3}.

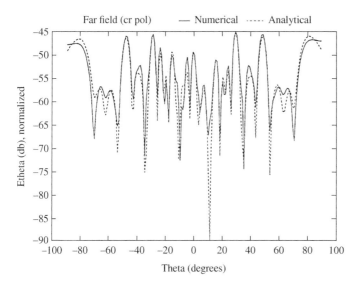

Figure 5.46 Cross-polarization characteristic for $\varphi = 90°$ cut for a 32×32 patch microstrip array using numerical and analytical results. Number of equally spaced source patches are 45×45 covering 1.9 m $\times 1.9$ m. Number of unknowns is 4802 and each patch is 0.46 λ apart. The rank of the Moment Matrix is 2968 with a cut off for the singular values to be 10^{-3}.

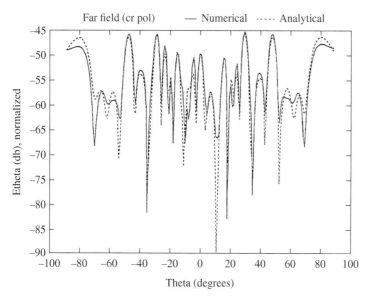

Figure 5.47 Cross-polarization characteristic for $\varphi = 90°$ cut for a 32×32 patch microstrip array using numerical and analytical results. Number of equally spaced source patches are 49×49 covering 1.9 m $\times$ 1.9 m. Number of unknowns is 4802 and each patch is 0.43 λ apart. The rank of the Moment Matrix is 2663 with a cut off for the singular values to be 10^{-3}.

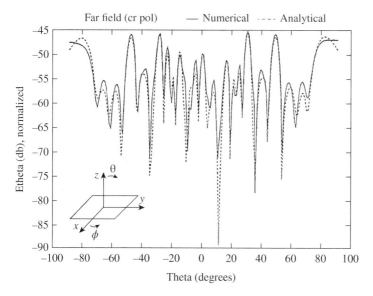

Figure 5.48 Cross-polarization characteristic for $\varphi = 90°$ cut for a 32×32 patch microstrip array using numerical and analytical results. Number of equally spaced source patches are 55×55 covering 1.9 m $\times$ 1.9 m. Number of unknowns is 4802 and each patch is 0.38 λ apart. The rank of the Moment Matrix is 3033 with a cut off for the singular values to be 10^{-3}.

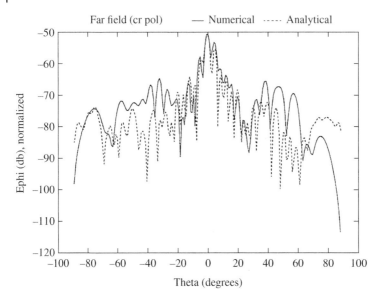

Figure 5.49 Cross-polarization characteristic for $\varphi = 90°$ cut for a 32 × 32 patch microstrip array using numerical and analytical results. Number of equally spaced source patches are 38 × 38 covering 1.9 m × 1.9 m. Number of unknowns is 4802 and each patch is 0.55 λ apart. The rank of the Moment Matrix is 1187 with a cut off for the singular values to be 10^{-3}.

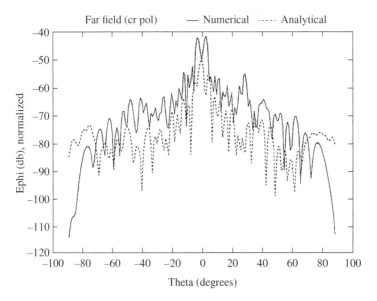

Figure 5.50 Cross-polarization characteristic for $\varphi = 90°$ cut for a 32 × 32 patch microstrip array using numerical and analytical results. Number of equally spaced source patches are 44 × 44 covering 1.9 m × 1.9 m. Number of unknowns is 4802 and each patch is 0.475 λ apart. The rank of the Moment Matrix is 2017 with a cut off for the singular values to be 10^{-3}.

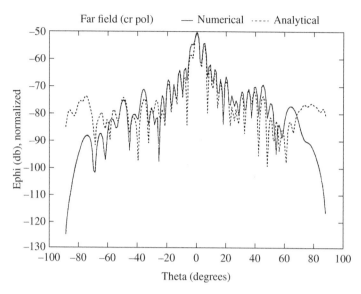

Figure 5.51 Cross-polarization characteristic for $\varphi = 90°$ cut for a 32 × 32 patch microstrip array using numerical and analytical results. Number of equally spaced source patches are 45 × 45 covering 1.9 m × 1.9 m. Number of unknowns is 4802 and each patch is 0.46 λ apart. The rank of the Moment Matrix is 2968 with a cut off for the singular values to be 10^{-3}.

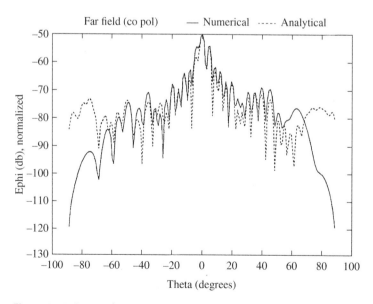

Figure 5.52 Cross-polarization characteristic for $\varphi = 90°$ cut for a 32 × 32 patch microstrip array using numerical and analytical results. Number of equally spaced source patches are 49 × 49 covering 1.9 m × 1.9 m. Number of unknowns is 4802 and each patch is 0.43 λ apart. The rank of the Moment Matrix is 2663 with a cut off for the singular values to be 10^{-3}.

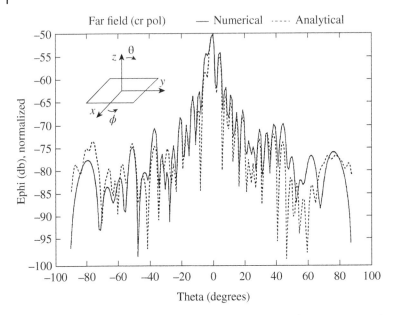

Figure 5.53 Cross-polarization characteristic for $\varphi = 90°$ cut for a 32×32 patch microstrip array using numerical and analytical results. Number of equally spaced source patches are 55×55 covering 1.9 m $\times$ 1.9 m. Number of unknowns is 4802 and each patch is 0.38 λ apart. The rank of the Moment Matrix is 3033 with a cut off for the singular values to be 10^{-3}.

5.5 Near-Field to Near/Far-Field Transformation for Arbitrary Near-Field Geometry Utilizing an Equivalent Electric Current

Presented here is a method for computing near- and far-field patterns of an antenna from its near-field measurements taken over an arbitrarily shaped geometry utilizing equivalent electric currents [21, 22]. This method utilizes near-field data to determine an equivalent electric current source over a fictitious surface which encompasses the antenna. This electric current, once determined, can be used to ascertain the near and the far field. This method demonstrates the concept of analytic continuity, i.e., once the value of the electric field is known for one region in space, from a theoretical perspective, its value for any other region can be extrapolated. It is shown that the equivalent electric current produces the correct fields in the regions in front of the antenna regardless of the geometry over which the near-field measurements are made. In this approach, the measured data need not satisfy the Nyquist sampling criteria, namely measurements are carried out at half a wavelength apart in space. An electric field integral equation is developed to relate the near field to the equivalent electric current. A moment method procedure is employed to solve the integral equation by transforming it into a matrix equation. A least-squares solution via singular value decomposition is used to solve the matrix equation. Computations with both synthetic and experimental data, where the near field of several antenna configurations are measured over various geometrical surfaces, illustrate the accuracy of this method.

In this approach, by using the equivalence principle [14, 15], an equivalent electric current replaces the radiating antenna. Furthermore, it is assumed that the near field is produced by the equivalent electric current and, therefore, via Maxwell's equation from the measured near-field data, the current source can be determined. Once this is accomplished, the near field and the far field of the radiating antenna in all regions in space in front of the radiating antenna can be determined directly from the equivalent electric current. It is consequently shown that from the knowledge of the field in one region of space measured over any geometry, the field values for any other region can be extrapolated, therefore confirming the concept of analytic continuity.

An electric field integral equation is developed which pertains to the measured near fields and the equivalent electric current. This integral equation has been solved for the unknown electric current source through a moment method procedure [16] with point matching, where the equivalent current is expanded as linear combinations of 2-D pulse basis functions and, therefore, the integral equation is then transformed into a matrix equation. In general, the matrix is rectangular whose dimensions depend on the number of field and source points chosen. The matrix equation is solved by the moment matrix, which is decomposed into a set of orthogonal matrices through the singular value decomposition. The set of orthogonal matrices can easily be inverted.

Another aspect of this approach is that the numerical integrations in the process of creating the moment matrix elements have been avoided by taking a limiting case. Since the field points and the source points are to never coincide and if their distances are much larger than the sizes of the current patches, then the pulse-basis functions expanding the current source can be approximated by the Hertzian dipoles.

The formulation and the theoretical basis for the equivalent electric current approach along with the formation of the corresponding matrix equation using the method of moments and its solution using the method of least squares via singular value decomposition is presented in the following sections. Near-field to far-field transformations are generally used for the far-field characterization of electrically large arrays and particularly phased arrays. The near-field measurement of a large antenna is not only very time consuming, but cumbersome because of measuring the fields over certain large canonical surfaces with high spatial accuracy. What this method provides is a methodology not only to minimize the number of measured data samples that needs to be collected from a measurement but also shows that it is not necessary to compute the fields over an entire canonical surface - imagine measuring over a cylindrical/spherical near-field measurement to deal with a 10-m-long antenna with submillimeter spatial accuracy! For planar near-field to far-field measurements, the method presented in this paper may be more efficient and accurate than conventional techniques as documented in the various references [12, 21, 22]. Also, in this method, the measured field points need not satisfy the Nyquist sampling criteria in space. Hence, this methodology is quite suitable for pattern measurements of large phased arrays.

5.5.1 Description of the Proposed Methodology

Consider an arbitrary-shaped antenna, as shown in Figure 5.3, which radiates into free-space. The aperture of the antenna is in a surface which separates all space into left-half and right-half spaces. The aperture of the antenna is placed in the plane and is facing the positive z axis. Since we are interested only in the electromagnetic field in the region where $z > 0$, we place a perfect magnetic conductor in front of the radiating antenna extending to infinity in the x and y directions on the xy plane of the antenna. This is denoted by S_∞. By the equivalence principle [14, 15], a surface electric current J' can be placed on a perfect magnetic conductor covering the aperture of the antenna. The value of J' is equal to the tangential value of the magnetic field on the xy-plane (same as in Figure 5.3 with the magnetic current replaced by an electric current)

$$J' = \hat{n} \times H \text{ on } S_\infty \qquad (5.39)$$

where H is the magnetic field on the xy-plane and $\hat{n}$ is the unit outward normal to the xy-plane pointing in the direction of the positive z-axis.

Using image theory [13], an electric current J radiating in free space (as seen in Figure 5.4 with the magnetic current replaced by an electric current), may be introduced on the xy-plane whose value is

$$J = 2J' \qquad (5.40)$$

Therefore

$$\hat{J} = 2\hat{n} \times \hat{H} \text{ on } S_\infty \qquad (5.41)$$

radiating in free-space (Figure 5.4 with the magnetic current replaced by the electric current) and producing exactly the same field as the original antenna in the region $z \geq 0$. Now the measured electric near field may be used to determine J from

$$E_{meas} = E(J) \qquad (5.42)$$

where E_{meas} is the electric near field measured over a geometry at a distance away from the aperture of the radiating antenna.

From Maxwell's equations, the electric field operator is expressed in terms of its electric current source in a homogeneous medium as

$$E(J) = -j \omega \mu A(J) + \frac{1}{j \omega \varepsilon} \nabla (\nabla \cdot A(J))$$ (5.43)

where $A(J)$ is the magnetic vector potential defined as

$$A(J) = -\frac{1}{4 \pi} \int_{S_\infty} J(r') G(r, r') ds'$$ (5.44)

where

$$\nabla \cdot = \frac{\partial}{\partial x} \hat{x} + \frac{\partial}{\partial y} \hat{y} + \frac{\partial}{\partial z} \hat{z}$$ (5.45)

and

$$G(r, r') = \frac{e^{-j k_0 |r - r'|}}{|r - r'|} = \frac{e^{-j k_0 R}}{R}$$ (5.46)

The primed variables correspond to the spatial locations of the source and the unprimed variables correspond to the locations of the field points. The region S_0 is where the equivalent electric current resides and is taken to be a rectangle in the xy-plane for which $- w_x/2 \leq x \leq w_x/2$ and $- w_y/2 \leq y \leq w_y/2$. S_0 is now divided into $N_x \times N_y$ equally spaced rectangular patches with dimensions Δx and Δy (as shown in Figure 5.5) given by

$$x_i = -\frac{w_x}{2} - \frac{\Delta x}{2} + i \Delta x$$
$$y_j = -\frac{w_y}{2} - \frac{\Delta y}{2} + j \Delta y$$ (5.47)

The components of the equivalent electric current J_x and J_y are approximated by equally spaced two-dimensional pulse basis functions as was done in the last section:

$$J_x(x', y') = \sum_{i=1}^{N_x} \sum_{j=1}^{N_y} \gamma_{ij} \Pi_{ij}(x', y')$$ (5.48)

$$J_y(x', y') = \sum_{i=1}^{N_x} \sum_{j=1}^{N_y} \zeta_{ij} \Pi_{ij}(x', y')$$ (5.49)

where γ_{ij} and ζ_{ij} are the unknown coefficients of the x- and the y-directed electric currents, respectively on the ijth patch. $\Pi_{ij}(x', y')$ is the two dimensional pulse basis function pertaining to the ijth patch, and was defined previously in equation (5.26) and is repeated here for convenience.

$$\Pi_{ij}(x',y') = \begin{cases} 1 & \text{if} \begin{cases} x_i - \dfrac{\Delta x}{2} \leq x' \leq x_i + \dfrac{\Delta x}{2} \\ y_j - \dfrac{\Delta y}{2} \leq y' \leq y_j + \dfrac{\Delta y}{2} \end{cases} \\ 0, & \text{otherwise} \end{cases}$$

Since it is assumed that the measured field points are somewhat away from the current carrying region S_0, the two dimensional pulse basis function may be further approximated in the integrals of (5.44) by Hertzian dipoles located at the center of the ijth patch. This is equivalent to utilizing a delta function as the expansion function for the source over each rectangular patch as shown in Figure 5.5. Therefore,

$$\Pi_{ij}(x',y') \approx \Delta x\, \Delta y\, \delta\left(x' - x_i; y' - y_j\right) \tag{5.50}$$

When (5.50) is substituted in (5.48) and (5.49), one obtains

$$J_x(x',y') = \sum_{i=1}^{N_x} \sum_{j=1}^{N_y} \gamma_{ij}\, \Delta x\, \Delta y\, \delta\left(x' - x_i; y' - y_j\right) \tag{5.51}$$

$$J_y(x',y') = \sum_{i=1}^{N_x} \sum_{j=1}^{N_y} \zeta_{ij}\, \Delta x\, \Delta y\, \delta\left(x' - x_i; y' - y_j\right) \tag{5.52}$$

The effect of the delta function in (5.51) and (5.52) results in replacing the integrals in (5.44) by their integrals evaluated at the location of the δ-function. Since the electric near field is known at discrete points on the geometry over which it has been measured, a point matching procedure [16] is chosen. Use of (5.51) and (5.52) in (5.44) leads to

$$\begin{bmatrix} E_{meas,\,\theta} \\ E_{meas,\,\phi} \end{bmatrix} = \begin{bmatrix} \mathbb{Q}_{11} & \mathbb{Q}_{21} \\ \mathbb{Q}_{12} & \mathbb{Q}_{22} \end{bmatrix} \begin{bmatrix} J_x \\ J_y \end{bmatrix} \tag{5.53}$$

where $E_{meas,\,\theta}$ and $E_{meas,\,\phi}$ are complex quantities whose elements are the θ- and ϕ-components of the electric near field, respectively, measured at discrete points. J_x and J_y are column vectors whose elements are the unknown coefficients γ_{ij} and ζ_{ij}, respectively. $\mathbb{Q}_{11}$, $\mathbb{Q}_{12}$, $\mathbb{Q}_{21}$, and $\mathbb{Q}_{22}$ are the submatrices of the entire moment matrix of (5.53) and are given by

$$[\mathbb{Q}_{11}]_{k,m} = \frac{\Delta x\, \Delta y}{4\pi} \left[\begin{aligned} &\cos\theta_k \cos\phi_k\, e^{-jk_0 R_{k,m}} \left\{ \left(x_k^f - x_m^s\right)^2 \left(\frac{j\omega\mu}{R_{k,m}^3} + \frac{3\eta}{R_{k,m}^4} + \frac{3}{j\omega\varepsilon R_{k,m}^5}\right) - \left(\frac{j\omega\mu}{R_{k,m}} + \frac{\eta}{R_{k,m}^2} + \frac{3}{j\omega\varepsilon R_{k,m}^3}\right) \right\} \\ &+ \cos\theta_k \sin\phi_k\, e^{-jk_0 R_{k,m}} \left\{ \left(x_k^f - x_m^s\right)\left(y_k^f - y_m^s\right) \left(\frac{j\omega\mu}{R_{k,m}^3} + \frac{3\eta}{R_{k,m}^4} + \frac{3}{j\omega\varepsilon R_{k,m}^5}\right) \right\} \\ &- \sin\theta_k\, e^{-jk_0 R_{k,m}} \left\{ \left(x_k^f - x_m^s\right)\left(z_k^f\right) \left(\frac{j\omega\mu}{R_{k,m}^3} + \frac{3\eta}{R_{k,m}^4} + \frac{3}{j\omega\varepsilon R_{k,m}^5}\right) \right\} \end{aligned} \right]$$

$$[\mathbb{Q}_{12}]_{k,m} = \frac{\Delta x\, \Delta y}{4\pi} \left[\begin{array}{l} \cos\theta_k\,\cos\phi_k\,e^{-j\,k_0 R_{k,m}}\left\{\left(x_k^f - x_m^s\right)\left(y_k^f - y_m^s\right)\left(\frac{j\omega\mu}{R_{k,m}^3} + \frac{3\eta}{R_{k,m}^4} + \frac{3}{j\omega\varepsilon R_{k,m}^5}\right)\right\} \\[2ex] + \cos\theta_k\,\sin\phi_k\,e^{-j\,k_0 R_{k,m}}\left\{\left(y_k^f - y_m^s\right)^2\left(\frac{j\omega\mu}{R_{k,m}^3} + \frac{3\eta}{R_{k,m}^4} + \frac{3}{j\omega\varepsilon R_{k,m}^5}\right) - \left(\frac{j\omega\mu}{R_{k,m}} + \frac{\eta}{R_{k,m}^2} + \frac{3}{j\omega\varepsilon R_{k,m}^3}\right)\right\} \\[2ex] - \sin\theta_k\,e^{-j\,k_0 R_{k,m}}\left\{\left(y_k^f - y_m^s\right)\left(z_k^f\right)\left(\frac{j\omega\mu}{R_{k,m}^3} + \frac{3\eta}{R_{k,m}^4} + \frac{3}{j\omega\varepsilon R_{k,m}^5}\right)\right\} \end{array} \right]$$

$$[\mathbb{Q}_{21}]_{k,m} = \frac{\Delta x\, \Delta y}{4\pi} \left[\begin{array}{l} \cos\phi_k\,e^{-j\,k_0 R_{k,m}}\left\{\left(x_k^f - x_m^s\right)\left(y_k^f - y_m^s\right)\left(\frac{j\omega\mu}{R_{k,m}^3} + \frac{3\eta}{R_{k,m}^4} + \frac{3}{j\omega\varepsilon R_{k,m}^5}\right)\right\} \\[2ex] - \sin\phi_k\,e^{-j\,k_0 R_{k,m}}\left\{\left(x_k^f - x_m^s\right)^2\left(\frac{j\omega\mu}{R_{k,m}^3} + \frac{3\eta}{R_{k,m}^4} + \frac{3}{j\omega\varepsilon R_{k,m}^5}\right) - \left(\frac{j\omega\mu}{R_{k,m}} + \frac{\eta}{R_{k,m}^2} + \frac{3}{j\omega\varepsilon R_{k,m}^3}\right)\right\} \end{array} \right]$$

$$[\mathbb{Q}_{22}]_{k,m} = \frac{\Delta x\, \Delta y}{4\pi} \left[\begin{array}{l} \cos\phi_k\,e^{-j\,k_0 R_{k,m}}\left\{\left(y_k^f - y_m^s\right)^2\left(\frac{j\omega\mu}{R_{k,m}^3} + \frac{3\eta}{R_{k,m}^4} + \frac{3}{j\omega\varepsilon R_{k,m}^5}\right) - \left(\frac{j\omega\mu}{R_{k,m}} + \frac{\eta}{R_{k,m}^2} + \frac{3}{j\omega\varepsilon R_{k,m}^3}\right)\right\} \\[2ex] - \sin\phi_k\,e^{-j\,k_0 R_{k,m}}\left\{\left(x_k^f - x_m^s\right)\left(y_k^f - y_m^s\right)\left(\frac{j\omega\mu}{R_{k,m}^3} + \frac{3\eta}{R_{k,m}^4} + \frac{3}{j\omega\varepsilon R_{k,m}^5}\right)\right\} \end{array} \right]$$

$$(5.54)$$

where θ_k and ϕ_k are the θ and ϕ coordinates, respectively, of the kth field measuring point. x_m^s and y_m^s are the x- and y-coordinates, respectively, of the mth source point. $R_{k,m}$ is the distance between the kth field point $(r_k^f)$and the mth source point (r_m^s) and is defined by

$$R_{k,m} = \sqrt{\left(x_k^f - x_m^s\right)^2 + \left(y_k^f - y_m^s\right)^2 + \left(z_k^f\right)^2} \qquad (5.55)$$

The superscript f denotes the field measuring point and the superscript s denotes the source point. Note that in (5.54) the two subscripts i and j have been replaced by the single subscript m. That is, $\left(x_m^s, y_m^s\right)$ is $\left(x_i, y_j\right)$ where i and j are determined by m. Δx and Δy represent the size of the patch on S_0 carrying the dipoles.

The resulting matrix (5.53) along with (5.54), when solved, determines the elements of J_x and J_y. This matrix equation is solved using the method of total least squares with singular value decomposition [20] as outlined in Chapter 1.

5.5.2 Numerical Results Using an Equivalent Electric Current

In this section we illustrate the accuracy of the equivalent electric current approach for near field to far field transformation. Experimentally measured data is utilized. Consider the same microstrip array as described in Section 5.4. The near field data is exactly the same as discussed in the previous section [19]. To demonstrate the accuracy of the electric current approach, fictitious planar surfaces on the xy-plane of several different dimensions, and number of unknown electric current patches were used to form an electric current sheet that would radiate the correct far field. First, a fictitious planar surface on the xy-plane of dimensions 1.6 m × 1.6 m is used to form an electric current sheet. This electric current sheet is divided into 45 × 45 equally spaced electric current patches, 0.36 λ apart.

Figure 5.54 compares the electric far field pattern obtained by the present method, with the result obtained analytically in Chapter 7. Figure 5.54 describes $20 \log_{10} |E_\varphi|$ for $\varphi = 0°$ and $-89° < \theta < 89°$. This is a principal plane or co-polarization pattern. As observed, the two results are in good agreement. To observe the effect of increasing the dimensions of the aperture, the same pattern as in Figure 5.54 is shown in Figure 5.55 where the simulated aperture's dimensions increase to 1.9 m × 1.9 m. As observed, the results are better as the size of the source plane is increased which, as explained in Section 5.1, is expected, Figures 5.56 and 5.57 describe $20 \log_{10} |E_\theta|$ for $\varphi = 90°$ utilizing the same measured data.

Figures 5.58 to 5.61 describe the cross-polarization patterns, and the same phenomenon again is observed. To demonstrate the effect of varying the cut-off for the singular values, the following experiments are performed. For the size

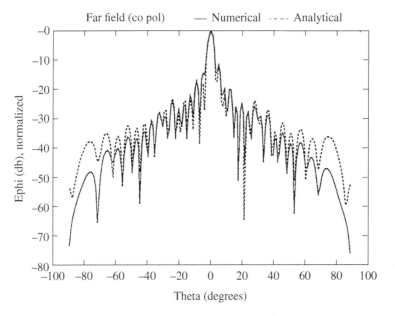

Figure 5.54 Co-polarization characteristic for $\varphi = 0°$ cut for a 32 × 32 patch microstrip array using numerical and analytical results. Number of equally spaced source patches are 45 × 45 covering 1.6 m × 1.6 m for the electric currents. Number of unknowns is 4050 and each patch is 0.39 λ apart. The rank of the Moment Matrix is 2367 with a cut off for the singular values to be 10^{-3}.

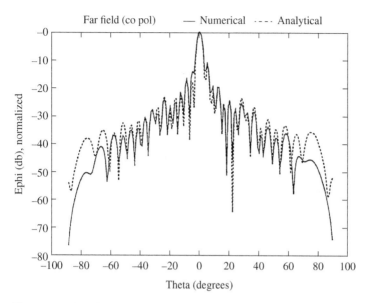

Figure 5.55 Co-polarization characteristic for $\varphi = 0°$ cut for a 32 × 32 patch microstrip array using numerical and analytical results. Number of equally spaced source patches are 45 × 45 covering 1.9 m × 1.9 m for the electric currents. Number of unknowns is 4050 and each patch is 0.46 λ apart. The rank of the Moment Matrix is 2964 with a cut off for the singular values to be 10^{-3}.

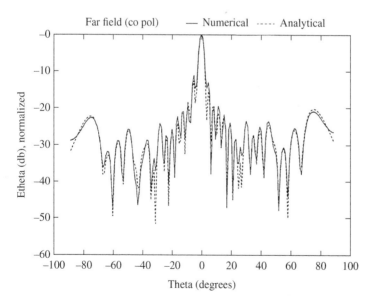

Figure 5.56 Co-polarization characteristic for $\varphi = 90°$ cut for a 32 × 32 patch microstrip array using numerical and analytical results. Number of equally spaced source patches are 45 × 45 covering 1.6 m × 1.6 m covering the electric currents. Number of unknowns is 4050 and each patch is 0.39 λ apart. The rank of the Moment Matrix is 2367 with a cut off for the singular values to be 10^{-3}.

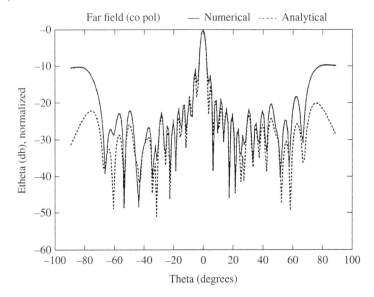

Figure 5.57 Co-polarization characteristic for $\varphi = 90°$ cut for a 32 × 32 patch microstrip array using numerical and analytical results. Number of equally spaced source patches are 45 × 45 covering 1.9 m × 1.9 m covering the electric current. Number of unknowns is 4050 and each patch is 0.46 λ apart. The rank of the Moment Matrix is 2964 with a cut off for the singular values to be 10^{-3}.

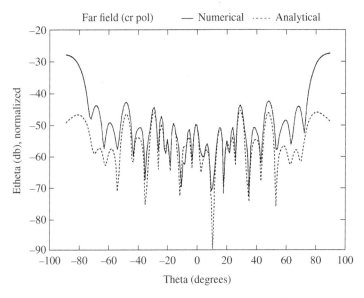

Figure 5.58 Cross-polarization characteristic for $\varphi = 0°$ cut for a 32 × 32 patch microstrip array using numerical and analytical results. Number of equally spaced source patches are 45 × 45 covering 1.6 m × 1.6 m covering the electric currents. Number of unknowns is 4050 and each patch is 0.39 λ apart. The rank of the Moment Matrix is 2367 with a cut off for the singular values to be 10^{-3}.

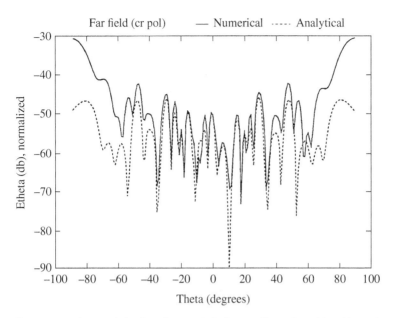

Figure 5.59 Cross-polarization characteristic for $\varphi = 0°$ cut for a 32 × 32 patch microstrip array using numerical and analytical results. Number of equally spaced source patches are 45 × 45 covering 1.9 m × 1.9 m. Number of unknowns is 4050 and each patch is 0.46 λ apart. The rank of the Moment Matrix is 2964 with a cut off for the singular values to be 10^{-3}.

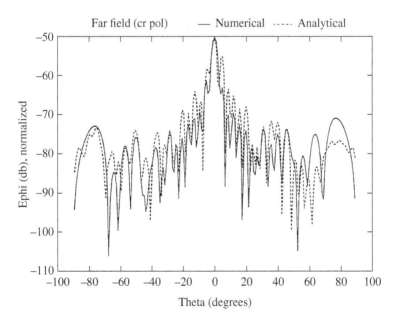

Figure 5.60 Cross-polarization characteristic for $\varphi = 90°$ cut for a 32 × 32 patch microstrip array using numerical and analytical results. Number of equally spaced source patches are 45 × 45 covering 1.6 m × 1.6 m of the electric currents. Number of unknowns is 4050 and each patch is 0.39 λ apart. The rank of the Moment Matrix is 2367 with a cut off for the singular values to be 10^{-3}.

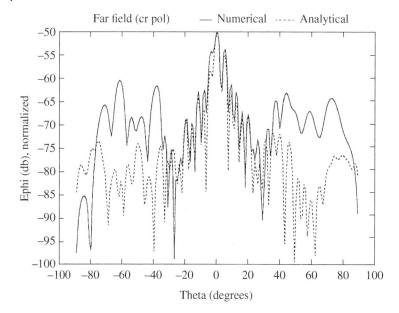

Figure 5.61 Cross-polarization characteristic for $\varphi = 90°$ cut for a 32 × 32 patch microstrip array using numerical and analytical results. Number of equally spaced source patches are 45 × 45 covering 1.9 m × 1.9 m of the electric currents. Number of unknowns is 4050 and each patch is 0.46 λ apart. The rank of the Moment Matrix is 2964 with a cut off for the singular values to be 10^{-3}.

of the source plane of 1.9 m × 1.9 m and the numbers of unknowns being of 45 × 45 square patches to represent the source plane, the cut-off for the singular values were varied from 10^{-2} to 10^{-3} and the result is shown in Figures 5.62 to 5.69. To demonstrate the effects of increasing the number of unknowns, the dimensions of the source plane and the cut-off for the singular values were kept constant and the number of unknown square patches representing the electric current in the source plane were varied from 45 × 45 to 60 × 60. The co- and cross-polarization patterns for these experiments are described in Figures 5.70 to 5.81. As observed the increase in the number of unknowns increases the accuracy of the results which is expected.

5.5.3 Summary

The method presented here determines the fields for $z > 0$ in front of the radiating antenna simply from the knowledge of the near field on any arbitrary geometry in space. Using various antenna configurations and near-field geometries, an investigation of the accuracy of this method was performed, when using equivalent electric currents as opposed to the use of magnetic currents outlined

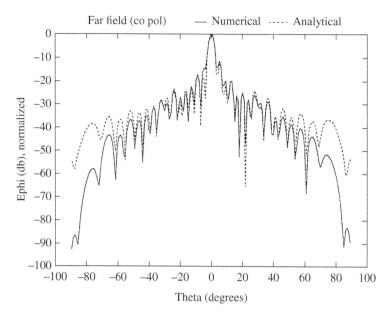

Figure 5.62 Co-polarization characteristic for $\varphi = 0°$ cut for a 32 × 32 patch microstrip array using numerical and analytical results. Number of equally spaced source patches are 45 × 45 covering 1.9 m × 1.9 m of the electric currents. Number of unknowns is 4050 and each patch is 0.46 λ apart. The rank of the Moment Matrix is 2763 with a cut off for the singular values to be 10^{-2}.

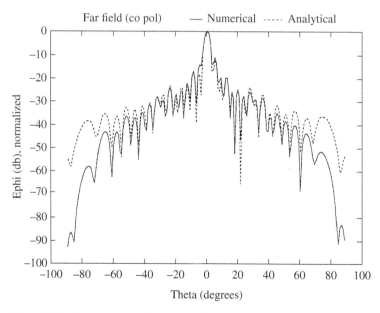

Figure 5.63 Co-polarization characteristic for $\varphi = 0°$ cut for a 32 × 32 patch microstrip array using numerical and analytical results. Number of equally spaced source patches are 45 × 45 covering 1.9 m × 1.9 m of the electric currents. Number of unknowns is 4050 and each patch is 0.46 λ apart. The rank of the Moment Matrix is 2964 with a cut off for the singular values to be 10^{-3}.

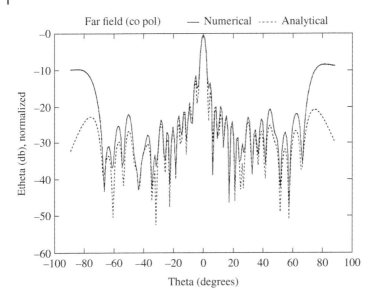

Figure 5.64 Co-polarization characteristic for $\varphi = 90°$ cut for a 32×32 patch microstrip array using numerical and analytical results. Number of equally spaced source patches are 45×45 covering 1.9 m $\times$ 1.9 m of the electric currents. Number of unknowns is 4050 and each patch is 0.46 λ apart. The rank of the Moment Matrix is 2763 with a cut off for the singular values to be 10^{-2}.

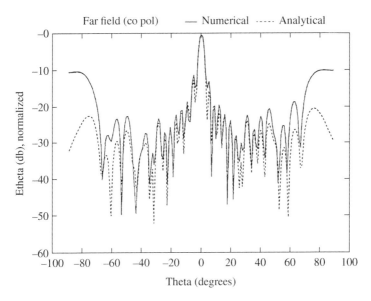

Figure 5.65 Co-polarization characteristic for $\varphi = 90°$ cut for a 32×32 patch microstrip array using numerical and analytical results. Number of equally spaced source patches are 45×45 covering 1.9 m $\times$ 1.9 m of the electric currents. Number of unknowns is 4050 and each patch is 0.46 λ apart. The rank of the Moment Matrix is 2964 with a cut off for the singular values to be 10^{-3}.

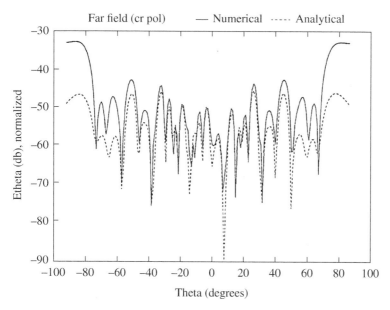

Figure 5.66 Cross-polarization characteristic for $\varphi = 90°$ cut for a 32 × 32 patch microstrip array using numerical and analytical results. Number of equally spaced source patches are 45 × 45 covering 1.9 m × 1.9 m of the electric currents. Number of unknowns is 4050 and each patch is 0.46 λ apart. The rank of the Moment Matrix is 2763 with a cut off for the singular values to be 10^{-32}.

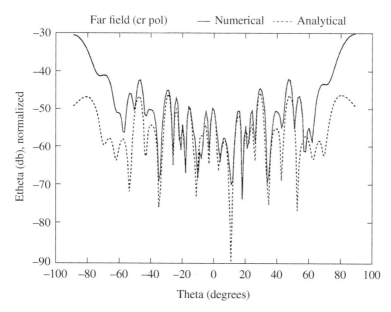

Figure 5.67 Cross-polarization characteristic for $\varphi = 0°$ cut for a 32 × 32 patch microstrip array using numerical and analytical results. Number of equally spaced source patches are 45 × 45 covering 1.9 m × 1.9 m of the electric currents. Number of unknowns is 4050 and each patch is 0.46 λ apart. The rank of the Moment Matrix is 2964 with a cut off for the singular values to be 10^{-3}.

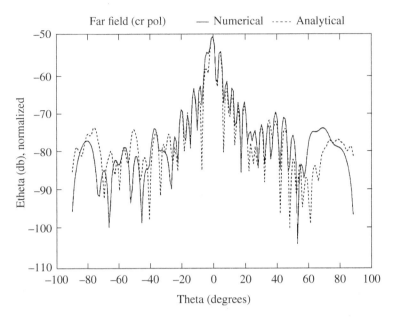

Figure 5.68 Cross-polarization characteristic for $\varphi = 0°$ cut for a 32 × 32 patch microstrip array using numerical and analytical results. Number of equally spaced source patches are 45 × 45 covering 1.9 m × 1.9 m of the electric currents. Number of unknowns is 4050 and each patch is 0.46 λ apart. The rank of the Moment Matrix is 2763 with a cut off for the singular values to be 10^{-2}.

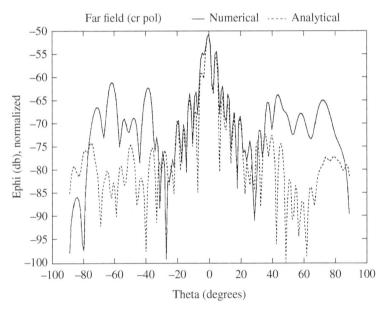

Figure 5.69 Cross-polarization characteristic for $\varphi = 90°$ cut for a 32 × 32 patch microstrip array using numerical and analytical results. Number of equally spaced source patches are 45 × 45 covering 1.9 m × 1.9 m of the electric currents. Number of unknowns is 4050 and each patch is 0.46 λ apart. The rank of the Moment Matrix is 2964 with a cut off for the singular values to be 10^{-3}.

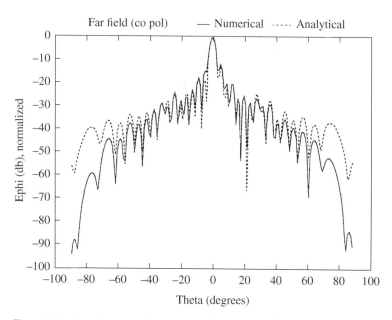

Figure 5.70 Co-polarization characteristic for $\varphi = 0°$ cut for a 32 × 32 patch microstrip array using numerical and analytical results. Number of equally spaced source patches are 45 × 45 covering 1.9 m × 1.9 m of the electric currents. Number of unknowns is 4050 and each patch is 0.46 λ apart. The rank of the Moment Matrix is 2964 with a cut off for the singular values to be 10^{-3}.

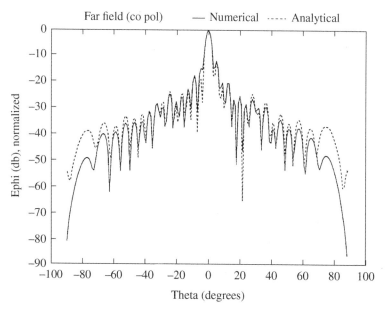

Figure 5.71 Co-polarization characteristic for $\varphi = 0°$ cut for a 32 × 32 patch microstrip array using numerical and analytical results. Number of equally spaced source patches are 55 × 55 covering 1.9 m × 1.9 m of the electric currents. Number of unknowns is 6050 and each patch is 0.38λ apart. The rank of the Moment Matrix is 3033 with a cut off for the singular values to be 10^{-3}.

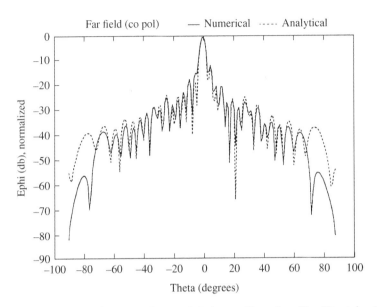

Figure 5.72 Co-polarization characteristic for $\varphi = 0°$ cut for a 32 × 32 patch microstrip array using numerical and analytical results. Number of equally spaced source patches are 60 × 60 covering 1.9 m × 1.9 m of the electric currents. Number of unknowns is 7200 and each patch is 0.348 λ apart. The rank of the Moment Matrix is 3047 with a cut off for the singular values to be 10^{-3}.

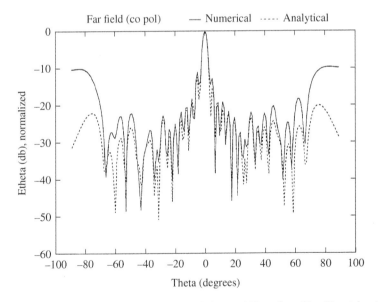

Figure 5.73 Co-polarization characteristic for $\varphi = 90°$ cut for a 32 × 32 patch microstrip array using numerical and analytical results. Number of equally spaced source patches are 45 × 45 covering 1.9 m × 1.9 m of the electric currents. Number of unknowns is 4050 and each patch is 0.46 λ apart. The rank of the Moment Matrix is 2964 with a cut off for the singular values to be 10^{-3}.

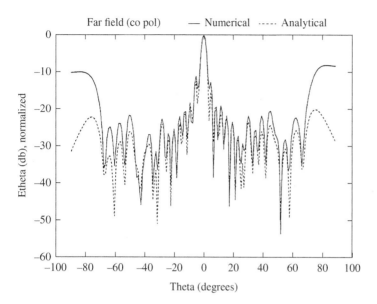

Figure 5.74 Co-polarization characteristic for $\varphi = 90°$ cut for a 32 × 32 patch microstrip array using numerical and analytical results. Number of equally spaced source patches are 55 × 55 covering 1.9 m × 1.9 m of the electric currents. Number of unknowns is 6050 and each patch is 0.38 λ apart. The rank of the Moment Matrix is 3033 with a cut off for the singular values to be 10^{-3}.

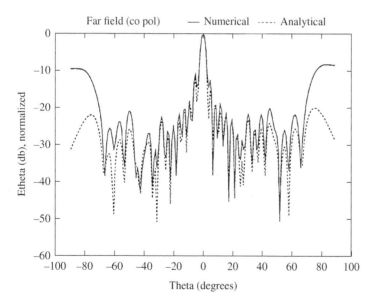

Figure 5.75 Co-polarization characteristic for $\varphi = 90°$ cut for a 32 × 32 patch microstrip array using numerical and analytical results. Number of equally spaced source patches are 60 × 60 covering 1.9 m × 1.9 m of the electric currents. Number of unknowns is 7200 and each patch is 0.348 λ apart. The rank of the Moment Matrix is 3047 with a cut off for the singular values to be 10^{-3}.

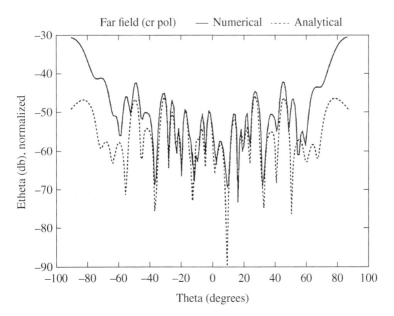

Figure 5.76 Cross-polarization characteristic for $\varphi = 0°$ cut for a 32 × 32 patch microstrip array using numerical and analytical results. Number of equally spaced source patches are 45 × 45 covering 1.9 m × 1.9 m of the electric currents. Number of unknowns is 4050 and each patch is 0.46 λ apart. The rank of the Moment Matrix is 2964 with a cut off for the singular values to be 10^{-3}.

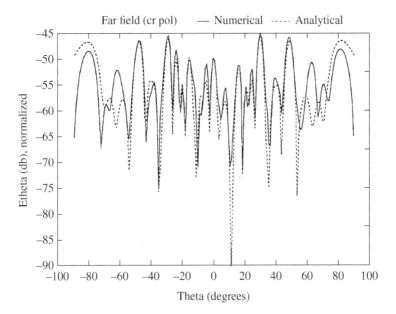

Figure 5.77 Cross-polarization characteristic for $\varphi = 0°$ cut for a 32 × 32 patch microstrip array using numerical and analytical results. Number of equally spaced source patches are 55 × 55 covering 1.9 m × 1.9 m of the electric currents. Number of unknowns is 6050 and each patch is 0.38 λ apart. The rank of the Moment Matrix is 3033 with a cut off for the singular values to be 10^{-3}.

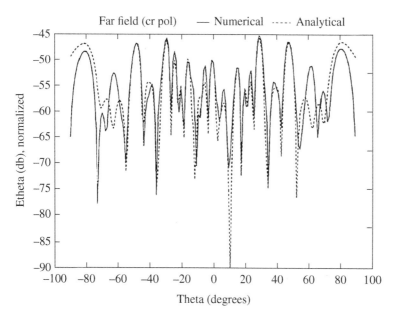

Figure 5.78 Cross-polarization characteristic for $\varphi = 0°$ cut for a 32 × 32 patch microstrip array using numerical and analytical results. Number of equally spaced source patches are 60 × 60 covering 1.9 m × 1.9 m of the electric currents. Number of unknowns is 7200 and each patch is 0.348 λ apart. The rank of the Moment Matrix is 3047 with a cut off for the singular values to be 10^{-3}.

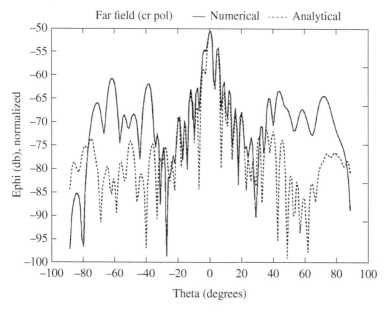

Figure 5.79 Cross-polarization characteristic for $\varphi = 90°$ cut for a 32 × 32 patch microstrip array using numerical and analytical results. Number of equally spaced source patches are 45 × 45 covering 1.9 m × 1.9 m of the electric currents. Number of unknowns is 4050 and each patch is 0.46 λ apart. The rank of the Moment Matrix is 2964 with a cut off for the singular values to be 10^{-3}.

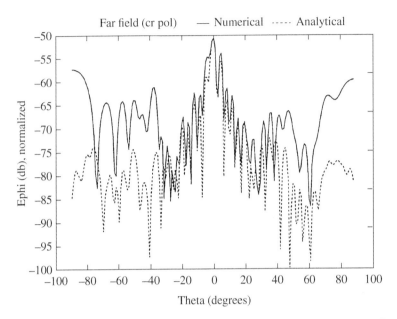

Figure 5.80 Cross-polarization characteristic for $\varphi = 90°$ cut for a 32 × 32 patch microstrip array using numerical and analytical results. Number of equally spaced source patches are 55 × 55 covering 1.9 m × 1.9 of the electric currents. Number of unknowns is 6050 and each patch is 0.38 λ apart. The rank of the Moment Matrix is 3033 with a cut off for the singular values to be 10^{-3}.

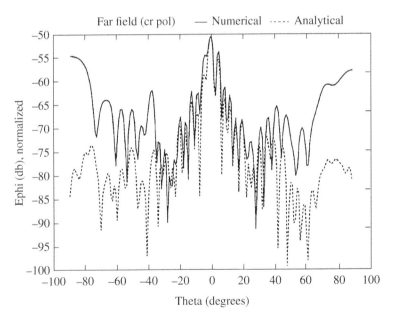

Figure 5.81 Cross-polarization characteristic for $\varphi = 90°$ cut for a 32 × 32 patch microstrip array using numerical and analytical results. Number of equally spaced source patches are 60 × 60 covering 1.9 m × 1.9 m of the electric currents. Number of unknowns is 7200 and each patch is 0.348 λ apart. The rank of the Moment Matrix is 3047 with a cut off for the singular values to be 10^{-3}.

in the previous section. When actual experimental sources were used, the far fields were compared with the conventional modal approach, which utilized the data from the analytical methodology outlined in Chapter 7. It has been our experience that the equivalent magnetic current approach provides always a better solution than the equivalent electric current approach. This may be due to the fact that the matrix arising from the electric field operator is more ill conditioned than the matrix involved with the magnetic field operator. In addition, the fields are decoupled for the magnetic field operator [4] for certain canonical geometries.

5.6 Evaluating Near-Field Radiation Patterns of Commercial Antennas

In this section, the source reconstruction technique from the measured near fields is applied to obtain a set of equivalent currents that will characterise the forward and backward radiation patterns of an antenna. This is particularly useful in characterizing the three dimensional near field data for an antenna as that information is generally not easily available. Once the equivalent sources are determined, the electromagnetic field at any aspect angle and distance from the antenna can be calculated. In this section, the method is applied to the evaluation of the radiation from commercial antennas at any observation point to determine the safety volume beyond which the hazardous field strength from the antenna is not visible. The electric field patterns of a DCS base station antenna at 1800 MHz and a horn antenna at 2500 MHz have been calculated and plotted at several distances from the antenna. The source reconstruction method is the only method that can be used in characterizing the "reference volumes" or exclusion zones for transmitting antennas dealing with the maximum levels of electromagnetic radiation safe for human exposure, as stated in many national and international regulations. This methodology is necessary as antenna manufacturers provide only the far field pattern and not the near field pattern which is necessary to characterize the exclusion zone.

5.6.1 Background

For most commercial antennas, only the far field pattern along the principal planes is specified. Although this can be useful for most cases, this does not provide information about the pattern at other aspect angles out of the main planes. Secondly, no phase information is available. Thirdly, one cannot derive how the antenna radiates at near distances except for the one provided. This makes it difficult to know what is the safe distance for human exposure for a given input power level to the antenna.

In addition to the possible interest of knowing how the E-field and H-field pattern of an antenna change with distance from the antenna, one application where the information at near field distances can be of great interest is the verification of compliance of radiofrequency stations, in general, and base stations of cellular radio communication services, in particular. The objective is to guarantee that the recommendations or regulations regarding limitations of exposure of the general public to Non Ionizing Radiation (NIR) of electromagnetic fields are maintained for the antenna under test (AUT).

The most known recommendations for human exposure to Radiofrequency (RF) electromagnetic fields are that of the Institute of Electrical and Electronics Engineers, Inc. (IEEE) [23] and that of the International Commission for non-ionizing radiation protection (ICNIRP) [24], which succeeded the International Non Ionizing Radiation Committee as a working group of the International Radiation Protection Association. Since its conception in 1974, they had been studying, jointly with the World Health Organization (WHO), the health effects of non-ionizing radiation. Their recommendations are mainly focused on short term effects of NIR, thus establishing maximum levels of specific absorption rate (e.g. SAR) or reference magnitudes (e.g. the electric field) as a function of frequency.

In the United States, the American National Standards Institute (ANSI), the IEEE and the National Council on Radiation Protection and Measurements (NCRP) have issued recommendations for human exposure to RF electromagnetic fields. The Federal Communications Commission (FCC) adopted, in 1996, the NCRP's recommendation of the Maximum Permissible Exposure limits for field strength and power density for the transmitters operating at frequencies of 300 kHz to 100 GHz, and the specific absorption rate (SAR) limits for devices operating within close proximity to the body as specified within the ANSI/IEEE C95.1-1992 guidelines. Certain applicants are required to routinely perform an environmental evaluation with respect to determining compliance with the FCC's exposure limits. In the event that an applicant determines the site is not within compliance, the submission of an Environmental Analysis is required. In the European Union (EU) [25–27], the Council of the previously named European Community established the Recommendation 1999/519/CE [25] that was based on the ICNIRP guidelines [24] and has been translated into national regulations by some of their members. Hence, it is necessary to know the radiated power from the antenna at various distances from it. Due to the development of second and third generation of digital mobile communications and the fast deployment of their Base Terminal Stations (BTS), the measurement and control of electromagnetic emissions due to these BTS need to be characterized.

Different methodologies exist in the literature regarding simulations and measurements for the characterization of electromagnetic fields surrounding the base station antennas. Among the proposed simulation techniques, the use of numerical methods to solve differential and integral equations as well as the use of propagation models for the analysis of base station antennas have been widely used [28–42]; even approximate formulas for fast estimation [35–36], combination techniques of dosimetric methods and simple simulations [43], and geometric methods [44] can also be found in the literature. The electromagnetic analysis of the antenna represented by an appropriate model in terms of its geometry and material composition could characterize the antenna completely in terms of its field at any arbitrary point. Letting aside the strategy of solving the electromagnetic problem and the selection of the numerical method, the main concern would be to obtain a representative geometric model of the antenna as well as an accurate characterization of the material properties of the elements that constitute the antenna. In many commercial antennas it would be difficult to predict accurately their radiation properties at any distance, mainly because of the difficulty in establishing a realistic model of the antenna since their geometric parameters and material characteristics may not be accessible or may even be unknown.

On the other hand, on-site measurements and dosimetric techniques for direct measurement of SAR and reference levels have been of common use for compliance testing with specific exposure limits [37–42]. However, these techniques carry inherent difficulties for accurate positioning and accurate level measurement, which depends on the configuration of the BTS and its environment, and do not allow a complete characterization of the antenna (estimation of the locations of the safeguard perimeters rather than a complete safeguard volume). Besides on-site measurements, the other way to characterize a commercial antenna is the direct measurement of its field pattern at a certain distance in an antenna measurement system with a canonical range (planar, cylindrical and spherical) in anechoic conditions (anechoic chamber as a test site). However, the measurement provides the field pattern at the measurement distance or the far-field pattern if a Near-Field to Far-Field (NF-FF) transformation software is used.

Some authors have addressed the calculation of field emissions at a specific distance when the measured electric field data at other distances are given, through NF to NF transformations over canonical surfaces. In [45] a spectral method is used to transform near-field measurements to any other distance using an acquisition over a cylindrical surface. In [46–50] the field pattern of a base station antenna over an arbitrary spherical surface is calculated from the knowledge of the measured electric field over a near-field spherical surface using a spherical range antenna measurement system and a transformation technique based on the spherical modal expansion. Those NF-NF transformations need measurement data over a canonical surface, requiring a minimum radius of the canonical surface where the field is to be calculated, and do not allow a straightforward calculation of the field over arbitrary surfaces conforming to the specified reference volumes, over which the antenna power pattern is desired.

In this section, we propose a technique to represent the electromagnetic behaviour of a real antenna (determination of its fields at any arbitrary point in space) through the calculation of an equivalent current density obtained from the near-field measurement of the antenna. The equivalent current densities representing the antenna are calculated making use of a source reconstruction method. Once the equivalent currents are determined over a surface close to the antenna, the electric and magnetic fields can be calculated at any arbitrary distances in a straightforward fashion using the free space Green's function. This technique can then be used to characterize antennas as those installed in BTS's of cellular services. This new method has some additional features:

- Arbitrary scanning can be used to get the known near-field data necessary to characterize the antenna electromagnetically through its equivalent currents.
- Once the antenna is characterized through its equivalent sources, the electric and magnetic fields can be calculated at any distance from the antenna, without restrictions of the minimum radius of canonical surface enclosing the antenna.
- For application of this methodology of using equivalent sources, it is not necessary to have available data over a complete "enclosing" surface covering 4π steradians.

Here, we use the source reconstruction technique to completely characterize the radiation fields (both near and far) of commercial antennas. Some practical results using the proposed technique are presented, both in terms of the evolution with distance of the near field patterns of a horn antenna, as well as the calculation and plot of a reference volume for a commercial 1800 MHz DCS base station antenna according to the regulations of an EU member state. Comparison using measured field patterns and computed data have been used to validate this methodology followed by some conclusions.

The main goal of the proposed technique is to determine the radiated fields of an antenna at any aspect angle and distance. The technique is based on a source reconstruction method, obtaining an equivalent current distribution which models the electromagnetic behaviour of the antenna. The optimum conditions for determining the equivalent current distribution of the antenna are to use the measured amplitude and phase data from a near-field pattern measurement system over a three-dimensional domain enclosing the antenna. Since the accuracy of the computation of the equivalent currents characterising the radiation characteristics of the antenna depends on the angular margin where field data is available, among the best known acquisition techniques using canonical ranges, is a near-field spherical scanning system. With the magnitude and phase information of the tangential electric field in the near-field zone over a spherical surface, the equivalent principle can be applied from a theoretical point of view to establish two equivalent problems for the original radiating antenna.

However, for computational purposes, it is not necessary to have the data over a complete enclosed surface.

5.6.2 Formulation of the Problem

In particular, considering that the domain of the equivalent currents "enclosing" the antenna is composed of two infinite planes located at, $z = z_1$ (e.g., the antenna aperture plane in the front) and $z = z_2$, (e.g., the antenna aperture plane in the back), the original problem can be decomposed into two equivalent problems: the "forward" one, with an Equivalent Magnetic Current (EqMC) distribution that radiates the known electric field data over the forward hemisphere, and the "backward" equivalent problem with a different EqMC distribution that radiates the known electric field data over the backward hemi-sphere. The original problem and the two external equivalent problems with their correspondent set of EqMC distributions are shown in Figure 5.82.

The source reconstruction technique must be applied to both the equivalent problems separately in order to obtain the "forward" and "backward" EqMC distributions. In each of the two equivalent problems the near electric field components and the Cartesian components of the EqMC distribution can be related through the free space Green's function (GF), leading to the integral equation:

$$\vec{E} = -\frac{1}{4\pi}\int\int_{S'} \nabla \times \left[\vec{M}_{se}(x',y',z')\frac{e^{j\beta R}}{R}\right]dS' = \frac{1}{4\pi}\int\int_{S'}\left(\hat{R} \times \vec{M}_{se}\right)\frac{1+j\beta R}{R^2}e^{-j\beta R}dS'$$

$$(5.56)$$

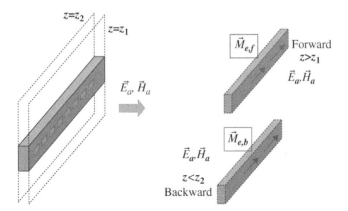

Figure 5.82 Original problem and external equivalent problems.

The tangential components of the near electric field (E_θ, E_φ) at the location (R_o, θ, φ) can be written as a function of the Cartesian components of the EqMC distribution $(\vec{M}_{se} = M_x \hat{x} + M_y \hat{y})$ over the surface S'. This results in:

$$E_\theta(R_o, \theta, \varphi) = -\frac{1}{4\pi} \iint_{S'} \left\{ \begin{array}{l} M_x[y' \sin\theta - (R_o - z' \cos\theta) \sin\varphi] + \\ + M_y[(R_o - z' \cos\theta) \cos\varphi - x' \sin\theta] \end{array} \right\} \left(\frac{1 + j\beta R}{R^3} \right) e^{-j\beta R} dS'$$

$$= \int_{S'_{AUT}} f_1 \{ M_x, M_y, GF \} dS'$$

$$(5.57)$$

$$E_\varphi(R_o, \theta, \varphi) = -\frac{1}{4\pi} \iint_{S'} \left\{ \begin{array}{l} M_x[(z' - R_o\cos\theta)\cos\varphi] + \\ + M_y[(z' - R_o\cos\theta)\sin\varphi] \end{array} \right\} \left(\frac{1 + j\beta R}{R^3} \right) e^{-j\beta R} dS'$$

$$= \int_{S'_{AUT}} f_2 \{ M_x, M_y, GF \} dS'$$

$$(5.58)$$

where GF represents the free space Green's function. The integral equations in (5.57) and (5.58) are coupled, as both are functions of M_x and M_y which are the unknowns to be determined. An alternate set of integral equations can be obtained for the source reconstruction technique, where both equations decouple. This can be done by relating the EqMC components to the Cartesian Electric Field components as:

$$E_x(R_o, \theta, \varphi) = \int_{S'_{AUT}} f_3 \{ M_y, GF \} dS' \tag{5.59}$$

$$E_y(R_o, \theta, \varphi) = \int_{S'_{AUT}} f_4 \{ M_x, GF \} dS' \tag{5.60}$$

Equations (5.59) and (5.60) can be used directly by neglecting the radial component of the Electric Field, E_r, and approximating the required Cartesian field components as:

$$E_x(R_o, \theta, \varphi) = E_\theta(R_o, \theta, \varphi)\cos\theta\cos\varphi - E_\varphi(R_o, \theta, \varphi)\sin\varphi + E_r(R_o, \theta, \varphi)\sin\theta\cos\varphi$$

$$E_y(R_o, \theta, \varphi) = E_\theta(R_o, \theta, \varphi)\cos\theta\sin\varphi + E_\varphi(R_o, \theta, \varphi)\cos\varphi + E_r(R_o, \theta, \varphi)\sin\theta\sin\varphi$$

$$(5.61)$$

This approach can be used on occasions if the acquisition is performed in the Fresnel region. For shorter acquisition distances, an iterative radial field retrieval procedure [47–50] should be implemented in the source reconstruction technique for an accurate determination of the Cartesian components of the fields in order to use the set of the decoupled equations (5.59) and (5.60).

Then, the solution for the EqMC distribution in the integral equations (5.59) and (5.60) can be done, given the values of the Cartesian components of the electric field over the acquisition domain. This can be accomplished through a matrix method scheme with an iterative optimization algorithm [46]:

$$E_x = T_a(M_y) \Rightarrow M_y = T_a^{-1}(E_x)$$
$$E_y = T_b(M_x) \Rightarrow M_x = T_b^{-1}(E_y) \tag{5.62}$$

With the calculated EqMC distributions representing the forward and backward radiation fields of the antenna, the complete E and H field pattern of the real antenna can be computed at any distance from the antenna.

Some practical considerations must be taken into account regarding the reconstruction of the EqMC distributions from electric field data at certain locations. Though a spherical scanning has been proposed, any scanning over an arbitrary surface covering a solid angle of 4π steradians could be used. In particular, the iterative procedure proposed in [49–50] to solve the integral equations involved in this type of equivalent problems fits very well to the case of scanning over arbitrary domains. Other typical scanning ranges (cylindrical and planar) could also be used. In these last cases, the extension of the scanning domain to provide accuracy in the source reconstruction process will depend on the directive gain function of the real antenna.

The knowledge of the near-field data over only a reduced part of a surface "enclosing" the antenna at a certain distance, does not guarantee the calculation of the correct values of the EqMC over the antenna. So neither diagnosis nor the characterization of the radiation of the antenna at any arbitrary observation point can be done with reliability. However, the computed EqMC over the antenna could be used to calculate the radiated field over a similar angular interval of the known field data, allowing some NF-FF transformation. As an example, if the near-field data is known over only the main planes (e.g. E-plane and H-plane), it is possible to characterize the antenna in terms of the calculation of the field at any distance in the same main planes of the antenna [32]. Regarding sampling of the measured fields, since the computational time is not usually a critical issue (CPU times are similar to those of an iterative MoM solution with source elements as unknowns and field data points as "testing points"), over sampling in the acquisition of electric field data is usually recommended to ensure a complete set of field information and to compensate for the measurement errors, increasing the overall accuracy of the method.

With regards to the choice of the region of support of the EqMC distribution, the planar domains for the "forward" and "backward" distributions placed at $z = z_1$ and $z = z_2$ respectively, should be selected as two parallel planes with a minimum distance of separation between these two planes such that they "enclose" the antenna under test. In practice, the selection of the planes tangential to the antenna (according to the established measurement setup) is the best choice since this ensures the most reliable representation of the fields over the antenna and thus, provides the possibility of performing diagnostic tasks. In addition, the finite dimensions of the antenna guarantee that both z_1 and z_2 cannot take the value zero simultaneously. Thus, the use of a spherical range for the field acquisition implies a slight reduction of the maximum angular interval ($|\theta_0| \leq 90°$ for the "forward" problem) in which the measured field data can be used. For example, to characterise an AUT, the probe can be placed at a distance of 5 meters and a possible choice of $z_1 = 0.5$ meters between the azimuth rotating axis (roll over azimuth spherical system) and the selected tangential plane to the AUT ($z = z_1$-plane) is a good choice. Under these circumstances, the measured field values beyond the angular value of $|\theta_0| \approx 84°$, cannot be used when using the external equivalent problem. This reduction in the angular domain for the measured field implies a similar reduction of the angular interval over which the far-field values can be computed using the measured near-field data. With regards to the region of support for the equivalent sources, for example for simple aperture antennas, it is usually sufficient to use a truncated plane matching the actual aperture plane. In most commercial antennas at microwave and millimetre wave bands, it is sufficient to consider, as the region of support for the equivalent sources, a planar domain tangential to the antenna which is approximately 50% larger than the projection of the radiating aperture to this tangential plane. For the examples presented in this section, the figures show exactly the regions of support with respect to the actual radiating source over which the equivalent sources have been considered.

5.6.3 Results for the Near-field To Far-field Transformation

5.6.3.1 A Base Station Antenna

In order to illustrate the application of the proposed method, the results measured for a commercial cellular base station antenna are presented. This sector antenna, designed for the DCS wireless system and operating in the 1800 MHz frequency band, was measured in an anechoic chamber, using a near field spherical range to obtain the amplitude and phase of the electric field all around the antenna. Figure 5.83 shows the experimental setup including the DCS base station antenna. To carry out the source reconstruction, the data is collected over a spherical surface ($-\pi < \theta < \pi; 0 < \varphi < \pi$). The complex field values are measured with angular steps of $\Delta\theta = 1°$, $\Delta\varphi = 2°$. (disregarding the data beyond the maximum angular value when dealing with two parallel planes located at $z_1 = 0$, and $z_2 = -0.2$ m). The distance between the AUT and the measurement probe was 5.40 m. A planar domain of size 2.0 m × 0.4 m (approx. dimensions of the antenna: 1.6 m × 0.3 m) was considered. The equivalent sources consisted of 120 × 40 Hertzian dipoles as basis functions for each component of the EqMC distribution, and a Conjugate Gradient technique was used to solve the integral equations for the EqMC. The near field measured data, which included data measured at the rear part of the antenna, was used as an input data to the

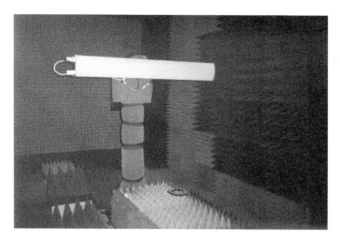

Figure 5.83 Near-field measurement setup and DCS base station antenna.

equivalent source reconstruction algorithm described in the previous section. Two equivalent problems, forward and backward, were actually established. Each equivalent problem with its own set of known near-field data was used to compute the corresponding set of equivalent magnetic currents. Figure 5.84 shows the M_y component of the forward and backward EqMC distributions calculated from their equivalent problems. Once the equivalent currents are known, the electric and magnetic-field patterns can be calculated at different distances form the antenna of interest. Figure 5.85 shows the evolution of the electric-field pattern associated with the DCS base station antenna at three different near-field distances: Figure 5.85(a) 0.5 m, 5.85(b) 1 m and 5.85(c) 10 m. The pattern at a distance of 1000 m, that can be considered as the far-field radiation pattern of the DCS antenna, has also been calculated and it is shown in Figure 5.85(d), where the typical sectoral radiation in the horizontal plane and the directive radiation in the vertical plane can be observed. In order to check the accuracy of the proposed method for this example, the reconstructed "forward" EqMC distribution has been used to calculate the far-field pattern over the range $|\theta| \leq 90°$ for the BTS antenna. The computed results have been compared to the far-field pattern obtained from the measured near-field data using a technique based on the spherical wave-modes expansion for a NF-FF transformation. The results for the co-polar (CP) and the cross-polar (XP) components of the radiation pattern in the plane $\varphi = 0°$ plane are depicted in Figure 5.86. It is seen that the two methods agree quite well upto $\pm 74°$ except at the end points as we have taken finite planar equivalent sources for this problem.

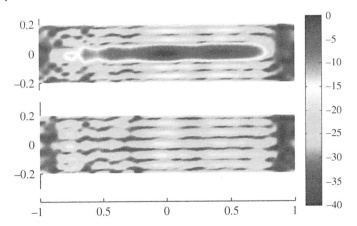

Figure 5.84 Normalized amplitude (dB) of the "forward" (up) and the "backward" (down) equivalent magnetic current distributions of the DCS base station antenna.

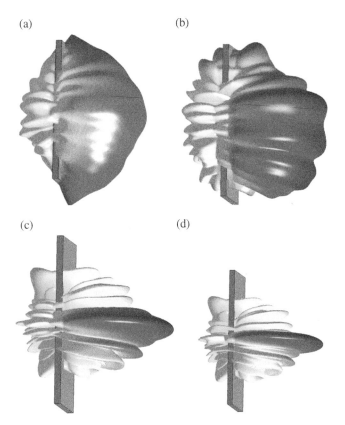

Figure 5.85 Amplitude (dB) of the E-field pattern of a DCS base station antenna at different distances: (a) 0.5 m; (b) 1.0 m; (c) 10 m; (d) 1000 m.

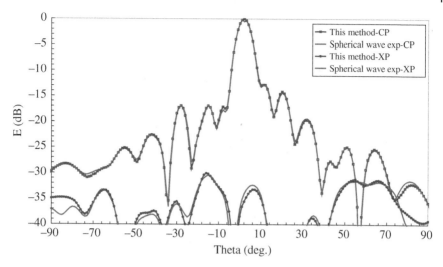

Figure 5.86 Comparison of the co-polarised and the cross-polarised components of the fields for the equivalent source method and using the spherical NF-FF transformations.

5.6.3.2 NF to FF Transformation of a Pyramidal Horn Antenna

The second example deals with a commercial pyramidal horn antenna operating at 2.5 GHz with the dimension 348.6 mm × 259.6 mm. The pyramidal horn antenna (see Figure 5.87) was measured using the same spherical near field range (seen in Figure 5.88) as for the DCS antenna, except for the difference in the frequency of operation. To carry out the source reconstruction, the data is collected over a spherical surface ($-\pi < \theta < \pi$; $0 < \varphi < \pi$). The complex field values are measured with angular steps of $\Delta\theta = 1°$, $\Delta\varphi = 2°$. (disregarding the data beyond the maximum angular value when dealing with two parallel planes located at $z_1 = 0.394$ m, and $z_2 = 0$ m). The distance between the AUT and the measurement probe was 4.86 m. A planar domain of size 520 mm × 400 mm was considered. The equivalent sources consisted of 57 × 57 Hertzian dipoles separated by 0.2 λ as basis functions for each component of the EqMC distribution, and a Conjugate Gradient technique solved the integral equations for the EqMC in 12 iterations. From the two equivalent problems, and the known sets of near-field data, the corresponding forward and backward EqMC distributions were calculated. Figure 5.89 shows the M_x component of the forward and backward EqMC distributions calculated for each of the equivalent problem. From the equivalent currents, the electric-field patterns can be calculated at different distances, showing the transition from Near-Field to Far-Field of the horn antenna. Figure 5.90 shows the field pattern at distances of a) 0.2 m, b) 0.5 m, c) 1 m, d) 100 m, where the last one can be already considered as the far-field radiation pattern of the horn antenna.

Figure 5.87 A pyramidal horn antenna with a 348.6 mm × 259.6 mm aperture operating at 2.5 GHz.

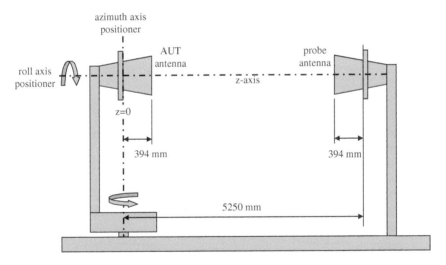

Figure 5.88 Measurement Setup using a spherical near field range.

In order to further verify the validity the results of the proposed methodology, a comparative study is presented by analyzing this horn antenna using a software tool based on the Finite Element (FE) method [49]. For the comparison, the tangential electric field close to the aperture, at $z = 444$ mm (to avoid

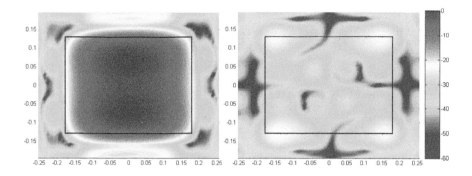

Figure 5.89 Normalized amplitude (dB) of the "forward" (up) and the "backward" (down) equivalent magnetic current distributions of the pyramidal horn antenna.

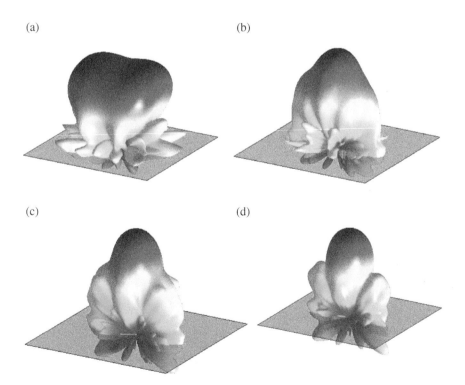

Figure 5.90 Amplitude (dB) of the E-field pattern of a pyramidal horn antenna at different distances: (a) 0.2 m; (b) 0.5 m; (c) 1 m; (d) 100 m.

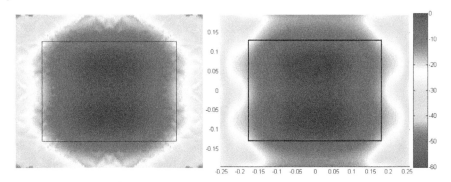

Figure 5.91 Comparison of the aperture fields produced by the finite element method and this novel method presented in this paper.

modeling the distortion near the edges of the aperture) had been simulated using a meshed model of the antenna and its surrounding radiation volume, and imposing the corresponding boundary conditions by considering the TE_{10} mode feeding the waveguide aperture (WR430). The computed results (left) for the vertically polarized tangential field over the $z = 444$ mm -plane are represented in Figure 5.91 together with the reconstructed EqMC distribution (M_x component–right) over the same plane for comparison. In addition, the results of the measured far-field pattern, the estimated far-field pattern from the FE analysis and the far-field pattern calculated from the reconstructed equivalent sources over the aperture plane, have been represented in Figure 5.92. It is seen that the agreement between the three is quite good.

5.6.3.3 Reference Volume of a Base Station Antenna for Human Exposure to EM Fields

Concerning the determination of reference volumes for compliance verification to recommendations and regulations as regards to the human exposure to electromagnetic fields, the DCS base station antenna of Figure 5.83 has been used to characterize its reference volume, specified in terms of a parallelepiped, where the total electric-field in the inner volume is below the allowed maximum level established by the regulations. In this case we are considering the maximum level of electric field at a frequency of 1800 MHz as stated in the regulation of an EU member state -Spain- [26]. According to Table 5.1, at 1800 MHz the corresponding maximum reference level is $E_{max} = 58.34$ V/m (35.3 dBV/m). Figure 5.93 shows the amplitude of the total electric field on the surface of a reference volume of dimensions 3 m × 2 m × 4 m, radiated by the DCS antenna for an EIRP value of 1300 Watts. One of the walls of the surface defining the reference volume has been removed for proper visualization.

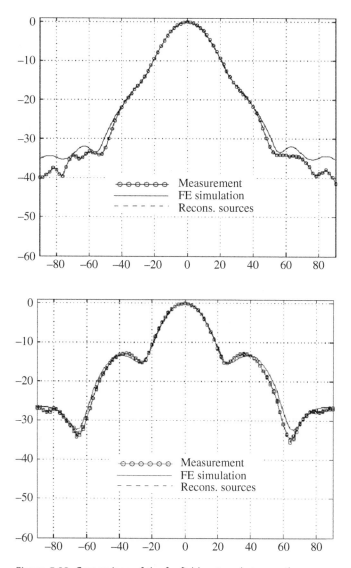

Figure 5.92 Comparison of the far field pattern between the measurement, finite element procedure and the proposed procedure based on the equivalent sources.

5.6.4 Summary

A technique, based on the establishment of two external equivalent problems of a radiating structure and the reconstruction of their respective sets of equivalent current distributions, has been proposed to characterize the radiation properties at any distance and aspect angle for commercial antennas. The technique

Table 5.1 Maximum emission levels for human exposure to EM fields, as regulated in Spain.

	Spain regulation [4] - Reference Emission Levels (rm s)			
Frequency	Electric field intensity E (V/m)	Magnetic field intensity H (A/m)	Flux density B (uT)	Equivalent power density (W/m^2)
400-2000MHz	$1,375 + f^{0,5}$	$0,0037 + f^{0,5}$	$0,0046 + f^{0,5}$	f/200
2-300GHz	61	0,16	0,20	10

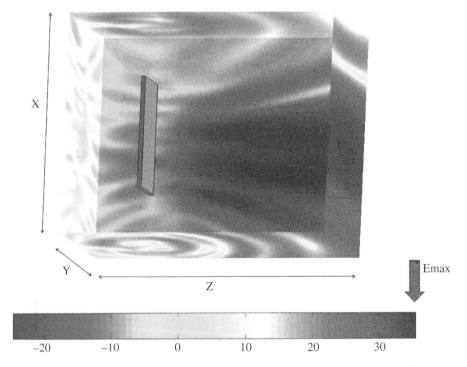

Figure 5.93 "Reference volume" of the DCS base station antenna. Amplitude (dBV/m) of the total E-field on its surface.

requires the measurement of the antenna at an arbitrary near-field distance using canonical or arbitrary scanning, and reconstructs the equivalent sources by solving the integral equations that relates sources and fields of the established unbounded equivalent problems. Once the equivalent sources are determined, the electromagnetic radiation of the antenna can be completely characterized at any point. This is particularly useful in commercial antenna characterization or verification since, usually, the geometry and material composition of the antenna are not completely available or easy to model and an electromagnetic simulation of the antenna can lead to inaccurate results. The method is very well suited for the establishment of near-field "reference volumes" or exclusion zones of arbitrary shape, for electromagnetic environmental studies or to comply with regulations and recommendations as regards to human exposure to non-ionizing electromagnetic fields.

5.7 Conclusions

The source reconstruction method (SRM) is presented in this chapter for antenna diagnostics and for carrying out near-field (NF) to far-field (FF) transformation. The SRM is based on the application of the electromagnetic Equivalence Principle, in which one establishes an equivalent current distribution that radiates the same fields as the actual currents induced in the antenna under test (AUT). The equivalent currents may be either magnetic or electric. The knowledge of the equivalent currents allows the determination of the antenna radiating elements, as well as the prediction of the AUT-radiated fields outside the equivalent currents domain. The unique feature of the novel methodology presented is that it can resolve equivalent currents that are smaller than half a wavelength in size, thus providing super-resolution. Furthermore, the measurement field samples can be taken at spacings greater than half a wavelength, thus going beyond the classical sampling criteria. These two distinctive features are possible due to the choice of a model-based parameter estimation methodology where the unknown sources are approximated by a basis in the computational Method of Moment (MoM) context and, secondly, through the use of the analytic free space Green's function. The latter condition also guarantees the invertibility of the electric field operator and provides a stable solution for the currents even when evanescent waves are present in the measurements. In addition, the use of the singular value decomposition in the solution of the matrix equations provides the user with a quantitative tool to assess the quality and the quantity of the measured data. Alternatively, the use of the iterative conjugate gradient (CG) method in solving the ill-conditioned matrix equations for the equivalent currents can also be implemented.

Finally, it is illustrated on how to characterize the radiation properties of an antenna at any distance and aspect angle from the actual physical structure. As before, once the equivalent sources are determined, the electromagnetic radiation of the antenna can be completely characterized at any point. This is particularly useful in commercial antenna characterization or verification since, usually, the geometry and material composition of the antenna are not completely available or easy to model and an electromagnetic simulation of the antenna can lead to inaccurate results. The method is very well suited for the establishment of near-field "reference volumes" or exclusion zones of arbitrary shape, for electromagnetic environmental studies or to comply with regulations and recommendations as regards to human exposure to non-ionizing electromagnetic fields.

References

1 Y. Rahmat-Samii, "Surface Diagnosis of Large Reflector Antennas Using Microwave Holographic Metrology: An Iterative Approach," *Radio Science*, Vol. 19, pp. 1205–1217, 1984.

2 A. D. Yaghjian, "An Overview of Near-Field Antenna Measurements," *IEEE Transactions on Antennas and Propagation*, Vol. 34, No. 1, pp. 30–45, 1986.

3 J. E. Hansen, J. Hald, F. Jensen, F. H. Larsen, and J. R. Wait, *Spherical Near-field Antenna Measurements*. Institution of Engineering and Technology (IET), London, England, 1988.

4 P. Petre and T. K. Sarkar, "Planar Near-Field to Far-Field Transformation Using an Equivalent Magnetic Current Approach," *IEEE Transactions on Antennas and Propagation*, Vol. 40, No. 11, pp. 1348–1356, 1992.

5 A. Taaghol and T. K. Sarkar, "Near-Field to Near/Far-Field Transformation for Arbitrary Near-Field Geometry Utilizing an Equivalent Magnetic Current," *IEEE Transactions on Electromagnetic Compatibility*, Vol. 38, No. 3, pp. 536–542, 1996.

6 F. Las-Heras, "Sequential Reconstruction of Equivalent Currents from Spherical Acquisition," *IEE Electronic Letters*, Vol. 35, No. 3, pp. 211–212, 1999.

7 F. Las-Heras and T. K. Sarkar, "A Direct Optimization Approach for Source Reconstruction and NF-FF Transformation Using Amplitude-Only Data," *IEEE Transactions on Antennas and Propagation*, Vol. 50, No. 4, pp. 500–510, 2002.

8 F. Las-Heras, M. R. Pino, S. Loredo, Y. Alvarez, and T. K. Sarkar, "Evaluating Near Field Radiation Patterns of Commercial Antennas," *IEEE Transactions on Antennas and Propagation*, Vol. 54, No. 8, pp. 2198–2207, 2006.

9 K. Persson and M. Gustafson, "Reconstruction of Equivalent Currents Using a Near-Field Data Transformation – with Radome Application," *Progress in Electromagnetic Research (PIER)*, Vol. 54, pp. 179–198, 2005.

10 Y. Alvarez, F. Las-Heras, and M. R. Pino, "Reconstruction of Equivalent Currents Distribution Over Arbitrary Three-Dimensional Surfaces Based on Integral Equation Algorithms," *IEEE Transactions on Antennas and Propagation*, Vol. 55, No. 12, pp. 3460–3468, 2007.

11 C. Cappellin, O. Breinbjerg, and A. Frandsen, "Properties of the Transformation from the Spherical Wave Expansion to the Plane Wave Expansion," *Radio Science*, Vol. 43, 2008.

12 Y. A. Lopez, F. Las-Heras, M. R. Pino, and T. K. Sarkar, "An Improved Super-Resolution Source Reconstruction Method," *IEEE Transactions on Instrumentation and Measurement*, Vol. 58, No.11, pp. 3855–3866, 2009.

13 P. Petre and T. K. Sarkar, "Differences Between Modal Expansion and Integral Equation Methods for Planar Near-Field to Far-Field Transformation," *Progress in Electromagnetics Research (PIER)*, J.A. Kong, editor, Vol. 12, EMW Publishing, Cambridge, MA, pp. 37–56, 1996.

14 S. A. Schelkunoff, "Some Equivalence Theorems of Electromagnetics and Their Application to Radiation Problems," *Bell System Technical Journal*, Vol. 15, pp. 92–112, 1936.

15 R. F. Harrington, *Time-Harmonic Electromagnetic Fields*, McGraw-Hill, New York, 1961.

16 R. F. Harrington, *Field Computation by Moment Methods*, Kreiger, Melbourne, FL, 1987.

17 G. H. Golub and C. F. Van Loan, *Matrix Computations*, Johns Hopkins University Press, Baltimore, MD, Second Edition, 1989.

18 J. Rahman and T. K. Sarkar, "Deconvolution and Total Least Squares in Finding the Impulse Response of an Electromagnetic Systems from Measured Data," *IEEE Transactions on Antennas and Propagation* Vol. 43, pp. 416–442, 1995.

19 Spherical near-field data obtained from C. Stubenrauch, NIST, Boulder, CO.

20 T. K. Sarkar, P. Petre, A. Taaghol, and R. F. Harrington, "An Alternate Spherical Near-Field to Far-Field Transformation," *Progress in Electromagnetic Research*, Vol. 16, pp. 268–284, EMW, Boston, MA, 1997.

21 A. Taaghol and T. K. Sarkar, "Near/Far Field Transformation for Arbitrary Near-Field Geometry Utilizing an Equivalent Electric Current and MOM," *IEEE Transactions on Antennas and Propagation*, Vol. 47, No. 9, pp. 566–571, 1999.

22 P. Petre and T. K. Sarkar, "Planar Near Field to Far Field Transformation Using an Array and Dipole Probes," *IEEE Transactions on Antennas and Propagation*, Vol. 42, pp. 534–537, 1994.

23 IEEE, "Standard for Safety Levels with Respect to Human Exposure to Radio-Frequency Electromagnetic Fields, 3kHz to 300 GHz," *IEEE C95.1-1991*, New York, 1992.

24 ICNIRP (International Commission on Non-Ionizing Radiation Protection), "Guidelines for Limiting Exposure to Time-Varying Electric, and Electromagnetic Fields (up to 300 GHz) – ICNIRP Guidelines", *Health Physics*, Vol. 74, No. 4, pp. 494–522, 1998.

25 EC, "Council Recommendation of 12 July 1999 on the Limitation of Exposure of the General Public to Electromagnetic Fields (0 Hz to 300 GHz)," *Official Journal of the European Community*, Vol. L199, p. 59, 1999.

26 Spain, "Real Decreto 1066/2001, de 28 de septiembre, por el que se aprueba el Reglamento que establece condiciones de protección del dominio público radioeléctrico, restricciones a las emisiones radioeléctricas y medidas de protección sanitaria frente a emisiones radioeléctricas" [*Regulation for the establishment of protection conditions of public radioelectric domain, limits to* *radioelectric emissions and health protection actions with respect to radioelectric emissions*]. BOE núm 234, 29-09-2001, 2001.

27 Italy, "Decree 10 September 1998, No.381. Regulations for the determination of ceiling values of radiofrequency compatible with human health," Gazzetta Ufficiale della Repubblica Italiana, n.257 del 3-11-1998, 1998.

28 Z. Altman, B. Begasse, C. Dale, A. Karwowski, J. Wiart, M. Wong, and L. Gattoufi, "Efficient Models for Base Station Antennas for Human Exposure Assessment," *IEEE Transactions on Electromagnetic Compatibility*, Vol. 44, pp. 588–592, 2002.

29 S. A. Hanna, "On Human Exposure to Radio-Frequency Fields Around Transmit Radio Sites," *Proceedings of the 1999 IEEE International Symposium Vehicular Technology Society*, Vol. 2, New Brunswick, NJ, pp. 1589–1593, 1999.

30 M. Barbiroli, C. Carciofi, G. Falciasecca, and M. Frullone, "Analysis of Field Strength Levels Near Base Station Antennas," *Proceedings of the 1999 IEEE International Symposium Vehicular Technology Society*, Amsterdam, The Netherlands, Vol. 1, pp. 156–160, 1999.

31 L. Catarinucci, P. Palazzari, and L. Tarricone, "Human Exposure to the Near-Field of Radiobase Antennas – A Fullwave Solution Using Parallel FDTD," *IEEE Transactions on Microwave Theory and Techniques*, Vol. 51, pp. 935–940, 2003.

32 G. Lazzi and O. P. Gandhi, "A Mixed FDTD-Integral Equation Approach for On-Site Safety Assessment in Complex Electromagnetic Environments," *IEEE Transactions on Antennas and Propagation*, Vol. 48, pp. 1830–1836, 2000.

33 D. Lautru, J. Wiart, W. Tabbara, and R. Mittra, "A MoMTD/FDTD Hybrid Method to Calculate the SAR Induced by a Base Station Antenna," *Proceedings of the 2000 IEEE International Symposium Antennas and Propagation Society*, Vol. 2, pp. 757–760, 2000.

34 P. Bernardi, M. Cavagnaro, S. Pisa, and E. Piuzzi, "Human Exposure to Radio Base-Station Antennas in Urban Environment," *IEEE Transactions on Microwave Theory and Techniques*, Vol. 48, pp. 1996–2002, 2000.

35 R. Cicchetti and A. Faraone, "Estimation of the Peak Power Density in the Vicinity of Cellular and Radio Base Station Antennas," *IEEE Transactions on Electromagnetic Compatibility*, Vol. 46, No. 2, pp. 275–290, 2004.

36 A. Faraone, R. Tay, K. Joyner, and Q. Balzano, "Estimation of the Average Power Density in the Vicinity of Cellular Base Station Antennas," *IEEE Transactions on Vehicular Technology*, Vol. 49, pp. 984–996, 2000.

37 *IEEE Recommended Practice for Measurements and Computations of Radio Frequency Electromagnetic Fields with Respect to Human Exposure to Such Fields, 100 kHz-300 GHz*, IEEE Std C95.3-2002.

38 *Federal Communications Commission Evaluating Compliance with FCC Guidelines for Human Exposure to Radio-Frequency Electromagnetic Fields*, 97–01 ed.: OET Bulletin 65, 1997.

39 A. M. Martinez-Gonzalez, A. Fernandez-Pascual, E. de los Reyes, W. Van Loock, C. Gabriel, and D. Sanchez-Hernandez, "Practical Procedure for Verification of

Compliance of Digital Mobile Radio Base Stations to Limitations of Exposure of the General Public to Electromagnetic Fields," *IEE Proceedings – Microwaves, Antennas and Propagation*, Vol. 149, pp. 218–228, 2002.

40 C. K. Chou, H. Bassen, J. Osepchuk, Q. Balzano, R. Petersen, M. Meltz, R. Cleveland, J. C. Lin, and L. Heynick, "Radio Frequency Electromagnetic Exposure: Tutorial Review on Experimental Dosimetry," *Bioelectromagnetics*, Vol. 17, No. 3, pp. 195–208, 1996.

41 O. P. Gandhi and M. S. Lam, "An On-Site Dosimetry System for Safety Assessment of Wireless Base Stations Using Spatial Harmonic Components," *IEEE Transactions on Antennas and Propagation*, Vol. 51, pp. 840–847, 2003.

42 J. Cooper, B. Marx, J. Buhl, and V. Hombach, "Determination of Safe Distance Limits for a Human Near a Cellular Base Station Antenna, Adopting the IEEE Standard or ICNIRP Guidelines," *Bioelectromagnetics*, Vol. 23, pp. 429–443, 2002.

43 C. Olivier and L. Martens, *A Practical Method for Compliance Testing of Base Stations for Mobile Communicatons with Exposure Limits*, Vol. 2, pp. 64–67, IEEE, Boston, MA, 2001.

44 W. T. Araujo Lopes, G. Glionna, and M. S. Alencar. "Generation of 3D Radiation Patterns: A Geometrical Approach," *IEEE Semiannual Vehicular Technology Conference*, Birmingham, AL, VTC Spring 2002.

45 A. Ziyyat, L. Casavola, D. Picard, and J. Ch. Bolomey "Prediction of BTS Antennas Safety Perimeter from NF to NF Transformation: An Experimental Validation," *AMTA* 2001, Denver, CO, 22–26 October 2001.

46 S. Blanch, J. Romeu, and A. Cardama, "Near Field in the Vicinity of Wireless Base-Station Antennas: An Exposure Compliance Approach," *IEEE Transactions on Antennas and Propagation*, Vol. 50, No. 5, pp. 685–691, 2002.

47 F. Las Heras Andrés, B. Galocha Iragüen, and J. L. Besada Sanmartín, "A Method to Transform Measured Fresnel Patterns to Far-Field Based on a Least-Squares Algorithm with Probe Correction," *Proceedings Antenna Measurement Techniques Association (AMTA), 17th Annual Meeting & Symposium*, Williamsburg, VA, pp. 352–357, November 1995.

48 F. Las-Heras and T. K. Sarkar, "Radial Field Retrieval in Spherical Scanning for Current Reconstruction and NF-FF Transformation," *IEEE Transactions on Antennas and Propagation*, Vol. 50, No. 6, pp.866–874, 2002.

49 F. Las-Heras, B. Galocha, and J. L. Besada. "Far-Field Performance of Linear Antennas Determined from Near-Field Data," *IEEE Transactions on Antennas and Propagation*, Vol. 50, No. 3, pp. 408–410, 2002.

50 F. Las-Heras, B. Galocha, and J. L. Besada, "Circular Scanning and Equivalent Magnetic Currents from Main Plane Near-Field to Far-Field Transformation," *International Journal of Numerical Modelling: Electronic Networks, Devices and Fields*, Vol. 15, No. 4, pp. 329–338, 2002.

6

Planar Near-Field to Far-Field Transformation Using a Single Moving Probe and a Fixed Probe Arrays

Summary

A fast and accurate method is presented for computing far-field antenna patterns from planar near-field measurements. The planar near-field data can be sampled either using a single moving dipole probe or using an array of dipole probes. The method then utilizes the measured near-field data to determine the equivalent magnetic current sources over a fictitious planar surface that encompasses the antenna, and these currents are then used to ascertain the far fields. Under certain approximations, the currents should produce the correct far fields in all regions in front of the antenna regardless of the geometry over which the near-field measurements are made. An electric field integral equation (EFIE) is developed using Maxwell's equations to relate the near fields measured using a single moving probe or using a fixed array of probes. In this methodology, use of small dipole probes does not require any probe corrections in the formulation. The Method of Moments (MOM) procedure is used to transform the EFIE into a matrix one to solve for the equivalent magnetic currents. The matrix equation is now solved using the iterative conjugate gradient method (CGM), and in the case of a rectangular matrix, a least-squares solution can still be found using this approach for the magnetic currents without explicitly computing the normal form of the equation. Near-field to far-field transformation for planar scanning may be efficiently performed under certain conditions by exploiting the block Toeplitz structure of the matrix and using CGM and the fast Fourier transform (FFT), thereby drastically reducing computation time and storage requirements. Numerical results are presented for several antenna configurations by extrapolating the far fields using synthetic and experimental near-field data.

It is also illustrated that a single scanning moving probe may be replaced by a fixed array of probes to sample the planar near field for the solution of the equivalent magnetic currents on the surface enclosing the AUT. It is also demonstrated that in this methodology a probe correction even when using an

Modern Characterization of Electromagnetic Systems and Its Associated Metrology, First Edition.
Tapan K. Sarkar, Magdalena Salazar-Palma, Ming Da Zhu, and Heng Chen.
© 2021 John Wiley & Sons, Inc. Published 2021 by John Wiley & Sons, Inc.

array of dipole probes is not necessary. The accuracy of this methodology is studied as a function of the size of the equivalent surface placed in front of the antenna under test and the error in the estimation of the far field along with the possibility of using a rectangular probe array which can efficiently and accurately provide the patterns in the principal planes. This can also be used when amplitude-only data is collected using an array of probes. Finally, it is shown that the probe correction can be useful when the physical lengths of the measurement dipole probes are comparable to a resonant antenna. And then it is shown under this circumstance how to carry it out the solution procedure followed by some conclusions and an abridged list of references.

6.1 Introduction

Near-field antenna measurements have become widely used in antenna testing since they allow for accurate measurements of antenna patterns in a controlled environment. The earliest works are based on the modal expansion method, in which the fields radiated by the test antenna are expanded in terms of planar, cylindrical, or spherical wave functions and the measured nearfields in any of these coordinate systems are then used to determine the coefficients of the appropriate wave functions [1–5]. Excellent overviews of the development of near-field scanning techniques are found in papers [6, 7]. The primary drawback of the modal expansion technique is that when the appropriate transform is used to compute the far fields, the near-fields outside the measurement region are assumed to be zero, particularly in the planar and cylindrical case. Consequently, the far fields are accurately determined only over a particular angular sector that is dependent on the separation between the plane containing the AUT and the measurement plane along with its size. Narasimhan and Kumar [8] have recently tried an approach similar to the one presented in this paper, in which measured near fields on a planar surface are used to synthesize source currents of planar arrays or apertures. However, the far fields were not ascertained from the equivalent sources. Furthermore, since the Fourier transform of the near fields was represented in terms of the finite Fourier transform of the measured near fields over a planar surface, all the truncation effects present in the modal expansion method are also found in their technique.

The equivalent current approach that represents an alternate method of computing far fields from measured near-fields has been recently explored [9–12]. In earlier papers, the radiating antenna was replaced by equivalent electric currents. The idea presented here is to replace the radiating antenna by equivalent magnetic currents and not electric currents which reside on a fictitious surface and encompass the antenna. Under certain approximation these equivalent currents should produce the correct far fields in all regions in front of the antenna.

In this approach the integral equation that pertains to the measured near fields and the equivalent magnetic currents is a decoupled one with respect to the coordinate axes for the planar scanning case. This means that two simple decoupled EFIE can be formulated, which contain only one component of the measured near fields. Each component of the magnetic current can be solved separately, in a parallel fashion.

A method of moments procedure [13] with point matching is used in which the equivalent magnetic currents are expanded in terms of two-dimensional pulse basis functions with unknown coefficients. The matrix elements of the resultant matrix equation represent the interaction between the probe and each of the current patches. The resultant MoM matrix may be rectangular or square depending on the number of data points and number of unknown currents chosen. The matrix equation is solved using CGM [14], and in the case of a rectangular matrix, a least-squares solution is found without explicitly computing the normal form of the equation. Near-field to far-field transformation for planar scanning thus may be efficiently performed under certain conditions. If the spacing between the measured field points and sizes of the unknown currents patches are chosen the same, the resultant system matrix is block Toeplitz. The structure of the matrix can be exploited by noting that a two-dimensional Fourier transform may be utilized to evaluate some expressions in CGM. This results in a tremendous decrease in storage and computation. The method has been called CGFFT and has previously been applied to analysis of radiation and scattering problems [15].

The other feature for the presented method is that the numerical integration in the process of creating the moment matrix elements can be avoided by taking the limiting case of the integral equation. For this special case, instead of using equivalent magnetic currents as equivalent sources, an equivalent magnetic dipole array is used to replace the aperture of the antenna. If the distance between the source and the field points is much larger than the sizes of the current patches, then these patches can be replaced by equivalent Hertzian dipoles. A detailed comparison of the accuracy and limitation of both equivalent currents and equivalent dipole array approximations is given in the present section.

The theoretical basis for the equivalent current approach is detailed in Section 6.2. The formulation of the EFIE is detailed in Section 6.3. The formulation of the matrix equation using MOM with the application of CGFFT is presented in Section 6.4. The idea of the equivalent magnetic dipole array approximation is given in Section 6.5. Numerical results for several antenna configurations are presented in Section 6.6 which is summarized in 6.7 followed by a detailed explanation on what is the difference between the classical use of the modal expansion method and the present use of equivalent sources to carry out the planar near-field to far-field transformation. In section 6.9 a direct optimization approach is presented to obtain the source reconstruction and a

Near-field (NF) to far-field (FF) transformation using amplitude-only data. In 6.10 it is shown how to extend the single moving probe methodology to deal with a fixed array of dipole probes to measure the NF. It is next shown in 6.11 on how to determine the FF patterns along principal planes using a rectangular probe array. The influence of the size of a square dipole probe array measurement plane on the accuracy of the NF to FF pattern. In 6.13 use of amplitude-only data is used to enhance the computational efficiency of the NF-FF transformation when utilizing a planar fixed probe array for measurement of the NF. Finally, in section 6.14 probe correction for using in electrically large probes is outlined followed by some conclusions and a list of references where additional materials may be found.

6.2   Theory

The methodology to be discussed next is quite similar to the one presented in section 5.2 but is repeated here as now it is specialized for a planar structure. Let us consider an arbitrary shaped antenna radiating into free space with the aperture of the antenna being a planar surface, which separates the entire space into left-half and right-half spaces.

Consider the general equivalent problem as shown in Figure 6.1. Because it is postulated that the electromagnetic fields, in the left-half space are zero, a perfect electric conductor can be placed on the xy-plane. If it is further assumed that for the general case, the tangential component of the electric field on S_∞ is zero except on S_0 then M' exists only on S_0. By the equivalence principle [16], a surface magnetic current M' may be placed on this perfect electric conductor (PEC) whose value is equal to the tangential value of the electric field on the xy-plane as

$$M' = E \times n \text{ on } S_\infty \tag{6.1}$$

where E is the electric field on the xy-plane, S_∞, which is the entire xy-plane at $z = 0$, and n is the unit outward normal to the xy-plane pointing in the direction of the positive z-axis.

Using image theory [1, 16], an equivalent magnetic current M can be introduced as seen in Figure 6.1 as

$$M = 2 M' \tag{6.2}$$

which is now radiating in free space, where M now replaces the source antenna and radiates into free space, producing exactly the same field as the original antenna in the region $z > 0$. If it is further assumed as for the general case that the tangential component of the electric field on S_∞ is zero except on S_0 then M exists only on S_0.

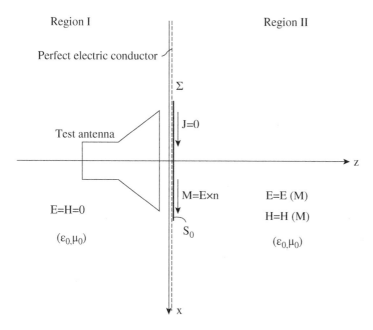

Figure 6.1 Original problem and the Equivalent problem introduced to produce the same external fields.

Therefore

$$M = 2\,E \times n \text{ on } S_0 \tag{6.3}$$

We can now use the measured NF to determine M. It is important to note that in the practical NF measurement process, usually only the electric near fields are measured. So we are focusing our attention to find the far fields from the measured electric near field only via the equivalent magnetic current approach, through

$$E_{meas} = E(M) \tag{6.4}$$

where E_{meas} is the electric near field measured over a geometry at a distance away from the aperture of the radiating antenna.

6.3 Integral Equation Formulation

From section 5.2, we find that

$$E(r) = \frac{1}{4\,\pi} \iint\limits_{S_\infty} [M(r') \times \nabla' G\,(r,r')]\,ds' \tag{6.5}$$

where $E(r)$ is the observed field at the observation point r. $M(r')$ is the equivalent magnetic current at the source point and r' is the source coordinate. ∇' is the gradient operator applied to the primed – or source coordinates. The free space Green's function $G(r, r')$ is given by

$$G(r,r') = \frac{e^{-j\,k_0\,|\,r-r'\,|}}{|\,r-r'\,|} \tag{6.6}$$

and $k_0 = \dfrac{2\,\pi}{\lambda}$ is the free space wave-number and λ is the wavelength of the radiated fields.

For the planar scanning case, the near-field measurement is performed over a planar surface assumed to be parallel with the source plane as shown in Figure 6.2. The aperture of the antenna (S_0) is assumed to be a rectangular plate in the x-y plane with dimensions W_x and W_y. The distance between the source plane (S_0) and the measurement plane is d. For the planar scanning the x and y components of the electric near-fields are usually measured.

Taking only the x and y components of the measured electric near fields into account in (6.5), the following integral equation can be obtained for the equivalent magnetic currents

$$\begin{bmatrix} E_{meas,\,x}(r) \\ E_{meas,\,y}(r) \end{bmatrix} = -\iint\limits_{S_0} \begin{bmatrix} 0 & \dfrac{\partial G(r,r')}{\partial z'} \\[2ex] -\dfrac{\partial G(r,r')}{\partial z'} & 0 \end{bmatrix} \times \begin{bmatrix} M_x(r') \\ M_y(r') \end{bmatrix} ds' \tag{6.7}$$

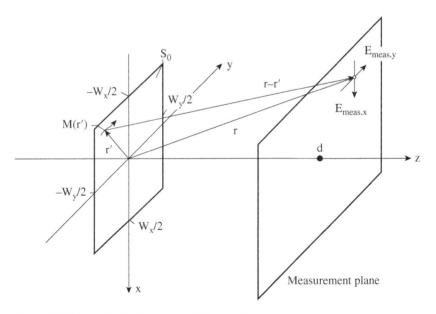

Figure 6.2 Schematic for Planar near field scanning.

It is evident from (6.7) that the resulting integral equation is a decoupled one with respect to the two components of the magnetic currents. So instead of solving (6.7), the two following simple decoupled integral equations

$$E_{meas,x}(\boldsymbol{r}) = -\iint\limits_{S_0} \frac{\partial G(\boldsymbol{r},\boldsymbol{r}')}{\partial z'} M_y(\boldsymbol{r}')ds' \tag{6.8}$$

$$E_{meas,y}(\boldsymbol{r}) = -\iint\limits_{S_0} \frac{\partial G(\boldsymbol{r},\boldsymbol{r}')}{\partial z'} M_x(\boldsymbol{r}')ds' \tag{6.9}$$

can be solved for separately. The presented method using the magnetic current approach is numerically more efficient than when using the equivalent electric current approach for planar scanning as the resulting integral equation decouples making solutions of very large matrix problems quite straightforward and simple.

6.4 Formulation of the Matrix Equation

After formulating the E-field integral equations, a MOM procedure is used to transform them into matrix equations. Both components of the equivalent magnetic currents (M_x and M_y) are approximated by equally spaced two-dimensional pulse basis functions as

$$M_x(x',y') = \sum_{i=1}^{N_x} \sum_{j=1}^{N_y} \alpha_{ij} \, \Pi_{ij}(x',y') \tag{6.10}$$

$$M_y(x',y') = \sum_{i=1}^{N_x} \sum_{j=1}^{N_y} \beta_{ij} \, \Pi_{ij}(x',y') \tag{6.11}$$

where α_{ij} and β_{ij} are the unknown coefficients of the x- and y-directed magnetic currents, respectively on the ijth patch. $\Pi_{ij}(x', y')$ is the two dimensional pulse basis function pertaining to the ijth patch and is defined by

$$\Pi_{ij}(x',y') = \begin{cases} 1 & \text{if} \quad \begin{cases} x_i - \dfrac{\Delta x}{2} \le x' \le x_i + \dfrac{\Delta x}{2} \\[2mm] y_j - \dfrac{\Delta y}{2} \le y' \le y_j + \dfrac{\Delta y}{2} \end{cases} \\[6mm] 0, & \text{otherwise} \end{cases} \tag{6.12}$$

For the above approximation the aperture plate (S_0) on which the equivalent magnetic current reside is assumed to be a rectangular one in the xy-plane of extension $-\dfrac{W_x}{2} \le x \le \dfrac{W_x}{2}$ and $-\dfrac{W_y}{2} \le y \le \dfrac{W_y}{2}$ as shown in Figure 6.2.

The region S_0 is now divided into $N_x N_y$ rectangular patches, each with the same dimensions Δx and Δy which are given by

$$\Delta x = \frac{w_x}{N_x}$$
$$\Delta y = \frac{w_y}{N_y} \tag{6.13}$$

In (6.12) x_i and y_j are the x- and y-coordinates of the center of the ijth patch and are given by

$$x_i = -\frac{w_x}{2} - \frac{\Delta x}{2} + i\,\Delta x$$
$$y_j = -\frac{w_y}{2} - \frac{\Delta y}{2} + j\,\Delta y \tag{6.14}$$

It is important to note that simple pulse basis functions can be used to approximate the magnetic currents because the integral equations do not contain any derivatives of these currents. This is a consequence of the previous assumption that the field points and the measurement points are not situated on the current carrying region S_0. Since it is assumed that the measured electric near fields are known at discrete points on the scanning plane, a point matching procedure is chosen.

Substituting (6.10) and (6.11) into (6.8) and (6.9) and utilizing point matching two decoupled matrix equations are obtained

$$E_{meas,x} = -GM_y$$
$$E_{meas,y} = -GM_x \tag{6.15}$$

where $E_{meas,\,x}$ and $E_{meas,\,y}$ are the vectors that contain the x and y components of the measured electric near fields, respectively. M_x and M_y are the column vectors that contain the unknown coefficients α_{ij} and β_{ij}, respectively. G is the moment matrix for the planar scanning case. The explicit expression for G is given by

$$G_{k,m} = \iint\limits_{\Omega_m} \frac{e^{-jk_0 R}}{R^2}(z_k - z')\left\{ jk_0 + \frac{1}{R} \right\}ds' \tag{6.16}$$

where Ω_m is the area of the mth patch and R is the distance between kth field point and the source point (r').

If the number of measured near-field points are the same as the number of current elements then the solution for (6.15) is unique. If the number of measured near-field points are larger than the number of current elements a least-squares solution is obtained. Gauss' quadrature formula is used to evaluate the two-dimensional integral numerically in (6.16) and CGM is used to solve the matrix equation $AX = Y$ resulting from (6.15).

The CG method [14, 15] starts with an initial guess X_1 and computes

$$R_1 = Y_1 - AX_1 \tag{6.17}$$

$$P_1 = A^*R_1 \tag{6.18}$$

and for $i = 1, 2, \ldots$, evaluate

$$a_i = \frac{\|A^*R_i\|}{\|AP_i\|} \tag{6.19}$$

$$X_{i+1} = X_i + a_i P_i \tag{6.20}$$

$$R_{i+1} = R_i - a_i AP_i \tag{6.21}$$

$$b_i = \frac{\|A^*R_{i+1}\|}{\|A^*R_i\|} \tag{6.22}$$

$$P_{i+1} = A^*R_{i+1} + b_i P_i \tag{6.23}$$

Here A^* is the conjugate transpose of A. Most of the computation in CGM occurs in the calculation of AP_i and $A*R_{i+1}$. Exploiting the block Toeplitz structure of the matrix A, these terms can be computed using a two-dimensional fast Fourier transform (FFT) [15, 17], by using

$$AP_i = \mathfrak{I}^{-1}\left\{\mathfrak{I}\left(A^c\right)\mathfrak{I}\left(P_i^c\right)\right\} \tag{6.24}$$

$$A^*R_i = f_c\left(\mathfrak{I}^{-1}\left\{\mathfrak{I}\left(A^c\right)\mathfrak{I}\left(R_{i+1}^{c^*}\right)\right\}\right)^* \tag{6.25}$$

Here $\mathfrak{I}$ denotes the two-dimensional discrete Fourier Transform, $\mathfrak{I}^{-1}$ the two-dimensional inverse discrete Fourier Transform, A^c is the convolution variation of the original matrix A. P_i^c and R_{i+1}^c are the convolution variations of the original vectors P_i and R_{i+1}, respectively, and $*$ denotes conjugate. For an $M \times N$ grid division the matrices A^c, P_i^c and R_{i+1}^c are $(2M - 1) \times (2N - 1)$, and therefore an FFT for nonpowers of two must be utilized to evaluate (6.24) and (6.25). A detailed description about the discrete convolution representation of a block Toeplitz system can be found in [15, 17].

Using the CGFFT method, for $M \times N$ measured field points, five matrixes of size $(2M - 1) \times (2N - 1)$ need to be stored, as opposed to one matrix of size $MN \times MN$ and four vectors of size $MN \times 1$ using CGM alone. Furthermore the required computation for the two dimensional FFT's is $[(2M - 1) \times (2N - 1)] \log 2 (M + N - 1)$ as opposed to $(MN)^2$, using CGM alone. If the number of measured data points is larger than the number of current elements chosen, a least-squares solution is found without explicitly computing the normal form of the equation. It is important to note that there is no change in the solution procedure when a least-squares formulation is applied. Details are available in [18].

6.5 Use of an Magnetic Dipole Array as Equivalent Sources

Instead of using equivalent magnetic currents as sources in this approach, equivalent magnetic dipole array can also be used to replace the aperture of the test antenna. This approximation can be explained from either a physical or mathematical point of view.

From a mathematical point of view, this equivalent magnetic dipole array approximation can be treated as a limit of the integral equation approach. For this limiting case, it can be assumed that the numerical integrations in the process of creating the moment matrix elements in (6.16) are executed using a one-point approximation, i.e.,

$$\iint_{\Omega_m} f\left(x_k, x', y_k, y'\right) ds' = \Omega_m f\left(x_k, x', y_k, y'\right) \tag{6.26}$$

In the above equation Ω_m is the area of the mth patch, $f(x_k, x', y_k, y')$ is the relevant function to be integrated, (x_k, y_k, z_k) are the coordinates of the kth field point and $\left(x'_m, y'_m, z'_m\right)$ are the coordinates of the center of the mth patch.

If the distance between the source and the field points are much larger than the sizes of the current patches, i.e.,

$$R \geq \max\left(\Delta x, \Delta y\right) \tag{6.27}$$

then the approximation given by equation (6.26) is valid. In our cases, the distance between the source plane and the scanning plane is always much greater than the largest size of the current patch.

Using this one-point approximation in equation (6.16) an equivalent magnetic dipole array approximation is developed because the right side in equation (6.26) can be treated as the appropriate electric field component of a magnetic dipole. From a physical point of view, this equivalent magnetic dipole array approximation can be treated as a special case of the general equivalent current approach. The advantage of this approximation is that the numerical integration in the process of creating the elements of the moment matrix can be avoided. This is a very time consuming procedure. The other advantage of using the magnetic dipole array as equivalent sources is that the moment matrix need not stored in an explicit form. This is because the elements of the moment matrix do not contain numerical integrations. This causes tremendous reduction in the storage requirements and makes the method computationally more efficient.

6.6   Sample Numerical Results

Far-field results for several planar antenna configurations are presented in this section. In all examples where near fields are measured on a planar surface, CGFFT has been employed. Experimental near-field data is utilized to ascertain far fields for a microstrip array. The extrapolated far fields are compared with those obtained using modal expansions. The experimental data was provided by Dr. Carl Stubenrauch from NIST [19].

In the examples, far fields obtained using planar measurements and the equivalent magnetic dipole array approximation are compared with those obtained using planar measurements and analyzing using planar modal expansions. Consider a microstrip array consisting of 32 × 32 uniformly distributed patches on a 1.5 m × 1.5 m surface. The operating frequency is 3.3 GHz. The array is considered to be in the x-y plane. For the planar modal expansion approach, the near fields are measured on a plane 3.24 m × 3.24 m at a distance of 35 cm from the array. There are 81 × 81 points measured 4 cm apart. The 81 × 81 data points are zero padded to produce 128 × 128 far-field points. Measurements are performed using a WR 284 waveguide.

For the first equivalent magnetic dipole array approximation, a 1.56 m × 1.56 m surface with 39 × 39 uniformly distributed magnetic dipoles are used for the source plane and a 3.24 m × 3.24 m surface with 81 × 81 measured near-field points are used for the field plane. So the solution obtained here is a least-squares one with exploiting the block Toeplitz structure of the matrix and utilizing CGFFT.

Figure 6.3 shows the normalized absolute values of the electric far-field component $|E_\phi|$ versus θ in dB for $\phi = 0°$ using the equivalent magnetic dipole array approximation and planar modal expansion. This is the copolarization pattern. The result obtained using the equivalent magnetic dipole array approximation and the result obtained using the planar modal expansion show excellent agreement up to ± 60° of $\phi = 0°$ and show acceptable agreement in the rest of the elevation range.

Figure 6.4 shows the normalized values of $|E_\theta|$ versus θ in dB for $\phi = 90°$ using the magnetic dipole array approximation and the planar modal expansion, respectively. This polarization is the main one. The agreement between the results obtained using the different approximations is excellent up to ± 60° of $\theta = 0°$ and acceptable in the rest of the elevation range.

Figure 6.5 shows the normalized values of $|E_\theta|$ versus θ in dB for $\phi = 0$ using the magnetic dipole array approximation and the planar modal expansion, respectively. This is the cross polarization. The result obtained using the equivalent magnetic dipole array approximation and the result obtained using the planar modal expansion show excellent agreement up to ± 60° of $\theta = 0°$ and acceptable agreement in the rest of the elevation range.

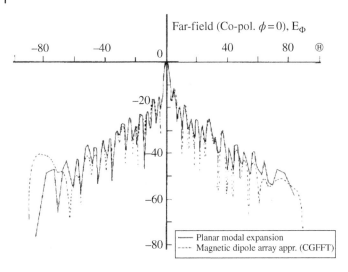

Figure 6.3 Copolarization characteristic for ϕ = 0° cut for a 32 × 32 patch microstrip array using planar modal expansion and equivalent magnetic dipole array approximation.

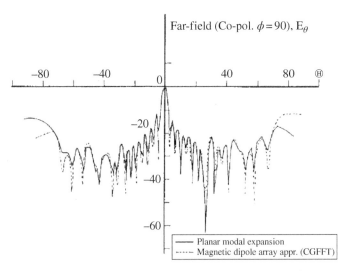

Figure 6.4 Copolarization characteristic for ϕ = 90° cut for a 32 × 32 patch microstrip array using planar modal expansion and equivalent magnetic dipole array approximation.

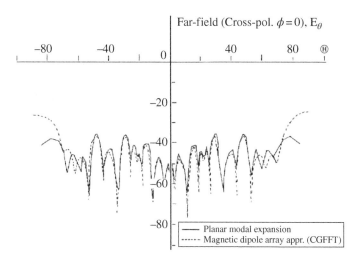

Figure 6.5 Cross-polarization characteristic for ɸ = 0° cut for a 32 × 32 patch microstrip array using planar modal expansion and equivalent magnetic dipole array approximation.

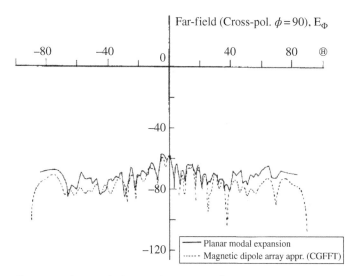

Figure 6.6 Cross-polarization characteristic for ɸ = 90° cut for a 32 × 32 patch microstrip array using planar modal expansion and equivalent magnetic dipole array approximation.

Figure 6.6 shows the normalized values of $|E_\phi|$ versus θ in dB for ɸ = 90 using the magnetic dipole array approximation and the planar modal expansion, respectively. This is the cross polarization. The shapes of the curves and the level of the sidelobes are the same for both approximations. Because the level of this cross-polarization characteristics is much lower than the level of the copolarization ones, largest absolute error may be allowed for these curves as for the main polarization characteristics. For Figures 6.3 to 6.6 it is known that the conventional planar modal expansion would provide acceptable results up to $\tan^{-1}(87/35) = 68°$ [20]. Perhaps that is the reason why the planar modal expansion results tend to deviate from our results beyond ± 60°.

For the second equivalent magnetic dipole array approximation, a 1.48 m × 1.48 m surface with 37 × 37 uniformly distributed magnetic dipoles is used for the source plane, and a 1.48 m × 1.48 m surface with 37 × 37 measured NF points is used for the field plane to enable use of CGFFT. It is important to note that for this case the conventional planar expansion method cannot be used because the area of the measured plane is equal to the area of the actual aperture of the test antenna. So the maximum angle for accurate far field (θ) when the planar modal expansion method is used is equal to zero [20].

Figure 6.7 shows the normalized absolute value of the electric far-field component $|E_\phi|$ versus θ in dB for φ = 0° using both equivalent magnetic dipole array approximations and a least squares solution. This is the co-polarization pattern. The result obtained using the first approximation and the result obtained using the second one show good agreement up to ± 35° for θ = 0° and show acceptable agreement in the rest of the elevation range. This formulation is more sensitive to the measurement errors than the least squares one.

Figure 6.8 shows the normalized values of $|E_\theta|$ versus θ in dB for φ = 90° using both equivalent magnetic dipole array approximations and a least squares

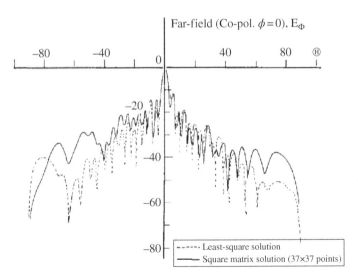

Figure 6.7 Copolarization characteristic for φ = 0° cut for a 32 × 32 patch microstrip array using least squares and square matrix solution.

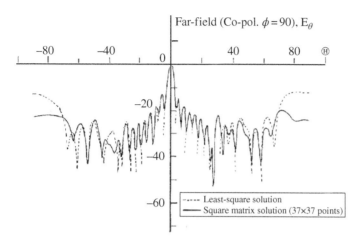

Figure 6.8 Copolarization characteristic for ϕ = 90° cut for a 32 × 32 patch microstrip array using least squares and square matrix solution.

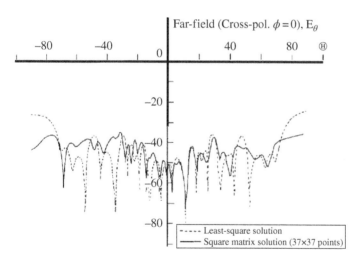

Figure 6.9 Cross-polarization characteristic for ϕ = 0° cut for a 32 × 32 patch microstrip array using least squares and square matrix solution.

solution. The polarization is the main one. The agreement between the results obtained using the different approximations is excellent up to ± 40° θ = 0° and acceptable in the rest of the elevation range, but the planar modal expansion cannot be applied to this data.

Figure 6.9 shows the normalized values of |E_θ| versus θ in dB for ϕ = 0° using both magnetic dipole array approximations and a least squares solution. This is

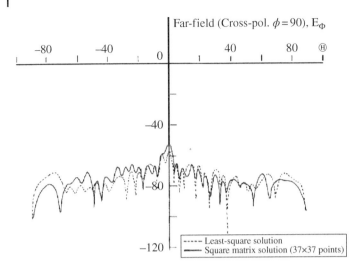

Figure 6.10 Cross-polarization characteristic for ϕ = 90° cut for a 32 × 32 patch microstrip array using least squares and square matrix solution.

the cross polarization. The result obtained using the first approximation and the result obtained using the second one show good agreement up to ±40°. Figure 6.10 shows the normalized values of $|E_\phi|$ versus θ in dB for ϕ = 90° using both magnetic dipole array approximations. This is the cross polarization. Again, the agreement between the results obtained using the different approximations is reasonable in the entire elevation range.

For the final example, we measure the planar near-field data in a narrow region (3.24 m × 0.76 m) and compute the copolar and the crosspolar pattern for ϕ = 0° as shown in Figure 6.11 and 6.12. The surface over which the equivalent currents are applied is approximated with 39 × 39 uniformly distributed magnetic dipoles on a 1.56 m × 1.56 m surface. The number of measurement points are 81 × 19, so a least-squares solution is obtained exploiting the block Toeplitz structure of the matrix and utilizing CGFFT. It is important to note that the modal expansion method does not allow us to approximate the source and the measurement plates in two different ways. Also in the figures, the results obtained using the modal expansion method are given for comparison.

Similarly, the copolar and the crosspolar patterns of Figures 6.13 and 6.14 for ϕ = 90° has been computed using data measured in the narrow region (0.76 m × 3.24 m).

Summarizing our computational results it can be concluded that the equivalent magnetic dipole array approximation gives accurate far-field patterns at angles beyond which the conventional planar modal expansion technique will fail. For example, for Figures 6.11 to 6.14, the conventional planar modal

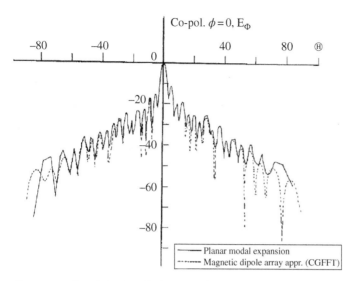

Figure 6.11 Copolarization characteristic for ϕ = 0° cut for the microstrip array when the near-field is measured in a narrow region (3.24 m × 0.76 m) as opposed to (3.24 m × 3.24 m).

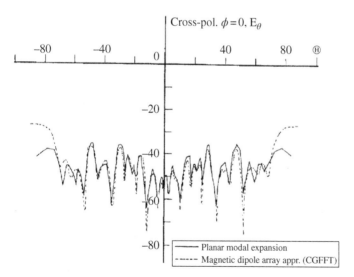

Figure 6.12 Cross-polarization characteristic for ϕ = 0° cut for the microstrip array with data measured on a narrow region (3.24 m × 0.76 m).

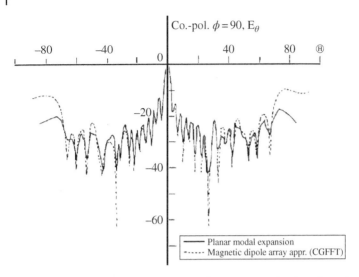

Figure 6.13 Copolarization characteristic for φ = 90° cut for the microstrip array when the near-field is measured in a narrow region (0.76 m × 3.24 m) as opposed to (3.24 m × 3.24 m).

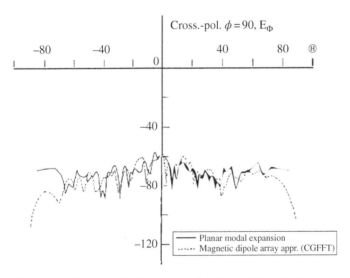

Figure 6.14 Cross-polarization characteristic for φ = 90° cut for the microstrip array with data measured on a narrow region (0.76 m × 3.24 m).

expansion will not work [18] as the sizes of the measurement plane and the surface covering the AUT are the same. The present approach provides accurate results up to ± 40° and the deterioration beyond that angle is graceful. The accuracy of the presented methodology depends mainly upon the accuracy of the measured near-field data and does not depend so much on the sizes of the measurement configuration. When a least-squares solution is obtained for the equivalent sources, the calculated far fields are not so sensitive to the errors in the noisy data compared to for the approximation when a square matrix solution is used.

6.7 Summary

A simple method is presented for computing far-field antenna patterns from near-field measurements. The method utilizes near-field data to determine equivalent magnetic source currents or an equivalent magnetic dipole array over the aperture of the antenna. Using this method it is possible to find the far fields of antennas that are not highly directive over large elevations and azimuthal ranges without using spherical scanning. Furthermore, the far field may be found with any desired resolution, and interpolation of the results is not required. Under certain conditions CGFFT may be used, thereby drastically reducing computation and storage requirements. Instead of using the equivalent magnetic current approach, a magnetic dipole array approximation can be applied, thereby eliminating the numerical integration in the process of creating the moment matrix elements. This method has a wider range of validity than the conventional modal expansion method.

6.8 Differences between Conventional Modal Expansion and the Equivalent Source Method for Planar Near-Field to Far-Field Transformation

6.8.1 Introduction

Near field antenna measurements are widely used in antenna testing since they allow for accurate measurements of antenna patterns in a controlled environment. The earliest works are based on the modal expansion method in which the fields radiated by the test antenna are expanded in terms of planar, cylindrical or spherical wave functions and the measured near fields are used to determine the coefficients in the expansion [4, 6, 7, 20, 21]. In this section, we focus our attention primarily on planar near-field measurements. A problem of the planar modal expansion technique is that the fields outside the measurement region are assumed to be zero. This assumption results in a systematic error in the computations involved in the planar modal expansion theory. If the measurement plane is of dimension $L \times L$ and the source plane (the plane which encompasses the sources) is of dimension $D \times D$, then the planar modal

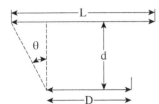

Figure 6.15 Relationship between scan length and the angular region of validity.

expansion theory can provide fields which are accurate up to an angle θ, which is given by

$$\theta = \tan^{-1} \frac{L-D}{2d} \tag{6.28}$$

where d is the separation distance between the two plane as shown in Figure 6.15 [20].

An alternate method of computing far-fields from measured near-fields has recently been explored [11, 18] utilizing an equivalent magnetic current approach which can go beyond the range of accuracy of the classical modal expansion technique. This method utilizes the measured near field data obtained on the $L \times L$ measurement plane to determine equivalent currents on the $D \times D$ source plane which encompass all sources as mentioned earlier. These equivalent currents may then be used to image the sources on the antenna if desired (for fault diagnosis) and also to compute the far-fields. This is in distinction to the planar modal expansion method where the far fields are computed first and then they are Fourier transformed to the source plane to image the sources [22].

The objective of this section is to demonstrate that:

A) When the measurement plane is infinite the planar modal expansion technique and the integral equation technique provide the same analysis equations. It will be shown that in that case the planar modal expansion solves the integral equation in the spectral domain [23] whereas the newly developed equivalent current approach solves the problem in the space domain.

B) The truncation error in the equivalent current approach is smaller as one transfers the data from the measurement plane to the source plane before Fourier transforming it to compute the far fields. Hence, the results for the equivalent current approach should hold over a larger azimuth angle θ than the planar modal expansion method.

Section 6.8.2 describes the planar modal expansion method. Section 6.8.3 presents the equivalent current approach using the integral equation and shows the equivalence between the two techniques. Limited numerical results are presented in Section 6.8.4 based on both numerically simulated and experimental data.

6.8.2 Modal Expansion Method

For the modal expansion method, two components of the electric field E_x and E_y are measured on the measurement plane, which is at a distance d from another plane located at $z = 0$, which is referred to as the source plane. All the sources are assumed to be behind this source plane as shown in Figure 6.16, i.e. $z \leq 0$.

The sources may not be planar sources. The measured fields are $E_{meas,x}$ and $E_{meas,y}$ at the measurement plane. Since an equivalent magnetic current on a plane is related to the tangential components of the electric fields, then one could say that on the measurement plane there are two equivalent magnetic currents, as given in equations (6.1)-(6.3). The far field from the two magnetic current sheets can be evaluated as [24]

$$E_{far,x}(r,\theta,\phi) = jk_0 \cos\theta \frac{e^{-jk_0 r}}{4\pi r} \int\int_{-\infty}^{\infty} M_y(x',y',z'=d)\, e^{j\vec{k}\cdot\vec{r}'}\, dx'dy'$$

(6.29)

where

$$\vec{k}\cdot\vec{r}' = k_0 \sin\theta\cos\phi\, x' + k_0 \sin\theta\sin\phi\, y' + k_0 \cos\theta\, z' = x'k_x + y'k_y + z'k_z$$

(6.30)

and

$$k_0^2 = k_x^2 + k_y^2 + k_z^2 = (2\pi/\lambda_0)^2$$

(6.31)

k_0 is the free space wavelength. The y-component of the far field is given by

$$E_{far,y}(r,\theta,\phi) = -jk_0 \cos\theta \frac{e^{-jk_0 r}}{4\pi r} \int\int_{-\infty}^{\infty} M_x(x',y',z'=d)\, e^{j\vec{k}\cdot\vec{r}'}\, dx'dy'$$

(6.32)

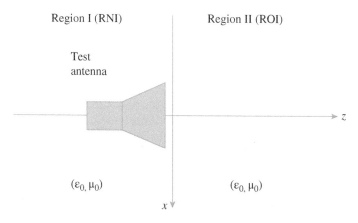

Region I (RNI)   Region II (ROI)

Test antenna

(ε_0, μ_0) (ε_0, μ_0)

Figure 6.16 All sources are located for $z < 0$.

If we consider $\tilde{M}_x\,(k_x,\,k_y,\,z^* = d)$ to be the two dimensional Fourier transform of $M_x(x^*,\,y^*,\,z^* = d)$, then

$$\tilde{M}_x\left(k_x, k_y, z' = d\right) = \int\int_{-\infty}^{\infty} M_x(x', y', z' = d)\, e^{\,j\,\left(x'k_x + y'k_y\right)} dx' dy' \qquad (6.33)$$

and

$$\tilde{M}_y\left(k_x, k_y, z' = d\right) = \int\int_{-\infty}^{\infty} M_y(x', y', z' = d)\, e^{\,j\,\left(x'k_x + y'k_y\right)} dx' dy' \qquad (6.34)$$

Using (6.33) and (6.34), (6.29) and (6.32) become

$$E_{far,x}(r,\theta,\phi) = C \cos\theta\,\tilde{M}_y\left(k_x, k_y, z' = d\right) e^{\,j\,d\sqrt{k_0^2 - k_x^2 - k_y^2}} \qquad (6.35)$$

$$E_{far,y}(r,\theta,\phi) = C \cos\theta\,\tilde{M}_x\left(k_x, k_y, z' = d\right) e^{\,j\,d\sqrt{k_0^2 - k_x^2 - k_y^2}} \qquad (6.36)$$

where C is some constant dependent on the parameter k_0.

If we consider $m_x(x^*, y^*, z^* = 0)$ and $m_y(x^*, y^*, z^* = 0)$ be two planar equivalent current sheets at the source plane $z = 0$, then $\tilde{m}_x\,(k_x,\,k_y,\,z^* = 0)$ and $\tilde{m}_y$ $(k_x,\,k_y,\,z^* = 0)$ will be their respective two dimensional Fourier transforms.

In that case

$$E_{far,x}(r,\theta,\varphi) = C \cos\theta\,\tilde{m}_y\left(k_x, k_y, z^* = 0\right) \qquad (6.37)$$

$$E_{far,y}(r,\theta,\varphi) = C \cos\theta\,\tilde{m}_x\left(k_x, k_y, z^* = 0\right) \qquad (6.38)$$

Hence the equivalent magnetic currents at the source plane and on the measurement plane are related in the spectral domain by

$$\tilde{m}_y\left(k_x, k_y, z' = 0\right) = \tilde{M}_y\left(k_x, k_y, z' = d\right) e^{\,j\,d\sqrt{k_0^2 - k_x^2 - k_y^2}} \qquad (6.39)$$

$$\tilde{m}_x\left(k_x, k_y, z' = 0\right) = \tilde{M}_x\left(k_x, k_y, z' = d\right) e^{\,j\,d\sqrt{k_0^2 - k_x^2 - k_y^2}} \qquad (6.40)$$

Therefore, to image the source plane, in the conventional planar modal expansion one takes the two dimensional Fourier transform of the measurement data to go to the far field and then the far field is directly related to the equivalent currents on the source plane by the inverse Fourier transform. The problem and the solution procedure is exact and no approximation is involved.

However, the problem is generated due to natural constraints. The limits for the double integrals in (6.33) and (6.34) are no longer infinity but finite. If the measurement plane is finite, then while taking the Fourier transforms of (6.29)–(6.34) it is clear that many of the relationships will be only approximate! The question is: under the constraint that the dimension of the measurement plane is finite which set of magnetic currents provide more accurate results for the far fields? Are they provided by $\tilde{M}_x$ and $\tilde{M}_y$ or by $\tilde{m}_x$ and $\tilde{m}_y$?

If the nonplanar sources are located behind the source plane and assuming Huygen's principle to hold, the source plane would capture more of the energy in the fields than the measurement plane if they are assumed to be of the same dimensions. This is because if the measurement plane, which is of the same dimension as the source plane, is situated at an additional distance d away from the sources it has to intercept less of the emanating fields. Hence the magnetic currents represented by M_x and M_y has a larger truncation error than m_x and m_y.

It is our intent to show that the integral equation approach utilizing the equivalent magnetic current approach solves for m_x (x^*, y^*, $z^* = 0$) and m_y (x^*, y^*, $z^* = 0$) and then performs a two dimensional Fourier transform to find the far fields.

To find E_θ and E_φ, the z-component of the electric field is also required. However, it is found from the divergence equation. Since in a source free region,

$$\vec{\nabla} \cdot \vec{E} = \frac{\partial E_x}{\partial x} + \frac{\partial E_y}{\partial y} + \frac{\partial E_z}{\partial z} = 0 \qquad (6.41)$$

If $\tilde{E}_z$ (k_x, k_y) denote the two dimensional Fourier transform of $E_z(x^*, y^*)$ then

$$\tilde{E}_z\left(k_x, k_y\right) = -\frac{k_x\tilde{E}_x\left(k_x, k_y\right) + k_y\tilde{E}_y\left(k_x, k_y\right)}{k_z} \qquad (6.42)$$

And from (5)

$$k_z = \sqrt{k_0^2 - k_x^2 - k_y^2} \qquad (6.43)$$

So once $\tilde{E}_z\left(k_x, k_y\right)$ is known from (6.42), the far fields are given by

$$E_\theta = E_{far,x} \cos\theta \cos\phi + E_{far,y} \cos\theta \sin\phi - E_{far,z} \sin\theta \qquad (6.44)$$

$$E_\phi = -E_{far,x} \sin\phi + E_{far,y} \cos\phi \qquad (6.45)$$

In the next section we look at the integral equation approach utilizing the equivalent magnetic current.

6.8.3 Integral Equation Approach

For the integral equation approach, a fictitious source plane is considered where the equivalent currents are located and is of the same dimension as the measurement plane but translated a distance d towards the sources. So the source plane is located at $z^* = 0$. As outlined in [18] an equivalent magnetic current of strength $2m$ is placed on the source plane at $z = 0$. The factor of 2 arises due to the application of the equivalence principle to a magnetic current radiating in the presence of a perfectly conducting ground plane. On this source plane, we put fictitious magnetic currents. The basic philosophy here is that if one knows the complex values of the magnetic currents on the source plane, one can evaluate the fields at the measurement plane. Conversely, if the measurement fields are known, then one can find the complex amplitudes of the magnetic currents m put on the source plane. Mathematically

$$E_{meas,x} = \iint_{-\infty}^{\infty} 2\,\tilde{m}_y(x',y',z'=0)\frac{\partial G(r,r')}{\partial z'}\,dx'\,dy' \qquad (6.46)$$

$$E_{meas,y} = \iint_{-\infty}^{\infty} 2\,\tilde{m}_x(x',y',z'=0)\frac{\partial G(r,r')}{\partial z'}\,dx'\,dy' \qquad (6.47)$$

where $G(r,r')$ is the free space Green's function. The explicit expressions is

$$\frac{\partial G(r,r')}{\partial z'} = \frac{e^{-j k_0 |r-r'|}}{4\pi|r-r'|^2}(z-z')\left[jk_0 + \frac{1}{|r-r'|}\right] \qquad (6.48)$$

where $|r - r'|$ is the distance between the source point and the field point. Observe here, $z^* = 0$ and $z = d$.

Now if we take the two dimensional Fourier Transform of both sides, this results in

$$\iint_{-\infty}^{\infty} E_{meas,x}(x,y,z=d)\,e^{j(k_x x + k_y y)}\,dxdy = 2\,\tilde{m}_y(k_x,k_y)\,\tilde{g}(k_x,k_y)$$

$$(6.49)$$

where $\tilde{m}_y$ is the two dimensional Fourier Transform of the magnetic currents located at the source plane and $\tilde{g}$ is two dimensional Fourier transform of the derivative of the Green's function

$$\tilde{G}(k_x,k_y,z,z') = \frac{-j}{2k_0}e^{j|z-z'|\sqrt{k_0^2 - k_x^2 - k_y^2}} \qquad (6.50)$$

with

$$\mathrm{Re}\left\{\sqrt{k_0^2 - k_x^2 - k_y^2}\right\} \geq 0 \qquad (6.51)$$

and

$$\mathrm{Im}\left\{\sqrt{k_0^2 - k_x^2 - k_y^2}\right\} < 0 \qquad (6.52)$$

In the transformed domain the derivative of the spectral domain with respect to z' yields

$$\frac{\partial}{\partial z'}\tilde{G}(k_x,k_y,z,z') = \frac{1}{2}\,\mathrm{sgn}\,(z-z')\,e^{j|z-z'|\sqrt{k_0^2 - k_x^2 - k_y^2}} \qquad (6.53)$$

where sgn(z) is the signum function. Since $z^* = 0$ and $z = d$

$$\tilde{g}(k_x,k_y) = \frac{1}{2}e^{jd\sqrt{k_0^2 - k_x^2 - k_y^2}} \qquad (6.54)$$

Utilizing (6.47) and (6.52) and comparing it with (6.29) and (6.30) it becomes clear that

$$\tilde{M}_y\left(k_x, k_y, z = d\right) = \tilde{m}_y\left(k_x, k_y, z = 0\right) e^{-jd\sqrt{k_0^2 - k_x^2 - k_y^2}} \tag{6.55}$$

and

$$\tilde{M}_x\left(k_x, k_y, z = d\right) = \tilde{m}_x\left(k_x, k_y, z = 0\right) e^{-jd\sqrt{k_0^2 - k_x^2 - k_y^2}} \tag{6.56}$$

Hence (6.53) and (6.54) are equivalent to (6.39) and (6.40). So if the information on the measurement plane were available, then the objective in the integral equation approach is basically to transfer the measurement data to the source plane and then take the Fourier transform to the far field. However, if the measurement plane is finite in size then the transfer of the data from $z = d$ plane to $z = 0$ plane utilizing the Fourier transform is not accurate because of the truncation error. Therefore, we utilize an alternate transformation to go to the source plane. This alternate transformation is utilized in the integral equation approach through the utilization of the Green's function. Theoretically, this reduces the truncation error problem introduced by the two dimensional Fourier transform.

When the measurement plane is finite in nature then (6.55) and (6.56) do not hold and one has to use (6.46) and (6.47) to solve for m_x and m_y. The classical method of moments has been utilized to solved for m_x and m_y. The unknowns are replaced by elementary dipoles. This is a good approximation as long as the source and the measurement planes are separated by a wavelength (i.e. $d = \lambda$) [24]. Secondly, the use of dipoles eliminates the need for integration of the matrix elements in the evaluation of the impedance matrix. The two decoupled equations (6.46) and (6.47) are solved as the two matrix equations:

$$\left[E_{meas,x}\right] = \left[Z_{yx}\right]\left[m_y\right] \tag{6.57}$$

$$\left[E_{meas,y}\right] = \left[Z_{xy}\right]\left[m_x\right] \tag{6.58}$$

These matrices can be very large. However these large equations can be solved very efficiently utilizing the FFT and the conjugate gradient method as outlined in [18]. Also the solution of (6.57) and (6.58) are decoupled and can be carried out simultaneously. Typically for 6400 unknowns for m_x or m_y the number of iterations taken to provide acceptable solution is about 10. Since per iteration, two two-dimensional FFT is computed, the integral equation method is about 20 times slower than the modal expansion method for 6400 unknowns. Typically on a VAX (a very old computer) workstation this amounts to a few seconds of CPU time for the modal expansion method as compared to several minutes of CPU time for the integral equation method to solve for the unknown magnetic currents m_x and m_y at the source plane. Once these magnetic currents are known, the far field can easily be computed utilizing the FFT.

6.8.4 Numerical Examples

In this section, we consider a theoretical simulation and a second example utilizing experimental data to illustrate the methodologies.

As a first example consider an array of 4 x-directed magnetic Hertzian dipoles placed on the corners of a 4 λ by 4 λ planar surface. This is considered as the test antenna. The measurement plane is considered of size 4.5 λ by 4.5 λ and is situated at a distance of 3 λ from the source plane. Near fields are computed at 15 equispaced points of 0.3 λ spacing (i.e., $\delta x = \delta y = 0.3 \lambda$). Utilizing the planar modal expansion to compute the far-field will be equivalent to taking the 2-dimensional Fourier transform of the 15×15 points. The normalized far fields E_θ ($\varphi = 0$) and E_φ ($\varphi = 90°$) for various values of θ are presented in Figures 6.17 and 6.18.

As expected the accuracy of the planar modal expansion will be valid up to $\tan^{-1}(0.25/3) \approx 10°$ [20]. The numerical simulations do illustrate that the calculated values for the far fields are acceptable up ± 12°.

For the integral equation approach, 15×15 magnetic dipole sources were considered at the source plane $z = 0$. The amplitudes of the 15×15 dipole sources were computed utilizing the integral equation approach utilizing magnetic currents for the same data over the 15 × 15 grid in the measurement plane.

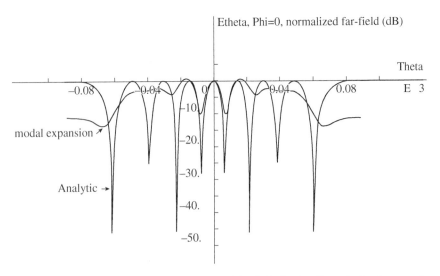

Figure 6.17 Comparison of exact and computed far-fields for φ = 0° cut for 2 × 2 magnetic dipoles on a 4λ × 4λ surface using modal expansion method.

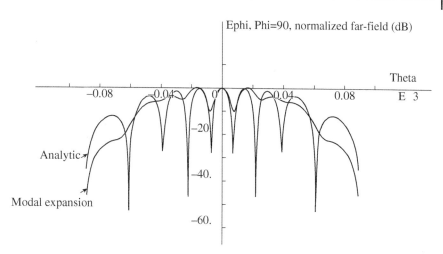

Figure 6.18 Comparison of exact and computed far-fields for ɸ = 90° cut for 2 × 2 magnetic dipoles on a 4λ × 4λ surface using modal expansion method.

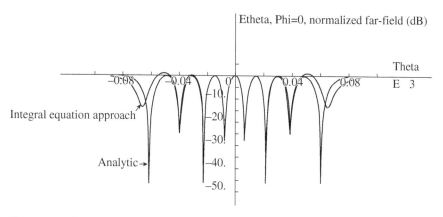

Figure 6.19 Comparison of exact and computed far-fields for ɸ = 0° cut for 2 × 2 magnetic dipoles on a 4λ × 4λ surface using integral equation method.

The 2D Fourier transform is then taken to find the far fields. The results are given in Figures 6.19 and 6.20. Figure 6.19 provides E_θ ($\varphi = 0$) and Figure 6.20 describes E_θ ($\theta = 90°$) for various angles of θ.

It is seen for this case the calculated far fields are accurate up to ± 50°. This is because the truncation error is reduced by transforming the measured fields to an equivalent plane much closer to the sources as discussed in the earlier section.

As a second example, we consider experimental data for a microstrip array. This microstrip array consists of 32 × 32 uniformly distributed dipoles over a

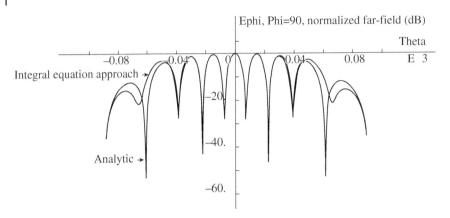

Figure 6.20 Comparison of exact and computed far-fields for $\phi = 90°$ cut for 2 × 2 magnetic dipoles on a $4\lambda \times 4\lambda$ surface using integral equation method.

1.5 m × 1.5 m surface. The operating frequency is 3.3 GHz. The array is considered in the *x-y* plane. The *x* and *y* components of the electric near fields are measured on a plane 3.24 m × 3.24 m at a distance of 35 cm from the array. There are 81× 81 measured points at a distance of 4 cm apart. This planar near field data was provided by Dr. Carl Stubenrauch of NIST [19].

If the planar modal expansion method is applied on the 81 × 81 measured data points, the computed far field will be accurate up to $\tan^{-1}(87/35) \approx 68°$ [20]. The results obtained using the modal expansion method were compared with the integral equation method as outlined in [18]. In [18], it was seen that the agreement between the two methods were good up to ± 60° and acceptable in the rest of the elevation range.

To demonstrate that by transferring the measured data to a fictitious surface, we are indeed reducing the truncation error, we take only 41×41 samples discarding 75% of the data. Therefore, we are considering the data only over a 1.64 m ×1.64 m measurement plane. By observing the magnitudes of the two components of the electric field (namely E_y in Figure 6.21 and E_x in Figure 6.22, we observe that the amplitude of the normalization component for E_x is 23.8 dB below that for E_y) it is clear that by taking 41× 41 points introduces significant truncation error if we were to take the direct Fourier transform of the data in the conventional modal expansion method. The Figure 6.21 and Figure 6.22 uses all the 81×81 data points. Hence by considering only 41× 41 data points in Figure 6.21 and Figure 6.22, one is incurring a large truncation error in the conventional modal expansion method.

If the modal expansion method is applied to the 41 × 41 data points, the calculated far fields would be accurate up to $\tan^{-1}(7/35) \simeq 11.3°$. For the integral equation approach, we consider a 41×41 magnetic dipole array uniformly distributed over a 1.64 m × 1.64 m surface at the $z = 0$ plane. From the 41×41

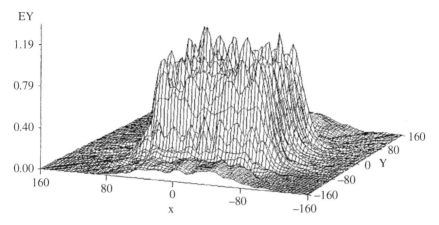

Figure 6.21 Amplitude of y component of measured electric near-field for a 32× 32 patch microstrip array.

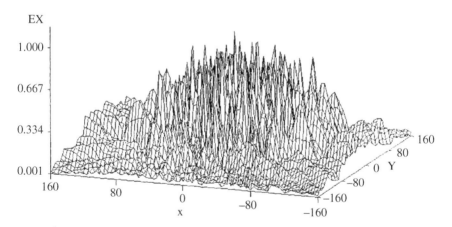

Figure 6.22 Amplitude of x component of measured electric near-field for a 32 × 32 patch microstrip array.

measured near-field sampled data, we compute the amplitudes of the 41×41 magnetic dipoles utilizing CGFFT [15].

The 2D Fourier transform was then utilized to find the far fields. Figure 6.23 provides E_φ ($\varphi = 0°$) and Figure 6.24 for E_φ ($\varphi = 90°$) by utilizing the modal expansion method with the original 81 × 81 points and the integral equation approach with only 41 × 41 data points. It is seen that for the two results agree till ± 40° and beyond that the agreement is reasonable. Therefore, by reducing the data by as much as 75% it is still possible to get reasonable results utilizing

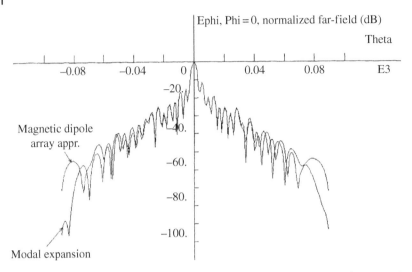

Figure 6.23 Co-polarization characteristic for $\Phi = 0°$ cut for a 32 × 32 patch microstrip array using planar modal expansion method (81 × 81 data points) and equivalent magnetic dipole array approximation (41×41data points).

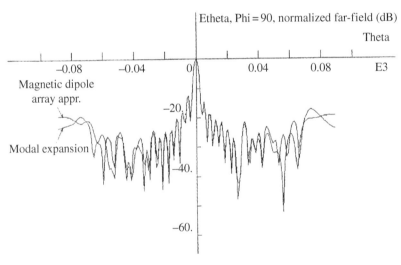

Figure 6.24 Co-polarization characteristic for $\Phi = 90°$ cut for a 32 × 32 patch microstrip array using planar modal expansion method (81 × 81 data points) and equivalent magnetic dipole array approximation (41×41) data points.

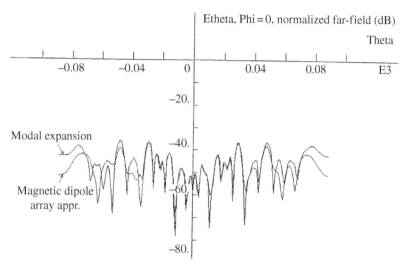

Figure 6.25 Cross-polarization characteristic for φ = 0° cut for a 32 × 32 patch microstrip array using planar modal expansion method (81 × 81 data points) and equivalent magnetic dipole array approximation (41×41) data points.

the integral equation technique. For this case 41× 41 data points, the CGFFT method takes about 10–12 iterations to provide an acceptable solution for the magnetic dipoles. Since the CGFFT method requires two 2D-FFT per iteration, the CGFFT method is slower than the modal expansion method by a factor 20–24 minutes as compared to seconds of the modal techniques.

Again, the two curves agree till ± 40° and the agreement is reasonable outside that region. These are the two principal planes co-polar pattern. Next we look at the cross-polar pattern. Figure 6.25 provides E_θ ($\varphi = 0°$) utilizing the modal expansion on the 81×81 points and the integral equation approach utilizing 41 × 41 data points. Even though the cross-polar pattern is 40 db down, reasonable agreement is seen upto 40°. Figure 6.26 provides E_θ ($\varphi = 90°$) utilizing the modal expansion and the integral equation approach. Because this pattern is 60 dB down, the accuracy of both the methods are being compared at a level which corresponds to the noise level. Even then, agreement is reasonable!

Limited experimentation utilizing these two examples (one utilizing synthetic data and other utilizing experimental data) illustrates that it is possible to go beyond the truncation error introduced by the measurement process. This is achieved by first transferring the measurement plane to a source plane containing equivalent sources on a plane closer to the source. It is important to point out that the integral equation method is not creating information that is not there in the measured data. The integral equation method has less truncation error as the processing is different from the conventional modal expansion

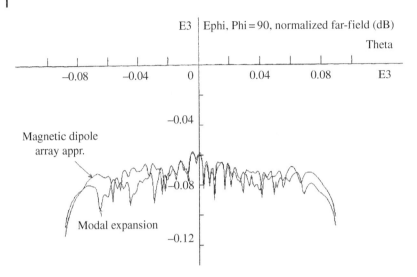

Figure 6.26 Cross-polarization characteristic for $\Phi = 90°$ cut for a 32 × 32 patch microstrip array using planar modal expansion method (81 × 81 data points) and equivalent magnetic dipole array approximation (41×41) data points.

method. Since the integral equation method is a model-based-parameter-estimation technique, it will always provide higher resolution than the conventional modal expansion method which is FFT based. Let us illustrate this phenomenon by an example. Consider two frequencies $f_2 > f_1 > 0$. Then we know in order to resolve them by the conventional Fourier Processing, we need a data record T which must be greater than $\dfrac{1.}{f_2 - f_1}$. This is the principle of uncertainty or the Rayleigh limit as it is popularly called. However, in order to solve for 2 frequencies by a model-based-parameter estimation technique one needs only 4 samples to solve the problem. There are 4 unknowns – 2 amplitudes and 2 frequencies and hence 4 samples should be sufficient to solve the problem. So through this model based parameter estimation technique we are not creating new information, but by assuming a model for the system we are significantly increasing the resolution. This is the motivation of our new approach based on the integral equation technique.

As shown in the block diagram of Figure 6.27, we have increased the resolution of the integral equation method over that of the conventional modal expansion method for processing the identical data set by simply reversing the order of the solution procedure.

It is seen that the modal expansion method first transforms the measured data to the far field by performing a 2D Fourier transform. From the far field, the data is transformed back to the source plane by performing a 2D inverse Fourier transform. For the integral equation approach, the measured data is first

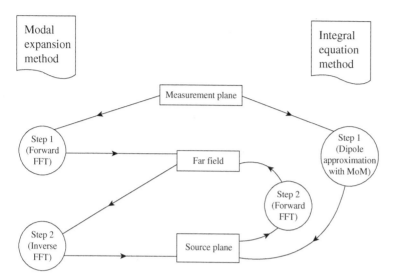

Figure 6.27 Flow diagrams for the processing of the measured data for the modal expansion method and for the integral equation method.

transformed to the source plane utilizing the dipole approximation for the equivalent sources in the source plane and then solving for their amplitudes by the method of moments. The results are then transformed from the source plane to the far field by performing a 2D-Fourier transform.

Also, it may be possible to significantly reduce the measurement time by as much as 75% if the integral equation approach is selected over the conventional modal expansion method to yield same degree of accuracy in the far field. However, the integral equation approach is computationally 20–24 times slower than the modal expansion method.

6.8.5 Summary

The integral equation approach can provide a reduction of the truncation error caused by performing measurements on a finite plane. This is achieved by transforming and shifting the measured data to a plane which is much closer to the sources. Once the equivalent magnetic currents are found on the source plane, the Fourier techniques can be used to compute the far fields from the magnetic currents. So the price one pays in utilizing the integral equation approach over the modal expansion method is that the integral equation approach takes about 20–25 times more CPU time (which may result to a few minutes on the modern computers) to produce the far fields for the same number of data points. This estimate has been obtained from limited simulation results. However, the integral equation approach requires fewer measured data points than the conventional modal expansion method to provide comparable numerical accuracy in the far fields when applied to the same near-field data. So the total measurement time in the integral equation method is less to achieve equivalent numerical accuracy in the far field result compared to the conventional modal expansion method.

6.9 A Direct Optimization Approach for Source Reconstruction and NF-FF Transformation Using Amplitude-Only Data

In this section a direct optimization procedure which utilizes phaseless electric field data over arbitrary surfaces for the reconstruction of an equivalent magnetic current density that represents the radiating structure or an antenna under test. Once the equivalent magnetic current density is determined, the electric field at any point can be calculated. Numerical results (both simulated and experimental) are presented to illustrate the applicability of this approach for non-planar near field to far field transformation as well as in antenna diagnostics.

6.9.1 Background

It is difficult to obtain complex (amplitude and phase) measurement at millimeter wave frequencies. Also it is economical to develop amplitude measurement equipments which are often the motivation of studying techniques that allow the calculation of the far-field pattern from amplitude measurements in the near-field.

Classical near field to far field (NF-FF) transformation methods using amplitude only data are based on phase retrieval techniques. Several approximations can be found in the literature using matrix methods with one and two measurement planes [25], or the classical two measurement planes technique with the Fast Fourier Transform (FFT) algorithm [26]. The main goal, in those techniques, is to reconstruct the phase of the near-field data. This usually leads to an iterative scheme in which the phase of the known amplitude near field data at two different planes is reconstructed in a forward-backward fashion. A typical phase retrieval scheme for two planes is shown in Figure 6.28(a).

Here, we propose a method that reconstructs directly the sources from the knowledge of the electric field amplitude data over some region (Figure 6.28 (b)). The sources are established in terms of an equivalent magnetic current (EMC) density and the equivalent principle has been used to represent the antenna under test (AUT) as an EMC distribution that encloses the AUT. From this representation, a relationship between sources and the amplitude field data

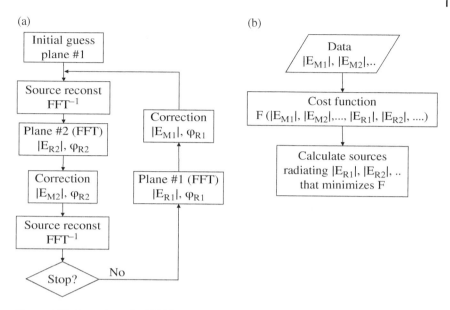

Figure 6.28 Flow charts for NF-FF reconstruction from only amplitude data: (a) classical phase retrieval algorithm, (b) proposed direct optimization.

over some domain can be established through a full wave integral equation. The technique of using equivalent currents and a matrix method to reconstruct them on arbitrary scanning surfaces from amplitude and phase information has been previously addressed in [18, 27–30]. In the case of amplitude only data, a cost function regarding the difference between the known amplitude data and the field radiated by the EMC distribution can be defined. The conjugate gradient method has been used to minimize the cost function and obtain the EMC distribution. Finally, the electric field data can be calculated at any point from this EMC distribution. NF-FF transformation has been considered as the main application, although the calculation of the EMC distribution can also be useful for diagnostics when the algorithm is applied to planar antennas.

The surface of the measured amplitude data (measurement domain) does not need to be restricted to the typical two planes usually defined but can be defined on arbitrary domains of non-planar measurement surfaces. Results are presented using one plane, two planes as well as other type of surfaces. Its accuracy, which depends on the amount of information given to the optimization procedure, are also discussed.

In Section 6.9.2, we present the concept of EMC in a succinct fashion. Then we illustrate how to extend this methodology to deal with amplitude only data over an arbitrary measurement plane utilizing an optimization procedure as outlined in section 6.9.3. Both simulation and experimental results are presented in Section 6.9.4 to illustrate how this methodology can be used for phaseless NF to FF transformation and antenna diagnostics for measurements made on arbitrary surfaces which need not conform to any coordinate system.

6.9.2 Equivalent Current Representation

According to the equivalence theorem, the field radiated by an antenna can be reproduced by the unbounded radiation from a surface distribution of electric and magnetic equivalent currents over a surface enclosing the original radiating antenna. Supposing that the enclosing surface is an infinite perfect electric conducting structure that coincides with the surface of the antenna, then image theory can be applied and only equivalent magnetic currents are needed to represent the radiated field due to the AUT in the forward region where the equivalence is valid (Figure 6.29) as also illustrated in [18, 30, 31].

The electric field at any point due to an EMC density over a planar surface can be obtained through the electric vector potential. When the EMC density is over the $z = 0$ plane, the Cartesian components of the electric field at the point (x, y, z) can be written as:

$$E_x(x,y,z) = \frac{-z}{4p} \times \int_{x'} \int_{y'} M_y(x',y') \times \frac{(1 + jkR)}{R^3} dx'dy' \tag{6.59}$$

$$E_y(x,y,z) = \frac{z}{4\pi} \cdot \int_{x'} \int_{y'} M_x(x',y') \cdot \frac{(1 + jkR)}{R^3} dx'dy' \tag{6.60}$$

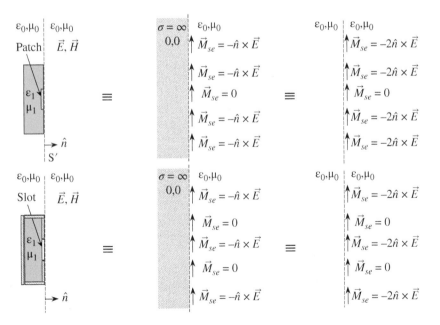

Figure 6.29 The equivalent problem for the EMC.

where R is the distance between each primed source point and the unprimed observation point. k is the wave number and M_x, M_y are the Cartesian components of the EMC density over the $z = 0$ plane. Equations (6.59) and (6.60) establish the relation between source and field components at any point where the equivalence is valid.

In order to manipulate numerically the above expressions, appropriate domain segmentation and an expansion of the EMC must be done. Using M subdomain-type basis functions, $f_m(x',y')$, weighted by unknown complex coefficients $C_{x,m}$, $C_{y,m}$; $m = 1,..,M$, each Cartesian component of the EMC distribution can be expressed as:

$$M_y(x',y') \cong \sum_{m=1}^{M} C_{y,m} \cdot f_m(x',y') \tag{6.61}$$

$$M_x(x',y') \cong \sum_{m=1}^{M} C_{x,m} \cdot f_m(x',y') \tag{6.62}$$

The numerical values of the field components at a point (x_n, y_n, z_0) radiated by the EMC distribution can be approximated by:

$$E_{x,n} = E_x(x_n, y_n, z_n) \cong \sum_{m=1}^{M} C_{y,m} \cdot \int_{x'}\int_{y'} \left(\frac{-z_n}{4\pi}\right) \cdot f_m(x',y') \cdot \frac{(1 + jkR_{n,m})}{R_{n,m}^3} \; dx' dy'$$

$$\tag{6.63}$$

$$E_{y,n} = E_y(x_n, y_n, z_n) \cong \sum_{m=1}^{M} C_{x,m} \cdot \int_{x'}\int_{y'} \left(\frac{z_n}{4\pi}\right) \cdot f_m(x',y') \cdot \frac{(1 + jkR_{n,m})}{R_{n,m}^3} \; dx' dy'$$

$$\tag{6.64}$$

If enough computing memory is available, the integral representations given by (6.63) and (6.64) can be used to calculate the EMC through (6.61) and (6.62) using measured Electric fields over arbitrary surfaces. In addition, because of equations (6.61)–(6.64) the measurement samples need not satisfy Nyquist sampling theory. This requirement is transferred to the equivalent source plane where the discretization through the choice of appropriate basis functions must meet certain criteria. Once the EMC is known, the far fields can easily be calculated. Another alternative is to take advantage of the translationally invariant properties of the Green function involved in equations (6.63),(6.64). For the case of planar scanning and the proposed configuration of the equivalent problem, the Green function involved in the electric field is a function of the absolute value of the differences between observation point and source point, becoming of the form $G_{n,m}\left(\vec{r} - \vec{r}'\right) = G\left(\left|x_n - x'_m\right|, \left|y_n - y'_m\right|\right)$. Hence, if equal spacing is selected both in the source domain as well as in the scanning domain, and the points in both domains are arranged so that all the distances $\left|x_n - x'_m\right|, \left|y_n - y'_m\right|$ are a multiple of a minimum separation, only a reduced number of matrix elements must be computed and stored.

More savings in CPU time can be achieved if the calculation of the field at the N points at each iterative step is performed using a Fast Fourier Transform (FFT) algorithm instead of a direct summation over the M elements for each of the N scanning points. In fact, at a constant distance from the plane of the antenna, equations (6.59) and (6.60) are of the convolutions type:

$$E(x,y) = \int_{x'} \int_{y'} M_{xy}(x',y') \cdot G(x - x', y - y') \, dx'dy' = M_{xy}{}^{*}G \qquad (6.65)$$

If the FFT of the unbounded medium Green function, $\widetilde{G}$, is calculated a priori and its elements stored, the N field values due to the estimated value of the EMC density M_{xy}^k, at the k-*th* iterative step of the minimization process, can be calculated by:

$$E(x,y) = FFT^{-1}\left(\widetilde{M}_{xy}^k \cdot \widetilde{G}\right) \qquad (6.66)$$

where $\widetilde{M}_{xy}^k$ corresponds to the FFT of the estimated EMC density at k-*th* iterative step. Therefore, information for both magnitude and phase for the EMC can be calculated using this procedure with amplitude only data.

6.9.3 Optimization of a Cost Function

Let us suppose we know the value of the amplitude of the electric field at N points. In a NF-FF application these points correspond to the measured data of the AUT. Let's term them $\left|E_{y,n}^{meas}\right|, \left|E_{x,n}^{meas}\right|, n = 1, .., N$ the measured values for the two Cartesian components of the electric field. In those data, values corresponding to one or more planes or values measured over any other arbitrary surface can be included.

Then, an optimization algorithm could be used to obtain the coefficients of the EMC distribution that produce the radiated field which best fits the measured amplitude near-field data. The cost function we propose to minimize to obtain each component of the EMC density is:

$$F = \sum_{n=1}^{N} \left[\left(E_{x,n}^{meas}\right) \cdot \left(E_{x,n}^{meas}\right)^{*} - \left(E_{x,n}\right) \cdot \left(E_{x,n}\right)^{*}\right]^2 \qquad (6.67)$$

In general, two independent functional are minimized, each one using one field component and one source component. Both components are used to calculate the complete EMC distribution and hence the far field pattern of the AUT.

With this reconstruction algorithm no assumption is made about the phase of the measured near-field data. In addition, it is not necessary to retrieve that phase first, but the EMC distribution is directly reconstructed and then, the field at any point can be calculated. A conjugate gradient algorithm with analytical derivatives has been used to calculate the coefficients $C_{x,m}$, $C_{y,m}$; $m = 1,.., M$ that represent the EMC distribution.

6.9.4 Numerical Simulation

This example corresponds to a two-dimensional magnetic current distribution, with M_y values located between 0 dB and −2 dB are arranged in a "chessboard" fashion as seen in Figure 6.30. The current distribution domain is a square of 4λ × 4λ with 12 × 12 elementary cells both for synthesizing and reconstruction purposes. Planar scanning was considered in this example. The near-field amplitude-only data at two different planes, 2λ and 4λ from the source plane, have been synthesized as seen in Figure 6.30. Each scanning plane is 8λ × 8λ and the spacing between samples is 0.05λ. Good results in the far-field data can

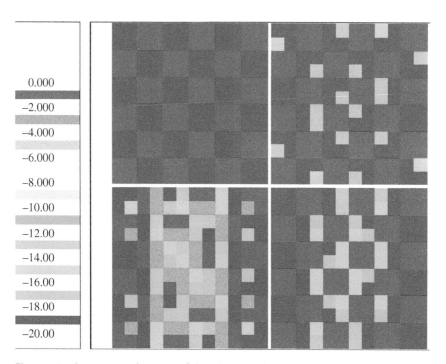

0.000
−2.000
−4.000
−6.000
−8.000
−10.00
−12.00
−14.00
−16.00
−18.00
−20.00

Figure 6.30 Reconstructed sources of the "chessboard" magnetic current distribution. Nominal (up-left), amplitude+phase (up-right), amplitude-1 plane (down-left), amplitude-2 planes (down-right).

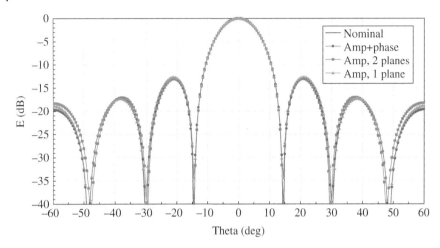

Figure 6.31 Far-field pattern of the "chessboard"-like magnetic current distribution. Planar scanning.

be obtained using only one plane or two planes of amplitude only data with a few iterations. The calculated far-field patterns for 50 iterations and different types of measured near-field information are shown in Figure 6.31. The rate of convergence is shown in Figure 6.32. However, more number of iterations in the minimization algorithm must be considered for an accurate reconstruction of the sources. The reconstructed magnetic current distribution using the information over one plane and two planes are shown in Figure 6.30. Also, the nominal distribution and the reconstructed one, using the amplitude and phase information over one plane, have been plotted in the same figure for comparison. Those results have been obtained with 1000 iteration of the minimization algorithm. The reconstructed magnetic current agrees well with the nominal one for all the configurations examined. The CPU time per iteration in a personal computer is less than one second in this example where 144 unknown sources and 25921 field data points over each scanning plane are used.

6.9.5 Results Obtained Utilizing Experimental Data

In this example, an aperture antenna is considered along with a main reflector of 40 cm diameter and a sub reflector in the aperture plane [31–33]. The measured data of this antenna were provided by the Laboratorio de Ensayos of El Casar de Talamanca, Spain. The antenna was measured at the operating frequency of 12.625 GHz at two scanning planes, 40 cm and 100 cm, far from the aperture of the antenna. The scanning domain at each plane is 0.88 m × 0.88 m with the spacing between field points of 1 cm and a total of 7921 field points. 100×100 source cells along a domain of 50 cm × 50 cm have been used for reconstruction.

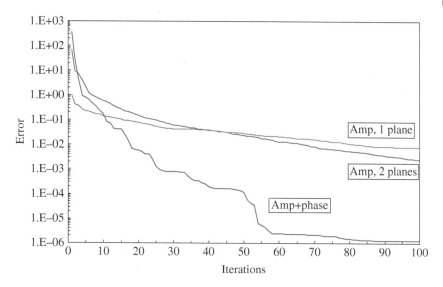

Figure 6.32 Error of the iterative optimization. "chessboard"-like magnetic current distribution.

The reconstructed EMC on the aperture of the antenna are shown in Figure 6.33 for the cases of amplitude data on two planes and amplitude data on only one plane. The reconstruction with amplitude and phase information is also plotted for comparison. Although the obstruction effect due to the sub-reflector can be observed in all cases, the diagnosis results from using amplitude in only one plane are far from the accuracy that can be obtained with the information from two planes. The calculated far-field pattern with this source reconstruction technique and amplitude only data without probe correction as well as the calculated result using a classical FFT technique with amplitude and phase data are shown in Figure 6.34 and Figure 6.35. The results using amplitude only data over two planes agree with those of using amplitude and phase information in the main lobe and in the first side lobe. However, in some differences appear in the first lobe level when using amplitude only data in one plane.

6.9.6 Summary

A source reconstruction technique is presented using measured amplitude-only field data through an equivalent magnetic current representation of the AUT. This approach provides the basis for NF-FF and diagnosis applications.

The possibility of measuring amplitude only data on arbitrary surfaces and the diagnosis capabilities makes this technique an alternative to the classical iterative phase retrieval schemes. The accuracy of this technique when using

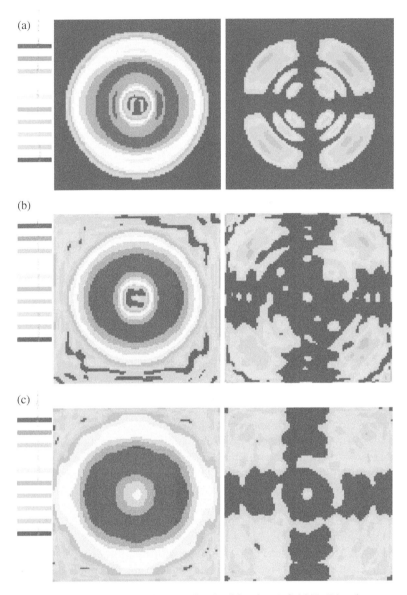

Figure 6.33 Reconstruction of the amplitude of the electric field (*Ex, Ey*) at the aperture of the "splas" antenna. (a) from amplitude and phase, (b) amplitude 2 planes, (c) amplitude 1 plane.

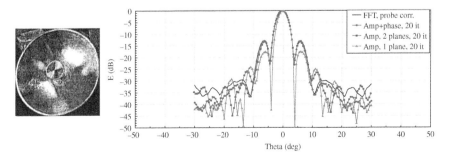

Figure 6.34 Far-field pattern of the splas antenna. $\varphi = 0$ –plane.

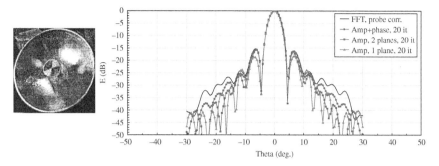

Figure 6.35 Far-field pattern of the splas antenna. $\varphi = 90°$ –plane.

measured amplitude only field data at one and two surfaces has been studied by means of synthesized data. Both pattern reconstruction as well as diagnosis applications have been presented. Some conclusions regarding the necessary information to improve accuracy are directly derived from the presented examples (for sufficient separation between the measurement surfaces, number of field points, maximum scanning angle, and so on). Amplitude measurements over two planes improves the accuracy both in the source reconstruction and in the calculated far-field pattern, while the quality of the results using amplitude information on one scanning surface can vary depending on the problem. The accuracy of the later is worse when both amplitude and phase information is available over one plane than of amplitude only information over two surfaces.

6.10 Use of Computational Electromagnetics to Enhance the Accuracy and Efficiency of Antenna Pattern Measurements Using an Array of Dipole Probes

The objective of this section is to illustrate that computational electromagnetics can be used very effectively to improve the accuracy and efficiency of planar antenna pattern measurements. This is accomplished by moving a single probe over the measurement plane to generate enhanced accuracy in planar near-field to far-field transformation than over the classical Fourier based modal expansion methods. It is also illustrated that this method provides reliable results for cases when the conventional method fails particularly for the case when the actual source plane and the measurement planes are approximately equal in size. Also in this approach there is no need to incorporate probe correction when using small dipole like probes, unlike in the existing approaches. In addition a methodology can be designed where one can use an array of dipole probes instead of moving a single probe over the measurement plane, thus improving the accuracy and efficiency of the measurements. In the use of the probe array there is also no need to perform probe correction. This proposed novel methodology is accomplished by solving for the equivalent magnetic current over a plane near the original source antenna under test and then employing the Method of Moments approach to solve for the equivalent magnetic currents on this fictitious surface. For this proposed methodology even though there is no need to satisfy the Nyquist sampling criteria in the measurement plane, a super resolution can be achieved in the solution of the equivalent magnetic current so that one can predict the status of the operation of each antenna elements in an array. Also the presence of evanescent fields in the measurements do not make this methodology unstable unlike in the conventional Fourier based techniques. The two components of the equivalent magnetic currents can be solved independently from the two measured components of the electric fields by solving the resultant Method of Moments matrix equation very efficiently and accurately by using the iterative conjugate gradient method enhanced through the incorporation of the Fast Fourier Transform (FFT) techniques. Sample numerical results are presented to illustrate the potential of a novel planar NF to FF transformation applied to the planar near field measurement technique.

6.10.1 Introduction

The source reconstruction method (SRM) is a recent computational technique developed for antenna diagnostics and for carrying out near-field (NF) to far-field (FF) transformation. The SRM is based on the application of the electromagnetic Equivalence Principle, in which one establishes an equivalent current distribution that radiates the same fields as the actual currents induced in the antenna under test (AUT). The knowledge of the equivalent currents allows the determination of the status of the radiating elements, as well as the prediction of the AUT-radiated fields outside the domain of the equivalent currents. The unique feature of the novel methodology has been illustrated that it has the potential to resolve equivalent currents patches that are smaller than half a wavelength in size, thus providing super-resolution. Furthermore, the measurement field samples can be taken at field spacing's greater than half a wavelength, thus going beyond the classical Nyquist sampling criteria. These two distinctive features are possible due to the incorporation of computational techniques into antenna measurement methodology thereby enhancing the latter's accuracy and efficiency. In the computational technique embedded in this measurement methodology, the unknowns are approximated by a sub sectional basis and, secondly, through the use of the analytic free space Green's function which is quite easy to compute numerically. The latter condition also guarantees the invertibility of the electric field operator and provides a stable solution for the currents even when evanescent waves are present in the measurements. In addition, the use of the singular value decomposition in the solution of the matrix equations can provide the user with a quantitative tool to assess the quality and the quantity of the measured data. Alternatively, the use of the iterative conjugate gradient (CG) method in solving the ill-conditioned matrix equations can also be implemented. Several examples have been presented to illustrate the applicability and accuracy of the proposed methodology.

6.10.2 Development of the Proposed Methodology

An equivalent magnetic current approach is a widely used method to calculate far field from near field data [18, 34]. Based on the equivalent principle, this method uses the near-field data to determine an equivalent magnetic current source on a fictitious planar surface that encompasses the antenna under test, and under certain approximations, the magnetic currents will produce the same field as the antenna under test in the region of interest. In our earlier works, the equivalent current approach of computing far fields from measured near fields using a single probe antenna measurement has been discussed [18] without incorporating probe correction. In this methodology, the introduction of a measurement probe array appears to have minimal mutual effects between the antenna under test (AUT) and the probe as shown in [34]. In addition, in this work, results are compared from two sets of the near field data obtained for two different scenarios. First, we use a 0.1 λ length dipole terminated in a 50 Ω load as a probe placed in front of the AUT and measure the voltage across the load. Then the probe is moved over a planar surface and measurements are taken with a 0.2 λ separation. Secondly, we replace a single probe antenna by an array of 0.1 λ length dipoles all terminated in 50 Ω loads and separated from each other in both directions by 0.2 λ. The advantage of choosing a probe array for measurement is that it can eliminate the inaccuracy of mechanical movement of the probe antenna over a large planar surface and can make the measurement methodology very efficient. This is more important particularly for measurements carried out in the high frequencies, say at M, V and W-bands. Also, one can obtain all the near field measurement information at once, thus making the entire measurement procedure very time-efficient and simple. Another important feature of the proposed approach that will be demonstrated is that in this methodology probe correction, which accounts for the mutual coupling between the AUT and the probe and also between the probes in the probe array, do not play a significant role.

6.10.3 Philosophy of the Computational Methodology

Consider an arbitrary shaped antenna radiating into free space with the aperture of the antenna being a planar surface (assumed for simplicity but this assumption can be relaxed), which separates the total space into two: left-half (Region I) and right-half (Region II) spaces as shown in Figure 6.36. Using the equivalence theorem [16], we postulate the electromagnetic fields in the RNI (Region of No Interest, i.e. Region I) to be zero and place a perfect electric conductor on the x-y plane (as shown in Figure 6.37).

We can further assume for the general case that the tangential component of the electrical field on the PEC surface is zero except over S_0, and then $\overline{M}$ exist only over a small region in front of the AUT characterized by S_0 as shown in Figure 6.37.

Then applying the image theory, the equivalent magnetic current $\overline{M}$ is obtained as

$$\overline{M} = 2\overline{E} \times \hat{n} \text{ on } S_0 \tag{6.68}$$

where now $\overline{M}$ radiates in the entire free space. Then determine $\overline{M}$ from the measured electric field components. For computational purposes the measurement plane and hence the source plane has to be truncated to a finite region S_0. As the fields decay almost exponentially on this plane as we go away from the radiating aperture one can truncate the surface to a small area without introducing any significant error in this approximation.

Now, the FF can be obtained from the measured electric near field via the equivalent magnetic current approach using a MoM formulation [13]. Furthermore,

$$\overline{E}_{meas} = \overline{E}\left(\overline{M}\right) \tag{6.69}$$

where $\overline{E}_{meas}$ is the measured electric NF, and $\overline{M}$ is the equivalent magnetic current that exists on S_0. After we solve for $\overline{M}$ using (2) we can calculate the far field.

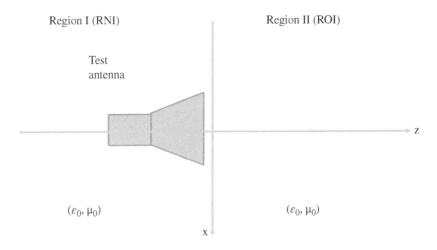

Figure 6.36 Original Problem.

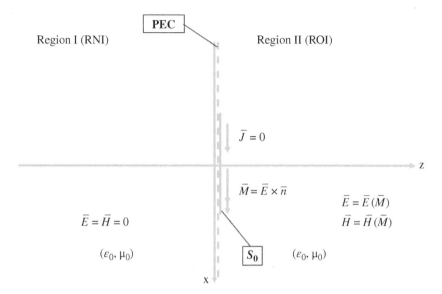

Figure 6.37 An Equivalent Problem.

6.10.4 Formulation of the Integral Equations

The electric field at any arbitrary point $P(r)$ can be found from

$$\overline{E}(\overline{r}) = -\iint_{S_0} \left[\overline{M}(\overline{r}') \times \nabla' g(\overline{r}, \overline{r}') \right] \mathrm{d}s' \tag{6.70}$$

where $\overline{E}(\overline{r})$ is the electric field. $\overline{M}(\overline{r}')$ is the equivalent magnetic current at the source point $\overline{r}'$. ∇' is the gradient operator with respect to the primed variables (sources), and $g(\overline{r}, \overline{r}')$ is the three-dimensional free space Green's function, given by

$$g(\overline{r}, \overline{r}') = \frac{e^{-jk_0|\overline{r} - \overline{r}'|}}{4\pi \, | \, \overline{r} - \overline{r}' \, |} \tag{6.71}$$

and k_0 is the free space wave number.

The near-field measurements are performed over a planar surface which is parallel with the source plane as shown in Figure 6.38. The source plane (S_0) is assumed to be a rectangular surface in the x-y plane with the dimensions w_x and w_y. The distance between the source plane and the measurement plane is d.

As we mentioned, there are two groups of measurements, for the first case, we use a 0.1 λ length Hertzian Dipole to estimate the sampled electric fields at 0.2 λ

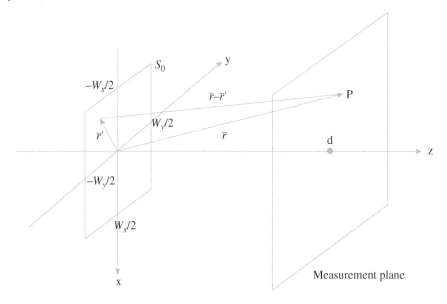

Figure 6.38 A planar scanning case.

separation on the measurement planar surfaces. The dipole probe is terminated in a 50 Ω load and the current across the load is measured.

We consider the measurement dipole to be x-directed and obtain the value of the current (I_x) at the center point of the dipole from the measured values of the voltage at each measurement point P. Then the dipole is rotated to be y-directed and we obtain the currents (I_y) from the measured voltages at the center of the dipole at each measurement point P. For the second case, we replace the single probe antenna by an array of 0.1 λ dipoles all terminated in 50 Ω loads and separated from each other by 0.2 λ. First, the array of dipoles are all x-directed to obtain the value of the current $[I_x]$ at the center of the dipole. Then they are rotated to be y-directed to obtain the current matrix $[I_y]$. Finally, we can multiply these current matrices by the diagonal impedance matrix $[Z]$ with the diagonal elements all being 50 Ω, to obtain $[V_x]$ *and* $[V_y]$, the voltages measured at the different orientations of the probes. It is estimated that the voltage V at the center of the dipole is proportional to the electric field $\overline{E}$ at that point. We can normalize the voltages induced at the center points of the dipoles and use that information to estimate the values for the sampled near field data. From that estimated near field data, the equivalent magnetic currents (M_x, M_y) on the source plane can be calculated. By using that equivalent magnetic currents, we calculate the far field. In the end, we compare the final far field results obtained from using the presented two methods, namely using a single probe or an array of probes with the results from an electromagnetic analysis code called HOBBIES [35].

We can use the x and y components of the measured electric near fields in (6.70), and obtain the following integro-differential equation as

$$\begin{bmatrix} E_{meas,x}(\bar{r}) \\ E_{meas,y}(\bar{r}) \end{bmatrix} = - \iint_{S_0} \begin{bmatrix} 0 & \dfrac{\partial g(\bar{r},\bar{r}')}{\partial z'} \\ -\dfrac{\partial g(\bar{r},\bar{r}')}{\partial z'} & 0 \end{bmatrix} \begin{bmatrix} M_x(\bar{r}') \\ M_y(\bar{r}') \end{bmatrix} ds' \tag{6.72}$$

Equation (6.72) shows that the integral equation is a decoupled one with respect to the two components of the magnetic currents. So the following two integral equations can be solved separately as

$$E_{meas,x}(\bar{r}) = - \iint_{S_0} \frac{\partial g(\bar{r},\bar{r}')}{\partial z'} M_y(\bar{r}') ds' \tag{6.73}$$

$$E_{meas,y}(\bar{r}) = \iint_{S_0} \frac{\partial g(\bar{r},\bar{r}')}{\partial z'} M_x(\bar{r}') ds' \tag{6.74}$$

6.10.5 Solution of the Integro-Differential Equations

Here we use MOM (Method of Moment) [13] to transform the E-field integral equations into matrix equations, so that they can be numerically calculated using the computer. Both M_x and M_y are approximated by equally spaced two-dimensional pulse basis functions

$$M_x(x',y') = \sum_{i=1}^{M} \sum_{j=1}^{N} \alpha_{ij} \Pi_{ij}(x',y') \tag{6.75}$$

$$M_y(x',y') = \sum_{i=1}^{M} \sum_{j=1}^{N} \beta_{ij} \Pi_{ij}(x',y') \tag{6.76}$$

where α_{ij} and β_{ij} are the unknown amplitudes of the x and y directed magnetic currents, respectively on the ij^{th} patch, and $\Pi_{ij}(x', y')$ is the two-dimensional pulse basis function of the ij^{th} patch and is defined by

$$\Pi_{ij}(x',y') = \begin{cases} 1 \text{ if } x_i - \dfrac{\Delta x}{2} \leq x' \leq x_i + \dfrac{\Delta x}{2} \\ y_i - \dfrac{\Delta y}{2} \leq y' \leq y_i + \dfrac{\Delta y}{2} \\ 0 \text{ otherwise} \end{cases} \tag{6.77}$$

For the above approximation, the source plane (S_0) is assumed to be a rectangular one in the x-y plane with extension to $-w_x/2 \leq x \leq w_x/2$ and $-w_y/2 \leq x \leq$

$w_y/2$ as shown in Figure 6.38. The source plane is divided into $M \cdot N$ equally spaced rectangular patches with dimensions Δx and Δy as

$$\Delta x = w_x/M \tag{6.78}$$
$$\Delta y = w_y/N \tag{6.79}$$

In (6.77) x_i and y_j are the x and y coordinates of the center of the ij^{th} patch and are given by

$$x_i = -\frac{w_x}{2} - \frac{\Delta x}{2} + i\Delta x \tag{6.80}$$

$$y_j = -\frac{w_y}{2} - \frac{\Delta y}{2} + j\Delta y \tag{6.81}$$

Since it is assumed that the measured electric near-field are known at discrete points on the scanning plane, a point matching procedure at the center of each plate is chosen. Substituting (6.75), (6.76) into (6.73), (6.74) and utilizing point matching the following two decoupled matrix equations are obtained,

$$E_{meas,x} = -GM_y \tag{6.82}$$
$$E_{meas,y} = GM_x \tag{6.83}$$

where G is the moment matrix for the planar scanning case. The explicit expressions for the elements of G is given by

$$G_{k,l} = \iint_{\Omega_i} \frac{e^{-jk_0R}}{4\pi R^2} (z_k - z') \left[jk_0 + \frac{1}{R} \right] ds' \tag{6.84}$$

where Ω_i is the area of the ith patch, and R is the distance between the kth field point (r_k) and l^{th} source point (r_l'), and $ds' = dx \cdot dy$.

These calculations can be very efficiently carried out under some specific conditions. If the spacing between the field points and the spacing between the current elements are chosen to be the same, the resultant matrix is block Toeplitz. The structure of the matrix can be exploited by noting that a two-dimensional Fourier transform may be utilized to evaluate the terms in the following CGM (Conjugate Gradient Method), which is called CGFFT (CGM and Fast Fourier Transform) [15, 18].

The CGM starts with an initial guess X_1 and computes

$$R_1 = Y_1 - AX_1 \tag{6.85}$$
$$P_1 = A^*R_1 \tag{6.86}$$

For $i = 1,2, \ldots$ let

$$a_i = \frac{\|A^*R_i\|^2}{\|AP_i\|^2} \tag{6.87}$$

$$X_{i+1} = X_i + a_iP_i \tag{6.88}$$

$$R_{i+1} = R_i - a_iAP_i \tag{6.89}$$

$$b_i = \frac{\|A^*R_{i+1}\|^2}{\|A^*R_i\|^2} \tag{6.90}$$

$$P_{i+1} = A^*R_{i+1} + b_iP_i \tag{6.91}$$

where A^* is the conjugate transpose of A. Most of the computational cost in CGM occurs in the calculation of AP_i and A^*R_{i+1}. These two calculations have to be performed inside a loop which needs to be carried out many times. This is the most time-consuming part if we multiply them directly. As we mentioned before, we can exploit the block Toeplitz structure of the matrix A, and these two terms can be computed using FFT. This would have a tremendous saving in computational time by using

$$AP_i = F^{-1}\{F(A^c)F(P_i^c)\} \tag{6.92}$$

$$A^*R_{i+1} = f_cF^{-1}\{F(A^c)F(R_{i+1}{}^{c*})\}^* \tag{6.93}$$

where F denotes the two-dimensional discrete Fourier Transform, F^{-1} denotes the two-dimensional inverse discrete Fourier Transform, A^c is the convolutional variation of the original matrix A, P_i^c and $R_{i+1}{}^c$ are the convolutional variations of the original vectors P_i and R_{i+1}, respectively, and $*$ denotes complex conjugate.

6.10.6 Sample Numerical Results

The comparisons of the far field calculated from the near field single probe measurement data and from the planar probe array measurement data are introduced next.

6.10.6.1 Example 1

A 2 λ by 2 λ pyramidal horn antenna is used as the antenna under test. A fictitious planar surface in the x-y plane of dimensions 3 λ by 3λ is used to form a planar magnetic current sheet. On the surface of the equivalent magnetic currents M_x and M_y are placed 15×15 current patches. This is the nature of the discretization. The near fields are sampled on a planar surface of the same dimensions and discretized to enable the use of CGFFT. The distance between the source plane and the scanning plane is 3 λ.

Figure 6.39 shows the x-directed single probe measurement system. Figure 6.40 shows the side view of the structure by using x-directed single probe as an example. Figure 6.41 shows the x-directed probe array measurement system. Figure 6.42 shows the side view of the structure by using the x-directed probe array as an example. The red lines in Figure 6.41 illustrate that the size of S_0 coincides exactly with the size of the measurement plane. The simulated results for the two methods mentioned above and the computed results for the

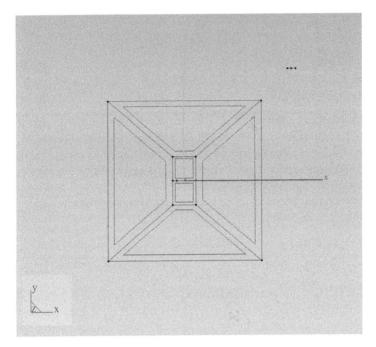

Figure 6.39 A x-directed single probe.

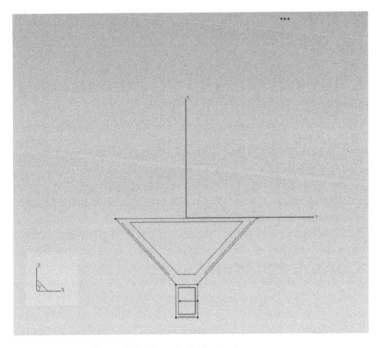

Figure 6.40 A x-directed single probe(side view).

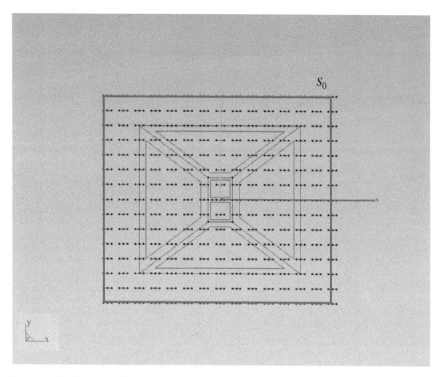

Figure 6.41 A *x*-directed probe array.

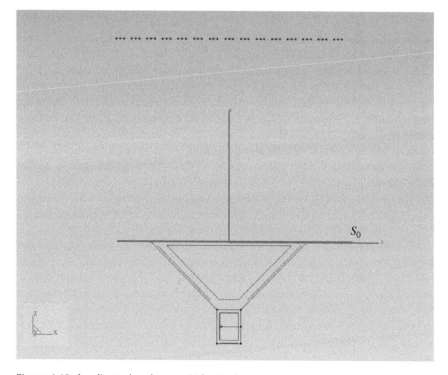

Figure 6.42 A *x*-directed probe array(side view).

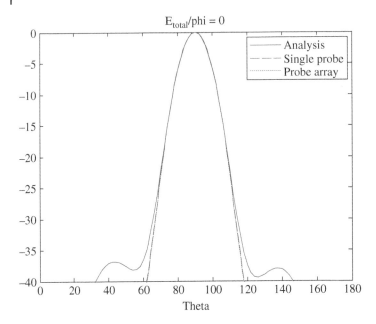

Figure 6.43 E_{total} when phi = 0° (dB Scale).

far fields are shown in Figure 6.43 and Figure 6.44. Figure 6.43 shows the normalized absolute value of the electric far field for φ = 0° in the dB scale. Figure 6.44 shows the normalized absolute value of the electric far field for φ = 90° in the dB scale. θ here is defined as the angle from the x axis to z axis and φ is the angle from the x axis to the y axis. This implies, φ equals 0° cut is the x-z plane and φ equals 90° is the y-z plane. The solid lines show the analytic results obtained using HOBBIES [35], dashed lines show the single probe measurement results and dotted lines show the measurement results using the probe array. We can see both the methods discussed above provides acceptable results. These results indicate that not incorporating probe correction and factoring it into the measurement have little effect on the accuracy of the final result. Hence this methodology is much simpler and more accurate than the classical modal based planar near-field to far-field transformation techniques.

It is important to point out that in terms of the absolute error this method does not perform well related to high accuracy essentially outside an angle of ± 40° in Figure 6.43. And we recognize this fact. But the real question is how this method performs as compared to the current state of the art. In the current state of the art, the fields outside the measurement plane is assumed to be zero and then a Fourier transform is taken to calculate the far field pattern. The accuracy of the current method has been stated in Figure 6.42 of [20] where the

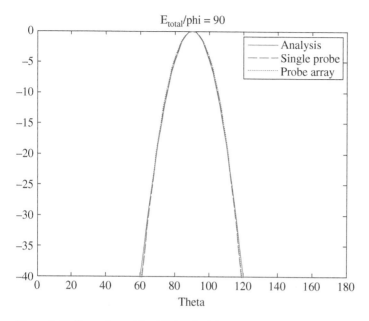

Figure 6.44 E_{total} when phi = 90° (dB Scale).

maximum angle θ_M for accurate far field region of validity is established using the relationship between the scan length L and size of the source plane S separated by d thereby yielding the expressions for the angle of validity of the classical modal techniques to be $\theta_M = \tan^{-1}\left[\dfrac{L-S}{2d}\right]$. In our examples using the equivalent magnetic current approach L and S are approximately of the same dimension. Therefore, the existing classical approach will not work at all for the example just discussed! Yet the equivalent current methodology has been able to accurately predict the pattern to ± 40°. It is important to note that in all the examples presented L and S are approximately the same and so the classical approach will not work for any of the examples presented in this specific section.

6.10.6.2 Example 2

For the next example, the AUT is made more complicated. We choose 16, 1.5 λ by 2 λ pyramidal horn antennas to form a 4 by 4 horn antenna array as the antenna under test. Each horn is separated from each other by 3 λ. A fictitious planar surface in the x-y plane of dimensions 10 λ by 10 λ is used to form a planar magnetic current sheet. On the surface, the equivalent magnetic currents M_x and M_y are divided into 50 × 50 current patches. The near fields are sampled

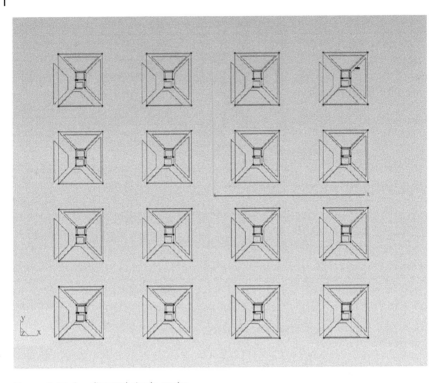

Figure 6.45 A *x*-directed single probe.

on a planar surface of the same dimensions and discretized to enable the use of the CGFFT technique for fast, efficient and accurate computation of the magnetic currents. The distance between the source plane and the scanning plane is 3λ. Figure 6.45 shows the *x*-directed single probe measurement system. Figure 6.46 shows the side view of the structure by using a *x*-directed single probe as an example. Figure 6.47 shows the side view of the *x*-directed probe array measurement structure. Figure 6.48 shows the side view of the structure by using a *x*-directed dipole probe array as an example. The red lines show that the size of S_0 coincides with the size of the measurement plane. Calculated results provided by the two methods from the simulated data are used. The two methods just mentioned and the analytical far field results computed using HOBBIES are shown in Figure 6.49 and Figure 6.50. Figure 6.49 shows the normalized absolute value of the electric far field for $\varphi = 0°$ in dB scale. Figure 6.50 shows the normalized absolute value of the electric far field for $\varphi = 90°$ in dB scale. θ here is defined as the angle from the *x*-axis to the *z*-axis and φ is the angle from the *x*-axis to the *y*-axis. This implies, φ equals 0° cut is the *x*-*z* plane and φ equals 90° is the *y*-*z* plane. The solid lines show the analytic results obtained

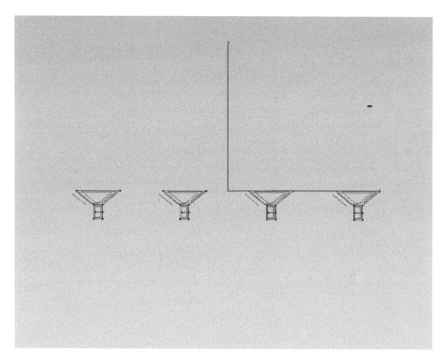

Figure 6.46 A x-directed single probe (side view).

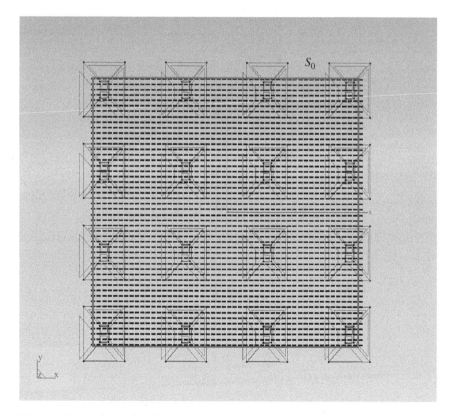

Figure 6.47 A x-directed probe array.

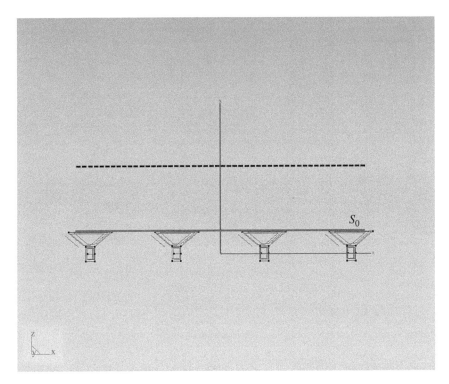

Figure 6.48 A x-directed probe array (side view).

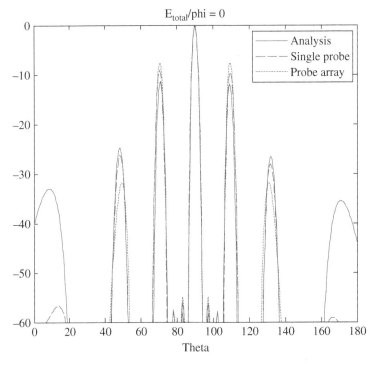

Figure 6.49 E_{total} when phi = 0° (dB Scale).

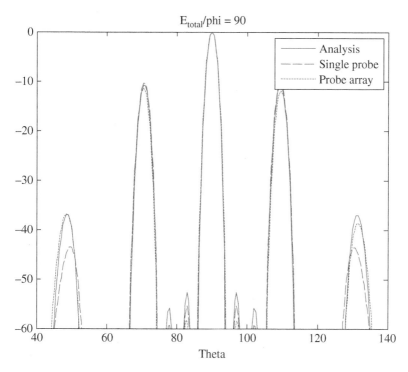

Figure 6.50 E_{total} when phi = 90° (dB Scale).

using HOBBIES, dashed lines show the single probe measurement results and the dotted lines show the probe array measurement results.

For both the methods, namely use of a single probe or an array of probes used in the simulation provide acceptable results. There are several observations that can be made from the results. First, the effect of mutual coupling between the probe and the array under test has little effect on the final result.

Even when a probe array is used it looks like the effect of mutual coupling is still not an issue. The other strength of this approach is that even though the size of the measurement plane barely covers the actual physical size of the antenna array, one can still obtain reliable results from 30° to 150°. Also, this computational methodology is quite fast and accurate. Finally, using this methodology the measurement plane can be deformed to any arbitrary shape and the Nyquist sampling criteria is not relevant for the measurement plane unlike in the Fourier transform based classical planar near-field to far-field transformation.

6.10.6.3 Example 3

For the third example a single three element Yagi-Uda antenna is selected as the AUT to illustrate the accuracy of this methodology. This antenna has a wide beam. Both the single probe method and the use of a probe array is used as

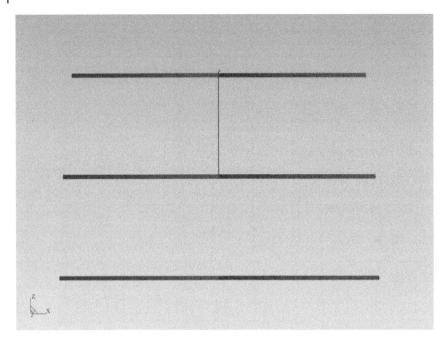

Figure 6.51 A three-element Yagi-Uda antenna.

samplers of the near field without any probe correction. The three-element Yagi-Uda antenna as shown in Figure 6.51 which consists of a driven element of length L = 0.47 λ, a reflector of length 0.482 λ, and a director of length 0.442 λ. They are all spaced 0.2 λ apart. The radius of the wire structure for all cases is 0.00425 λ. A fictitious planar surface in the *x-y* plane of dimensions 10λ by 10λ is used to approximate the equivalent source which is going to radiate the same fields in the desired region as the original antenna. On this surface equivalent magnetic currents M_x and M_y are applied. These two current components are discretized into 50 × 50 current patches. The two planar components of the near fields are measured on a planar surface of the same dimensions and are discretized to an equivalent value as of the same size as the equivalent current sources so as to make it possible to apply the CGFFT method to solve these large systems of equations using modest computational resources and using minimal CPU time. The distance between the source plane and the measurement plane is assumed to be 3λ.

The measurement methodology for this Yagi-Uda antenna is quite similar to the measurement system used for the horn antenna as described in section 6.10.6.1 illustrating Example 1. The calculated results of the two methods described earlier are used to generate the far-field along with the use of an accurate numerical electromagnetic analysis tool called HOBBIES [35] so as to assess the accuracy for the computed results obtained by the proposed methods.

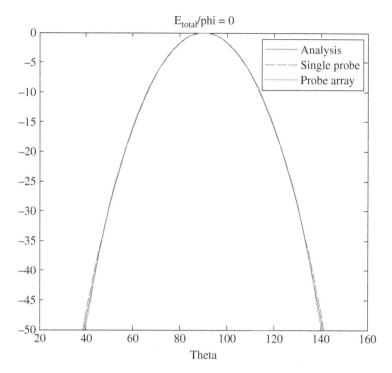

Figure 6.52 E_{total} when $\varphi = 0°$ (dB Scale).

All the three results are presented in Figure 6.52 and Figure 6.53. Figure 6.52 shows the normalized absolute value of the electric far field for $\varphi = 0°$ in a dB scale. Figure 6.53 shows the normalized absolute value of the electric far field for $\varphi = 90°$ in the dB scale. θ here is defined as the angle from x-axis to z-axis and φ is the angle from the x-axis to the y-axis, which implies φ equals $0°$ cut is the x-z plane and φ equals $90°$ cut is the y-z plane. θ equals $0°$ means $+x$ direction and θ equals $180°$ means $-x$ direction. The solid lines show the analytic results obtained using HOBBIES, dashed lines show the single probe measurement results and the dotted lines show the probe array measurement results.

We can see both the methods (namely using a single probe and an array of probes for measurement) we discussed above provides acceptable results further emphasizing that probe correction has little impact on this novel measurement procedure.

6.10.6.4 Example 4

For the final example, we deal with an antenna array under test. The array consists of 9 Yagi-Uda antennas to form a 3 by 3 antenna array as the AUT. Each element of the Yagi-Uda array has been described in example 3 and they are separated from each other by 2 λ. A fictitious planar surface in the x-y plane

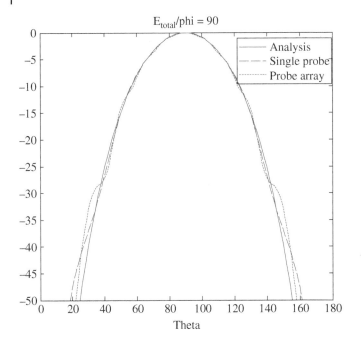

Figure 6.53 E_{total} when $\varphi = 90°$ (dB Scale).

of dimensions 5 λ by 5 λ is used to form a planar magnetic current sheet to approximate the fields that will be generated by the actual array in the desired region. On this surface, the applied equivalent magnetic currents M_x and M_y are divided into 25 × 25 current patches to approximate the measured electric fields on the measurement plane. The measurement plane is assumed to have the same size as that of the equivalent planar surface on which the magnetic currents are applied so as to be able to use the CGFFT method to solve the matrix equations containing the complex amplitudes of the unknown currents. The distance between the source plane and the measurement plane is 3 λ.

Figure 6.54 shows the x-directed single probe measurement system. Figure 6.55 shows the side view of the structure by using a x-directed single probe for measurement. Figure 6.56 shows the x-directed probe array measurement set up. Figure 6.57 shows the side view of the structure by using a x-directed probe array as an example. The red lines show that the size of S_0 coincides with the size of the measurement plane. The calculated results for the far field obtained by the two methods described in this paper, namely sliding a single probe and using a probe array along with the results computed by a numerical electromagnetics code HOBBIES invoking the electric field integral equation are illustrated in Figure 6.58 and Figure 6.59. Figure 6.58 shows the normalized absolute value of the electric far field for $\varphi = 0°$ in a dB scale.

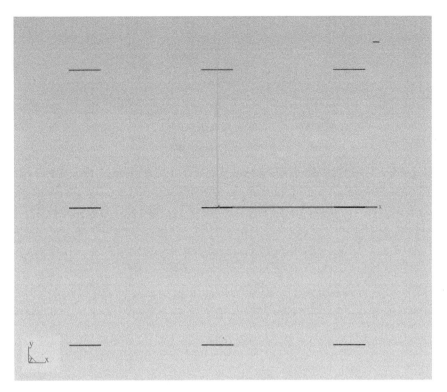

Figure 6.54 A x-directed single probe.

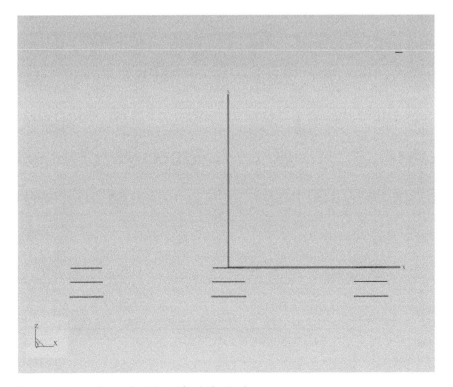

Figure 6.55 A x-directed single probe (side view).

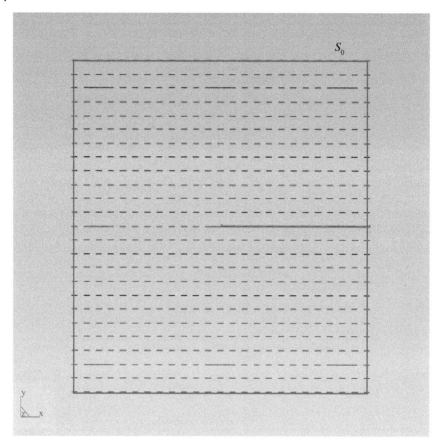

Figure 6.56 A x-directed probe array.

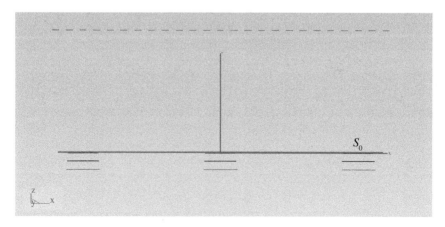

Figure 6.57 A x-directed probe array (side view).

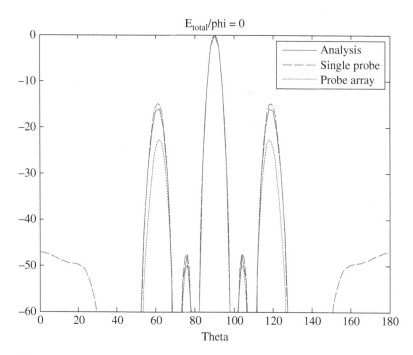

Figure 6.58 E_{total} when phi = 0° (dB Scale).

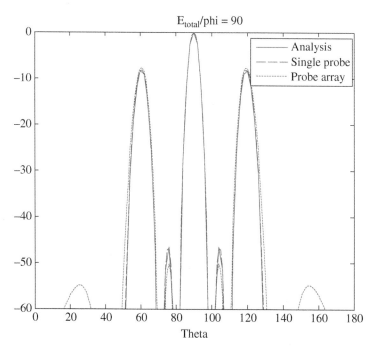

Figure 6.59 E_{total} when phi = 90° (dB Scale).

Figure 6.59 shows the normalized absolute value of the electric far field for $\varphi = 90°$ in a dB scale. θ here is defined as the angle from the x-axis to the z-axis and φ is the angle from the x-axis to the y-axis. This implies that φ equals $0°$ cut is the x-z plane and φ equals $90°$ is the y-z plane. θ equals $0°$ implies $+x$ direction and θ equals $180°$ implies $-x$ direction. The solid lines show the analytic results obtained using HOBBIES, dashed lines show the single probe measurement result and the dotted lines show the results from the probe array measurement.

We can see both methods, using a single moving dipole probe or a fixed array of dipole probes discussed above provides acceptable results and that a probe correction is not at all a requirement for this methodology. In this case, acceptable results are obtained from $40°$ to $140°$. For the classical approach of planar modal expansion, it will not have been possible to solve this problem for the given data as in this case the source and the measurement planes are of the same size!

6.10.7 Summary

A comparison is made between the calculated antenna pattern by moving a single probe across the entire measurement plane as opposed to using a planar fixed dipole array as the probe to equivalently scan the entire surface of the measurement plane just once. The equivalent magnetic current approach is used to calculate the far field for both measurement systems. For the results presented, both systems can obtain acceptable results for the far field even when the equivalent source plane and the measurement plane are of equal size. The remarkable point to note is that in this novel methodology probe correction is not deemed necessary when a small dipole is used as a measurement probe or a fixed array of dipoles used as a probe.

6.11 A Fast and Efficient Method for Determining the Far Field Patterns Along the Principal Planes Using a Rectangular Probe Array

In this section it is illustrated that using a planar rectangular shaped dipole probe array to sample the near field of an antenna under test (AUT) over the measurement plane can be used to calculate the far-field pattern along the principal planes only [36]. Also compensation for mutual coupling between the AUT and the probe array is not required. This thus provides a fast and efficient methodology to determine the E-plane and H-plane far field patterns of an antenna using partial near-field data. Conventional classical Fourier based methods cannot provide any meaningful results using partial near-field data. The current methodology requires placing the probe array over two rectangular planes near the original antenna source individually and measuring the two components of the electric fields and employing the Method of Moments approach to solve for the equivalent magnetic currents on some fictitious planes located in front of the AUT. The Method of Moments matrix is solved by using the iterative conjugate gradient method enhanced through the incorporation of the Fast Fourier Transform (FFT) techniques. Hence, in this methodology, the total time for measurement is dramatically reduced as compared to the previously discussed method of using a square dipole array for measurement. Sample numerical results are presented to illustrate the potential of a novel rectangular probe array measurement of near-field to far-field transformation technique.

6.11.1 Introduction

For a linearly polarized antenna, performance is often described in terms of its principal E-plane and H-plane patterns. Under this circumstance, we don't have to apply a square probe array measurement methodology to get the equivalent magnetic currents necessary on a square region of the source plane to get the entire radiation pattern. Instead we can utilize a smaller sized rectangular probe array to obtain the E-plane and H-plane patterns separately. For the computational technique to be used in conjunction with this measurement methodology, the unknowns in the method of moments scheme are approximated by a sub sectional basis and, secondly, the analytic free space Green's function is used which is quite easy to compute numerically. The unique feature of the novel methodology has been illustrated in the previous section and here it is illustrated that for this methodology to compute the principal plane patterns, one can further reduce the number of measurement samples required as compared to the square dipole array approach.

6.11.2 Description of the Proposed Methodology

In this methodology, a dipole array is used for the measurement of the near-field of the AUT. The probe array consist of 0.1λ length dipoles which are all terminated in 50 Ω loads. Each element in the array is separated from each other by 0.2λ in both directions to estimate the sampled electric fields on a rectangular measurement planar surfaces. To carry out the measurements of the near-field the following steps are used. First, the dipoles are all rotated by 90° to measure the field in another direction. Then, rotate the rectangular measurement plane by 90° and measure the field information for those two directions. The advantage of choosing a probe array rather than a single probe for measurement is that it can eliminate the inaccuracy of mechanical movement of the probe antenna over a large planar surface and can make the measurement methodology very efficient. This is more important particularly for measurements carried out in the high frequencies, say at M, V and W-bands. Another important

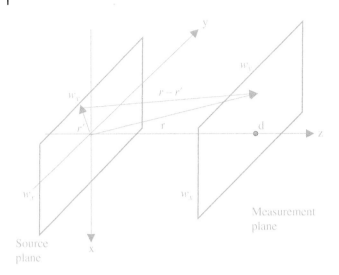

Figure 6.60 Planar scanning for the *x-z* plane.

feature of the proposed approach that will be demonstrated is that in this methodology probe correction, which accounts for the mutual coupling between the AUT and the probe and also between the probes in the probe array, do not play a significant role.

First, as shown in Figure 6.60, we make all the dipoles in the array to be *x*-directed and obtain the values of $[V_{x1}]$ across all the terminals of all the loads connected to the dipole array. Then, keeping the parameters of the array the same, each dipole is now rotated by 90° so that they allof them now become *y*-directed. This yields the value of $[V_{y1}]$ across the loads of the dipole probes. It is estimated that the voltage *V* at the center of the dipole is proportional to the electric field E at that point. We can normalize the voltages induced at the center points of the dipoles and use that information to estimate the values for the sampled near field data. From that estimated near field data, the equivalent magnetic currents (M_{x1}, M_{y1}) on the source plane can be calculated as outlined in the previous section. By using that equivalent magnetic currents, we calculate the far-field which is expected to be accurate only for the *x-z* cut of the pattern.

Next, as shown in Figure 6.61, all the dipoles in the array are now rotated to be *x*-directed and this will yield the values of $[V_{x2}]$ across the loads connected to the dipole probe array. Then, keeping the parameters of the array to be the same, each dipole is rotated by 90° to be *y*-directed and this will yield the values of $[V_{y2}]$ across the loads of the dipoles in the array. Again, it is estimated that the voltage *V* at the center of the dipole is proportional to the electric field E at that point. We can normalize the voltages induced at the center points of the dipoles and use that information to estimate the values for the sampled near

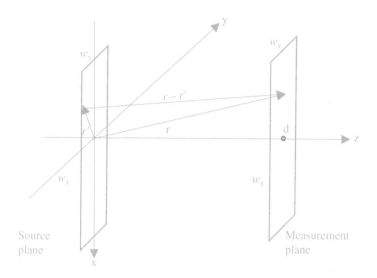

Figure 6.61 Planar scanning for y-z plane.

field data. From that estimated near field data, the equivalent magnetic currents (M_{x2}, M_{y2}) on the source plane can be calculated. By using these equivalent magnetic currents, one can calculate the far field which is expected to be accurate for the *y-z* cut.

6.11.3 Sample Numerical Results

The comparisons of the E-plane and the H-plane patterns are calculated using source and measurements over a planar slice of space covering a portion of the AUT. The assumed two principal source and the measurement planes are a slice of the planar space as described in the following examples.

6.11.3.1 Example 1

Consider a 2λ by 2λ pyramidal horn antenna under test whose principal plane patterns are desired. To generate those patterns first measurements are made using a probe array and are illustrated by the following sequence of calculations.

Step 1: The probes for this case is an array of Hertzian Dipoles of 0.1 λ length and are all terminated by 50 Ω loads. The individual elements in the array are separated from center to center along *x*-direction by 0.2 λ and along the *y*-direction also by the same amount. So the spacing between the two linear probe arrays in the *y*-direction is 0.2 λ. The induced voltages in the terminated loads are used to estimate the sampled electric fields on a planar slice of the principal measurement planes formed by the probe array. First, we make all

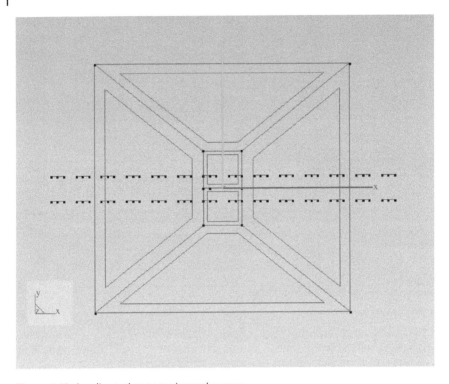

Figure 6.62 A *x*-directed rectangular probe array.

the dipoles to be *x*-directed and choose the dimensions of the array to be 2.7λ by 0.2λ, which means 28 dipoles (14 × 2 dipoles) in total for the measurement probe array as shown in Figure 6.62. The voltages across the dipoles are now measured to obtain the Matrix $[V_{x1}]$. They are now used to estimate the unknown magnetic currents on the source plane. The source plane is of the size 2.8λ by 0.4λ. It consists of 28 square patches of size $0.2\ \lambda$ each. The separation between the source plane and the measurement plane is $3\ \lambda$ as shown in Figure 6.63. Next, all the dipoles are individually rotated so that they are now *y*-directed. The measurement plane in this case is of size 2.6λ by 0.3λ to calculate the other component of the magnetic current placed on the same source plane as shown in Figure 6.64, which implies that the measurements are carried out using 28 dipoles (14 × 2 dipoles). The size of the source plane is the same as in the previous case. The measured voltages across the terminated loads of the dipoles are used to obtain the Matrix $[V_{y1}]$. The plane of the equivalent magnetic current is chosen to be of the same dimensions and discretized to enable the use of CGFFT method as outlined in the previous sections. This yields two components of the magnetic currents on the source plane providing the values

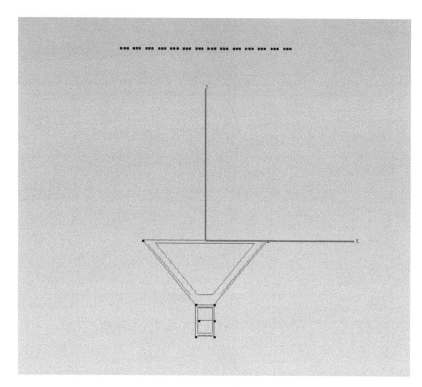

Figure 6.63 A x-directed probe array (side view).

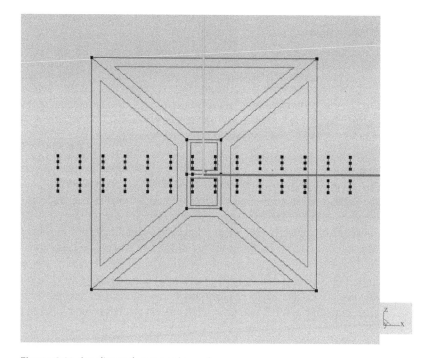

Figure 6.64 A y-directed rectangular probe array.

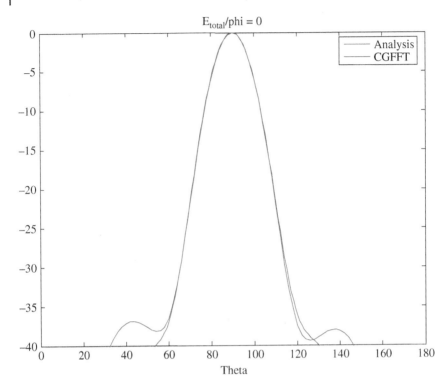

Figure 6.65 *E* total when $\phi = 0$ (dB Scale).

for M_{x1} and M_{y1} which are defined on the 14×2 current patches. The far field is now calculated in this principal plane using these two sets of magnetic currents. Figure 6.65 provides the principal plane pattern for this case. The pattern is accurate from 60° to 120°. The classical planar near to far field transformation method will not provide any result for this set of measurements [20].

Step 2: To obtain the pattern for the other cut, we place a rectangular probe array with 0.1 λ length dipole as probes with all of them terminated in 50 Ω loads. They are separated from each other in both directions by 0.2λ to estimate the sampled electric fields on the measurement plane. Next, the rectangular measurement plane is rotated by 90° when compared to that in Step 1, as shown in Figure 6.66 and 6.67. First, we make all the dipoles in the probe array to be x-directed and choose the dimensions of the array to be 0.3λ by 2.6λ, which means 28 dipoles (2 × 14 dipoles) in total as shown in Figure 6.66. The voltages across the dipoles are measured to obtain the voltage matrix $[V_{x2}]$ which estimates the near fields of the AUT Next, we make all the dipoles to be y-directed as shown in Figure 6.67. The dipole array is now of size 0.2λ by 2.7λ as shown.

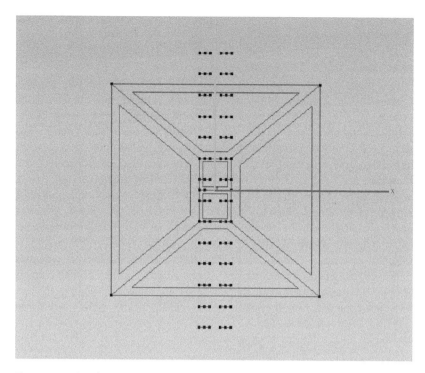

Figure 6.66 A *x*-directed probe array (step 2).

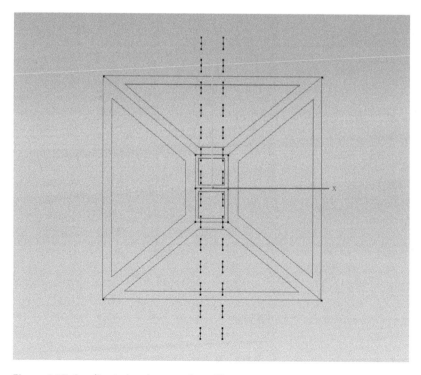

Figure 6.67 A *y*-directed probe array (step 2).

The source plane in both cases are the same 0.4λ by 2.8λ. We measure the voltages across the 28 dipoles to obtain the matrix $[V_{y2}]$ estimating the near fields from the AUT. The equivalent magnetic current plane is chosen to be of the same dimensions and discretized to enable the use of CGFFT. Which means the equivalent magnetic currents M_{x2} and M_{y2} are placed into 2×14 current patches. The equivalent magnetic currents M_{x2} and M_{y2} are first calculated and then they are used to obtain the far field pattern as shown in Figure 6.68.

It is seen that by placing the equivalent magnetic current over a planar sector covering the plane across which the far field pattern is to be computed, results of engineering accuracy can be obtained using such an approximate procedure. The other interesting point is that the mutual coupling between the measurement dipoles is not taken into account. The simulated results for the two steps we mentioned above, and the analytic results using a numerical electromagnetics code for the evaluation of the far-fields are shown in Figures 6.65 to Figure 6.68 for the two principal plane cuts. Figure 6.65 shows the normalized

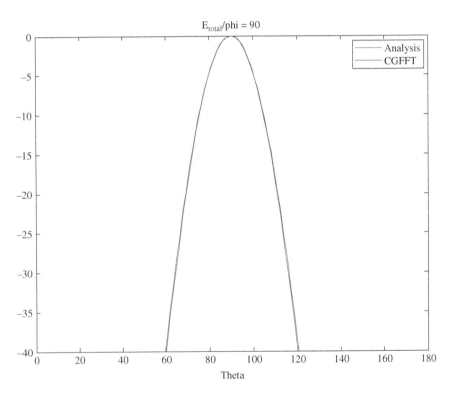

Figure 6.68 **E** total when $\phi = 90°$ (dB Scale) in step 2.

absolute value of the electric far field for $\varphi = 0°$ in the dB scale of step1. Figure 6.68 shows the normalized absolute value of the electric far field for $\varphi = 90°$ in dB scale of step1. θ is defined as the angle from x-axis to z-axis and φ is the angle from the x-axis to the y-axis. This implies, φ equals $0°$ cut is the x-z plane (*E*-plane) and φ equals $90°$ is the y-z plane (*H*-plane). θ equals $0°$ means $+x$ direction and θ equals $180°$ implies the x direction. The blue lines show the analytic results obtained by using HOBBIES, red lines show the rectangular probe array simulated results using the probe array data as the starting point over a sector. This provides a possibility of obtaining a quick solution of engineering accuracy for the principal plane patterns in a short time.

6.11.3.2 Example 2

For the next example, the AUT is made more complicated. We choose 16, 1.5λ by 2λ pyramidal horn antennas to form a 4 by 4 horn antenna array as the AUT. Each horn antenna is separated from each other by 3 λ. Next we follow the two steps described before in section 6.11.3.1.

Step 1: Put a rectangular dipole array of 0.1 λ length dipoles which are all terminated in 50 Ω loads. They are separated from each other in both directions by 0.2λ to estimate the sampled electric fields at 0.2 λ separation on the rectangular measurement plane. First, make all the measurement dipoles in the probe array to be x-directed and choose the dimensions of the array to be 9.9λ by 1.4λ, which means 400 dipoles (50 × 8 dipoles) in total in the array as shown in Figure 6.69. The difference between the source plane and the measurement plane is 3λ as shown in Figure 6.70. The voltages across the loads on the dipoles are measured to obtain the matrix $[V_{x1}]$. Next, all the dipoles are rotated to be y-directed and choose the dimensions of the array to be 9.8λ by 1.5λ, which means 400 dipoles (50 × 8 dipoles) in total as shown in Figure 6.71. Measure the voltages across the dipoles to obtain the matrix $[V_{y1}]$. The equivalent magnetic current plane is chosen to be 10λ by 1.6λ so that we can use the CGFFT method. Which means the equivalent magnetic currents M_{x1} and M_{y1} are placed into 50 × 8 current patches. From the computed equivalent magnetic currents M_{x1} and M_{y1}, the far field pattern for the principal plane is shown in Figure 6.72. Even though all the peaks in the pattern are located at the same place the amplitude is slightly off.

Step 2: Next, put a rectangular dipole array each of which is 0.1 λ and all terminated in 50 Ω loads. They are separated from each other in both directions by 0.2λ to estimate the sampled electric fields at 0.2 λ separation on the rectangular measurement plane. (This time, the rectangular measurement plane is rotated by 90° degree compared to that in Step 1, as shown in Figure 6.73 and Figure 6.74.)

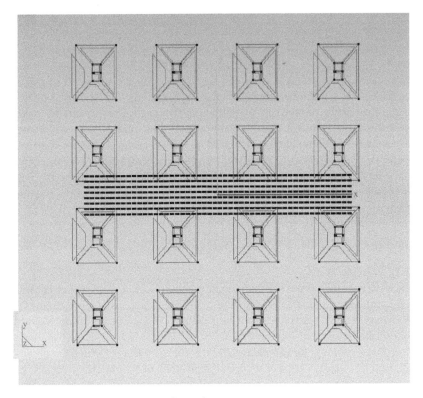

Figure 6.69 A x-directed rectangular probe array.

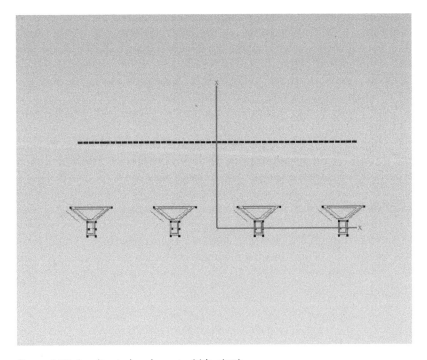

Figure 6.70 A x-directed probe array (side view).

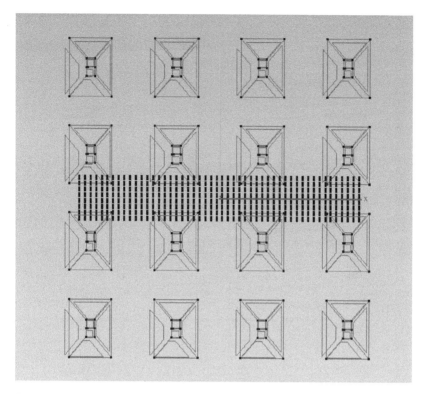

Figure 6.71 A *y*-directed rectangular probe array.

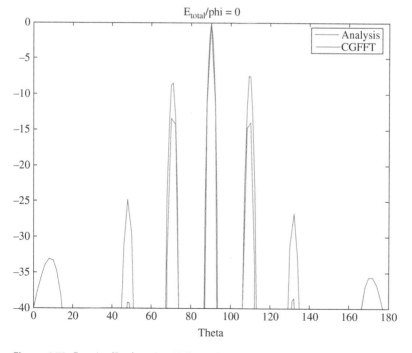

Figure 6.72 E_{total} in dB when $\phi = 0°$ (step 1).

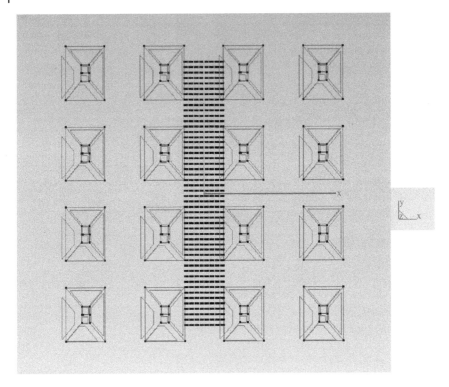

Figure 6.73 A x-directed probe array (step 2).

First, make all the dipoles to be x-directed and choose the dimensions of the array to be 1.5λ by 9.8λ, which means 400 dipoles (8 × 50 dipoles) in total as shown in Figure 6.73. Measure the voltages across the dipoles to obtain the matrix $[V_{x2}]$. Secondly, we make all the dipoles to be y-directed and choose the same dimensions of the array to be 1.4λ by 9.9λ, which means 400 dipoles (8 × 50 dipoles) in total as shown in Figure 6.74. Measure the voltages across the dipoles to obtain the matrix $[V_{y2}]$. The equivalent magnetic current plane is chosen to be 1.6λ by 10λ consisting of 400 square patches of dimensions $0.2\,\lambda$ so that the CGFFT method can be applied to compute the equivalent magnetic currents M_{x2} and M_{y2} that are placed into 8 × 50 current patches. From the equivalent magnetic currents M_{x2} and M_{y2}, the far field pattern can be computed as shown in Figure 6.75.

The simulated results for the two steps mentioned above, and the analytic results are shown in Figure 6.72 and Figure 6.75. Figure 6.72 shows the normalized absolute value of the electric far field for $\varphi = 0°$ in the dB scale of step1. Figure 6.75 shows the normalized absolute value of the electric far field for

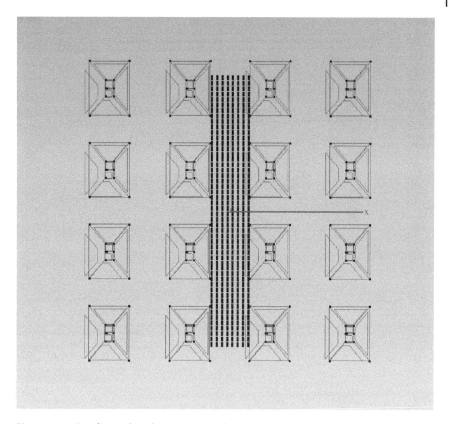

Figure 6.74 A y-directed probe array (step 2).

$\varphi = 90°$ in dB scale of step2. The blue lines show the analytic results obtained by using HOBBIES, and the red lines show the computed results using the rectangular probe array measurements.

6.11.3.3 Example 3

For the third example a single three element Yagi-Uda antenna is selected as the AUT to illustrate the reliability and the accuracy of the computed results using this methodology. This antenna has a wide beam. The three-element Yagi-Uda antenna as shown in Figure 6.76 consist of a driven element of length $L = 0.47 \lambda$, a reflector of length 0.482 λ, and a director of length 0.442 λ.They are all spaced 0.2 λ apart. The radius of the wire structure for all cases is 0.00425 λ. The measurement methodology for this Yagi-Uda antenna is quite similar to the measurement system used for the horn antenna as described in section 6.11.3.1.

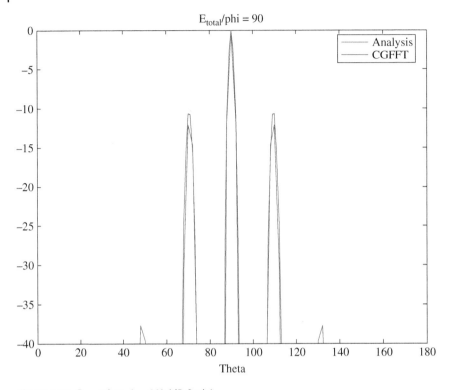

Figure 6.75 E_{total} when $\phi = 90°$ (dB Scale).

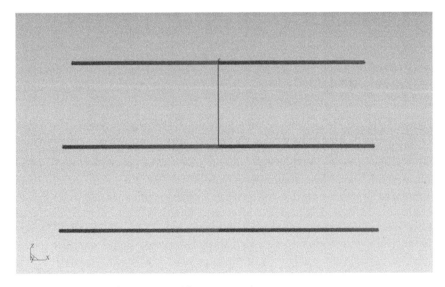

Figure 6.76 A three-element Yagi-Uda antenna.

Step 1: Put a rectangular dipole array consisting of elements with 0.1 λ length and all terminated in 50 Ω loads. The elements in the array are separated from each other in both directions by 0.2λ to estimate the sampled electric fields on a rectangular measurement plane. First, make all the dipoles to be x-directed and choose the dimensions of the array to be 1.7λ by 0.4λ, which means 27 dipoles (9 × 3 dipoles) in total. Measure the voltages across the loads on the dipoles to obtain the matrix $[V_{x1}]$. Secondly, make all the dipoles to be y-directed and choose the dimensions of the array to be 1.6λ by 0.5λ, which means 27 dipoles (9 × 3 dipoles) in total. Measure the voltages across the loads of the dipoles to obtain the matrix $[V_{y1}]$. The source plane over which the equivalent magnetic currents M_{x1} and M_{y1} are placed resolves into 9 ×3 current patches. The equivalent magnetic currents are now solved using the CGFFT method. From those currents the far field pattern is obtained as shown in Figure 6.77.

Step 2: Next, orient the rectangular dipole array where all the elements are of length 0.1 λ are all terminated in 50 Ω loads. They are separated from each other in both directions by 0.2λ to estimate the sampled electric fields at 0.2 λ separation on the rectangular measurement plane. (This time, the rectangular measurement plane is rotated by 90° compared to that in Step 1). First, make all

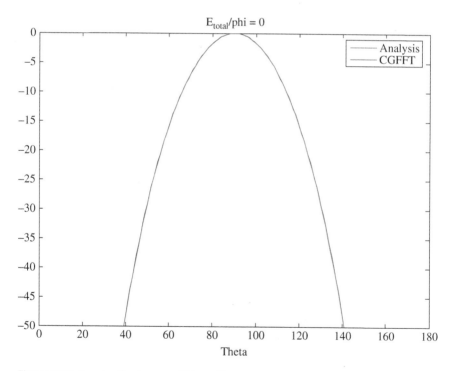

Figure 6.77 E_{total} in dB when $\varphi = 0°$ (step 1).

the dipoles to be x-directed and choose the dimensions of the array to be 0.5λ by 1.6λ, which means 27 dipoles (3×9 dipoles) in total are used. We measure the voltages across the dipoles to obtain the matrix $[V_{x2}]$. Secondly, make all the dipoles to be y-directed and choose the dimensions of the array to be 0.4λ by 1.7λ, which means 27 dipoles (3×9 dipoles) in total are used. We measure the voltages across the dipoles to obtain the matrix $[V_{y2}]$. The plane of the equivalent magnetic current is chosen to be of the same dimensions and is discretized so as to be able to apply the fast and accurate CGFFT method. From the computed equivalent magnetic currents M_{x2} and M_{y2} the far field pattern is obtained and shown in Figure 6.78.

The simulated results for the two steps as described above, and the computations using an electromagnetic analysis code are shown in Figure 6.77 and Figure 6.78. Figure 6.77 shows the normalized absolute value of the electric far field for $\varphi = 0°$ in the dB scale of step1. Figure 6.78 shows the normalized absolute value of the electric far field for $\varphi = 90°$ in dB scale of step 2. The blue lines show the analytic results obtained by using HOBBIES, and red lines show the results computed from the rectangular probe array measurements.

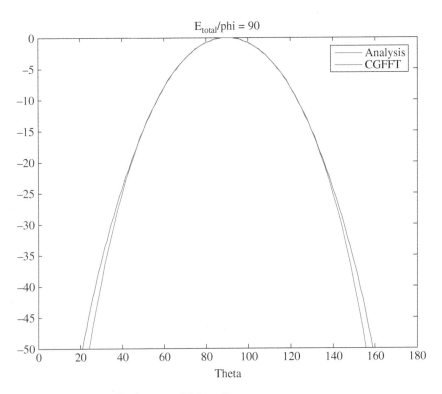

Figure 6.78 E_{total} in dB when $\varphi = 90°$ (step 2).

6.11.3.4 Example 4

For the final example, we deal with an antenna array of 9 Yagi-Uda antenna placed in a 3 by 3 array. Each element of the Yagi-Uda array has been described in 6.11.3.3 and the elements in the array are separated from each other by 2 λ.

Step 1: Put a rectangular dipole array each of 0.1 λ in length and all terminated in 50 Ω loads The dipole elements in the array are separated from each other in both directions by 0.2λ to estimate the sampled electric fields at 0.2 λ separation on the rectangular measurement plane. First, make all the dipoles to be x-directed and chose the dimensions of the array to be 4.9λ by 0.4λ, which means 75 dipoles (25 × 3 dipoles) in total are chosen as shown in Figure 6.79. Measure the voltages across the dipoles to obtain the matrix $[V_{x1}]$. Secondly, make all the dipoles to be y-directed and choose the dimensions of the measurement array to be 4.8λ by 0.5λ, which means 75 dipoles (25 × 3 dipoles) in total are required as shown in Figure 6.80. The measurement plane is 3λ away from the AUT as shown in Figure 6.81. Measure the voltages across the loads connected to the dipoles to obtain the matrix $[V_{y1}]$. The equivalent magnetic current plane is

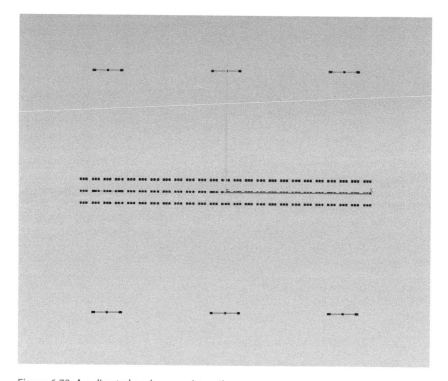

Figure 6.79 A x-directed probe array (step 1).

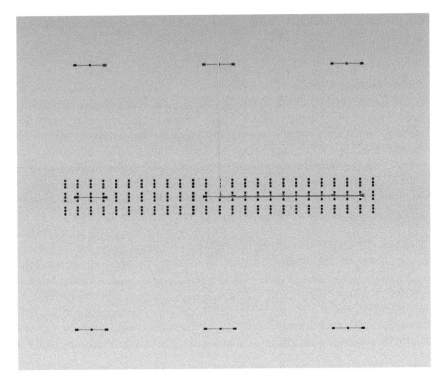

Figure 6.80 A *y*-directed probe array (step 1).

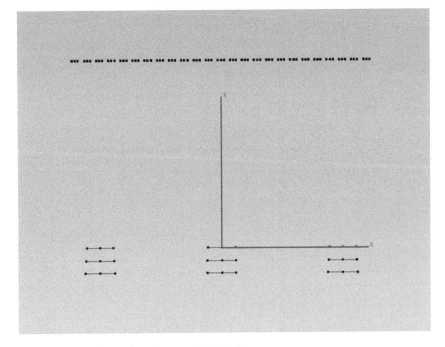

Figure 6.81 A *x*-directed probe array (side view).

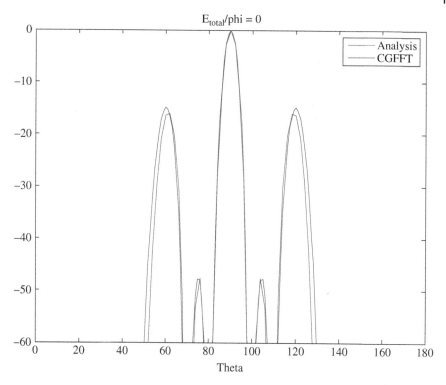

Figure 6.82 E_{total} in dB when $\phi = 0°$ (step 1).

chosen to be of the same dimensions and discretized to enable the use of CGFFT method in the computations. Which means the equivalent magnetic currents M_{x1} and M_{y1} are placed into 25×3 current patches. From the computed equivalent magnetic currents M_{x1} and M_{y1} utilizing the voltages induced in the probe array, we obtain the far-field pattern as shown in Figure 6.82.

Step 2: Now, place choose a rectangular dipole array where the length of each of the elements are 0.1 λ and all the dipoles are terminated in 50 Ω loads. They are separated from each other in both directions by 0.2λ to estimate the sampled electric fields at 0.2 λ separation on the rectangular measurement plane. (This time, the rectangular measurement plane is rotated by 90° compared to that in Step 1, as shown in Figure 6.83 and Figure 6.84) First, make all the dipoles to be x-directed and choose the dimensions of the array to be 0.5λ by 4.8λ, which means 75 dipoles (3 × 25 dipoles) in total are chosen as shown in Figure 6.83. Measure the voltages across the dipoles to obtain the matrix

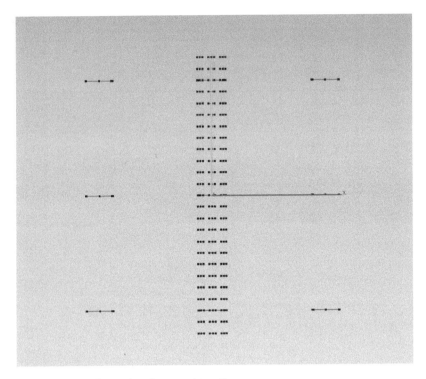

Figure 6.83 A *x*-directed probe array (step 2).

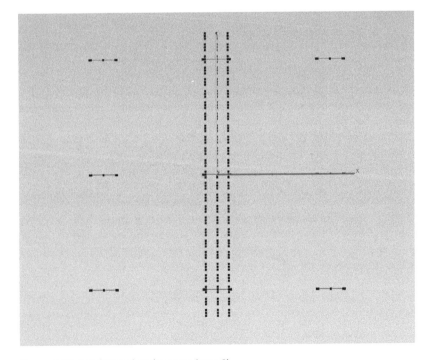

Figure 6.84 A *y*-directed probe array (step 2).

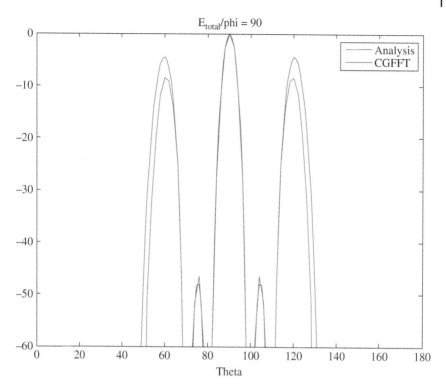

Figure 6.85 E_{total} in dB when $\phi = 90°$ (step 2).

$[V_{x2}]$. Secondly, make all the dipoles to be y-directed and choose the dimensions of the array to be 0.4λ by 4.9λ, which means 75 dipoles (3×75 dipoles) in total are used as shown in Figure 6.84. Measure the voltages across the loads of the dipoles to obtain the matrix $[V_{y2}]$. The equivalent magnetic current plane is chosen to be 0.6λ by 5λ in order to use the CGFFT method. The equivalent magnetic currents M_{x2} and M_{y2} are discretized into 3×25 current patches. From the computed equivalent magnetic currents M_{x2} and M_{y2}, the far field can be obtained as shown in Figure 6.85. The simulated results for the two steps we mentioned above, and the analytic results are shown in Figure 6.82 and Figure 6.85. Figure 6.82 shows the normalized absolute value of the electric far field for $\varphi = 0°$ in the dB scale of step1. Figure 6.85 shows the normalized absolute value of the electric far field for $\varphi = 90°$ in dB scale of step 1. The blue lines show the analytic results obtained by using HOBBIES, red lines show the rectangular probe array measurement results.

6.11.4 Summary

A fast and efficient planar near field measurement methodology along the principal planes is proposed for an AUT. It is shown that a dipole probe array can be used to sample the near fields on a planar sector of the measurement plane and there is no need to cover the entire antenna under test and yet compute the principal plane patterns with engineering accuracy. Conventional planar near field transformation techniques require the measurement over a large plane covering at least the front face of the antenna under test. Mutual coupling compensation between the elements of the probe under test seems not to be necessary in this methodology. Simulation results indicate that the fields along the principal cuts of the antenna under test can be obtained very quickly and efficiently using this methodology.

6.12 The Influence of the Size of Square Dipole Probe Array Measurement on the Accuracy of NF-FF Pattern

The objective of this section is to illustrate the influence of the size of the measurement plane on the accuracy of the far-field pattern computed using the near-field obtained from using a planar rectangular dipole probe array measurement system. The error is defined as the difference between the computed far-field using the equivalent magnetic currents computed form using a probe array to sample the near fields with the results from an electromagnetic analysis code called HOBBIES [35] for different sizes of the measurement plane keeping the size of the source plane fixed. Analysis is made between the relation of the size of the rectangular measurement plane and the accuracy of the computed far-field pattern using this data. Here we define a relative error as follows,

$$\Xi = \sum_{\theta = 0°}^{\theta = 180°} \left(E_{Theory} - E_{Cal} \right)^2 \tag{6.94}$$

where, Ξ the relative error, E_{Theory} is the theoretical far-field result simulated by HOBBIES at one cut, and E_{Cal} is the result obtained from the presented NF-FF approach.

As compared to the classical Fourier based modal expansion methods, use of a square dipole probe array to generate the near-field data provides reliable results for cases when the classical method fails for the case when the actual source plane and the measurement plane are approximately equal in size [20]. Also, in this approach there is no need to incorporate probe correction unlike in the existing approaches. In addition, the methodology of using probe array instead of moving a single probe over the measurement plane improved the accuracy and efficiency of the whole process which is accomplished by solving for the equivalent magnetic current over a plane near the original source antenna under test and then employing the Method of Moments approach to solve for the equivalent magnetic currents on this fictitious surface. It is expected that the larger the size of the measurement plane is chosen, the more accurate are the computed results. So the question is what is the relation between the accuracy and the size of the measurement plane so that one can make a smart choice to get the result that is accurate enough in an efficient way. Sample numerical results are presented to illustrate how accurate it can be and what the relation between the size of the near-field measurement plane and the accuracy of the final result is.

6.12.1 Illustration of the Proposed Methodology Utilizing Sample Numerical Results

The objective of this section is to show the influence of different sizes of measurement planes on the accuracy of the computed far-field pattern. Thus, when we have an AUT whose near-field are to be measured, one can then have a clear idea of what size of the measurement plane one needs to choose. The far field calculated by using different sizes of the measurement planes and the relative error with respect to the sizes of the measurement planes are introduced next.

6.12.1.1 Example 1

A 2 λ by 2 λ pyramidal horn antenna is used as the antenna under test. The distance between the source plane and the scanning plane is 3 λ. In this case the size of the actual source plane of the antenna under test is 2 λ by 2 λ. We use 0.1 λ length Hertzian Dipoles in the array to estimate the sampled electric fields at certain separations on the measurement planar surfaces. The dipole probes are all terminated in 50 Ω loads and the voltages across the loads are measured.

The number of the measurement dipoles starts from 4 by 4 and end up with 100 by 100. In order to keep the symmetry, we increase the number of dipoles on each side by a factor of 2 at a time, and we choose the separation between dipoles in both directions to be 0.2 λ. Which means the measurement plane start from the dimensions of 0.8 λ by 0.8λ and increases to 20 λ by 20λ. On this surface, the equivalent magnetic currents M_x and M_y are distributed and discretized to enable the use of CGFFT to solve the integral equations.

Figure 6.86 shows a 10 by 10 x-directed probe array with 0.2 λ separations in both directions of the structure. Figure 6.87 displays the side view of the same structure. The objective here is to study the relationship between the error in the final result between the simulated results using an electromagnetic numerical analysis tool HOBBIES and the results predicted by this methodology as a

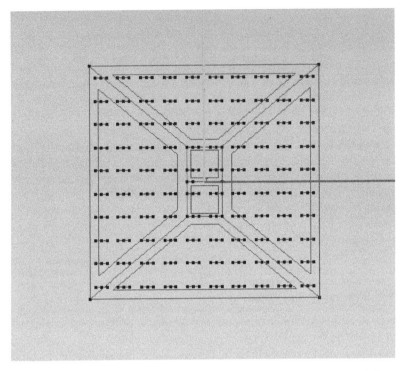

Figure 6.86 A 10 by 10 x-directed probe array with 0.2 λ separations in both directions.

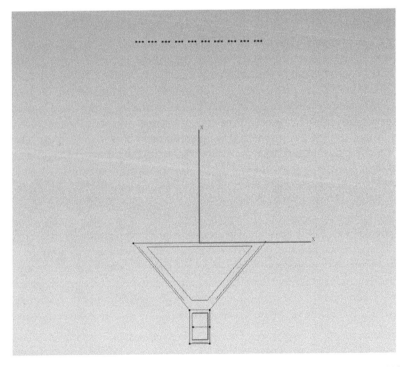

Figure 6.87 The same x-directed probe array with 0.2 λ separations in both directions (side view).

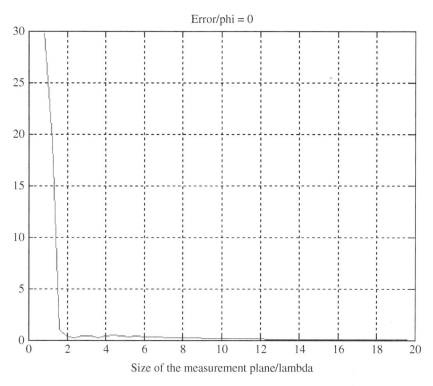

Figure 6.88 Relative error for different sizes of the measurement planes with 0.2 λ separations between the elements of the probe in both directions ($\varphi = 0°$).

function of the size of the measurement plane. The various errors are plotted in Figure 6.88 and Figure 6.89 for all the sizes of the measurement planes from 0.4 λ by 0.4λ to 20 λ by 20λ. We observe in the relative error plots of (6.94) for both the cuts, that the error goes down approximately to zero when the size of the measurement becomes larger than 2 λ. It is important to note that 2 λ is also the size of the actual source plane of the antenna under test. Also, the far field results obtained using the sizes of the measurement planes which are larger than 2 λ are as shown in Figure 6.90 and Figure 6.91. θ here is defined as the angle from the x-axis to the z-axis and φ is the angle from the x-axis to the y-axis. This implies, φ equals 0° cut is the x-z plane and φ equals 90° is the y-z plane. The solid blue lines show the analytic results obtained using HOBBIES, dashed red lines show the results obtained using different sizes of square dipole probe array measurement. We can see that all the red dashed lines provide accurate estimates.

We can see that after the size of the measurement plane chosen to be larger than the size of the actual source plane of the antenna under test, the presented NF-FF approach provides acceptable results.

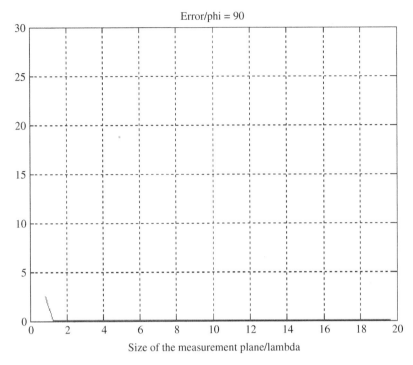

Figure 6.89 Relative error for different sizes of the measurement planes with 0.2 λ separations between the elements of the probe in both directions ($\varphi = 90°$).

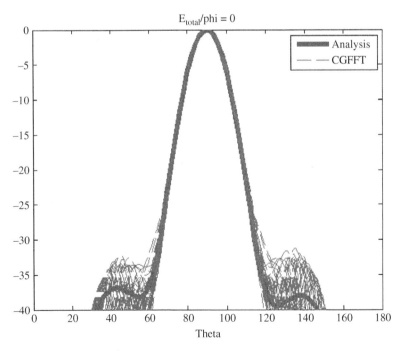

Figure 6.90 E_{total} when $\varphi = 0°$ (dB Scale) for all sizes of the measurement planes from 2 λ by 2 λ to 20 λ by 20λ with 0.2 λ separations in both directions.

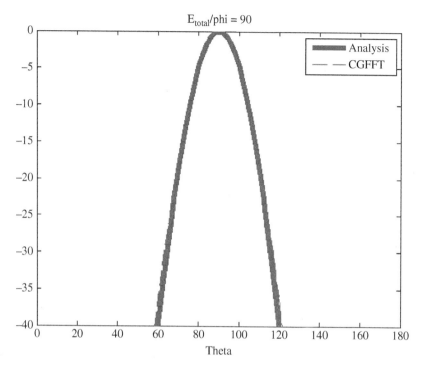

Figure 6.91 E_{total} when $\varphi = 90°$ (dB Scale) for all sizes of the measurement planes from 2 λ by 2 λ to 20 λ by 20λ with 0.2 λ separations in both directions.

6.12.1.2 Example 2

For the next example, the AUT is made more complicated by choosing 16, 1.5 λ by 2 λ pyramidal horn antennas to form a 4 by 4 horn antenna array as the AUT. Each horn antenna is separated from each other by 3 λ. The distance between the source plane and the scanning plane is 3λ. In this case, the size of the source plane at the AUT is 10.5 λ by 11 λ.

We perform 3 different groups of measurements:

Group 1: The number of the measurement dipoles starts from 4 by 4 and end up with 100 by 100. In order to keep the symmetry, the number of dipoles on each side is increased by a factor of 2 at a time, and the separation between the dipoles in both directions is chosen to be 0.2 λ. This implies that the measurement plane starts from the dimensions of 0.4 λ by 0.4λ and increases to 20 λ by 20λ. On the surface of the equivalent magnetic currents M_x and M_y are placed to be of the same dimensions and discretized so as to enable one to use the CGFFT.

Group 2: The number of the measurement dipoles starts from 4 by 4 and end up with 50 by 50. In order to keep the symmetry, we increase the number of dipoles on each side by a factor of 2 at a time, and we choose the separation between dipoles in the array to be 0.2 λ along both directions. This implies that the measurement plane starts from the dimensions of 1.6 λ by 1.6λ and increases to 20 λ by 20λ. On the surface, the equivalent magnetic currents M_x and M_y are made of the same dimensions and discretized so as to enable one to use the CGFFT method.

Group 3: The number of the measurement dipoles starts from 4 by 4 and end up with 40 by 40. In order to keep the symmetry, we increase the number of dipoles on each side by a factor of 2 at a time, and we choose the separation between the dipoles in both directions to be 0.5 λ. This means that the measurement plane starts from the dimensions of 2 λ by 2λ and increases to 20 λ by 20λ. On the surface of the equivalent magnetic currents M_x and M_y are placed into the same sizes and discretized so as to enable one to use the CGFFT method.

Figure 6.92 shows the 50 by 50 x-directed probe array with 0.2 λ separations in both directions of the structure. Figure 6.93 shows the side view of the structure

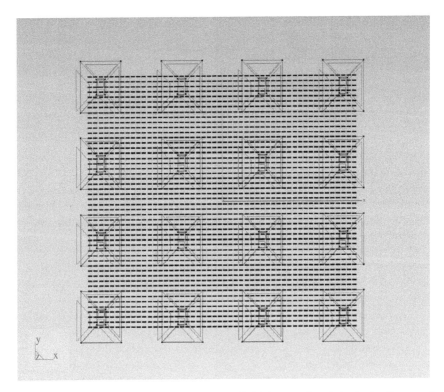

Figure 6.92 A 50 by 50 x-directed probe array with 0.2 λ separations in both directions used as a measurement probe array.

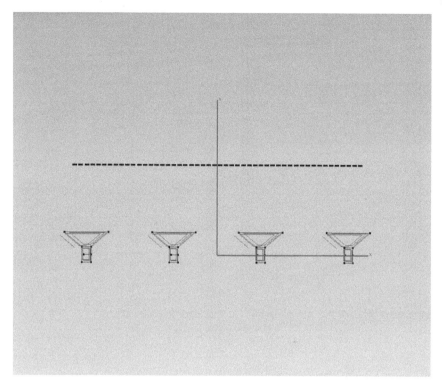

Figure 6.93 A 50 by 50 *x*-directed probe array with 0.2 λ separations in both directions (side view) as the probe array.

by displaying the same 50 by 50 *x*-directed probe array with 0.2 λ separations in both directions. It's necessary for us to analyze the relation between the relative error of (6.94) as a function of the size of the measurement plane. The relations are shown in Figure 6.94 and Figure 6.95.

We observe that plots of the relative error for both the pattern in the principal planes goes down close to zero after the size of the measurement plane becomes larger than 10 λ. The remarkable point is that the relative error goes down close to zero for all the 3 groups of measurements with different separations (0.2 λ, 0.4 λ, 0.5 λ) and occurs near the same size of the measurement plane. Notice that 11 λ is about the size of the actual source plane of the AUT, where this occurs. Also, the far-field results obtained using different sizes of the measurement planes with using 0.2 λ separations in both directions between the dipoles in the probe which are larger than 10 λ are as shown in Figure 6.96 and Figure 6.97. The far-field results obtained using different sizes of the measurement planes with 0.4 λ separations in both directions and whose total

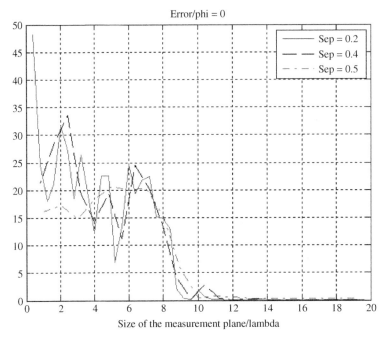

Figure 6.94 Relative error for different sizes of measurement planes with 0.2 λ, 0.4 λ, 0.5 λ separations between the dipole elements in both directions ($\varphi = 0°$).

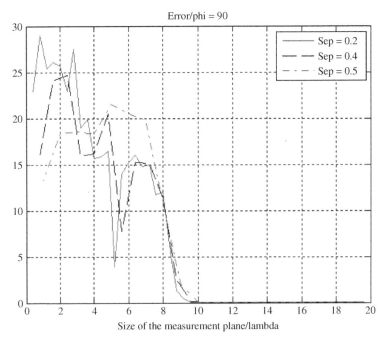

Figure 6.95 Relative error for different sizes of measurement planes with 0.2 λ, 0.4 λ, 0.5 λ separations between the dipole elements in both directions ($\varphi = 90°$).

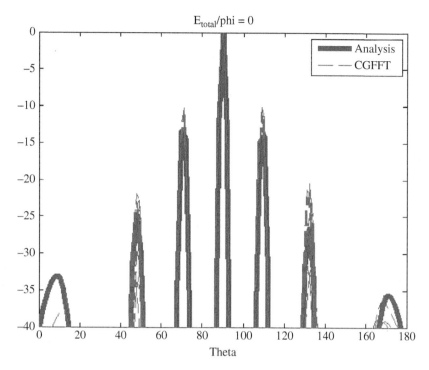

Figure 6.96 E_{total} when $\varphi = 0°$ (dB Scale) for all sizes of the measurement planes from 10 λ by 10λ to 20 λ by 20λ with 0.2 λ separations between the dipole elements in both directions.

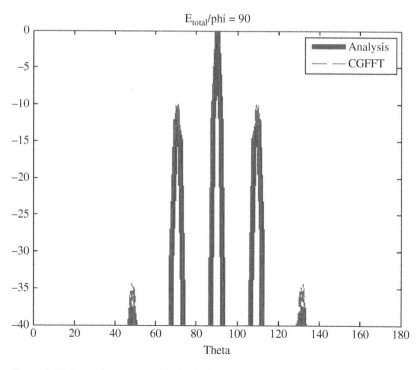

Figure 6.97 E_{total} when $\varphi = 90°$ (dB Scale) for all sizes of the measurement planes from 10 λ by 10λ to 20 λ by 20λ with 0.2 λ separations between the dipole elements in both directions.

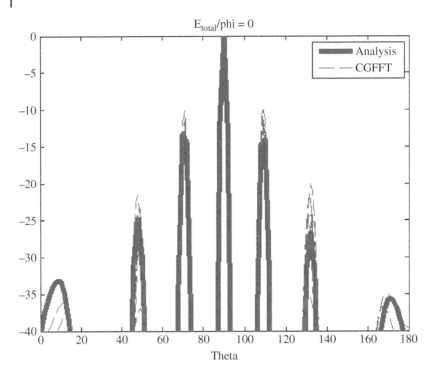

Figure 6.98 E_{total} when $\varphi = 0°$ (dB Scale) for all sizes of the measurement planes from 10 λ by 10λ to 20 λ by 20λ with 0.4 λ separations between the dipole elements in both directions.

dimensions are larger than 10 λ are shown in Figure 6.98 and Figure 6.99. The far field results obtained for the sizes of measurement planes with the dipoles in the probe are separated from each other by 0.5 λ in both directions. Their total sizes are larger than 10 λ and are shown in Figure 6.100 and Figure 6.101. θ here is defined as the angle from x axis to z axis and φ is the angle from the x-axis to the y-axis. This implies, φ equals 0° cut is the x-z plane and φ equals 90° is the y-z plane. The solid blue lines show the analytic results obtained using HOB-BIES, and the dashed red lines show the computed results obtained using measurements from different sizes of square dipole probe array measurements. We can see that all the red dashed lines are accurate.

Hence, it is clear that after the size of the measurement plane is chosen to be larger than the size of the actual source plane of the AUT, the presented NF-FF transformation provides acceptable results.

6.12.1.3 Example 3

For the third example a single three element Yagi-Uda antenna is selected as the AUT to illustrate the validity of this methodology. This antenna has a wide beam. The three-element Yagi-Uda antenna is shown in Figure 6.76. It consists

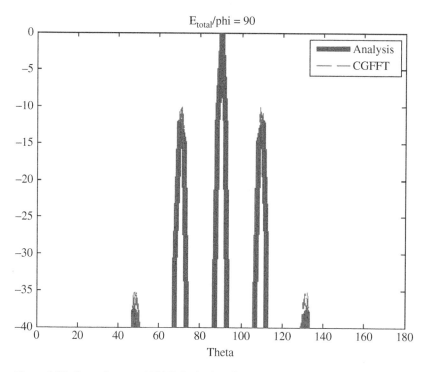

Figure 6.99 E_{total} when $\varphi = 90°$ (dB Scale) for all sizes of the measurement planes from 10 λ by 10λ to 20 λ by 20λ with 0.4 λ separations between the dipole elements in both directions.

of a driven element of length $L = 0.47$ λ, a reflector of length 0.482 λ, and a director of length 0.442 λ. They are all spaced 0.2 λ apart. The radius of the wire structure for all cases is 0.00425 λ. The distance between the source plane and the measurement plane is assumed to be 3λ. The measurement methodology for this Yagi-Uda antenna is quite similar to the measurement system used for the horn antenna as described in 6.12.1.1.

The number of the measurement dipoles starts from 4 by 4 and end up with 100 by 100. In order to keep the symmetry, we increase the number of dipoles on each side by a factor of 2 at a time, and we choose the separation between the dipoles in the probe to be separated in both directions by 0.2 λ. This implies that the measurement plane starts from the dimensions of 0.8 λ by 0.8λ and increases to 20 λ by 20λ. The equivalent magnetic currents M_x and M_y are chosen to be of the same dimensions and discretized so as to enable one to use the CGFFT method. The simulated results using all the sizes of the measurement planes from 0.8 λ by 0.8λ to 20 λ by 20λ and the analytic results for the far fields are shown in Figure 6.102 and Figure 6.103. Figure 6.102 shows the normalized absolute value of the electric far field for $\varphi = 0°$ in the dB scale. Figure 6.103 shows the normalized absolute value of the electric far field for $\varphi = 90°$ in

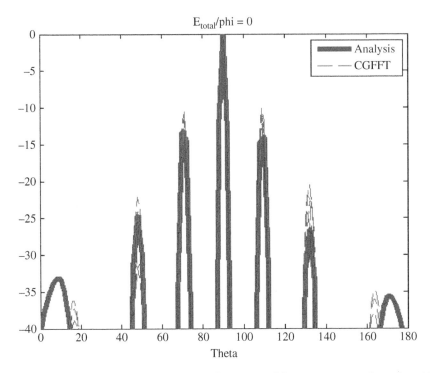

Figure 6.100 E_{total} when $\varphi = 0°$ (dB Scale) for all sizes of the measurement planes from 10 λ by 10λ to 20 λ by 20λ with 0.5 λ separations between the dipole elements in both directions.

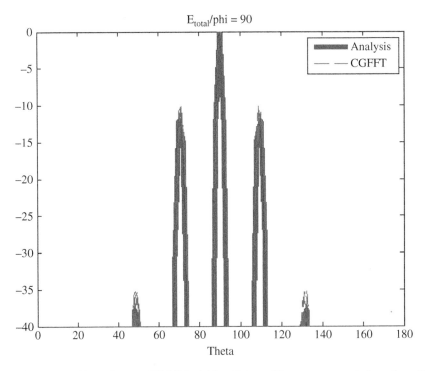

Figure 6.101 E_{total} when $\varphi = 90°$ (dB Scale) for all sizes of the measurement planes from 10 λ by 10λ to 20 λ by 20λ with 0.5 λ separations between the dipole elements in both directions.

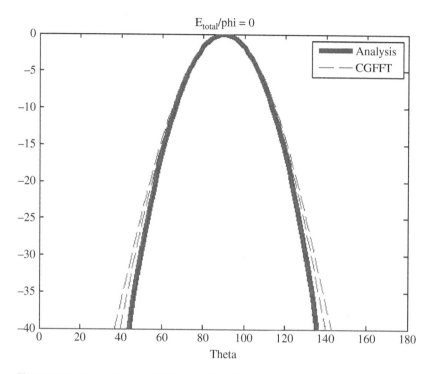

Figure 6.102 E_{total} when $\varphi = 0°$ (dB Scale) for all sizes of the measurement planes from 0.8 λ by 0.8λ to 20 λ by 20λ with 0.2 λ separations between the dipole elements in both directions.

the dB scale. The solid blue lines show the analytic results obtained using HOB-BIES, and the dashed red lines show the results obtained using different sizes of the plane containing the square dipole probe array. From Figure 6.102 and Figure 6.103 it is seen that the dashed red lines are very close with respect to the solid blue line. It's also necessary to analyze the relation between the relative error defined in (6.94) with the size of the measurement plane. The relations are shown in Figure 6.104 and Figure 6.105. The error goes down to zero after the size of the measurement is larger than 0.8 λ. Notice that 0.5 λ is the size of the actual source plane of the AUT. After the size of the measurement plane is chosen to be larger than the size of the actual source plane of the AUT, the presented NF-FF approach provides acceptable results.

6.12.1.4 Example 4

For the final example, an antenna array of 9 Yagi-Uda antennas in a 3 by 3 configuration is chosen as the AUT. Each element of the Yagi-Uda antenna has been described in 6.11.3.4 and they are separated from each other by 2 λ. The distance between the source plane and the measurement plane is 3λ.

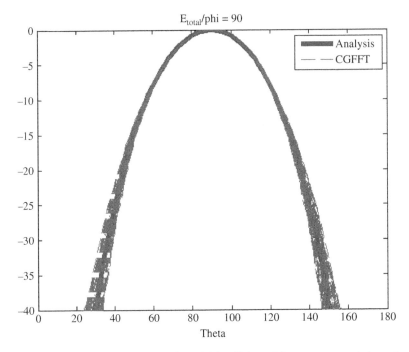

Figure 6.103 E_{total} when $\varphi = 90°$ (dB Scale) for all sizes of the measurement planes from 0.8 λ by 0.8λ to 20 λ by 20λ with 0.2 λ separations between the dipole elements in both directions.

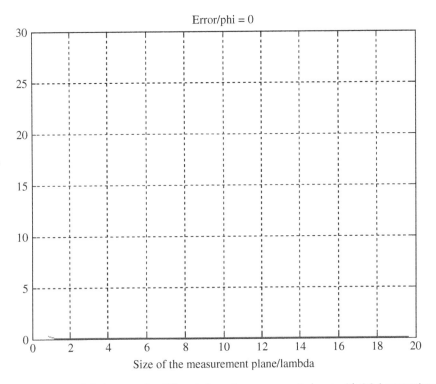

Figure 6.104 Relative error for different sizes of measurement planes with 0.2 λ separations between the dipole elements in both directions ($\varphi = 0°$).

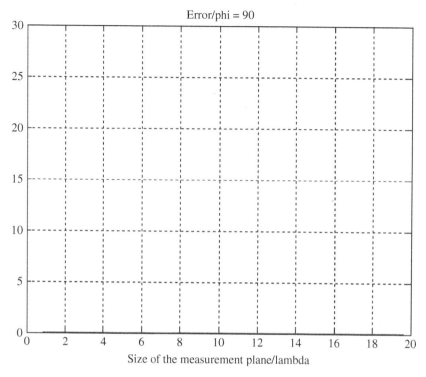

Figure 6.105 Relative error for different sizes of measurement planes with 0.2 λ separations between the dipole elements in both directions ($\varphi = 90°$).

We study 3 different groups of measurements:

Group 1: The number of the measurement dipoles starts from 4 by 4 and end up with 100 by 100. In order to keep the symmetry, we increase the number of dipoles on each side by a factor of 2 at a time, and we choose the separation between the dipoles in both directions to be 0.2 λ. This implies that the measurement plane starts from the dimensions of 0.8 λ by 0.8λ and increases to 20 λ by 20λ. The equivalent magnetic currents M_x and M_y on the source plane are structured so as to enable one to use the CGFFT method.

Group 2: The number of the measurement dipoles starts from 4 by 4 and end up with 50 by 50. In order to keep the symmetry, we increase the number of dipoles on each side by a factor of 2 at a time, and we choose the separation between the dipoles in both directions to be 0.4 λ. This implies that the measurement plane starts from the dimensions of 1.6 λ by 1.6λ and increases to 20 λ by 20λ. The equivalent magnetic currents M_x and M_y on the source plane are structured so that the CGFFT method can be implemented.

Group 3: The number of the measurement dipoles starts from 4 by 4 and end up with 40 by 40. In order to keep the symmetry, we increase the number of dipoles on each side by a factor of 2 at a time, and we choose the separation between the dipoles in both directions to be 0.5 λ. This implies that the measurement plane starts from the dimensions of 2 λ by 2λ and increases to 20 λ by 20λ. The equivalent magnetic currents M_x and M_y on the source plane are structured so as to enable one to use the CGFFT method.

Figure 6.106 shows the 24 by 24 x-directed probe array with 0.2 λ separations in both the directions. Figure 6.107. shows the side view of the structure by from the 24 by 24 x-directed probe array with 0.2 λ separations in both directions. The relation between the relative error of (6.94) as a function of the size of the measurement plane is now presented in Figure 6.108 and Figure 6.109. It is seen the error for both the principal planes goes down close to zero after the size of the measurement plane becomes larger than 5 λ. The remarkable point is that the relative error goes down to zero for all the 3 groups of measurements with

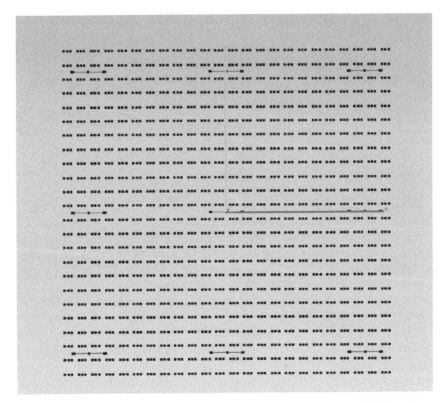

Figure 6.106 A 24 by 24 x-directed probe array with 0.2 λ separations in both directions between the dipoles in the probe array.

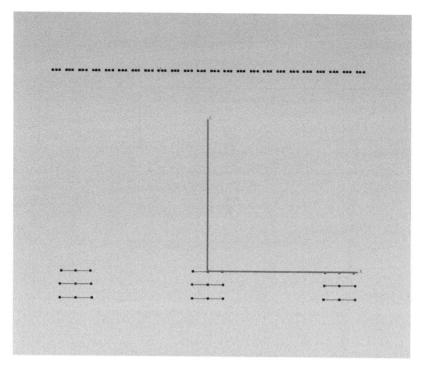

Figure 6.107 A 24 by 24 x-directed probe array with 0.2 λ separations in both directions (side view). between the dipoles in the probe array.

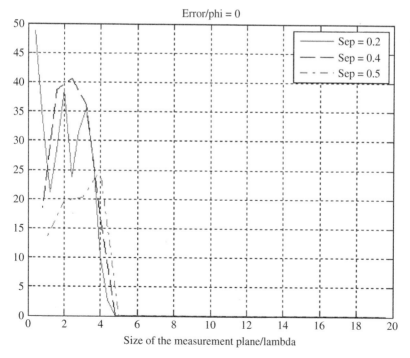

Figure 6.108 Relative error for different sizes of measurement planes with 0.2 λ, 0.4 λ, 0.5 λ separations in both directions ($\varphi = 0°$) between the dipoles in the probe array.

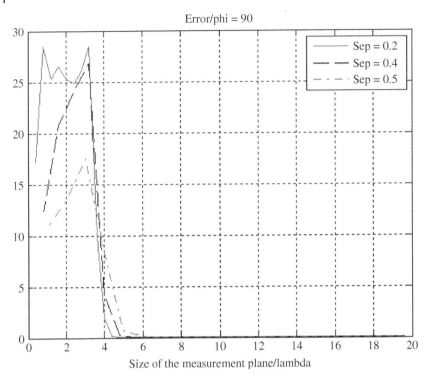

Figure 6.109 Relative error for different sizes of measurement planes with 0.2 λ, 0.4, 0.5 λ separations in both directions (φ = 90°) between the dipoles in the probe array.

different separations (0.2 λ, 0.4 λ, 0.5 λ) between the dipoles in the probe array at about the same size of the measurement plane. Observe that 5 λ is also about the size of the actual source plane of the AUT. Also, the far field results obtained using different sizes of the measurement planes with 0.2 λ separations between the dipoles in the probe array along both directions which are larger than 5 λ are shown in Figure 6.110 and Figure 6.111. The far field results obtained from the sizes of measurement planes with 0.4 λ separations in both directions which are larger than 4.8 λ are as shown in Figure 6.112 and Figure 6.113. The far field results obtained using different sizes of the measurement planes with 0.5 λ separations along both directions which are larger than 5 λ are shown in Figure 6.114 and Figure 6.115. θ here is defined as the angle from the x-axis to the z-axis and φ is the angle from the x-axis to the y-axis. This implies, φ equals 0° cut is the x-z plane and φ equals 90° is the y-z plane. The solid blue lines show the analytic results obtained using HOBBIES, and the dashed red lines show the computed results for different sizes of the measurement planes

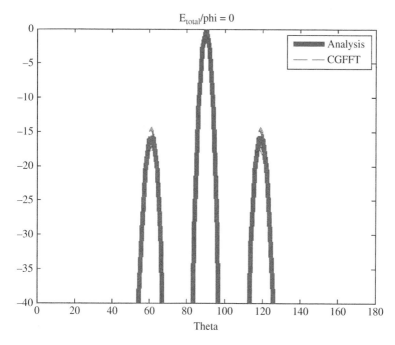

Figure 6.110 E_{total} when $\varphi = 0°$ (dB Scale) for all sizes of the measurement planes from 5 λ by 5λ to 20 λ by 20λ with 0.2λ separations in both directions between the dipoles in the probe array.

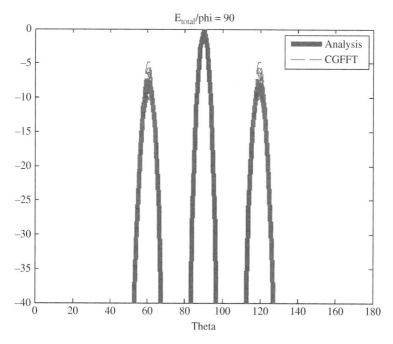

Figure 6.111 E_{total} when $\varphi = 90°$ (dB Scale) for all sizes of the measurement planes from 5λ by 5λ to 20 λ by 20λ with 0.2λ separations in both directions between the dipoles in the probe array.

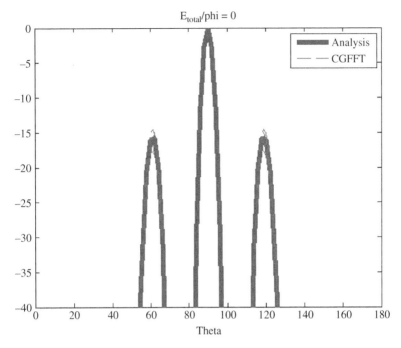

Figure 6.112 E_{total} when $\varphi = 0°$ (dB Scale) for all sizes of the measurement planes from 4.8 λ by 4.8 λ to 20 λ by 20λ with 0.4 λ separations in both directions between the dipoles in the probe array.

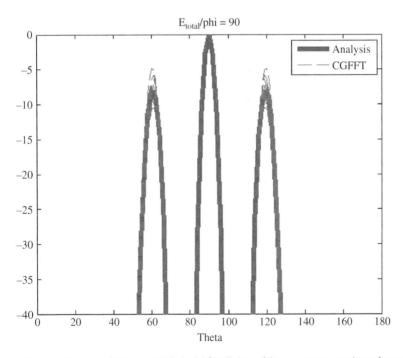

Figure 6.113 E_{total} when $\varphi = 90°$ (dB Scale) for all sizes of the measurement planes from 4.8 λ by 4.8λ to 20 λ by 20λ with 0.4 λ separations in both directions between the dipoles in the probe array.

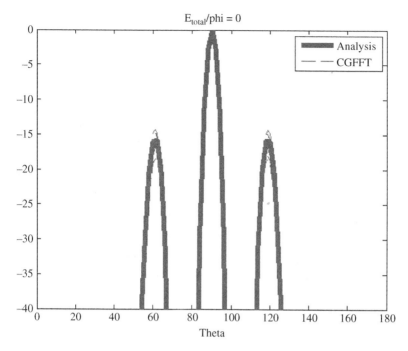

Figure 6.114 E_{total} when $\phi = 0°$ (dB Scale) for all sizes of the measurement planes from 5 λ by 5 λ to 20 λ by 20λ with 0.5 λ separations in both directions between the dipoles in the probe array.

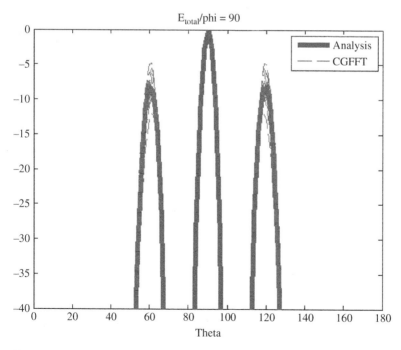

Figure 6.115 E_{total} when $\phi = 90°$ (dB Scale) for all sizes of the measurement planes from 5 λ by 5 λ to 20 λ by 20λ with 0.5 λ separations in both directions between the dipoles in the probe array.

containing the dipole probe array. One can see that the results from all the red dashed lines are quite accurate.

Hence, one can see that after the size of the measurement plane is selected to be larger than the size of the actual source plane of the AUT, the results from the presented NF-FF transformation provides acceptable results.

6.12.2   Summary

The equivalent magnetic current approach is used to calculate the far-field for all measurement systems. The far field results calculated by using near field square dipole probe array method using different sizes of the measurement planes are compared, also for complicated structured AUT. For the results presented, if the sizes of the measurement planes are chosen to be larger than or approximately equal in size of the actual source plane of the AUT, one can obtain acceptable results, where the classical approach of planar modal expansion fails.

6.13   Use of a Fixed Probe Array Measuring Amplitude-Only Near-Field Data for Calculating the Far-Field

The section utilizes computational electromagnetics techniques to obtain the far-field from the measured amplitude-only near-field data using an array of dipole probes. This is accomplished by placing a square array composed of dipoles located at two different distances from the AUT measuring amplitude-only near-field data on two separate planes. The amplitude-only voltages are measured at the loads of the probe dipoles and collected on two separate planes. We start with an initial guess of the phase information and after sufficient number of iterations recover the correct phases, so that one can obtain the final far field result. In every iteration, we need to solve the equivalent magnetic current over a plane near the original source AUT and then employing the Method of Moments approach solve for the equivalent magnetic currents on this fictitious surface. The two components of the equivalent currents can be solved independently from the two measured components of the voltages by solving the resultant Method of Moments matrix equation very efficiently and accurately by using the iterative conjugate gradient method enhanced through the incorporation of the Fast Fourier Transform (FFT) techniques. Also, in the use of the probe array there is no need to perform probe correction unlike in the existing approaches. For this proposed methodology even though there is no need to satisfy the Nyquist sampling criteria in the measurement plane, till a super resolution can be achieved in the solution of the equivalent magnetic current. Sample numerical results are presented to illustrate the accurate transformed far field result calculated from the near field measurement of amplitude data only. It is also important to note that for probes one can now use RFID tags to generate amplitude only data for the near fields as illustrated in [37].

In summary, one can still obtain acceptable far-field results by using the amplitude-only near-field measurement data. Thus, a square array of dipole probes is used to sample the near-field amplitude from the AUT. In this methodology, the far-field patterns obtained from the near-field amplitude-only data measured over two different planes, can be used to obtain acceptable far-field results.

6.13.1 Proposed Methodology

In this amplitude-only methodology, instead of measuring the complex voltages $[V_x]$ & $[V_y]$, the near-field amplitude only measurements are performed over two planar surfaces which are both parallel with the source plane as shown in Figure 6.116. The source plane (S_0), the measurement plane 1 (P_1) and the measurement plane 2 (P_2) are all assumed to be rectangular surfaces in the x-y plane with the same dimensions w by w.

The amplitude of the voltages measured on the plane P_1 are defined as $[A_{1x}]$ & $[A_{1y}]$ where one performs the measurements in a similar way as outlined in the previous sections by following similar procedures. First orient all the dipoles in the probe array to be all x-directed and then the structure is rotated so that the dipoles of the array become all y-directed. The separation distance between the planes S_0 and P_1 is defined to be d_1. The amplitude of the voltages measured

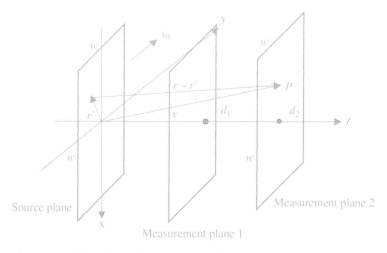

Figure 6.116 Two planes of measurement for amplitude-only data.

on the plane P_2 are defined to be $[A_{2x}]$ & $[A_{2y}]$. The separation distance between the planes S_0 and P_2 is d_2. The computations for the solution for the phase starts with an iterative process by providing an initial guess for the phases associated with the amplitudes measured on P_1 to be 0, so that the voltages $[V_{1x}]$ & $[V_{1y}]$ on the plane P_1 are assumed to be

$$[V_{1x}] = [A_{1x} \times e^{j0}] \tag{6.95}$$

$$[V_{1y}] = [A_{1y} \times e^{j0}] \tag{6.96}$$

Utilizing these initial values for the voltages $[V_{1x}]$ & $[V_{1y}]$, one can calculate the equivalent magnetic currents on the source plane M_x^1 & M_y^1 (where the number on the upper right hand corner represents the values computed after so many iterations). Next, these equivalent magnetic currents M_x^1 & M_y^1 are used to calculate the complex values of the voltages on plane P_2. One can now replace the inaccurate amplitudes for the voltages by the measured amplitudes ($[A_{2x}]$ & $[A_{2y}]$) observed on the plane P_2. Using these correct amplitudes for the voltages and concatenated with the earlier calculated phases ($[\varphi_{2x}]$ & $[\varphi_{2y}]$), one can combine the measured amplitudes ($[A_{2x}]$ & $[A_{2y}]$) and the calculated phases ($[\varphi_{2x}]$ & $[\varphi_{2y}]$) to represent the voltages on plane P_2, as

$$[V_{2x}] = [A_{2x} \times e^{j\varphi_{2x}}] \tag{6.97}$$

$$[V_{2y}] = [A_{2y} \times e^{j\varphi_{2y}}] \tag{6.98}$$

By using this new set of complex voltages $[V_{2x}]$ & $[V_{2y}]$, one can obtain the equivalent magnetic currents M_x^2 & M_y^2 on S_0, which presumably is are more accurate to represent the actual equivalent sources. This type of bootstrapping iterative computations is continued to update the phases of the voltages on P_1 by combing the measured amplitudes ($[A_{1x}]$ & $[A_{1y}]$) and the calculated phases ($[\varphi_{1x}]$ & $[\varphi_{1y}]$) to obtain the new voltages $[V_{1x}]$ & $[V_{1y}]$ on P_1. By carrying out enough number of iterations, it has been our experience that one can obtain the equivalent magnetic currents on the source plane with engineering accuracy which will provide the far field. In the end, the final iterated far-field results obtained from the use of an electromagnetic analysis code called HOBBIES is used to compute the differences.

6.13.2 Sample Numerical Results

6.13.2.1 Example 1
A 2 λ by 2 λ pyramidal horn antenna is used as the antenna under test. The near-field amplitude only measurements are first performed over the measurement plane 1 (P_1) and then performed over the measurement plane 2

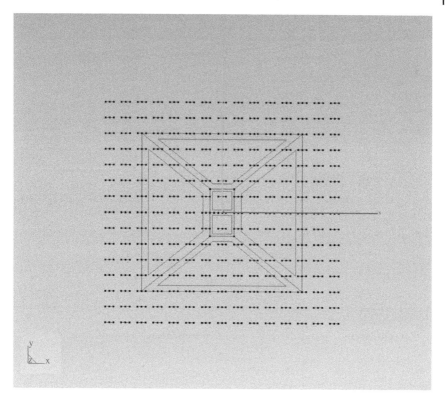

Figure 6.117 A x-directed probe array.

(P_2) by using an array of 15 × 15 dipoles of length 0.1 λ as probes. All the dipoles are terminated in 50 Ω loads and separated from each other by 0.2 λ in both directions. The two planar surfaces P_1 and P_2 are both parallel with the source plane (S_0), as shown in Figure 6.117 and Figure 6.118. In this case, S_0, P_1 and P_2 are all square surfaces in the x-y plane with the same dimensions 3λ by 3λ. On S_0, the equivalent magnetic currents M_x and M_y are approximated by 15 × 15 current patches and similar discretizations are made on P_1 & P_2 for the measurements so that one can apply the CGFFT method. The distance between S_0 and P_1 is 2 λ, and the distance between S_0 and P_2 is 3 λ. Then one can obtain the far field results by using the procedure outlined using amplitude only data measured on P_1 and P_2. Figure 6.117 shows the x-directed probe array measurement structure. Figure 6.118 shows the side view of the structure by using the x-directed probe array as an example. The simulated results for the method mentioned above and the analytic results

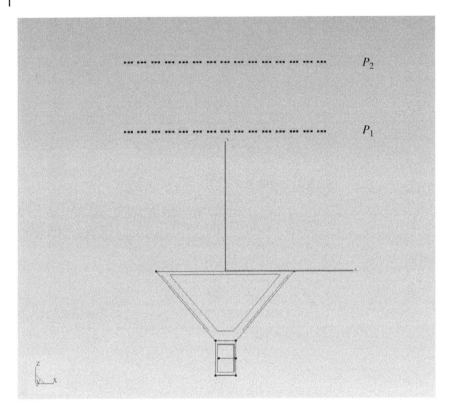

Figure 6.118 A x-directed probe array (side view).

for the far fields are shown in Figure 6.119 and Figure 6.120. Figure 6.119 shows the normalized absolute value of the electric far field for $\varphi = 0°$ in the dB scale. Figure 6.120 shows the normalized absolute value of the electric far field for $\varphi = 90°$ in the dB scale. θ here is defined as the angle from the x-axis to the z-axis and φ is the angle from the x-axis to the y-axis. This implies, φ equals 0° cut is the x-z plane and φ equals 90° is the y-z plane. The blue lines show the analytic results obtained using HOBBIES, and the red lines show the results obtained from processing amplitude only data obtained from the probe array. In conclusion, one can observe that the presented method provides acceptable results. This indicates that using measured amplitude-only data and using a probe array consisting of dipoles and without performing any probe correction, one can still obtain results for the far-field pattern with engineering accuracy.

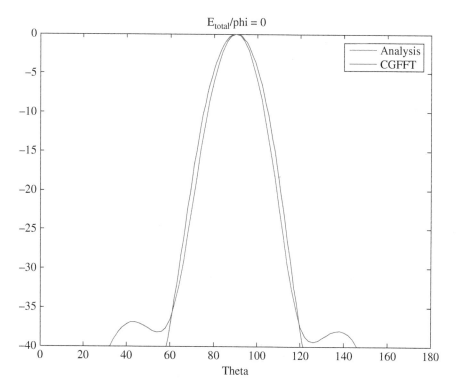

Figure 6.119 E_{total} when $\varphi = 0°$ (dB Scale).

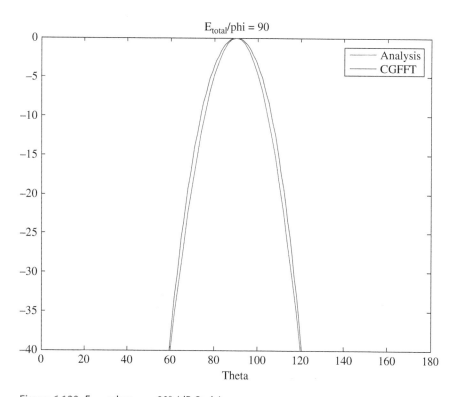

Figure 6.120 E_{total} when $\varphi = 90°$ (dB Scale).

6.13.2.2 Example 2

For the next example, the AUT is more complicated as it consists of 16, 1.5 λ by 2 λ pyramidal horn antennas to form a 4 by 4 horn antenna array. Each horn is separated from each other by 3 λ.

The near-field amplitude-only measurements are first performed over the measurement plane 1 (P_1) and then performed over the measurement plane 2 (P_2) by using an array of 40 by 40 dipoles of length 0.1 λ and they are all terminated in 50 Ω loads. The elements in the probe array are separated from each other by 0.5 λ in both directions. The two planar surfaces P_1 and P_2 are both parallel to the source plane (S_0), as shown in Figure 6.121 and Figure 6.122. In this case, S_0, P_1 and P_2 are all square surfaces in the x-y plane with the same dimensions of 20λ by 20λ.

On S_0, the equivalent magnetic currents M_x and M_y are discretized into 40×40 equivalent magnetic current patches and on planes P_1 & P_2 the amplitude-only data are measured with similar spacings between the dipole elements in the probe so as to be able to use the CGFFT method. The distance between S_0

Figure 6.121 A *x*-directed probe array.

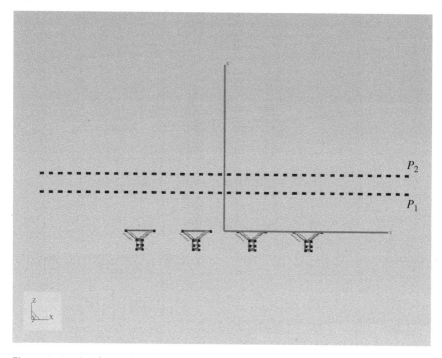

Figure 6.122 A x-directed probe array (side view).

and P_1 is 2 λ, and the distance between S_0 and P_2 is 3 λ. Then one can obtain the far field results by following the presented recipe on how to process the amplitude only data measured on planes P_1 and P_2. Figure 6.121 shows the x-directed probe array measurement structure. Figure 6.122 shows the side view of the structure by using the x-directed probe array as an example. The simulated results for the method mentioned above and the analytic results for the far-fields are shown in Figure 6.123 and Figure 6.124. Figure 6.123 shows the normalized absolute value of the electric far field for $\varphi = 0°$ in the dB scale. Figure 6.124 shows the normalized absolute value of the electric far field for $\varphi = 90°$ in the dB scale. θ represents the angle from x axis to z axis and ϕ is the angle from the x axis to the y axis. This implies, φ equals $0°$ cut is the x-z plane and φ equals $90°$ is the y-z plane. The blue lines show the analytic results obtained using HOBBIES, and the red lines show the results obtained using amplitude only data from the probe array. One can observe that the method presented provides results of engineering accuracy by utilizing amplitude-only data obtained over two planes. Also, in this method probe correction is not necessary as long as the dipoles in the array are small in dimensions so as not to affect the current distribution on the AUT.

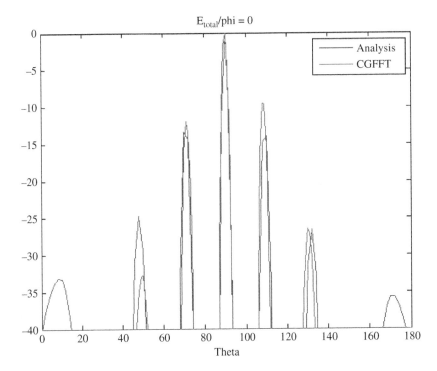

Figure 6.123 E_{total} when $\varphi = 0°$ (dB Scale).

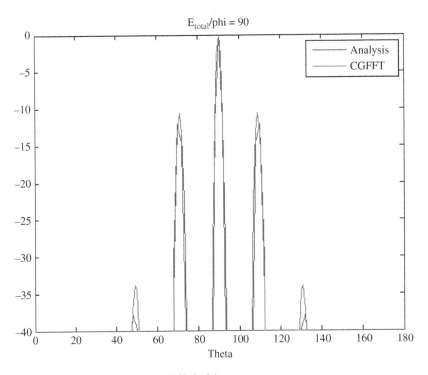

Figure 6.124 E_{total} when $\varphi = 90°$ (dB Scale).

6.13.2.3 Example 3

For the third example a single three element Yagi-Uda antenna is selected as the AUT to illustrate the efficiency and accuracy of this methodology. This antenna has a wide beam. The three-element Yagi-Uda antenna is shown in Figure 6.76. It consists of a driven element of length $L = 0.47 \lambda$, a reflector of length 0.482λ, and a director of length 0.442λ. They are all spaced 0.2λ apart. The radius of the wire structure for all the cases is 0.00425λ.

The measurement methodology for this Yagi-Uda antenna is quite similar to the measurement system used for the horn antenna as described in section 6.11.3.3. The near-field amplitude-only measurements are first performed over the measurement plane 1 (P_1) and then performed over the measurement plane 2 (P_2) by using an array of 25 by 25 dipoles of length 0.1λ. All the dipoles are terminated in 50 Ω loads and separated from each other by 0.2λ in both directions. The two planar surfaces P_1 and P_2 are both parallel with the source plane (S_0). In this case, S_0, P_1 and P_2 are all rectangular surfaces in the x-y plane with the same dimensions $5\lambda \times 5\lambda$. On S_0, the equivalent magnetic currents M_x and M_y are placed over 25×25 current patches and the measurements are carried out in the planes P_1 & P_2 which are of similar spacings between the dipole elements so as to be able to use the CGFFT method. The distance between S_0 and P_1 is 2λ, and the distance between S_0 and P_2 is 3λ. Then one can obtain the far-field results by using the method mentioned above using amplitude-only data measured on P_1 and P_2. The simulated results for the method mentioned above and the analytic results for the far-fields are shown in Figure 6.125 and Figure 6.126. Figure 6.125 shows the normalized absolute value of the electric far field for $\varphi = 0°$ in the dB scale. Figure 6.126 shows the normalized absolute value of the electric far field for $\varphi = 90°$ in the dB scale. The blue lines show the analytic results obtained using HOBBIES, and the red lines show the results obtained using amplitude only data using a probe array. One can observe that the presented method provides acceptable results even when utilizing amplitude-only data and not incorporating probe correction. This method provides results of engineering accuracy under these circumstances.

6.13.2.4 Example 4

For the final example, we deal with an array of 9 Yagi-Uda antennas to form a 3 by 3 antenna array as the AUT. Each element of the Yagi-Uda array has been described in section 6.13.2.3 and they are separated from each other by 2λ. The near-field amplitude-only measurements are first performed over the measurement plane 1 (P_1) and also over the measurement plane 2 (P_2) by using an array of 40×40 dipoles of length 0.1λ and all terminated in 50 Ω loads. The dipole elements in the probe array are separated from each other by 0.5λ in both directions. The two planar surfaces P_1 and P_2 are both parallel with the source

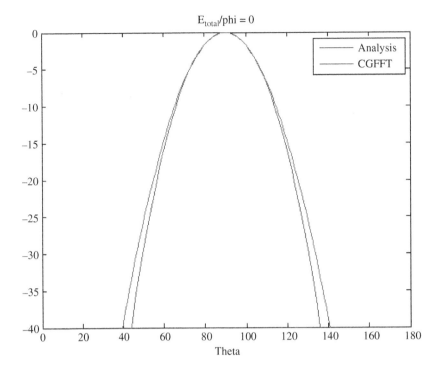

Figure 6.125 E_{total} when $\varphi = 0°$ (dB Scale).

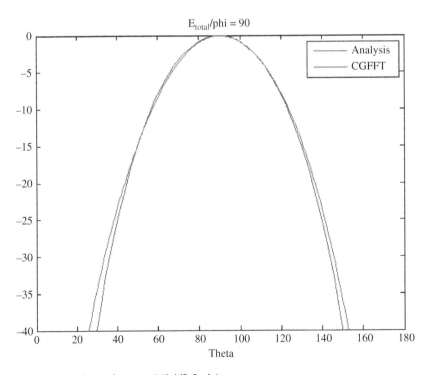

Figure 6.126 E_{total} when $\varphi = 90°$ (dB Scale).

plane (S_0), as shown in Figure 6.127 and Figure 6.128. In this case, S_0, P_1 and P_2 are all rectangular surfaces in the x-y plane with the same dimensions $20\lambda \times 20\lambda$. On S_0, the equivalent magnetic currents M_x and M_y are discretized into 40×40 current patches and P_1 & P_2 are same discretized to enable the use of CGFFT. The distance between S_0 and P_1 is 2 λ, and the distance between S_0 and P_2 is 3 λ. Then one can obtain the far-field results by using the method mentioned above from the amplitude-only data measured on P_1 and P_2. Figure 6.127 shows the x-directed probe array measurement structure. Figure 6.128 shows the side view of the structure while using the x-directed probe array. The simulated results for the method mentioned above and the analytic results for the far fields are shown in Figure 6.129 and Figure 6.130. Figure 6.129 shows the normalized absolute value of the electric far field for $\varphi = 0°$ in the dB scale. Figure 6.130 shows the normalized absolute value of the electric far field for $\varphi = 90°$ in the dB scale. θ here is defined as the angle from the x-axis to the z-axis and φ is the angle from the x-axis to the y-axis.

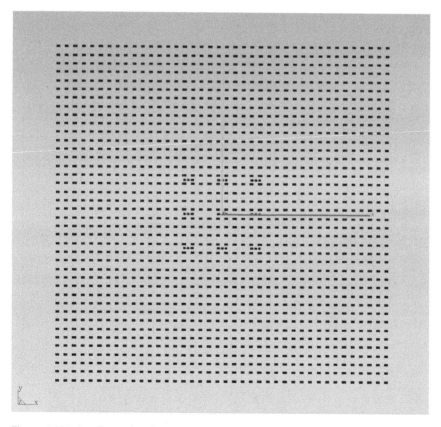

Figure 6.127 A x-directed probe array.

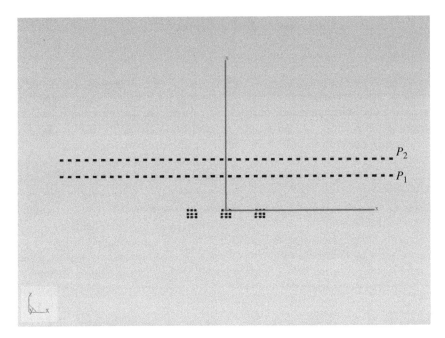

Figure 6.128 A x-directed probe array (side view).

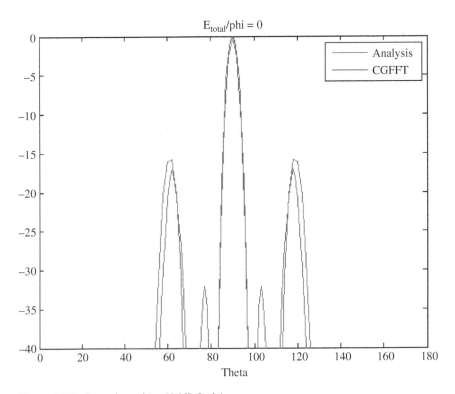

Figure 6.129 E_{total} when phi = 0° (dB Scale).

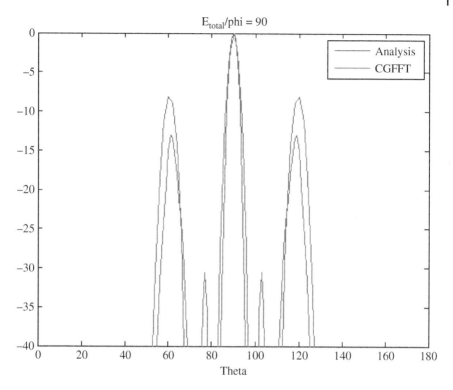

Figure 6.130 E_{total} when phi = 90° (dB Scale).

This implies, φ equals 0° cut is the x-z plane and φ equals 90° is the y-z plane. The blue lines show the analytic results obtained using HOBBIES, and the red lines show the computations using amplitude-only data obtained from the measured data obtained using the probe array. One can observe that this method provides acceptable results even when using amplitude-only data and without incorporating any probe correction even when using an array of dipoles as probes.

6.13.3 Summary

A method is presented using an estimated equivalent magnetic current evaluated from the amplitude-only data measured at the loads of a probe array. The results indicate that one obtains solutions of engineering accuracy even when not incorporating any probe correction which deal with the mutual coupling between the various elements of the antennas located in the probe array. The data is measured on two planes at different distances from the antenna. The interesting point to make here is that this methodology can be a very efficient option for measurements carried out in high frequencies where it may be difficult to obtain the phase information.

6.14 Probe Correction for Use with Electrically Large Probes

In the previous sections it has been illustrated that for most of the cases when using a reasonably sized dipole as a probe, it is seen that probe correction is not necessary. When an array of dipoles as probes is deployed to replace a moving single probe, it is seen that even in those cases it is not necessary to take into account the electromagnetic coupling effects between the AUT and the probes. However, if the probes are chosen to be large resonant sized structures then a probe correction is absolutely essential as the probes then influence the original current distribution of the AUT. In this section, it is illustrated how to accomplish this following the presentation ion [38].

In this section, a method is presented for computing the far-field antenna patterns from measured near-field data measured by an array of planar resonant sized dipole probes. The method utilizes the near-field data to determine some equivalent magnetic current sources over a fictitious planar surface which encompasses the antenna. These currents are then used to find the far-fields. The near-field measurement is carried out by terminating each resonant dipole with 50 Ω load impedances and measuring the complex voltages across the loads. An electric field integral equation (EFIE) is developed to relate the measured complex voltages to the equivalent magnetic currents. The mutual coupling between the array of probes and the test antenna modeled by magnetic dipoles is taken into account. The method of moments with a Galerkin's type testing solution procedure is used to transform the integral equation into a matrix one. The matrix equation is solved with the conjugate gradient-fast Fourier transformation (CG-FFT) method exploiting the block Toeplitz structure of the matrix.

Planar near-field antenna measurements have become widely used in antenna testing since they allow for accurate measurements of antenna patterns in a controlled environment. In the equivalent current approach which represents an alternate method of computing far fields from measured near fields has been presented in the previous sections. The basic idea for the equivalent current approach is to replace the radiating antenna by an equivalent magnetic and/or electric currents which reside on a fictitious surface and encompass the antenna. Under certain conditions, these equivalent currents produce the correct far fields in all regions in front of the antenna. For the equivalent magnetic current approach where only magnetic currents are used as equivalent sources, the obtained EFIE equation is a decoupled one with respect to the coordinate axes for the planar scanning case. This means that two simple decoupled EFIE's can be formulated which contain only one component of the measured near fields and equivalent currents and which can be solved separately, in a parallel fashion.

In this work, one replaces a small probe antenna in a probe array by resonant length dipoles which are separated from each other by approximately half a wavelength. The use of an array of probes is quite advantageous as it eliminates cumbersome precise mechanical movement of the probe antenna over a large planar surface. Secondly, the utilization of a probe array also eliminates the necessity of knowing the physical location of the probes accurately as it moves across the test antenna. These considerations become more important at the millimeter-wave frequencies.

In this method, the array of dipoles is planar and are of resonant length. They are all terminated in a 50 Ω load, and a network analyzer or any other noninvasive probing (e.g., optoelectronic devices) may be used to measure the voltages across the 50 Ω loads. Hence, in principle, this method [38] is quite different from [39] where an end-fire array of dipole probes has been used to modulate the response and other noninvasive probing (e.g., optoelectronic devices) may be used to measure the voltages across the 50 Ω loads.

In this section, to compute the mutual interaction between the array of probes and the test antenna, the method of moments with Galerkin's type solution procedure has been used to transform the EFIE to a matrix equation. For this particular case, the matrix has a block Toeplitz structure, and so the CGFFT method can be utilized to solve such matrix equations without even explicitly forming the normal equations, thereby drastically reducing computation and storage requirements.

6.14.1 Development of the Proposed Methodology

Consider an arbitrarily shaped antenna radiating into free space. A planar array of probes consisting of thin dipoles is located in front of the test antenna as shown in Figure 6.131. The equivalence principle is applied on an infinite plane in front of the test antenna, so that for $z > 0$, this equivalent source provides the correct fields. The planar array of probes is considered to be parallel to this fictitious plane in front of the test antenna on which some equivalent currents have been applied to predict the correct fields in the region $z > 0$. The distance between the plane placed in front of the test antenna and the planar array is d. For illustration purposes, consider the elements of the probe array to consist of x-directed dipoles (typically a half wavelength long), and are uniformly distributed along both the x and the y directions. The spacing between the probes is considered to be Δx and Δy in the x and y directions, respectively. The dipoles are terminated by the impedance z_L (typically 50 Ω) to match the input impedance requirement of the measurement device.

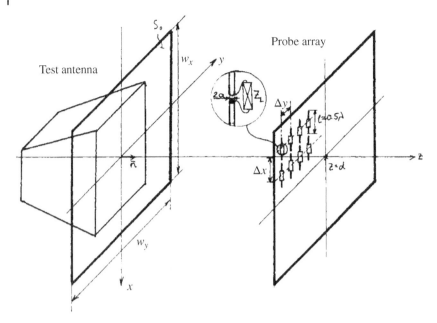

Figure 6.131 Planar near-field measurement using a phased array of resonant dipoles as a probe antenna.

Invoking the surface equivalence principle by using an equivalent magnetic current M on the plane in front of the test antenna, as illustrated in the previous sections, one can predict the proper electric field for $z > 0$. Utilizing the image theory, one can evaluate the radiated electric field as

$$E_{meas,x}(r) = -\iint_{S_0} \frac{\partial G(r, r')}{\partial z'} M_y(r')ds' \qquad (6.99)$$

$$E_{meas,y}(r) = -\iint_{S_0} \frac{\partial G(r, r')}{\partial z'} M_x(r')ds' \qquad (6.100)$$

The explicit expression for $\dfrac{\partial G}{\partial z'}$ is given by

$$G_{k,m} = \iint_{\Omega_m} \frac{e^{-jk_0 R}}{R^2}(z_k - z')\left\{jk_0 + \frac{1}{R}\right\}ds' \qquad (6.101)$$

where Ω_m is the area of the mth patch and R is the distance between kth field point and the source point (r').

The total electric field in front of the test antenna is produced by the equivalent magnetic current sources M and by the induced electric currents J_{Pr} on the probes so that

$$E_{total} = E_m^s(M) + E_{Pr}^s(J_{Pr}) \tag{6.102}$$

Enforcing the tangential component of the total electric field to be zero on the probes results in the following electric field integral equation:

$$E_m^s(M) + E_{Pr}^s(J_{Pr}) = 0 \quad \text{on probe dipoles} \tag{6.103}$$

In (6.103), the unknown currents are M and J_{Pr}. Since the nearfield measurements are carried out by terminating the dipole elements in the probe array by z_L (which is typically 50 Ω) and we measure the complex voltages across the load, the current distribution J_{Pr} is known. This is because, for the present approach, on every resonant size probe, only one entire domain basis function has been applied for the numerical calculation. Hence, (6.103) becomes

$$E_m^s(M)_{\tan} = -E_{Pr}^s(V_{meas}) \tag{6.104}$$

where V_{meas} are the known values of the measured voltages.

To obtain the far fields, we need to find the equivalent source M' "on the test antenna surface" when the probes are not present. If we assume that the same excitation on the test antenna is present with and without the probes, then

$$H(M')_{\tan} = -H_{\tan}'(V_{mag}) \tag{6.105}$$

where $H_{\tan}'$ are the known magnetic fields across the excitation gaps of the test antenna produced by the magnetic voltage generators V_{mag}. However, when the probes are present then (6.105) becomes

$$\{H(J_{Pr}) + H(M)\}_{\tan} = -H_{\tan}'(V_{mag}) \quad \text{on the source plane} \tag{6.106}$$

where $H(J_{Pr})$ is the magnetic field produced by the electric currents on the probes.

By combining (6.105) with (6.106), the following equation is obtained for the unknown magnetic current distribution M' as

$$H(M')_{\tan} = \{H(J_{Pr}) + H(M)\}_{\tan} \quad \text{on the source plane} \tag{6.107}$$

where M is the magnetic current distribution when the probes are present and J_{pr} is the known electric current distribution on the probes. So the two decoupled integral equations to be solved for are (6.104) and (6.107). It is important to note that the numerical effort required to solve (6.107) is the same as discussed earlier. This is because the computation of the term $E_{Pr}^s(V_{meas})$ requires only a simple preprocessing of the measured voltage data. The solution of (6.107) is a well-behaved one because the moment matrix has dominant diagonal elements, and therefore does not cause any numerical difficulty.

It is important to point out that this formulation takes all the mutual couplings into account, i.e., the mutual coupling between the test antenna and the dipole probes, and the mutual coupling between the dipole probes.

6.14.2 Formulation of the Solution Methodology

For use of computation, on the planar equivalent surface of dimension $w_x \times w_y$ are placed an array of P magnetic dipoles in the x-directions and Q in the y-directions so the total number of dipoles is PQ. This is a good approximation if d is greater than a few wavelengths. Also, this assumption does away with the numerical integration required in the evaluation of the matrix elements as carried out and mentioned earlier in the previous sections.

Next, on the probe dipoles an entire domain sinusoidal basis functions is assumed to be

$$
f_{ij}(x) = \left\{
\begin{array}{ll}
\dfrac{\sin\left[k_0(x - x_c + h/2)\right]}{\sin\left(k_0 h/2\right)} & \text{for } x_c - h/2 \le x \le x_c \\[3mm]
\dfrac{\sin\left[k_0(x_c + h/2 - x)\right]}{\sin\left(k_0 h/2\right)} & \text{for } x_c \le x \le x_c + h/2
\end{array}
\right\}
\tag{6.108}
$$

where h and a are the length and the radius of the dipole probes, and x_c and y_c are the coordinates of the center of the a dipole probe.

Applying the discretization procedure to (6.104) and enforcing Galerkin's method one obtains

$$
HM_y = \frac{1}{z_L}\left(G_{xx}^j + \langle z_L \rangle\right) V
\tag{6.109}
$$

where matrix H represents the interaction between the probes and the magnetic current elements. The explicit expression for the elements of H is given by

$$
H_{(ij),(kl)} = \int\limits_{P_{kl}} \frac{\partial G\left(r_{kl}, r'_{ij}\right)}{\partial z'} f_{kl}(x)\, dx
\tag{6.110}
$$

Here, r_{kl} defines the surface of the (r_k)th probe antenna of the phased array and r'_{ij} defines the (ij)th magnetic dipole on the source plane. M_y is the unknown vector containing the y-component of the equivalent magnetic source current elements. z_L is the load impedance terminating the probes. The matrix G_{xx} represents the interactions between the probes and whose elements are given by

$$
G_{(ij),(kl)} = \frac{1}{j\,\omega\,\varepsilon_0} \int\limits_{P_{ij}}\int\limits_{P_{kl}} \left\{\left[k_0^2 f_{ij}(x') f_{kl}(x) - \frac{\partial f_{ij}(x')}{\partial x'}\frac{\partial f_{kl}(x)}{\partial x}\right] G\left(x, x', y_l, y'_j\right)\right\} dx\, dx'
\tag{6.111}
$$

6.14.3   Sample Numerical Results

In order to illustrate this methodology, some typical numerical results are presented. As an example, consider a half-wave dipole antenna of radius 0.007 λ located at the center of the coordinate system as shown in Figure 6.132. The planar near fields are measured by an array of probes displayed in Figure 6.132. In this example, the input impedance of the half-wave dipole is calculated from the measured near fields. The input impedance directly depends on the current of the source dipole in free space. The total number of probes sampling the near field is varied from $N = 1$ to 961 (= 31 × 31). The input impedance of a half-wave dipole is about 73 Ω in free space. If the proposed methodology is meaningful, then even when the near-fields are measured by an array of 961 resonant length probes (which significantly affects the current distribution on the test dipole), one should still be able to obtain reliably the input impedance of the single dipole under test in isolation. This would illustrate that it is possible to de-embed the effects of the probe array and obtain the proper current (or input impedance) on the test antenna in isolation. Once the currents are known, the fields anywhere in space can easily be computed.

Table 6.1 contains the calculated value of the input impedance of the test antenna in isolation as a function of $N = 1$ to 961. It is seen that the largest value of the error is 2.4% in the real part and 5.6% in the imaginary part of the input impedance.

It has been our experience that, by utilizing this new method, if $d > 3\lambda$, then instead of solving two equations (6.104) and (6.107), it is possible to drop (6.107) and just solve (6.104) with M replaced by M'. Our limited experimentation has proved that this is valid irrespective of the size of the planar probe array.

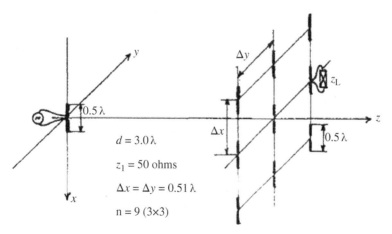

Figure 6.132 A planar near-field measurement configuration using an array of probes. The test antenna is a resonant size electric dipole. The probe antenna is a phased array ($N = 9$).

Table 6.1 Input impedance of the test antenna as a function of the number of elements of phased array probes. (Number, Real part of the Input Impedance, Imaginary part of the input impedance).

1	73.4	j 41.6
9	74.4	j 39,7
25	71.6	j 39.3
49	73.2	j 40.5
81	72.1	j 39.7
121	73.1	j 39.7
169	72.9	j 40.4
225	72.3	j 40.3
441	72.9	j 39.7
961	72.8	j 40.4
Single	73.1	j 41.8

In summary, a method is described for computing the fields anywhere in space of a test antenna from the measured near-field data obtained by a planar phased array of probe dipoles. In this approach, there is no need to move the probe mechanically, nor is there any necessity to measure the spatial probe position accurately as it is carrying out the measurements. This method could be very useful, particularly at millimeter waves where it may be difficult to move a probe mechanically with millimeter accuracy. The surface equivalence principle along with the method of moments is utilized to solve the integral equation for equivalent unknown magnetic currents on a planar surface located in front of the test antenna. A numerically very efficient CG-FFT scheme has been implemented to solve the integral equation for the magnetic currents, thereby drastically reducing both computation time and storage requirements. The interesting point in the section is that the measurement probes can be of resonant lengths and therefore probe correction can be implemented with ease to yield very accurate results.

This procedure takes into account not only the mutual interaction between the dipoles in the planar probe array into account, but also the interaction between the test antenna and the planar probe array consisting of probes of resonant lengths. However, it has been our experience that if d (the separation distance between the assumed source plane and the measurement plane) > 3λ, then the second mutual interaction in this procedure can be neglected without affecting the final results significantly. The validity of this approach has been shown through limited examples, and this method has a wider range of validity than that of the contemporary modal expansion methods.

6.15 Conclusions

A fast and accurate method has been presented for computing far-field antenna patterns from planar near-field measurements. The method utilizes near-field data to determine equivalent magnetic current sources over a fictitious planar surface that encompasses the antenna, and these currents are used to ascertain the far fields. An electric field integral equation has been developed to relate the near fields to the equivalent magnetic currents. The Method of Moments procedure has been used to transform the integral equation into a matrix one. The matrix equation is solved using the iterative conjugate gradient method, and in the case of a rectangular matrix, a least-squares solution is implemented. Numerical results are presented for several antenna configurations by extrapolating the far fields using synthetic and experimental near-field data.

It is also illustrated that a single probe can be replaced by an array of probes to compute the equivalent magnetic currents on the surface enclosing the AUT and it has been demonstrated that in this methodology a probe correction even when using an array of dipoles is not necessary. The accuracy of this methodology is studied as a function of the size of the equivalent surface placed in front of the antenna under test and the error in the estimation of the far field along with the possibility of using a rectangular probe array which can efficiently and accurately provide the patterns in the principal planes. This can also be used when amplitude-only data is collected using an array of probes. Finally, it is shown that the probe correction can be useful when the size of the probes are of resonant lengths. It has been illustrated on those circumstances on how to address that issue.

References

1 J. Brown and E. V. Jull, "The Prediction of Aerial Radiation Patterns for Near-Field Measurements," *Proceedings of the IEE – Part B: Electronic and Communication Engineering*, Vol. 108B, pp. 635–644, 1961.

2 D. M. Kerns, "Plane-Wave Scattering-Matrix Theory of Antennas and Antenna-Antenna Interactions," *NBS Monograph 162*, U.S. Govt. Printing Office, Washington, DC, 1981.

3 F. Jensen, *Electromagnetic Near-Field Correlations*, Ph.D. Dissertation, Technical University of Denmark, July 1970.

4 W. M. Leach, Jr. and D. T. Paris, "Probe-Compensated Near-Field Measurements on Cylinder," *IEEE Transactions on Antennas and Propagation*, Vol. AP-21, pp. 435–445, 1973.

5 D. T. Paris, W. M. Leach, and E. B. Joy, "Basic Theory of ProbeCompensated Near-Field Measurements," *IEEE Transactions on Antennas and Propagation*, Vol. AP-26, pp. 373–379, 1978.

6 A. D. Yaghjian, "An Overview of Near-Field Antenna Measurements," *IEEE Transactions on Antennas and Propagation*, Vol. AP-34, pp. 30–45, 1986.

7 J. J. H. Wang, "An Examination of the Theory and Practice of Planar Near-Field Measurement," *IEEE Transactions on Antennas and Propagation*, Vol. AP-36, pp. 746–753, 1988.

8 M. S. Narasimhan and B. P. Kumar, "A Technique of Synthesizing the Excitation Currents of Planar Arrays or Apertures," *IEEE Transactions on Antennas and Propagation*, Vol. AP-38, pp. 1326–1332, 1990.

9 T. K. Sarkar, S. Ponnapolli, and E. Arvas, "An Accurate Efficient Method of Computing a Far-Field Antenna Patterns from Near-Field Measurements," *Proceedings of the International Conference on Antennas and Antennas & Propagation*, Dallas, TX, May 1990.

10 S. Ponnapolli, *The Computation of Far-Field Antenna Patterns from Near-Field Measurements Using an Equivalent Current Approach*, Ph.D. Dissertation, Syracuse University, December 1990.

11 S. Ponnapolli, "Near-Field to Far-Field Transformation Utilizing the Conjugate Gradient Method," in Application of Conjugate Gradient Method in Electromagnetics and Signal Processing, PIER, T. K. Sarkar, editor, VNU Sci. Press, New York, Vol. 5, Chapter 11, 1990.

12 S. Ponnapolli, T. K. Sarkar, and P. Petre, "Near-Field to Far-Field Transformation Using an Equivalent Current Approach," *Proceedings of the International Conference on Antennas and Antennas & Propagation*, London, Ontario, June 1991.

13 R. F. Harrington, Field Computation by Moment Methods, Robert E. Kreiger Publishing, Malabor, 1968.

14 T. K. Sarkar and E. Arvas "On a Class of Finite Step Iterative Methods (Conjugate Directions) for the Solution of an Operator Equation Arising in Electromagnetics," *IEEE Transactions on Antennas and Propagation*, Vol. AP-33, pp. 1058–1066, 1985.

15 T. K. Sarkar, E. Arvas, and S. M. Rao, "Application of FFT and Conjugate Gradient Method for the Solution of Electromagnetic Radiation from Electrically Large and Small Conducting Bodies," *IEEE Transactions on Antennas and Propagation*, Vol. AP-34, pp. 635–640, 1986.

16 R. F. Harrington, Time-Harmonic Electromagnetic Fields, McGraw-Hill, New York, 1961.

17 A. V. Oppenheim and R. W. Shafer, Digital Signal Processing, Prentice-Hall, Englewood Cliffs, NJ, 1975.

18 P. Petre and T. K. Sarkar, "Planar Near-Field to Far-Field Transformation Using an Equivalent Magnetic Current Approach," *IEEE Transactions on Antennas and Propagation*, Vol. 40, No. 11, pp. 1348–1356, 1992.

19 C. Stubenrauch, [private communication]

20 A. C. Newell, "Error Analysis Techniques for Planar Near-Field Measurements," *IEEE Transactions on Antennas and Propagation*, Vol. AP-36, pp. 754–768, 1988.

21 Joy, E. B., W. M. Leach, Jr., G. P. Rodriguez, and D. T. Paris, "Applications of Probe-Compensated Near-Field Measurements," *IEEE Transactions on Antennas and Propagation*, Vol. AP-26, No. 3, pp. 379–389, 1978.

22 Y. Rahmat-Samii, "Surface Diagnosis of Large Reflector Antennas Using Microwave Holographic Metrology: An Iterative Approach," *Radio Science*, Vol. 19, pp. 1205–1217, 1984.

23 P. C. Clemmow, The Plane Wave Spectrum Representation of Electromagnetic Fields, Pergamon Press, 1966.

24 P. Petre and T. K. Sarkar, "Differences Between Modal Expansion and Integral Equation Methods for Planar Near-Field to Far-Field Transformation," *Progress in Electromagnetic Research* 12, pp. 37–56, 1996.

25 R. G. Yaccarino and Y. R. Samii, "Phase-Less Bi-Polar Planar Near-Field Measurements and Diagnostics of Array Antennas," *IEEE Transactions on Antennas and Propagation*, Vol. 47, pp. 574–583, 1999.

26 R. G. Yaccarino and Y. R. Samii, "Phaseless Near-Field Measurements Using the UCLA Bi-Polar Planar Near Field Measurement System," *Proceedings AMTA Meeting*, Long Beach, CA, October 1994.

27 J. R. Mautz and R. F. Harrington, "Computational Methods for Antenna Pattern Synthesis," *IEEE Transactions on Antennas and Propagation*, Vol. AP-23, pp. 507–510, 1975.

28 T. Isernia, G. Leone, and R. Pierri, "Radiation Pattern Evaluation from Near-Field Intensities on Planes," *IEEE Transactions on Antennas and Propagation*, Vol. 44, pp. 701–710, 1996.

29 T. Isernia, G. Leone, and R. Pierri, "Phaseless Near Field Techniques: Formulation of the Problem and Field Properties," *Journal of Electromagnetic Waves and Applications*, Vol. 8, pp. 871–878, 1994.

30 A. Taaghol and T. K. Sarkar, "Near-Field to Near/Far-Field Transformation for Arbitrary Near Field Geometry, Utilizing an Equivalent Magnetic Current," *IEEE Transactions On Electromagnetic Compatibility*, Vol. 38, No. 3, pp. 536–541, 1996.

31 F. Las-Heras and T. K. Sarkar, "A Direct Optimization Approach for Source Reconstruction and NF-FF Transformation Using Amplitude-Only Data," *IEEE Transactions on Antennas and Propagation*, Vol. 50, No. 4, pp. 500–510, 2002.

32 F. Las-Heras, "Sequential Reconstruction of Equivalent Currents from Cylindrical Near Field," *Electronics Letters*, Vol. 35, No. 3, pp. 211–212, 1999.

33 F. Las-Heras, B. Galocha, and J. L. Besada, "Far-Field Performance of Linear Antennas Determined from Near-Field Data," *IEEE Transactions on Antennas and Propagation*, Vol. 50, No. 3, pp. 408–410, 2002.

34 H. Chen, T. K. Sarkar, M. Zhu, and M. Salazar-Palma, "Use of Computational Electromagnetics to Enhance the Accuracy and Efficiency of Antenna Pattern Measurements," *IEEE Journal on Multiscale and Multiphysics Computational Techniques*, Vol. 3, pp. 214–224, 2018.

35 Y. Zhang, T. K. Sarkar, X. Zhao, D. Garcia-Donoro, W. Zhao, M. Salazar-Palma, and S. Ting, Higher Order Basis Based Integral Equation Solver (HOBBIES), John Wiley & Sons, Piscataway, NJ, 2012.

36 H. Chen and T. K. Sarkar, "Influence of the Probe When Computing Far Field from Near Field Measurements," *IEEE Conference on Antenna Measurement & Applications*, Syracuse, NY, 2016.

37 M. W. William, N. K. S. Abdelhadi, S. A. S. Elmeadawy, Y. A. Zaghloul, and H. F. Hammad, "DIY Antenna Characterization System Using Universal UHF RFID Hemispherical Dome," *2019 IEEE International Symposium on Antennas and Propagation and USNC-URSI Radio Science Meeting*, German University, Cairo, Student Design Contest, Atlanta, GA, 2019.

38 P. Petre and T. K. Sarkar, "Planar Near-Field to Far-Field Transformation Using an Array of Dipole Probes," *IEEE Transactions on Antennas and Propagation*, Vol. 42, No. 4, pp. 534–537, 1994.

39 J.-C. Bolomey, B. J. Cown, G. Fine, L. Jofre, M. Mostafavi, D. Picard, J. P. Estrada, P. G. Friederich, and F. L. Cain, "Rapid Near-Field Antenna Testing Via Arrays of Modulated Scattering Probes," *IEEE Transactions on Antennas and Propagation*, Vol. 36, No. 6, pp. 804–814, 1988.

7

Spherical Near-Field to Far-Field Transformation

Summary

In this chapter two methods for spherical near-field to far-field transformation is presented. The first methodology presents an exact explicit analytical formulation for transforming near-field data generated over a spherical surface to the far field radiation pattern. The results are validated with experimental data. A computer program involving this method is provided at the end of the chapter. The second method presents the equivalent source formulation that was described earlier and illustrates how it can be deployed to the spherical scanning case where one component of the field is missing from the measurements, when the tangential components of the fields on a spherical surface is given. Again the methodology is validated using other techniques and also with experimental data.

7.1 An Analytical Spherical Near-Field to Far-Field Transformation

In this section an analytical spherical near-field to far-field transformation is presented. In this technique all the parameters of interest in the spherical wave expansion can be a priori controlled to a prespecified degree of accuracy.

7.1.1 Introduction

A large volume of literature exists on the spherical near field to far field transformation [1–3]. In this section, the approach similar to [4] is taken to expand the fields in terms of TM and TE modes to r as described by [5–7]. The final expressions of this section are somewhat different and simpler than the expressions of [1, 4] and [6].

Modern Characterization of Electromagnetic Systems and Its Associated Metrology, First Edition.
Tapan K. Sarkar, Magdalena Salazar-Palma, Ming Da Zhu, and Heng Chen.
© 2021 John Wiley & Sons, Inc. Published 2021 by John Wiley & Sons, Inc.

7.1.2   An Analytical Spherical Near-Field to Far-Field Transformation

Consider a sphere of radius a over which the tangential components of the electric field, E_θ and E_φ, are known. So

$$E_\theta\,(a,\theta,\varphi) = f_1\,(\theta,\varphi) \tag{7.1}$$

$$E_\varphi\,(a,\theta,\varphi) = f_2\,(\theta,\varphi) \tag{7.2}$$

From this near field given in equations (7.1) and (7.2) one can determine the far field. The complete expression for the field external to the sphere is given by [4, 7] using the following vector potentials

$$A_r = j\omega\,\varepsilon \sum_{n=0}^{\infty}\sum_{m=0}^{n} r\,h_n^{(2)}(kr)\,P_n^m(\cos\theta)\,[\alpha_{mn}\,\cos\,m\phi + \beta_{mn}\,\sin\,m\phi] \tag{7.3}$$

$$F_r = j\omega\,\mu \sum_{n=0}^{\infty}\sum_{m=0}^{n} r\,h_n^{(2)}(kr)\,P_n^m(\cos\theta)\,[\gamma_{mn}\,\cos\,m\phi + \delta_{mn}\,\sin\,m\phi] \tag{7.4}$$

where $h_n^{(2)}(kr)$ is the spherical Bessel function of the second kind of order n and argument kr, $P^m(\cos\theta)$ is the associated Legendre function of the first kind of argument $\cos\theta$, and $\alpha_{mn}, \beta_{mn}, \gamma_{mn}, \chi_{mn}$ are the four constants to be determined from the boundary value problem specified by (7.1) and (7.2). Here, k is the free space wave number and ε and μ are the permittivity and permeability of free space. The θ and φ field components are then given by [4] as

$$E_\theta = \frac{-1}{r\,\sin\theta}\frac{\partial F_r}{\partial\phi} + \frac{1}{j\omega\,\varepsilon\,r}\frac{\partial^2 A_r}{\partial r\,\partial\theta} = \sum_{n=0}^{\infty}\sum_{m=0}^{n}\frac{\partial}{r\,\partial r}\left[r\,h_n^{(2)}(kr)\right]\frac{dP_n^m(\cos\theta)}{d\theta}$$

$$[\alpha_{mn}\,\cos\,m\phi + \beta_{mn}\,\sin\,m\phi] + j\omega\,\mu \sum_{n=0}^{\infty}\sum_{m=0}^{n}\frac{m\,h_n^{(2)}(kr)}{\sin\theta}\,P_n^m(\cos\theta)$$

$$[\gamma_{mn}\,\sin\,m\phi - \delta_{mn}\,\cos\,m\phi] \tag{7.5}$$

$$E_\phi = \frac{1}{r}\frac{\partial F_r}{\partial\theta} + \frac{1}{j\omega\,\varepsilon\,r\,\sin\theta}\frac{\partial^2 A_r}{\partial r\,\partial\phi} = -\sum_{n=0}^{\infty}\sum_{m=0}^{n}\frac{\partial}{r\,\partial r}\left[r\,h_n^{(2)}(kr)\right]$$

$$\frac{m\,P_n^m(\cos\theta)}{\sin\theta}\,[\alpha_{mn}\,\sin\,m\phi - \beta_{mn}\,\cos\,m\phi]$$

$$+ j\omega\,\mu \sum_{n=0}^{\infty}\sum_{m=0}^{n} h_n^{(2)}(kr)\frac{dP_n^m(\cos\theta)}{d\theta}\,[\gamma_{mn}\,\cos\,m\phi + \delta_{mn}\,\sin\,m\phi] \tag{7.6}$$

$$H_\theta = \frac{1}{r \sin \theta} \frac{\partial A_r}{\delta \phi} + \frac{1}{j \omega \mu r} \frac{\partial^2 F_r}{\partial r \partial \theta} = \sum_{n=0}^{\infty} \sum_{m=0}^{n} \frac{\partial}{r \partial r} \left[r h_n^{(2)}(kr) \right]$$

$$\frac{dP_n^m(\cos \theta)}{d\theta} \left[\gamma_{mn} \cos m\phi + \delta_{mn} \sin m\phi \right]$$

$$- j\omega \varepsilon \sum_{n=0}^{\infty} \sum_{m=0}^{n} h_n^{(2)}(kr) \frac{m P_n^m(\cos \theta)}{\sin \theta} \left[\alpha_{mn} \sin m\phi - \beta_{mn} \cos m\phi \right]$$

$$(7.7)$$

$$H_\phi = \frac{1}{r} \frac{\partial A_r}{\partial \theta} + \frac{1}{j \omega \mu r \sin \theta} \frac{\partial^2 F_r}{\partial r \partial \phi} = - \sum_{n=0}^{\infty} \sum_{m=0}^{n} \frac{\partial}{r \partial r} \left[r h_n^{(2)}(kr) \right]$$

$$\frac{m P_n^m(\cos \theta)}{\sin \theta} \left[\gamma_{mn} \sin m\phi - \delta_{mn} \cos m\phi \right]$$

$$- j\omega \varepsilon \sum_{n=0}^{\infty} \sum_{m=0}^{n} h_n^{(2)}(kr) \frac{dP_n^m(\cos \theta)}{d\theta} \left[\alpha_{mn} \cos m\phi + \beta_{mn} \sin m\phi \right]$$

$$(7.8)$$

By replacing the field components E_θ and E_φ in equations (7.1) and (7.2) with their expressions described by equations (7.5) and (7.6), respectively, and then using orthogonality relationships at $r = a$, the coefficients α_{mn}, β_{mn}, γ_{mn} and δ_{mn} can be determined. Therefore, from equations (7.1) and (7.5) one can obtain

$$f_1(\theta, \phi) = j\omega \mu \sum_{n=0}^{\infty} \sum_{m=0}^{n} \frac{m h_n^{(2)}(ka)}{\sin \theta} P_n^m(\cos \theta) \left[\gamma_{mn} \sin m\phi - \delta_{mn} \cos m\phi \right]$$

$$+ \sum_{n=0}^{\infty} \sum_{m=0}^{n} \left\{ \frac{\partial}{r \partial r} \left[r h_n^{(2)}(kr) \right] \right\}_{r=ka} \frac{dP_n^m(\cos \theta)}{d\theta} \left[\alpha_{mn} \cos m\phi + \beta_{mn} \sin m\phi \right]$$

$$(7.9)$$

and from equations (7.2) and (7.6) one can obtain,

$$f_2(\theta, \phi) = j\omega \mu \sum_{n=0}^{\infty} \sum_{m=0}^{n} h_n^{(2)}(ka) \frac{dP_n^m(\cos \theta)}{d\theta} \left[\gamma_{mn} \cos m\phi + \delta_{mn} \sin m\phi \right]$$

$$- \sum_{n=0}^{\infty} \sum_{m=0}^{n} \left\{ \frac{\partial}{r \partial r} \left[r h_n^{(2)}(kr) \right] \right\}_{r=ka} \frac{m P_n^m(\cos \theta)}{\sin \theta} \left[\alpha_{mn} \sin m\phi - \beta_{mn} \cos m\phi \right]$$

$$(7.10)$$

From equations (7.9) and (7.10) one can obtain,

$$f_1(\theta, \phi) = \sum_{n=0}^{\infty} \sum_{m=0}^{n} \frac{D\, m}{\sin \theta} P_n^m(\cos \theta) \left[\gamma_{mn} \sin m\phi - \delta_{mn} \cos m\phi \right]$$

$$+ \sum_{n=0}^{\infty} \sum_{m=0}^{n} \frac{N}{a} \frac{dP_n^m(\cos \theta)}{d\theta} \left[\alpha_{mn} \cos m\phi + \beta_{mn} \sin m\phi \right] \qquad (7.11)$$

$$f_2(\theta, \phi) = \sum_{n=0}^{\infty} \sum_{m=0}^{n} D \frac{dP_n^m(\cos \theta)}{d\theta} \left[\gamma_{mn} \cos m\phi + \delta_{mn} \sin m\phi \right]$$

$$- \sum_{n=0}^{\infty} \sum_{m=0}^{n} \frac{N}{a} \frac{m\, P_n^m(\cos \theta)}{\sin \theta} \left[\alpha_{mn} \sin m\phi - \beta_{mn} \cos m\phi \right] \qquad (7.12)$$

where

$$D = j\omega\mu\, h_n^{(2)}(ka) \qquad \text{and} \quad N = ka\, h_n^{(2)'}(ka) + h_n^{(2)}(ka) \qquad (7.13)$$

At this point orthogonality between functions is invoked to determine the four unknown coefficients. From (7.11) 'one can obtain

$$\int_0^{2\pi} d\phi \int_0^{\pi} d\theta\, f_1(\theta, \phi) \frac{dP_{n'}^{m'}(\cos \theta)}{d\theta} \sin \theta \cos m'\phi$$

$$= \sum_{n=0}^{\infty} \sum_{m=0}^{n} D\, m \int_0^{\pi} P_n^m(\cos \theta) \frac{dP_n^{m'}(\cos \theta)}{d\theta} d\theta$$

$$\times \int_0^{2\pi} \cos m'\phi \left[\gamma_{mn} \sin m\phi - \delta_{mn} \cos m\phi \right] d\phi$$

$$+ \sum_{n=0}^{\infty} \sum_{m=0}^{n} \frac{N}{a} \int_0^{\pi} \frac{dP_n^m(\cos \theta)}{d\theta} \frac{dP_n^{m'}(\cos \theta)}{d\theta} \sin \theta\, d\theta$$

$$\times \int_0^{2\pi} \cos m'\phi \left[\alpha_{mn} \cos m\phi + \beta_{mn} \sin m\phi \right] d\phi \qquad (7.14)$$

since

$$\int_0^{2\pi} \cos m'\phi\, \cos m\phi\, d\phi = \begin{cases} 0 & \text{for } m \neq m' \\ \dfrac{2\pi}{\varepsilon_m} & \text{for } m = m' \end{cases} \qquad (7.15)$$

where

$$\varepsilon_m = \begin{cases} 1 & \text{for } m = 0 \\ 2 & \text{for } m \neq 0 \end{cases} \tag{7.16}$$

and

$$\int_0^{2\pi} \cos m'\phi \, \sin m\phi \, d\phi = 0 \text{ for all } m \text{ and } m' \tag{7.17}$$

Equation (7.14) may be written as

$$\int_0^{2\pi} d\phi \int_0^{\pi} d\theta f_1(\theta, \phi) \frac{dP_{n'}^{m'}(\cos \theta)}{d\theta} \sin \theta \cos m'\phi$$

$$= \sum_{n=0}^{\infty} \sum_{m=0}^{n} \frac{2\pi D m'}{\varepsilon_{m'}} \int_0^{\pi} (-\delta_{m'n}) P_n^{m'}(\cos \theta) \frac{dP_{n'}^{m'}(\cos \theta)}{d\theta} d\theta$$

$$+ \sum_{n=0}^{\infty} \sum_{m=0}^{n} \frac{2\pi N}{\varepsilon_{m'} a} \int_0^{\pi} (\alpha_{m'n}) \frac{dP_n^{m'}(\cos \theta)}{d\theta} \frac{dP_{n'}^{m'}(\cos \theta)}{d\theta} \sin \theta \, d\theta$$

$$\tag{7.18}$$

and using (7.12) one can obtain

$$\int_0^{2\pi} d\phi \int_0^{\pi} d\theta \, m' f_2(\theta, \phi) \, P_{n'}^{m'}(\cos \theta) \sin m'\phi$$

$$= \sum_{n=0}^{\infty} \sum_{m=0}^{n} D m' \int_0^{\pi} P_{n'}^{m'}(\cos \theta) \frac{dP_n^{m}(\cos \theta)}{d\theta} d\theta$$

$$\times \int_0^{2\pi} \sin m'\phi \, [\gamma_{mn} \cos m\phi + \delta_{mn} \sin m\phi] \, d\phi \tag{7.19}$$

$$- \sum_{n=0}^{\infty} \sum_{m=0}^{n} \frac{N m}{a} \int_0^{\pi} P_{n'}^{m'}(\cos \theta) \frac{m \, P_n^{m}(\cos \theta)}{\sin \theta} d\theta$$

$$\times \int_0^{2\pi} \sin m'\phi \, [\alpha_{mn} \sin m\phi - \beta_{mn} \cos m\phi] \, d\phi$$

where D and N are described by equation (7.13). By using equation (7.17) and

$$\int_0^{2\pi} \sin m'\phi \, \sin m\phi \, d\phi = \begin{cases} 0 & \text{for } m \neq m' \\ \dfrac{2\pi}{\varepsilon_m} & \text{for } m = m' \end{cases}$$
(7.20)

one can rewrite (7.19) as

$$\int_0^{2\pi} d\phi \int_0^{\pi} d\theta m' f_2(\theta,\phi) P_{n'}^{m'}(\cos\theta)\sin m'\phi$$

$$= \sum_{n=0}^{\infty}\sum_{m=0}^{n} \frac{2\pi D\,m'}{\varepsilon_{m'}} \int_0^{\pi} \delta_{m'n} P_{n'}^{m'}(\cos\theta)\frac{dP_n^{m'}(\cos\theta)}{d\theta} d\theta$$
(7.21)

$$- \sum_{n=0}^{\infty}\sum_{m=0}^{n} \frac{2N\,(m')^2\pi}{\varepsilon_{m'}\,a} \int_0^{\pi} \alpha_{m'n} P_{n'}^{m'}(\cos\theta)\frac{dP_n^{m'}(\cos\theta)}{\sin\theta} d\theta$$

Now by subtracting (7.21) from (7.18) results in

$$\frac{-2\pi D\,m'}{\varepsilon_{m'}} \delta_{m'n} \int_0^{\pi} \left[P_n^{m'}(\cos\theta)\frac{dP_{n'}^{m'}(\cos\theta)}{d\theta} + P_{n'}^{m'}(\cos\theta)\frac{dP_n^{m'}(\cos\theta)}{d\theta} \right] d\theta$$

$$+ \frac{2N}{\varepsilon_{m'}\,a}\alpha_{mn} \int_0^{\pi} \sin\theta \left[\frac{dP_n^{m'}(\cos\theta)}{d\theta}\frac{dP_{n'}^{m'}(\cos\theta)}{d\theta} + \frac{(m')^2}{\sin^2\theta}P_{n'}^{m'}(\cos\theta)P_n^{m'}(\cos\theta) \right] d\theta$$

$$= \int_0^{2\pi}\int_0^{\pi} \left[f_1(\theta,\phi)\frac{dP_{n'}^{m'}(\cos\theta)}{d\theta}\sin\theta\cos m\phi - m f_2(\theta,\phi) P_n^m(\cos\theta)\sin m\phi \right] d\theta\, d\phi$$
(7.22)

And using the following orthogonality relationship

$$\int_0^{\pi} \sin\theta\, d\theta \left[\frac{dP_n^{m'}(\cos\theta)}{d\theta}\frac{dP_{n'}^{m'}(\cos\theta)}{d\theta} + \frac{(m')^2}{\sin^2\theta}P_{n'}^{m'}(\cos\theta)P_n^{m'}(\cos\theta) \right]$$

$$= \begin{cases} 0 & \text{for } n \neq n' \\ \dfrac{2\,n(n+1)+(n+m)!}{(2\,n+1)(n-m)!} & \text{for } n = n' \end{cases}$$
(7.23)

$$\int_0^{\pi} d\theta \left[P_n^{m'}(\cos\theta)\frac{dP_{n'}^{m'}(\cos\theta)}{d\theta} + P_{n'}^{m'}(\cos\theta)P_n^{m'}(\cos\theta) \right] = 0$$
(7.24)

(7.22) can be written as

$$\frac{2N}{\varepsilon_m a} \alpha_{mn} \left[\frac{2\,n\,(n+1)\,(n+m)!}{(2\,n+1)\,(n-m)!} \right]$$

$$= \int_0^{2\pi} \int_0^{\pi} \left[f_1(\theta,\phi)\, \frac{dP_n^m(\cos\theta)}{d\theta}\, \sin\theta\, \cos m\phi - m f_2(\theta,\phi)\, P_n^m(\cos\theta)\, \sin m\phi \right] d\theta\, d\phi$$

$$(7.25)$$

Therefore

$$\alpha_{mn} = \frac{\psi\, a}{N} \int_0^{2\pi} \int_0^{\pi} \left[f_1(\theta,\phi)\, \frac{dP_n^m(\cos\theta)}{d\theta}\, \sin\theta\, \cos m\phi - m f_2(\theta,\phi)\, P_n^m(\cos\theta)\, \sin m\phi \right] d\theta\, d\phi$$

$$(7.26)$$

where

$$\psi = \frac{1}{\pi}\, \frac{(2\,n+1)\,(n-m)!}{2\,n\,(n+1)\,(n+m)!}\, \frac{\varepsilon_m}{2}$$

$$(7.27)$$

To determine β_{mn} one can rewrite equations (7.11) and (7.12) in the following form

$$\int_0^{2\pi} d\phi \int_0^{\pi} d\theta\, f_1(\theta,\phi)\, \frac{dP_{n'}^{m'}(\cos\theta)}{d\theta}\, \sin m'\phi\, \sin\theta$$

$$= \sum_{n=0}^{\infty} \sum_{m=0}^{n} D\,m \int_0^{\pi} P_n^m(\cos\theta)\, \frac{dP_{n'}^{m'}(\cos\theta)}{d\theta}\, d\theta$$

$$\times \int_0^{2\pi} \sin m'\phi\, [\gamma_{mn}\, \sin m\phi - \delta_{mn}\, \cos m\phi]\, d\phi \qquad (7.28)$$

$$- \sum_{n=0}^{\infty} \sum_{m=0}^{n} \frac{N}{a} \int_0^{\pi} \frac{dP_n^m(\cos\theta)}{\sin\theta}\, \frac{dP_{n'}^{m'}(\cos\theta)}{\sin\theta}\, \sin\theta\, d\theta$$

$$\times \int_0^{2\pi} \sin m'\phi\, [\alpha_{mn}\, \cos m\phi + \beta_{mn}\, \sin m\phi]\, d\phi$$

and

$$\int_0^{2\pi} d\phi \int_0^{\pi} d\theta \, m'f_2(\theta,\phi) \, P_{n'}^{m'}(\cos\theta) \, \cos m'\phi$$

$$= \sum_{n=0}^{\infty} \sum_{m=0}^{n} \frac{2D\,m'}{\varepsilon_{m'}} \int_0^{\pi} \gamma_{mn} P_{n'}^{m'}(\cos\theta) \, \frac{dP_n^{m'}(\cos\theta)}{d\theta} \, d\theta \qquad (7.29)$$

$$- \sum_{n=0}^{\infty} \sum_{m=0}^{n} \frac{2N\,(m')^2}{\varepsilon_{m'}\,a} \int_0^{\pi} \beta_{mn} P_{n'}^{m'}(\cos\theta) \, P_n^{m'}(\cos\theta) \, d\theta$$

And via a similar approach the unknown coefficient β_{mn} are determined as

$$\beta_{mn} = \frac{\psi\,a}{N} \int_0^{2\pi} \int_0^{\pi} \left[f_1(\theta,\phi) \, \frac{dP_n^m(\cos\theta)}{d\theta} \, \sin\theta \, \sin m\phi + m f_2(\theta,\phi) \, P_n^m(\cos\theta) \, \cos m\phi \right] d\theta \, d\phi$$

$$(7.30)$$

Similarly γ_{mn} and δ_{mn} may be determined as

$$\gamma_{mn} = \frac{\psi\,a}{N} \int_0^{2\pi} \int_0^{\pi} \left[m f_1(\theta,\phi) \, P_n^m(\cos\theta) \, \sin m\phi + f_2(\theta,\phi) \, \frac{dP_n^m(\cos\theta)}{d\theta} \, \cos m\phi \, \sin\theta \right] d\theta \, d\phi$$

$$(7.31)$$

$$\delta_{mn} = \frac{\psi\,a}{N} \int_0^{2\pi} \int_0^{\pi} \left[f_2(\theta,\phi) \, \frac{dP_n^m(\cos\theta)}{d\theta} \, \sin\theta \, \sin m\phi - m f_1(\theta,\phi) \, P_n^m(\cos\theta) \, \cos m\phi \right] d\theta \, d\phi$$

$$(7.32)$$

To now determine the far field components related to E_θ and E_ϕ, one invokes

$$E_\phi = -Z_0 \, H_\theta \qquad (7.33)$$

$$E_\theta = Z_0 \, H_\phi \qquad (7.34)$$

Here Z_0 is the characteristic impedance of free space which is $120\pi = 377\,\Omega$. The electric far fields are related to the orthogonal components of the magnetic far fields through the characteristic impedance of free space. To compute the far field results the large argument approximation of the spherical Hankel function is used. It is

$$h_n^{(2)}(kr) \approx \frac{j^{n+1} \, e^{-jkr}}{kr} \quad \text{for } r \gg \lambda \qquad (7.35)$$

From (7.35) one can obtain

$$\frac{1}{r}\frac{\partial}{\partial r}\left[r\,h_n^{(2)}(kr)\right] \approx \frac{j^{n+1}(-jk)e^{-jkr}}{kr} = \frac{j^n\,e^{-jkr}}{r} \tag{7.36}$$

Substituting (7.33) and (7.36) into (7.7) results in

$$H_\theta(r,\theta,\phi) = \sum_{n=0}^{\infty}\sum_{m=0}^{n} -j\omega\varepsilon\frac{j^{n+1}e^{-jkr}}{kr}\frac{P_n^m(\cos\theta)\,m}{\sin\theta}\,[\alpha_{mn}\sin m\phi - \beta_{mn}\cos m\phi]$$

$$+ \sum_{n=0}^{\infty}\sum_{m=0}^{n}\frac{j^n e^{-jkr}}{r}\frac{dP_n^m(\cos\theta)}{d\theta}\,[\gamma_{mn}\cos m\phi + \delta_{mn}\sin m\phi] \tag{7.37}$$

After some simplification

$$H_\theta(r,\theta,\phi) = \sum_{n=0}^{\infty}\sum_{m=0}^{n}\omega\varepsilon\frac{j^n e^{-jkr}}{kr}\frac{P_n^m(\cos\theta)\,m}{\sin\theta}\,[\alpha_{mn}\sin m\phi - \beta_{mn}\cos m\phi]$$

$$+ \sum_{n=0}^{\infty}\sum_{m=0}^{n}\frac{j^n e^{-jkr}}{r}\frac{dP_n^m(\cos\theta)}{d\theta}\,[\gamma_{mn}\cos m\phi + \delta_{mn}\sin m\phi] \tag{7.38}$$

Substituting from equations (7.26), (7.30), (7.31) and (7.32) for the quantities α_{mn}, β_{mn}, γ_{mn}, δ_{mn}, respectively, into equation (7.38) yields:

$$H_\theta(r,\theta,\phi) = \frac{e^{-jkr}}{4\,\pi k\,r\,\eta}\sum_{n=0}^{\infty}\frac{(2n+1)}{n(n+1)}j^n\ \times$$

$$\left[\frac{j}{h_n^{(2)}(ka)}\int_0^{2\pi}\int_0^{\pi}\left\{f_1(\theta',\phi')\frac{d^2P_n(\xi)}{d\theta\,d\phi} + f_2(\theta',\phi')\sin\theta'\frac{d^2P_n(\xi)}{d\theta\,d\theta'}\right\}d\theta'd\phi'\right.$$

$$\left.+ \frac{a}{N}\int_0^{2\pi}\int_0^{\pi}\left\{f_1(\theta',\phi')\frac{\sin\theta'}{\sin\theta}\frac{d^2P_n(\xi)}{d\theta'd\phi} + \frac{f_2(\theta',\phi')}{\sin\theta}\frac{d^2P_n(\xi)}{d\phi\,d\phi'}\right\}d\theta'd\phi'\right] \tag{7.39}$$

Utilizing

$$P_n(\xi) = \sum_{m=0}^{n}\varepsilon_m\frac{(n-m)!}{(n+m)!}P_n^m(\cos\theta')\cos m(\phi-\phi') \tag{7.40}$$

and via a similar approach one obtains

$$
H_\phi(r,\theta,\phi) = \frac{e^{-jkr}}{4\pi r\eta} \sum_{n=0}^{\infty} \frac{(2n+1)}{n(n+1)} j^n \times
$$

$$
\left[
\frac{j}{k\,h_n^{(2)}(ka)} \int_0^{2\pi}\int_0^\pi \left\{ \frac{f_1(\theta',\phi')}{\sin\theta} \frac{d^2 P_n(\xi)}{d\phi'd\phi} + f_2(\theta',\phi') \frac{\sin\theta'}{\sin\theta} \frac{d^2 P_n(\xi)}{d\theta\,d\phi'} \right\} d\theta'd\phi'
\right.
$$

$$
\left.
+ \frac{a}{N} \int_0^{2\pi}\int_0^\pi \left\{ f_1(\theta',\phi')\sin\theta' \frac{d^2 P_n(\xi)}{d\theta'd\theta} + f_2(\theta',\phi') \frac{d^2 P_n(\xi)}{d\theta\,d\phi'} \right\} d\theta'd\phi'
\right]
$$

$$(7.41)$$

It is interesting to note that both (7.39) and (7.41) do not contain any summation over m, which has been eliminated in the present formulation by utilizing the addition theorem for Legendre polynomials introduced through (7.40). Also observe that the second derivatives of the Legendre polynomials can be evaluated, for example, as

$$
\frac{\partial^2 P_n(\xi)}{\partial\theta\,\partial\phi} = \frac{\partial^2 P_n(\xi)}{\partial\xi^2} \frac{\partial\xi}{\partial\phi} \frac{\partial\xi}{\partial\theta} + \frac{\partial P_n(\xi)}{\partial\xi} \frac{\partial^2\xi}{\partial\theta\,\partial\phi}
\tag{7.42}
$$

where

$$
\frac{\partial P_n(\xi)}{\partial\xi} = \frac{n+1}{1-\xi^2} [\xi P_n(\xi) - P_{n+1}(\xi)]
\tag{7.43}
$$

and

$$
\frac{\partial^2 P_n(\xi)}{\partial\xi^2} = \frac{n+1}{(1-\xi^2)^2} [\{(2+n)\xi^2 - n\} P_n(\xi) - 2\xi P_{n+1}(\xi)]
\tag{7.44}
$$

Furthermore, for a given prespecified ka, one could precompute the summation over n, in terms of the four "pseudo" Green's functions and store them. Under these conditions, one then needs to perform only an integral over θ and φ as

$$
H_\theta(r,\theta,\phi) \simeq \frac{e^{-jkr}}{4\pi r\eta} \int_0^{2\pi}\int_0^\pi \{ f_1(\theta',\phi') G_1(\theta,\phi,\theta',\phi') + f_2(\theta',\phi') G_2(\theta,\phi,\theta',\phi') \} d\theta'd\phi'
$$

$$(7.45)$$

$$
H_\phi(r,\theta,\phi) \simeq \frac{e^{-jkr}}{4\pi r\eta} \int_0^{2\pi}\int_0^\pi \{ f_1(\theta',\phi') G_3(\theta,\phi,\theta',\phi') + f_2(\theta',\phi') G_4(\theta,\phi,\theta',\phi') \} d\theta'd\phi'
$$

$$(7.46)$$

where

$$G_1(\theta,\phi,\theta',\phi') = \sum_{n=1}^{\infty} \frac{(2n+1)}{n(n+1)} j^n \left[\frac{j}{h_n^{(2)}(ka)} \frac{d^2 P_n(\xi)}{d\theta\, d\phi} + \frac{ka}{N} \frac{\sin\theta'}{\sin\theta} \frac{d^2 P_n(\xi)}{d\theta'd\phi} \right]$$

(7.47)

$$G_2(\theta,\phi,\theta',\phi') = \sum_{n=1}^{\infty} \frac{(2n+1)}{n(n+1)} j^n \left[\frac{j\sin\theta'}{h_n^{(2)}(ka)} \frac{d^2 P_n(\xi)}{d\theta\, d\theta'} + \frac{ka}{N\sin\theta} \frac{d^2 P_n(\xi)}{d\phi'd\phi} \right]$$

(7.48)

$$G_3(\theta,\phi,\theta',\phi') = \sum_{n=1}^{\infty} \frac{(2n+1)}{n(n+1)} j^n \left[\frac{j\sin\theta}{h_n^{(2)}(ka)} \frac{d^2 P_n(\xi)}{d\phi'd\phi} + \frac{ka\sin\theta'}{N} \frac{d^2 P_n(\xi)}{d\theta'd\theta} \right]$$

(7.49)

$$G_4(\theta,\phi,\theta',\phi') = \sum_{n=1}^{\infty} \frac{(2n+1)}{n(n+1)} j^n \left[\frac{j\sin\theta'}{h_n^{(2)}(ka)\,\sin\theta} \frac{d^2 P_n(\xi)}{d\theta'd\phi'} + \frac{ka}{N} \frac{d^2 P_n(\xi)}{d\theta\, d\phi'} \right]$$

(7.50)

The integrals in (7.45) and (7.46) can efficiently and accurately be carried out using the conventional Fast Fourier Transformation technique. The functions $G_1 - G_4$ are called "pseudo" Green's functions because for a true Green's functions, f_1 and f_2 would be convolved with the Green's functions, but here it is an integral. These equations indicate that if for a fixed ka, the Green's functions are precomputed and stored, then the actual computations of (7.45) and (7.46) can be done even on a laptop PC. If the quality of the measured data, i.e., $f_1(\theta, \varphi)$ and $f_2(\theta, \varphi)$ are good (which implies that quite a few significant bits are accurate), the derivatives in (7.39) and (7.41) can be transferred from P_n to f_1 and f_2. This may enhance the rate of convergence of the summation over n.

In summary, the present approach offers the following features:

1) The transformation is expressed in an analytic form.
2) Therefore, for a given radius of the sphere where the fields are computed it is not known a priori as to how many spherical modes need to be computed. Through an explicit representation this uncertainty can be removed and the summation can be carried out to a pre-specified degree of accuracy over the spherical modes.
3) There is only one summation, i.e., over n. To obtain a relative numerical accuracy of 10^{-7} in the computation of the fields, the limit of the summation over n should be $n = 1.27\ ka$ for $ka > 60$.
4) The derivatives in the expressions for the far field can be transferred to the data (if the quality is good) to further enhance the rate of convergence. Or equivalently the data can be expanded in a Fourier series as is conventionally done (at least in the first step) and the derivatives can be carried out in an analytic fashion utilizing the FFT.

5) For a fixed *ka*, all the summations over *n* can be precomputed and stored in the memory of a PC. This NF/FF transformation procedure is equivalent to synthesizing a plane wave region using an infinite number of point sources on a sphere having radius *a* and each individual point source having a complex amplitude is given by the "Pseudo" Green's functions G_1-G_4.

A computer program incorporating this analytical spherical near-field to far-field transformation is given in the Appendix 7A.

7.1.3   Numerical Simulations

7.1.3.1   Synthetic Data
As a first example consider a four dipole array. The dipoles are located at the corners of a $4\lambda \times 4\lambda$ planar surface which is in the *x* - *y* plane. The center of the $4\lambda \times 4\lambda$ square surface is located at $x = 0.22\lambda$ and $y = 0.22\lambda$. The plane of the array is the *x* - *y* plane. So the four dipoles are not located symmetrically about the origin. A spherical surface is drawn with the center defined above and with a radius of 10 λ. On that spherical surface of 20 λ diameter encapsulating the four offset dipoles located on the *x* - *y* plane at the corners of a $4\lambda \times 4\lambda$ surface both the electric field components are computed. E_θ and E_φ are computed analytically. Next, the two field components are used in conjunction with (7.39) and (7.41) to evaluate the far fields. Figure 7.1 presents E_φ in dB for $\varphi = 0°$

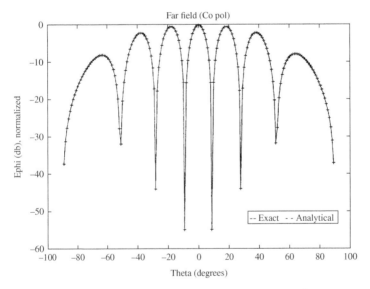

Figure 7.1 Comparison of exact and computed far field for $\varphi = 0°$ cut for a 4 dipole array each located at the corners of a $4\lambda \times 4\lambda$ plate.

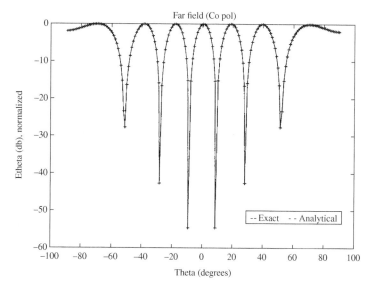

Figure 7.2 Comparison of exact and computed far field for $\varphi = 90°$ cut for a 4 dipole array located at the corners of a $4\lambda \times 4\lambda$ plate.

as a function of θ. Both the exact analytical far field and the far field computed by using the present theory are presented in Figure 7.1. They are visually indistinguishable. In Figure 7.2, E_θ is presented in dB for $\varphi = 90°$ as a function of θ. Again, the analytical far fields from the four off centered dipoles and the computed far fields are visually indistinguishable. The cross polar components in both the figures are quite negligible.

7.1.3.2 Experimental Data

Next, measured data is utilized. Consider a microstrip array antenna consisting of 32×32 uniformly distributed patches on a 1.5m × 1.5m surface. The near fields are measured on a spherical surface at a distance of 1.23 m away from the array antenna at a frequency of 3.3 GHz. The data is taken every 2° in φ and every 1° in θ. Measurements have been performed using an open ended cylindrical WR284 waveguide fed with the TE_{11} mode. The measured data was provided by Dr. Carl Stubenrauch of NIST [8]. Figures 7.3–7.6 compare the far-field patterns obtained by the present analytical method with the far field patterns obtained by the numerical technique described in [9]. These numerically computed far-field patterns employ the same measured data utilizing an equivalent magnetic current approach for near field to far field transformation [9]. Figure 7.3 describes $20 \log_{10} E_\varphi$ for $\varphi = 0°$ and for various angles of θ. Figure 7.4 represents $20 \log_{10} E_\theta$ for $\varphi = 90°$ and for $-90° < \theta < 90°$. These are the principal plane patterns. As observed, the agreement is good. Figures 7.5

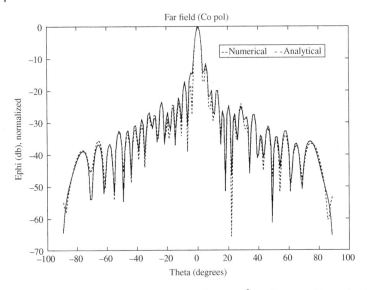

Figure 7.3 Co-polarization characteristic for $\varphi = 0°$ cut for a 32×32 patch microstrip array using analytical and numerical results.

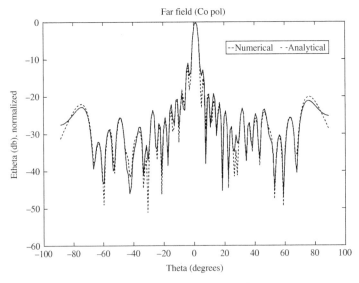

Figure 7.4 Co-polarization characteristic for $\varphi = 90°$ cut for a 32×32 patch microstrip array using analytical and numerical results.

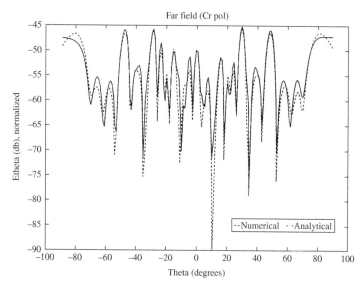

Figure 7.5 Cross-polarization characteristic for $\varphi = 0^\circ$ cut for a 32×32 patch microstrip array.

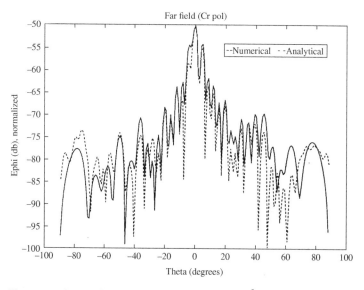

Figure 7.6 Cross-polarization characteristic for $\varphi = 90^\circ$ cut for a 32 × 32 patch microstrip array using analytical and numerical results.

and 7.6 show the cross polar patterns. Figure 7.5 depicts $20 \log_{10} E_\theta$ for $\varphi = 0^\circ$ and for different values of θ. Figure 7.6 presents $20 \log_{10} E_\varphi$ for $\varphi = 90^\circ$ and for $-90^\circ < \theta < 90$. The agreement between the approach presented in this section and the numerical approach in evaluating the cross polar pattern is reasonable for pattern levels above -70 dB.

7.1.4 Summary

An alternate method is described and have been presented in [10] for spherical near field to near/far field transformation without probe correction. The advantage of this approach is that one of the summations over m has been eliminated by utilizing the addition theorem for Legendre polynomials. Hence the presented expressions are more concise and easier to visualize. This method is accurate, as illustrated by the computed results and comparing it with other techniques. The performance of this method has been illustrated using both synthetic and real experimental data.

7.2 Radial Field Retrieval in Spherical Scanning for Current Reconstruction and NF–FF Transformation

In this section another alternate spherical near-field to far-field (NF–FF) transformation using equivalent magnetic currents (EMCs) and matrix methods is addressed. It is based on the decoupling of the field components and the iterative retrieval of the radial component of the electric field. The technique is applied for far-field calculation as well as for the estimation of the current distribution of the antenna under test (AUT) using spherical near-field facilities. Results from measured near-field data of several antennas are presented and compared to those of the analytical solution via a spherical wave mode expansion method.

7.2.1 Background

Near-field scanning over canonical surfaces—planar, cylindrical, and spherical—is often used to obtain the far-field pattern characteristics of an antenna under test (AUT) via near-field to far-field (NF–FF) transformation algorithms. Spherical scanning is most versatile in the sense that it makes it possible to obtain the radiation pattern over the entire angular span of any arbitrary antenna. Spherical NF–FF transformation techniques based on a modal expansion in spherical waves has been widely used [10, 11]. In these techniques, the convergence of the series representation makes it necessary to have a knowledge of the near field over an angular interval sufficiently greater (depending on the maximum AUT dimension and acquisition distance) than the desired far-field angular interval. Diagnosis tasks in those techniques, can be done through the Fourier transform to obtain the aperture field from the far field.

Finally, NF–FF transformation [10] and diagnosis [12] can be accomplished using arbitrary scanning surfaces with a matrix method formulation based on the reconstruction of some equivalent current distribution over that arbitrary surface. One characteristic of the matrix techniques is that it gives the best far-field solution (in the least squares sense) for a given near-field data set. In this way, it has been experimentally observed that near-field scanning is only required in the desired far-field angular interval. The intermediate step of source reconstruction inherent in a matrix method formulation yields the possibility of performing diagnosis tasks in some type of antennas, such as planar arrays and aperture antennas, since the method gives the best solution of the equivalent sources (field at the aperture) for a given near-field data set. However, a large system of equations appears when directly applying numerical matrix methods to the resultant integral equations relating source components and the electric field components.

In fact, in a general equivalent problem for spherical near-field scanning, two components (tangential to the surface enclosing the AUT) of both the equivalent magnetic and equivalent electric currents must be used. When reconstructing over an infinite plane, only one type of equivalent currents can be used, but still two components of this equivalent currents are needed. So, finally, two coupled integral equations arise when relating electric field components (angular components over the spherical scanning surface) and the equivalent source components along an infinite plane (the antenna plane). When solving numerically the integral equations, it is necessary to solve a large system of equations.

In this section, after establishing the equivalent problem with only an equivalent magnetic current (EMC) distribution, an alternate formulation is presented. The formulation is based on the decoupling of the two integral equations under an assumption of a zero value of the radial component of the electric field. Then, each integral equation can be solved independently for each component of the EMC. The numerical solution of each integral equation is performed by minimizing a least squares error functional. This approach works well when measurements are performed in the Fresnel region. For a more general acquisition distance and in order to overcome the limitations resulting from neglecting the radial electric field, an iterative algorithm that retrieves the radial component of the electric field is proposed [13].

In order to compare this technique with the modal expansion technique, a numerical code utilizing the NF-FF transformation based on a wave-mode expansion [14] has been used to calculate the radiation patterns from synthesized and measured near-field data. Synthesized data from radiating configurations, where the radial component of the electric field cannot be neglected, have been used to assess the accuracy of this method. Results using both the radial field retrieval technique and the modal expansion technique have been presented for most of the examples.

Special attention has been paid to experimental validation. The measured near-field data from a reflector antenna has been used to calculate the far-field pattern and to reconstruct the field at the antenna aperture. Also, the analytical solution for the far-field pattern with a wave expansion method has also been calculated for comparison purposes. Other measured data have been used to illustrate the applicability of the proposed radial field retrieval technique.

7.2.2 An Equivalent Current Reconstruction from Spherical Measurement Plane

In this section, the principle of the equivalent current approach is presented. Here, an equivalent magnetic current distribution over an infinite plane is made coincident with the plane of the antenna (i.e. the aperture plane is replaced by a larger of equivalent impressed radiating elements in the case of two dimensional arrays or a virtual plane "enclosing" the antenna). For computational purposes, the domain of this EMC distribution is truncated to some region where the tangential electric field away from this region can be neglected. To reach the final matrix formulation, an equivalent problem is established with an EMC distribution over an infinite plan that coincides with the antenna plane (aperture plane, layer of impressed radiating elements in the case of two-dimensional (2-D) arrays, or a virtual plane "enclosing" the antenna). For numerical purposes, the domain of this EMC distribution is truncated at some distance from the antenna where the tangential electric field can be neglected. Then, for the external equivalent problem, a relation between the field radiated by the original AUT and the EMC density distributed over the plane can be written as the vector integral equation [13]

$$\vec{E}(x,y) = \Im\left\{\vec{M}_{su}(x',y')\right\} \tag{7.51}$$

where $\Im$ is an integral operator that includes the free space Green function. Considering the particular case of spherical scanning, two scalar integral equations are generated in such a way that each angular component of the electric field depends on both the Cartesian components of the equivalent magnetic current density, i.e.,

$$E_\theta(R_0,\theta,\phi) = \Im\left(M_x, M_y\right) \tag{7.52}$$

$$E_\phi(R_0,\theta,\phi) = \Im\left(M_x, M_y\right) \tag{7.53}$$

Under this representation, a large number of unknowns representing the EMC distribution are used in a typical method of moments solution procedure to solve (7.52) for the unknown EMC. The Nyquist sampling is transformed form the measurement plane to the source plane and here roughly 10 unknowns per wavelength in a linear direction is necessary for accurate computation of the EMC. This result in a large number of unknowns resulting in approximately 100 unknowns per square wavelength of surface area is necessary to represent each component of the EMC. However, two independent integral equations can be obtained if one neglects the radial component of the near field. Under this assumption, the Cartesian components of the electric field can be obtained from the measured angular components of the electric field using the following approximation:

$$
\begin{bmatrix} E_x \\ E_y \end{bmatrix} = \begin{bmatrix} \cos\theta\,\cos\phi & -\sin\phi \\ \sin\theta\,\sin\phi & \cos\phi \end{bmatrix} \begin{bmatrix} E_\theta \\ E_\phi \end{bmatrix} \tag{7.54}
$$

Now an integral equation relating the Cartesian EMC components and the Cartesian field components can be set up. This can significantly reduce the solution time as the number of unknowns is now reduced by half. If the observation point is represented by the scanning spherical coordinates (R_0, θ, ϕ) and the source points are represented by the Cartesian components (x', y', z_0'), then the field due to an EMC distribution over a surface S can be written as:

$$
E_x(R_0, \theta, \phi) = -\int_S M_y(x', y')\, G\left(\lambda, R_0, \theta, \phi, x', y', z_0'\right) ds' \tag{7.55}
$$

$$
E_y(R_0, \theta, \phi) = -\int_S M_x(x', y')\, G\left(\lambda, R_0, \theta, \phi, x', y', z_0'\right) ds' \tag{7.56}
$$

where

$$
G\left(\lambda, R_0, \theta, \phi, x', y', z_0'\right) = (R_0\cos\theta - z_0')\left(\frac{1 + jkR}{4\pi R^3}\right) e^{-jkR} \tag{7.57}
$$

and λ is the wavelength and k is the wave number given by $\dfrac{2\pi}{\lambda}$. R is the distance between the source and the observation point.

The advantage with this representation is that two decoupled integral equations are now obtained, each of them relating one Cartesian component of the electric field with one Cartesian component of the EMC distribution. Therefore, only half the number of unknowns at a time of the conventional matrix representation [derived from equation (7.52)] is involved in each solution of each of the integral equation. In addition, the conjugate gradient method can be used to efficiently solve the matrix equation, dealing with each component of the EMC. The decoupled integral equations (7.54) and (7.55), relating the measured field components to the EMC, are solved numerically for the EMC, which are expanded in terms of subdomain-type basis functions in the usual Method of

Moment context. A cost function then is minimized to obtain the desired solution.

Represent the scalar measured data corresponding to one component of the electric field by the vector Y and the corresponding computed values at the same points predicted by the EMC over the equivalent surface [values obtained using equations (7.54) or (7.55)] by the vector E; where $E = G\,M_s$ where M_s is the vector representing the discretized EMC components over the surface S. Through this definition a least squares error function can be defined as

$$\Xi = (G\,M_s - Y)^T (G\,M_s - Y)^* \tag{7.58}$$

Where the superscript T represents transpose and the $*$ represents a complex conjugate. From (7.58) a quadratic functional can be derived given by

$$\Theta = M_s^T\,\mathbf{H}\,M_s + \left(B\,M_s + B^*\,M_s^*\right) + C \tag{7.59}$$

where $\mathbf{H}$ is defined by the Hessian and M_s^T and M_s^* define the transpose and the conjugate of M_s, respectively of the vector representing the EMC distribution. Then the optimization procedure to obtain M_s reduces to finding the minimum of a quadratic form. Taking the first derivative and setting it to equal to zero results in the optimum value given by:

$$M_{s,optimum} = \mathbf{H}^{-1}\,B^* \tag{7.60}$$

Next, the iterative reconstruction of the radial field component using the surface EMC is presented

7.2.3 The Radial Electric Field Retrieval Algorithm

The approximation used in (7.52) and (7.53) can provide good results in certain cases when measuring the fields in the Fresnel region of a test antenna in a near-field spherical scanning range. However, that formulation is not general and is not accurate under general conditions. If due to the electric size of the antenna and the scanning distance, it can so happen that the values of the angular components of the electric field are no longer greater than the radial component, and then, the radial component of the electric field cannot be neglected and one cannot use directly the formulation established in (7.53) to (7.56).

In order to evaluate the radial component of the electric field when transforming to the Cartesian components from its angular components, an iterative algorithm for the radial field retrieval is proposed. As a starting point, the value of the radial component of the electric field is set to zero and an initial EMC reconstruction step is performed according to the formulation previously presented. With this first estimate of the EMC distribution, the electric field over the hemispherical scanning surface is calculated, replacing the zero values of the radial component by the calculated ones at each scanning point. Now the Cartesian components of the measured field are calculated using both the angular and radial field components. A new EMC is then computed. The iterative procedure continues until the predefined accuracy of the solution is achieved. The steps of the algorithm are summarized as follows.

1) Start with an initial guess for the EMC components M_x^0 and M_y^0.
2) Set to zero initially the value of the radial component of the electric field. The Cartesian components of the electric fields E_x^0 and E_y^0 are obtained from the scanned angular components of E_θ and E_ϕ using the approximation of (7.52) and (7.53).
3) Reconstruct each component of EMC by solving the integral equations of (7.54) and (7.55) for M_x^k and M_y^k at the kth iteration.
4) Calculate the radial component of the electric field E_r^k from the magnetic current components M_x^k and M_y^k.
5) Calculate of the Cartesian components of the electric field E_x^k and E_y^k from the scanned angular components of E_θ and E_ϕ and using the estimate for the radial component E_r^k.
6) Start the next iteration from (7.52) and (7.53) and continue the iteration till the stopping criteria which is given by the error functional (7.58) is reduced to a small value.

7.2.4 Results Obtained Using This Formulation

First some results are shown using simulated data and then using actual measured data.

7.2.4.1 Simulated Data

As a first example consider near-field data from some known EMC distributions given in Table 7.1. An acquisition distance, close enough to the radiating structure, has been selected so that the radial component of the electric field is significant. Then scanning is performed over a spherical surface with increments of $\Delta\theta = 10°$ and $\Delta\varphi = 10°$ have been considered in generating the near-field data. A fictitious surface containing the antenna under test (AUT) is considered as the source plane. Here, the EMC is discretized using sub sectional basis functions in the MoM context using 10×10 cells. This discretization is used for the source reconstruction.

In Figures 7.7 and 7.8 the reconstructed EMC distribution (amplitude and phase of the M_y component) for various iterations of the radial field retrieval algorithm are compared to the nominal EMC distribution. A comparison between the far-field pattern due to the nominal EMC distribution and the far-field pattern calculated from the reconstructed EMC distribution for two

Table 7.1 Nominal Values of 2-D EMC Distribution.

Frequency	1,5 GHz
Dimensions, LX x LY	1m x 1m ($5\lambda \times 5\lambda$)
M_y (amplitude)	$1/(1 + x^2 + y^2)$
M_y (phase)	$45\left(\sqrt{x^2 + y^2}/LX\right)$, deg
M_x (amplitude)	$0.01 + x^2 + y^2$
M_x (phase)	0, deg
Acquisition distance	0,8 m (4λ)

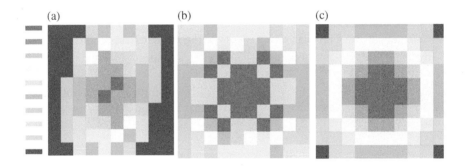

Figure 7.7 Amplitude (dB) of the equivalent magnetic current M_y: reconstructed from the results of 1 iteration (a), reconstructed from the results of 20 iterations (b), nominal (c). Color scale:(−3, 0).

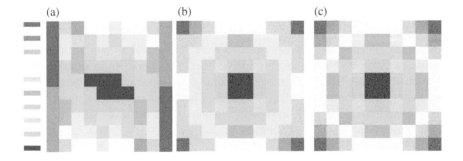

Figure 7.8 Phase (deg.) of the equivalent magnetic current M_y: reconstructed from the results of 1 iteration (a), reconstructed from the results of 20 iterations (b), nominal (c). Color scale:(5, 60).

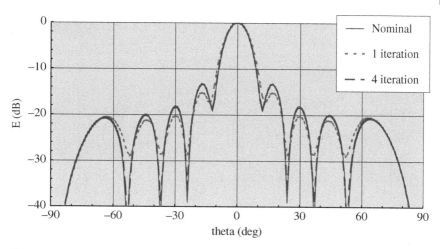

Figure 7.9 Two-dimensional EMC distribution. Far-field pattern ($\varphi = 0°$) from nominal and reconstructed sources.

different iterations of the radial field retrieval algorithm is shown in Figure 7.9. It is seen that good convergence is obtained after only 4 iterations.

Next we illustrate the results using actual measured data.

7.2.4.2 Using Measured Data

The experimental data was obtained at Laboratorio de Ensayos of El Casar de Talamanca, and was taken by Pr. J. L. Besada and P. Caballero.

Linear Patch Array: A commercial base-station antenna for mobile communications of 1.5 m length was measured at the near-field spherical facility of Grupo de Radiacion-Universidad Politecnica de Madrid, at a frequency of 1.8 GHz, with a scanning radius of 3.3 m and scanning with an angular increments of $\Delta\theta = 1°$ and $\Delta\varphi = 1°$. The calculated co-polar and cross-polar components of the far field in the E-plane (polarization along the array direction) using the proposed numerical algorithm based on the equivalent currents as well as using the spherical wave expansion of [10] and the SNIFTD software [14] are shown in Figures 7.10 and 7.11. In this case, no iterative process for the calculation of the radial field is needed (at iteration 1, zero value of the radial component is considered) to obtain results using the equivalent current algorithm. Near-field data in the front hemisphere have been used in the EMC algorithm while complete spherical near-field data are considered for the spherical wave expansion results. Under these premises, the calculation of the far field in the front region (containing 32,851 points) with the equivalent current radial field retrieval algorithm is 17 times faster than with the spherical wave expansion method [13].

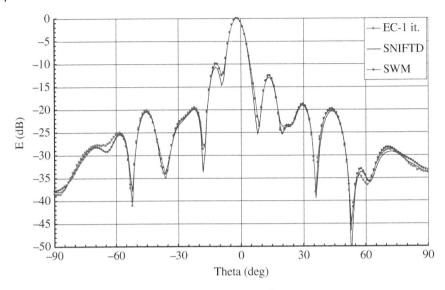

Figure 7.10 Linear patch array. Far-field pattern ($\varphi = 0°$). Co-polar pattern.

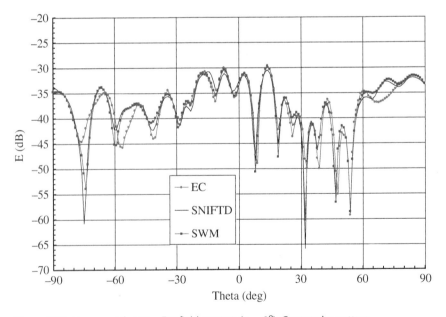

Figure 7.11 Linear patch array: Far-field pattern ($\varphi = 0°$). Cross-polar pattern.

As a second example, consider measured near field components from a 32x32 microstrip patch array. This data was given to us by Carl Stubenrauch from NIST [8]. The near-field data of an square planar array of 32 × 32 patches, at 3.3 GHz, measured over a spherical surface of radius 1.23 m is used to validate the proposed algorithm. This data set is now used in the iterative radial field retrieval algorithm, the field at the antenna plane (EMC distribution), and the far-field pattern in the main planes were calculated. The results of the copolar far field in the main planes are compared to the results of the analytical spherical wave expansion technique in Figures 7.12 and 7.13. The results of the EMC technique have been calculated with four iterations of the radial field retrieval algorithm. Results with both techniques are undistinguishable except close to grazing angles, where discrepancies are due to the truncation of the equivalent current domain that imposes a zero value of the tangential electric field outside this domain in the EMC technique. The degree of agreement in the cross-polar results with both the techniques can be observed in Figures 7.14 and 7.15. The reconstructed electric field at the plane of the antenna, in both amplitude and phase, are shown in Figure 7.16, where a 1.6 m ×1.6 m domain and 64 × 64 cell elements were used to reconstruct the EMC. In Figure 7.16, the outlines of the radiating structure are visible. For a complete far-field pattern in the front of the antenna with 1° increment for both the angular coordinates, the equivalent current–radial field retrieval algorithm results are achieved in a CPU time five times faster than the CPU time used with the spherical wave mode expansion algorithm using 134 modes in the expansion.

For the third and final example, consider a 90 cm reflector antenna. The near field data from a 90 cm reflector antenna was measured on the spherical near-field range of the Laboratorio Ensayos (El Casar de Talamanca, Madrid). The

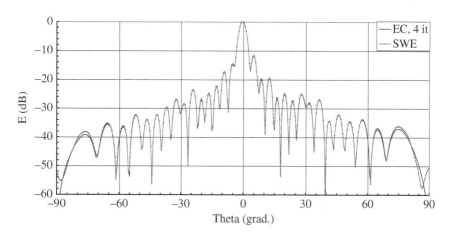

Figure 7.12 32 × 32 microstrip patch array. Far-field pattern ($\varphi = 0°$). Co-polar pattern.

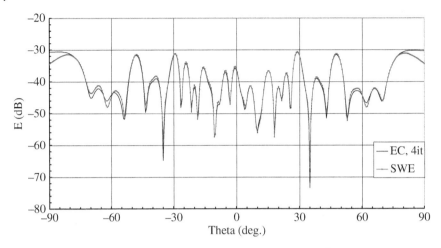

Figure 7.13 32 × 32 microstrip patch array. Far-field pattern ($\varphi = 0°$). Cross-polar pattern.

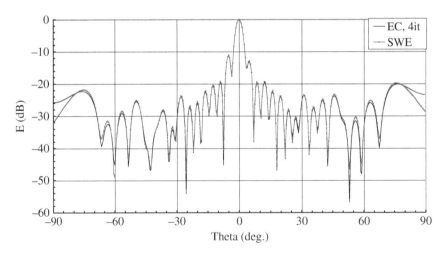

Figure 7.14 32 × 32 microstrip patch array. Far-field pattern ($\varphi = 90°$). Co-polar pattern.

radial distance of the spherical scanning was 126 cm and the frequency of operation was 10.7 GHz. For this configuration, the AUT generates an electric field with some non-negligible radial component. Only scanned near-field data in the range $\theta \in [0°, 40°]$, with $\varphi \in [0°, 360°]$, with angular increments $\Delta\theta = 0.5°$ and $\Delta\varphi = 0.75°$, have been used as input for the source reconstruction-radial field retrieval algorithm. No special care was taken to adjust the feeding element for

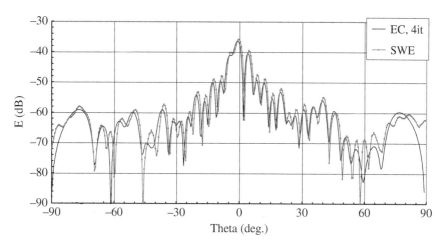

Figure 7.15 32 × 32 microstrip patch array. Far-field pattern (φ = 90°). Cross-polar pattern.

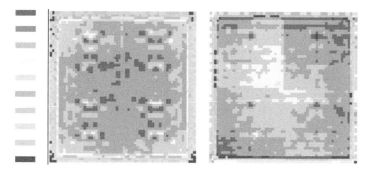

Figure 7.16 32 × 32 microstrip patch array. Amplitude (–30dB to 0dB) and phase (–180° to 180°) of the electric field at the plane of the antenna.

generating linear polarization at the focus of the parabolic reflector and the goal was to compare the NF-FF transformation results using the proposed method with those of spherical wave expansion method using the SNIFTD software [14]. Both the copolar and the cross polar far-field patterns are shown in Figures 7.17–7.20. Even for this acquisition at very short distance, it is seen that accurate far field results are obtained for the copolar φ = 0° plane (Figure 7.17) without performing radial component retrieval (iteration 1). This is also true in Figure 7.18 for the cross –polar (However, for some cuts in the pattern, the zero radial field results must be corrected to obtain accurate results. Figure 7.19 provides the copolar (far field pattern in the φ = 90° –plane) but for the cross polar

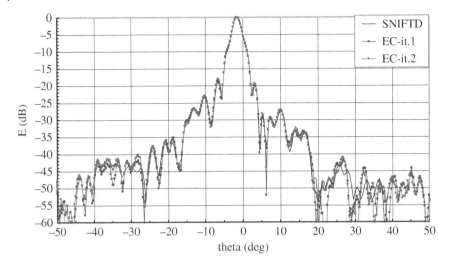

Figure 7.17 Reflector antenna of size 90 cm. Far-field pattern ($\varphi = 0°$). Copolar pattern.

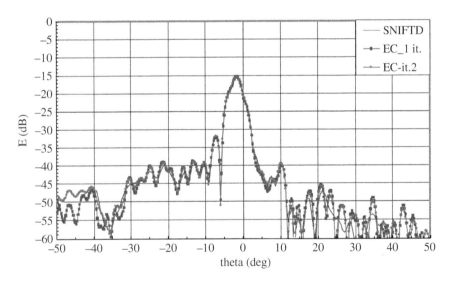

Figure 7.18 Reflector antenna of size 90 cm. Far-field pattern ($\varphi = 0°$). Cross-polar pattern.

case in Figure 7.20 there are differences. In that plane, if the radial field values are neglected, the proposed EMC reconstruction gives a far-field result that differs from the one obtained with spherical wave expansion at levels around –30dB near the broadside $\theta = 0°$ direction. However, after a complete iteration

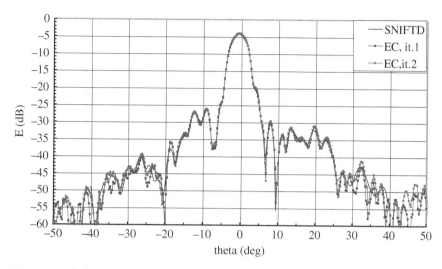

Figure 7.19 Reflector antenna of size 90 cm. Far-field pattern ($\varphi = 90°$). Copolar pattern.

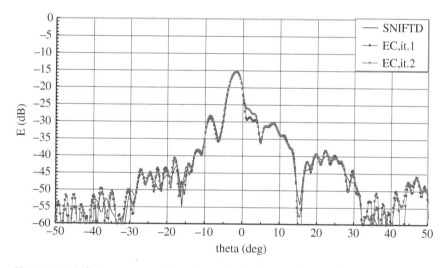

Figure 7.20 Reflector antenna of size 90 cm. Far-field pattern ($\varphi = 90°$). Cross-polar pattern.

of the radial field retrieval algorithm the calculated numerical results agree perfectly with those of the conventional wave expansion. Segmentation of the source domain using 80 × 80 cells was used for the EMC reconstruction. The intermediate results of the amplitude of the EMC distribution (components of the tangential electric field) over a domain of 1 m × 1.2 m in the aperture

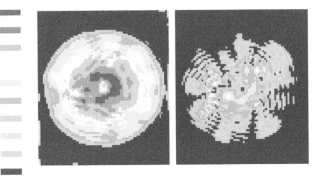

Figure 7.21 Reflector antenna of size 90 cm. Electric field components at the aperture of size (100 cm × 120 cm).

plane are shown in Figure 7.21. The circular profile along with the feed blockage can be clearly observed in this figure for the EMC on the reflector antenna.

In short, an equivalent source reconstruction and a NF–FF transformation method have been presented for the case of spherical scanning. The method makes use of a radial component retrieval algorithm in an iterative fashion. This algorithm is based on the decoupling of the integral equations appearing in the source reconstruction algorithm, reducing to half the number of unknowns that must be handled in the matrix solution and increasing the computational efficiency over traditional NF–FF matrix methods for the spherical scanning case. Just a few iterations are typically required for accurate NF–FF transformation while antenna diagnosis can require more refinement. A comparison with the technique based on the spherical wave-mode expansion has been performed in terms of far-field results. The results of this comparison together with the computational efficiency and the possibility of performing direct antenna operational diagnosis tasks, make the described technique an interesting alternative in the area of near-field antenna measurements.

7.3 Conclusion

Two methods have been presented for spherical near-field to far-field transformation. The first method presents an exact explicit analytical formulation for transforming near-field data generated over a spherical surface to the far field radiation pattern. A computer program involving this method is provided at the end of the chapter. The second method describes the equivalent source formulation that was described earlier and illustrates how it can be deployed to the spherical scanning case where one component of the field is missing from the measurements, when the tangential components of the fields on a spherical surface is given. The validity of both of these techniques has been illustrated using both synthetic and experimental data.

Appendix 7A   A Fortran Based Computer Program for Transforming Spherical Near-Field to Far-Field

```
program neartofar
    use omp_lib
    USE MM
    implicit none
    real(8) x,r,freq,cc
    complex(8) CEP,CET,temp
    real(8) CMPr,CMTr,CMPi,CMTi
    real(8) et(1:361),ep(1:361)
    real(8) bt
    integer i,j,ipff,ipf,aTF,ai,itff,ii

    cc=299792458.0d0
    pi=2.0d0*dasin(1.0d0)
    qw=pi/180.0d0
    maxM=500
    freq=1.55d9    ! The frequency
    r=0.774d0      ! Distance away from the antenna
    aka=2.0d0*pi*freq/cc*r

    call BES(aka)
  numi=1

    open(unit=1,file="./input/Horn_0.txt")
    do j=1,360,numi
        do i=1,180,numi
            read(1,*) CMTr,CMTi,CMPr,CMPi
            CMT(j,i)=dcmplx(CMTr,CMTi);
            CMP(j,i)=dcmplx(CMPr,CMPi);
        end do
    end do

    do ipff=1,1
    !do ipff=91,91
        ipf=ipff-1
        bt=0.0d0
```

```
          !$OMP PARALLEL FIRSTPRIVATE(itff,aTF,CET,CEP)
          !$OMP DO
          do itff=1,361,2
              aTF=(itff-181)
              call field(aTF,IPF,CET,CEP)
              et(itff )=cdabs(CET)
              ep(itff )=cdabs(CEP)
              if(et(itff )>bt) then
                  bt=et(itff )
              end if
              if(ep(itff )>bt) then
                  bt=ep(itff )
              end if
              print *,atf,ipf,20.0d0*dlog10(et(itff )),
20.0d0*dlog10(ep(itff ))
          end do
          !$OMP END DO NOWAIT
          !$OMP END PARALLEL
          print*,'bt=',bt,'bt in db=',20.0d0*dlog10(bt)
          open(unit=19,file="./Elog.txt")!!!
          open(unit=18,file="./E.txt")!!!
          do ii=1,361,2
              ai=(ii-181)
              print*,ai,ipf,20.0d0*dlog10(et(ii)/
bt),20.0d0*dlog10(ep(ii)/bt)
              write(19,*)ai,20.0d0*dlog10(et(ii)/
bt),20.0d0*dlog10(ep(ii)/bt)
              write(18,*)ai,(et(ii)),(ep(ii))
          end do
          close(18,status='SAVE')!!!
          close(19,status='SAVE')!!!
      end do

      stop
end program neartofar

!--------------------------------------------------------
!--------------------------------------------------------

module MM
    complex(8) CMP(361,181),CMT(361,181)
    complex(8) ch(1000),chd(1000)
    real(8) pi,qw,aka
```

```fortran
    integer maxM, numi
end module MM

!--------------------------------------------------
!--------------------------------------------------

subroutine field(aTF,IPF,CET,CEP)
    use MM
    implicit none
    integer,intent(in)::aTF,IPF
    complex(8),intent(out)::CET,CEP

    real(8) P(1000)
    real(8) CTF,STF,AQ,ACP,ASP,dlt,dlp,X,err
    real(8) ea1,ea2,eb1,eb2,ec1,ec2,ed1,ed2,CTH,STH
    real(8) o1,o2,o3,o4,FF1,FF2
    integer IPSS,IPS,ITSS,ITS,n
    complex(8) csum1,csum2,c1,c2,c3,c4,ct,ca,cb
    complex(8) chn,chnd,cd1,cd2

    err=1.0d-12

    CTF=dcos(qw*aTF)
    STF=dsin(qw*aTF)
    CET=0.0d0
    CEP=0.0d0
    dlt=qw
    dlp=qw
    do IPSS=1,360,numi
        IPS=IPSS-1
        AQ=dfloat(IPF-IPS)
        ACP=dcos(qw*AQ)
        ASP=dsin(qw*AQ)
        do ITSS=1,180,numi
            ITS=ITSS-1
            CTH=dcos(qw*dfloat(ITS))
            STH=dsin(qw*dfloat(ITS))
            X=CTF*CTH+STF*STH*ACP
            CALL LEG(X,P)
            ea1=(-CTF*STH+STF*CTH*ACP)*(-STF*CTH
+CTF*STH*ACP)
            ea2=STF*STH+CTF*CTH*ACP
            eb1=(-STF*CTH+CTF*STH*ACP)*(STF*ASP)
```

```
                eb2=CTF*ASP
                ec1=(-STF*STH*ASP)*(ASP)
                ec2=ACP
                ed1=(-CTF*STH+STF*CTH*ACP)*(STH*ASP)
                ed2=CTH*ASP
                ca=CMT(IPSS,ITSS)
                cb=CMP(IPSS,ITSS)
                csum1=0.0d0
                csum2=0.0d0
                do1:do n=1,maxM
                    if (cdabs(ca)==0.0d0 .and. cdabs(cb)
==0.0d0) then
                        exit do1
                    end if
                    call FF(n,x,P,FF1,FF2)
                    o1=FF2*eb1+FF1*eb2
                    o2=FF2*ea1+FF1*ea2
                    o3=FF2*ed1+FF1*ed2
                    o4=FF2*ec1+FF1*ec2
                    c1=ca*o2+cb*o1
                    c2=ca*o4+cb*o3
                    c3=-ca*o3+cb*o4
                    c4=-ca*o1+cb*o2
                    chn=aka*ch(n+1)
                    chnd=aka*chd(n+1)+ch(n+1)
                    ct=(0.0d0,1.0d0)**n*(2.0d0+1.0d0/dfloat(n))
                    cd1=ct*(c1/chnd+(0.0d0,1.0d0)*c2/chn)
                    cd2=ct*(c3/chnd+c4*(0.0d0,1.0d0)/chn)
                    csum1=csum1+cd1
                    csum2=csum2+cd2
                    if(cdabs(csum2)<err .and. cdabs(csum1)
                        exit do1
                    end if
                if(cdabs(csum2)<err .and. cdabs(cd1/csum1)
<err) then
                        exit do1
                    end if
                if(cdabs(csum1)<err .and. cdabs(cd2/csum2)
<err) then
                        exit do1
                    end if
            if(cdabs(cd1/csum1)<err .and. cdabs(cd2/csum2)
<err) then
                        exit do1
                    end if
```

```fortran
            end do do1
        if (n==maxM) then
            print *,n,'maxM'
        end if
        CEP=CEP+sth*csum2*dlt*dlp
        CET=CET+sth*csum1*dlt*dlp
        end do
    end do
    return
end subroutine


!--------------------------------------------------
!--------------------------------------------------


subroutine FF(n,x,P,FF1,FF2)
    use MM
        implicit none
    integer,intent(in)::n
    real(8),intent(in)::x
    real(8),intent(in)::P(1000)
    real(8),intent(out)::FF1,FF2

    if ((1.0d0-dabs(x))
        FF1=n*P(n+1)/(2.0d0*x)
        FF2=n*(n-1)*(n+2)*P(n+1)/(12.0d0*x*x-4.0d0)
    else
        FF1=1.0d0/(1.0d0-x**2)*(x*P(n+1)-P(n+2))
        FF2=1.0d0/(1.0d0-x**2)**2*(P(n+1)*((2.0d0+n)
*x**2-n)-2.0d0*x*P(n+2))
    end if

    return
end subroutine FF


!--------------------------------------------------
!--------------------------------------------------


subroutine LEG(Z,P)
    use MM
        implicit none
    real(8),intent(in)::Z
    real(8),intent(out)::P(1000)
    integer II,N
    real(8) A1,A2
    P(1)=1.0d0
```

```
    P(2)=Z
    P(3)=0.5d0*(3.0d0*Z*Z-1.0d0)
    P(4)=0.5d0*Z*(5.0d0*Z*Z-3.0d0)
    do II=5,(maxM+2)
        N=II-2
        A1=dfloat(2*N+1)
        A2=dfloat(N+1)
        P(II)=(P(II-1)*Z*A1-dfloat(N)*P(II-2))/A2
        !print *,N+1,P(II),Z
    end do
return
end subroutine LEG

!----------------------------------------------------
!----------------------------------------------------

subroutine BES(X)
    use MM
        implicit none
    real(8),intent(in)::X
    integer I,N
    real(8) AJ,AY

    AJ=dsin(X)/X
    AY=-dcos(X)/X
    ch(1)=dcmplx(aj,-ay)
    AJ=dsin(X)/(X*X)-dcos(X)/X
    AY=-dcos(X)/(X*X)-dsin(X)/X
    ch(2)=dcmplx(AJ,-AY)
    chd(1)=-ch(2)

    DO I=2,(maxM+2)
        N=I-1
        ch(I+1)=dfloat(2*N+1)/X*ch(I)-ch(I-1)
    end do

    DO I=2,(maxM+2)
        N=I-1
        chd(I)=(dfloat(N)*ch(I-1)-dfloat(N+1)*ch(I+1))/
dfloat(2*N+1)
    end do

    return
end subroutine BES
```

References

1 R. Laroussi and G. I. Costache, "Far-Field Predictions from Near-Field Measurements Using an Exact Integral Solution," *IEEE Transactions on Electromagnetic Compatibility*, Vol. 36, pp. 189–195, 1994.

2 J. J. H. Wang, "An Examination of the Theory and Practices of Planar Near-Field Measurements," *IEEE Transactions on Antennas and Propagation*, Vol. 36, No. 6, pp. 746–753, 1988.

3 A. J. Poggio and E. K. Miller, "Integral Equation Solutions of Three Dimensional Scattering Problems," *Computer Techniques for Electromagnetics*, R. Mittra, editor, Elsevier, 1973.

4 A. C. Ludwig, "Near-Field Far-Field Transformations Using Spherical Wave Expansions," *IEEE Transactions on Antennas and Propagation*, Vol. 19, No. 2, pp. 214–220, 1971.

5 R. F. Harrington, *Time Harmonic Electromagnetic Fields*, McGraw-Hill, New York, 1961.

6 J. A. Stratton, *Electromagnetic Theory*, McGraw-Hill, New York, 1941.

7 L. L. Bailin and S. Silver, "Exterior Electromagnetic Boundary Value Problems for Spheres and Cones," *IRE Transactions on Antennas and Propagation*, Vol. 4, No. 1, pp. 5–16, 1956. See correction by the authors in *IRE Transactions on Antennas and Propagation*, Vol. 5, No. 3, p. 313, 1957.

8 C. Stubenrauch, NIST, Private communication.

9 A. Taaghol and T. K. Sarkar, "Near Field to Near/Far Field Transformation for Arbitrary Near Field Geometry Utilizing an Equivalent Magnetic Current," *IEEE Transactions on Electromagnetic Compatibility*, Vol. 38, No. 3, pp. 536–548, 1995.

10 J. E. Hansen, J. Hald, F. Jensen, F. H. Larsen, and J. R. Wait, *Spherical Near-field Antenna Measurements*. Institution of Engineering and Technology (IET), London, England, 1988.

11 T. K. Sarkar, P. Petre, A. Taaghol, and R. F. Harrington, "An Alternative Spherical Near Field to Far Field Transformation," *Progress in Electromagnetics Research*, Vol. PIER 16, pp. 269–284, 1997.

12 P. Petre and T. K. Sarkar, "Planar Near-Field to Far Field Transformation Using an Equivalent Magnetic Current Approach," *IEEE Transactions on Antennas and Propagation*, Vol. 40, pp. 1348–1356, 1992.

13 F. Las-Heras, "Radial Field Retrieval for Current Reconstruction from Spherical Acquisition," *Electronics Letters*, Vol. 36, No. 10, pp. 867–869, 2000.

14 TICRA Engineering Consultants, "SNIFTD: Software Package for Spherical Near-Field Far-Field Transformations with Full Probe Correction," *Kron Prinsens Gade 13*, DK-1114 Copenhagen K, Denmark.

8

Deconvolving Measured Electromagnetic Responses

Summary

Two deconvolution techniques is presented to illustrate how the ill-posed deconvolution problem has been regularized. Depending on the nature of regularization utilized based on the given data one can obtain a reasonably good approximate solution. The two techniques presented here have the built in self-regularizing schemes. This implies that the regularization process, which depends highly on the data, can be automated as the solution procedure continues. The first method is based on solving the ill-posed deconvolution problem by the iterative conjugate gradient method. The second method uses the method of total least squares implemented through the singular value decomposition (SVD) technique. The methods have been applied to measured data to illustrate the nature of their performance.

8.1 Introduction

For a given system, if the input to the system is represented by $x(t)$ and the output of the system is represented by $y(t)$ and the system impulse response is given by $h(t)$, then by the terms of the problem,

$$y(t) = \int_{-\infty}^{\infty} x(\tau) h(t-\tau) d\tau = \int_{-\infty}^{\infty} x(t-\tau) h(\tau) d\tau \; y(t) \qquad (8.1)$$

Deconvolution [1–4] is referred to obtaining the impulse response $h(t)$ from a given input $x(t)$ and the output $y(t)$.

To show that this problem is ill-posed [5, 6] consider two system impulse responses $h_1(t)$ and $h_2(t)$ of the same system. Let the exact impulse response be

$$h_1(t) = h(t) \qquad (8.2)$$

Modern Characterization of Electromagnetic Systems and Its Associated Metrology, First Edition.
Tapan K. Sarkar, Magdalena Salazar-Palma, Ming Da Zhu, and Heng Chen.
© 2021 John Wiley & Sons, Inc. Published 2021 by John Wiley & Sons, Inc.

and a perturbed impulse response be

$$h_2(t) = h(t) + C \sin \omega t \tag{8.3}$$

where the amplitude C of the sinusoid and its frequency ω can be chosen arbitrarily. Now if we consider the difference in the outputs $y_2(t)$ and $y_1(t)$ due to the two different system impulse responses $h_2(t)$ and $h_1(t)$ for the same input to the system x(t), then

$$y_2(t) - y_1(t) = \int_{-\infty}^{\infty} x(t-\tau)\left[h_2(\tau) - h_1(\tau)\right] d\tau = C \int_{-\infty}^{\infty} x(t-\tau) \sin \omega \tau \, d\tau \tag{8.4}$$

If we consider the input to be realizable and hence bounded, then

$$|x(t)| \leq M \text{(a constant independent of } t) \tag{8.5}$$

Furthermore, if we consider a real system then the upper limit of the integration cannot be ∞ but some finite value T. In addition, if the system is physical, then it must be causal, i.e.,

$$h(t) = 0 \quad \text{for} \quad t < 0 \tag{8.6}$$

and so the lower limit in (8.4) is not $-\infty$ but 0. Hence (8.4) becomes under a constant input $x(t) = M$

$$|y_2(t) - y_1(t)| \leq C M \int_0^T \sin \omega \tau \, d\tau \leq \frac{2 C M}{\omega} \tag{8.7}$$

From (8.7) we conclude that, by selecting ω to be sufficiently large, the difference between $y_2(t)$ and $y_1(t)$ can be made arbitrarily small. The ill-posedness [6] of this example is evidenced by the fact that small difference in y can map into large differences in h. This is a serious problem because, in practice measurement of y will be accompanied by a nonzero measurement error (or representation error in a finite dimensional digital system). Use of the "noisy data" can yield solutions for the impulse responses $h_2(t)$ and $h_1(t)$ which may be significantly different from the desired solution. For example if $M = 0.5$ and $\omega = 10^6$ and $C = 1$, then the difference between

$$|y_2(t) - y_1(t)| \leq 10^{-6} \tag{8.8}$$

This implies, that one can have totally different impulse responses, but the output from each of them due to the same input may be nearly equal. The question is: what is the mathematical problem? This has been explained in [6].

In summary, one can represent (8.1) symbolically as an operator equation

$$\mathbf{X}\,\mathbf{h} = \mathbf{Y} \tag{8.9}$$

where the unknown $\mathbf{h}$ is to be solved for the given right hand side $\mathbf{Y}$ and the operator $\mathbf{X}$. Let us expand $\mathbf{h}$ in terms of the eigenfunctions of the convolution operator $\mathbf{X}$. An operator may not have eigenvalues and eigenvectors for an infinite dimensional space, but all operators have eigenvalues and eigenvectors in a finite dimensional space. If λ_i and $\mathbf{e}_i$ are the respective eigenvalues and eigenfunctions of the operator $\mathbf{X}$, then

$$\mathbf{X}\,\mathbf{e}_i = \lambda_i\,\mathbf{e}_i \tag{8.10}$$

and one can write an expansion for the unknown $\mathbf{h}$ by

$$h(t) = \sum_{j=1}^{N} \alpha_j\, e_j\,(t) \tag{8.11}$$

where $e_j(t)$ are the known eigenfunctions with the independent variable t and α_j are the unknown constants to be solved for. Equation (8.11) assumes that the solution $h(t)$ can be adequately represented by a linear sum of N eigenfunctions $e_j(t)$.

By substituting (8.11) into (8.9), one obtains

$$\mathbf{X}\,\mathbf{h} = \sum_{j=1}^{N} X(\alpha_j\, e_j) = \sum_{j=1}^{N} \alpha_j X(e_j) = \sum_{j=1}^{N} \alpha_j\, \lambda_j\, e_j = \mathbf{Y} \tag{8.12}$$

Multiplying both sides of (8.12) by the eigenvectors $e_k(t)$ and taking the usual Hilbert inner product over the interval 0 to T over which the eigenfunctions $e_k(t)$ are orthogonal one obtains

$$< \mathbf{X}\,h;\, e_k > = \sum_{j=1}^{N} \alpha_j\, \lambda_j < e_j;\, e_k > = < \mathbf{Y};\, e_k > \tag{8.13}$$

If we repeat (8.13) for all the N eigenfunctions, $k = 1, \ldots N$, then one obtains a set of N equations. Since

$$< e_j;\, e_k > = \int_{0}^{T} e_j\,(t)\, e_k(t)\, dt = \partial_{jk} \tag{8.14}$$

where

$$\partial_{jk} = \begin{cases} 1 & \text{for } j = k \\ 0 & \text{for } j \neq k \end{cases} \tag{8.15}$$

Then

$$\alpha_k \lambda_k = \ <Y; e_k> \tag{8.16}$$

from which the unknown α_k is obtained as

$$\alpha_k = \frac{<Y; e_k>}{\lambda_k} \tag{8.17}$$

and the complete solution is explicitly given by

$$h(t) = \sum_{j=1}^{N} \frac{<Y; e_j>}{\lambda_j} \ e_j \tag{8.18}$$

However, the problem is as follows: Since the operator X is square integrable and bounded, i.e.,

$$\int_0^T dt \int_0^T d\tau \ x^2(t - \tau) \leq W \ [\text{a constant independent of } T \text{ and } x(t)] \tag{8.19}$$

the convolution operator X is Hilbert-Schmidt which implies that its eigenvalues has a limit point of zero. This implies that as $k \to \infty$,

$$\lim_{k \to \infty} \lambda_k \to 0 \tag{8.20}$$

If λ_k goes to zero the sum in (8.18) starts blowing up for the small eigenvalues and hence the solution of the problem becomes unstable. So the key step in solving (8.18) is to truncate the sum before the contribution from the small eigenvalues enters into the picture. If the truncation process can be accomplished in an efficient fashion one would have a stable solution [6].

Various deconvolution methods essentially attempts to truncate the sum in (8.18) in many different ways. Since the eigenvalues are data dependent the truncation process cannot be determined a priori as the results will depend significantly on the data.

Many books [1–4] exist in the published literature that deal with this deconvolution problem. These methods regularize the ill-pose problem in various ways but are not quite amenable to a complete automatic solution to the problem without any manual intervention. In this section, two techniques will be presented that solve the deconvolution problem. The unique feature for these techniques are that they can be somewhat automated and an a priori criteria can be established on how to truncate the sum in (8.18). The first technique to be described is the conjugate gradient method applied to solve the operator equation defined in (8.1) and the second technique is the singular value decomposition based on the total least squares approach of solving the discretized version of (8.1).

8.2 The Conjugate Gradient Method with Fast Fourier Transform for Computational Efficiency

8.2.1 Theory

In this method, the basic philosophy is that instead of solving for $\mathbf{h} = \mathbf{X}^{-1}\mathbf{Y}$ directly, the following functional $\mathbf{F}(\mathbf{h})$ is minimized which is given by

$$\mathbf{F}(\mathbf{h}) = \; < \mathbf{R}; \mathbf{R} > \; = \; < \mathbf{Xh} - \mathbf{Y}, \mathbf{Xh} - \mathbf{Y} > \tag{8.21}$$

where $\mathbf{X}$ is the convolution operator

$$\mathbf{X} = \int_0^\infty x\,(t-\tau)\,(\bullet)\,d\tau \tag{8.22}$$

and the inner product is defined as

$$< F; G > \; = \int_0^\infty F(t)\,G(t)\,d\,t \tag{8.23}$$

and the norm of G is given by

$$\|G\|^2 = \; < G; G > \; = \int_0^\infty |G(t)|^2\,dt \tag{8.24}$$

The conjugate-gradient method starts with an initial guess $\mathbf{h}_0$ and generates

$$\mathbf{P}_0 = -\,\mathbf{b}_0\,\mathbf{X}^*(\mathbf{X}\,\mathbf{h}_0 - \mathbf{Y}) \tag{8.25}$$

where $\mathbf{X}^*$ represents the adjoint operator. With reference to (8.25), the adjoint operator $\mathbf{X}^*$ is the advanced convolution operator, which is defined as

$$\mathbf{X}^*\mathbf{z} = \int_0^\infty x\,(t-\tau)\,z(t)\,dt = \int_0^\infty x\,(t)\,z(t+\tau)\,dt \tag{8.26}$$

In defining (8.26), it has been assumed that $x(t)$ is causal. The conjugate-gradient method then develops according to the following:

$$\mathbf{h}_{k+1} = \mathbf{h}_k + \alpha_k\,\mathbf{P}_k \tag{8.27}$$

$$\mathbf{R}_{k+1} = \mathbf{R}_k + \alpha_k\,\mathbf{X}\mathbf{P}_k \tag{8.28}$$

$$\alpha_k = \frac{1}{\|\mathbf{X}\,\mathbf{P}_k\|^2} \tag{8.29}$$

$$\mathbf{P}_{k+1} = \mathbf{P}_k - b_{k+1} \mathbf{X}^* \mathbf{R}_{k-1} \qquad (8.30)$$

$$b_k = \frac{1}{\|\mathbf{X}^* \mathbf{R}_k\|^2} \qquad (8.31)$$

The above procedure, from (8.25) - (8.31), is somewhat different from the conventional Hestenes and Stiefel algorithms [7, 8]. The above method brings about a certain saving in memory. This is because in [7] one needs to store the four vectors $\mathbf{h}_k$, $\mathbf{P}_k$, $\mathbf{R}_k$ and $\mathbf{XP}_k$; whereas in the modified presented algorithm it is not necessary to store $\mathbf{XP}_k$.

The numerical computation is done in the following way. One starts with an array of elements for $\mathbf{h}_0$, which the initial guess for the solution to this problem. One next convolves $\mathbf{h}_0$ with $\mathbf{X}$, to yield another linear array, say $\mathbf{Q}$. Then subtract $\mathbf{Y}$ from the array $\mathbf{Q}$, resulting in $\mathbf{R}_0$. The computation of the convolution can be speeded up utilizing the fast Fourier transform (FFT) [9]. To utilize the FFT, it is necessary to pad the arrays $\mathbf{h}_0$ and $\mathbf{X}$ with an approximately equal number of zeros. Then take the FFT of both $\mathbf{h}_0$ and $\mathbf{X}$, and multiply them. Then take the inverse FFT to yield the array $\mathbf{B}$. Truncate the array $\mathbf{B}$ to form the array $\mathbf{Q}$, so that $\mathbf{Q}$ has a dimension less than $\mathbf{X}$, as the original array was padded up with zeros before taking FFT.

Next, the advance convolution between $\mathbf{X}$ with $\mathbf{R}_0$, is performed, resulting in the array $\mathbf{X}^* \mathbf{R}_0$. Again the FFT can be utilized to speed up the computation. The remaining computations are done in the fashion as described earlier.

The conjugate-gradient method always converges for any initial guess and for any functional equation for a bounded operator as long as (8.25) - (8.31) are implemented. The solution is unique as the local minimum of the functional is also the global one as the functional is quadratic. The rate of convergence is given in [7, 10]. Even when $\mathbf{Y}$ is not in the range of the operator $\mathbf{X}$ (this may happen when $\mathbf{Y}$ is contaminated by noise), the conjugate-gradient method still yields the minimum norm solution. Another important question in this numerical technique is how to terminate the iteration. Since the conjugate gradient method terminates in at most M steps (in absence of round-off error), where M is the number of independent eigenvalues of the operator in the space in which the problem is being solved. The conjugate gradient method seeks the solution, at each iteration, along the direction of the eigenvectors corresponding to first large and then the small eigenvalues in descending order. For an ill-posed problem, there are some small eigenvalues due to noise.

Therefore, to obtain an acceptable solution, it is necessary to stop the iterative process before the method seeks a solution along the direction of the eigenvector corresponding to small eigenvalues. Hence, too few iterations may give a solution that is not good enough, and for ill-posed problems, too many iterations might introduce spurious oscillations in the solution as the method seeks the solution along the eigenvectors corresponding to the small noise eigenvalues. Hence the conjugate gradient method essentially performs a singular value decomposition without actually computing the eigenvalues and the criteria that one utilizes to stop the iterative process is that the magnitude of the largest element in the residual must be below a certain value, that is [10],

$$|\mathbf{R}_k(t)| \leq W \text{ (a prefixed constant)} \tag{8.32}$$

The method of choosing W is now discussed. For most problems of interest, the elements of the input and output are accurate up to the quantization errors associated with each sample of $x(t)$ and $y(t)$. Assume that the quantization errors $\Delta x(t_0)$ and $\Delta y(t_0)$ are independent, t_0 being the sampling time. Further assume that

$$|\Delta x(t_n)| \leq \alpha \tag{8.33}$$
$$|\Delta y(t_n)| \leq \alpha \tag{8.34}$$

where α is a constant dependent on the number of bits used in the quantization process to represent the data. In the current experiments, eight-bit quantization has been used in recording the digital form of the signal. Hence in our computations, the effective number of bits ε_p to chosen to be eight, which then yields the constant α as

$$\alpha = 2^{-\varepsilon_p} = 2^{-8} = 0.00390625 \tag{8.35}$$

Now observe that the error in the solution after n-iteration would be

$$\mathbf{R}_n(t) = \int_0^\infty x(t-\tau)\, h_n(\tau)\, d\tau - y(t) \tag{8.36}$$

As $n \to \infty$, one would have a different unstable solution other than the exact solution due to the accumulation of the discretization errors. Let us term the true impulse response by $h_0(t)$. Now define a discretization error $\mathbf{R}_D$ due to sampling and express it as

$$\mathbf{R}_D(t) = \int_0^\infty [x(t-\tau) - \Delta x(t-\tau)]\, h_e(\tau)\, d\tau - y(t) - \Delta y(t)$$
$$\simeq \int_0^\infty \Delta x(t-\tau)\, h_e(\tau)\, d\tau - \Delta y(t) \leq \left\{ \int_0^\infty |h(\tau)|\, d\tau + 1 \right\} \tag{8.37}$$

$h(\tau)$ is now replaced by $h_n(\tau)$ in (8.37) and check at each iteration whether

$$\mathbf{R}_D(t) \leq \alpha \left\{ \int_0^T |h_n(\tau)|\, d\tau + 1 \right\}, \text{ for all } t \tag{8.38}$$

i.e., the residuals $\mathbf{R}_n(t)$ are of the same order of magnitude as the discretization errors is satisfied.

Then check at each iteration whether

$$\alpha \left\{ \int_0^T |h_n(\tau)| \, d\tau + 1 \right\} \leq |\mathbf{R}_n(t)| \qquad (8.39)$$

It was shown by Oettli [10] that the above inequality is a necessary and sufficient condition for $h(t)$ to be a solution of (8.3). Note an upper limit is defined in (8.38) by T. This is because in practice the data is recorded up to a finite time T.

8.2.2 Numerical Results

Consider determination of the impulse response of a conducting sphere of 1.5 inch in diameter. The incident electric field on the sphere is denoted by $x(t)$ which is obtained from the scattering from a plane conducting sheet and $x(t)$ is shown in Figure 8.1. The measured response from the target is presented in Figure 8.2. This is $y(t)$ [11].

It is seen that both signals are relatively noise free. The conjugate gradient method is now applied to the set $x(t)$ and $y(t)$ to obtain the impulse response of the sphere. With $\varepsilon_p = 8$ in (8.35), it took eight iteration (for 256 data samples for the input $x(t)$ and the output $y(t)$ waveforms) to obtain the impulse response $h(t)$ shown in Figure 8.3. It is noticed that the deconvolved response has a rather noisy tail. This is because the signal-to-noise ratio deteriorates as one progresses in time. In order to minimize the effect of the noisy tail of the waveform, the definition of the inner product is changed by introducing an exponential weighting. Since the impulse response exists over the first 300 picoseconds of the data it is necessary to modify the error minimization criterion by defining

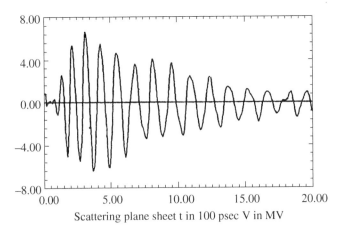

Figure 8.1 Transient waveform scattered from a plane conducting sheet.

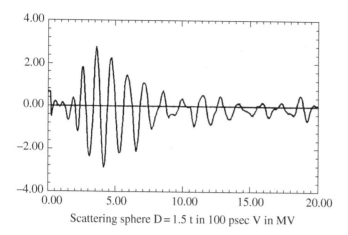

Figure 8.2 Transient waveform scattered from a plane conducting sphere of 1.5 inch in diameter.

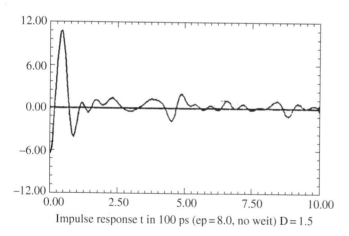

Figure 8.3 Impulse response of a conducting sphere of 1.5 inch in diameter with $\varepsilon_p = 8.0$.

a weighted inner product in (8.25)-(8.31). Hence the modified error criterion is [11, 12].

$$< \mathbf{R}_n; \mathbf{R}_n > \; = \; \int_0^T \mathbf{R}_n \, (t) \, \mathbf{R}_n(t) \, \exp \, (-t/t_0) \, dt; T = 2 \text{ nanoseconds}$$

$$(8.40)$$

where $t_0 = 500$ picoseconds. The exponential weighting function tends to suppress the effect of the noisy tail of the evolving impulse response on the deconvolution process. With this modified functional, the conjugate gradient method is applied to $x(t)$ and $y(t)$ to obtain the impulse response for an effective number of bits $\varepsilon_p = 8$, which required nine iterations for a 256 point discretized signal. The impulse response is shown in Figure 8.4 and its spectra in Figure 8.5. The resonant frequencies are obtained as 6.6 and 9.0 GHz which are close to the theoretical values of 6.9 and 9.3 GHz, respectively.

This algorithm has also been implemented in hardware for real time processing [13–15].

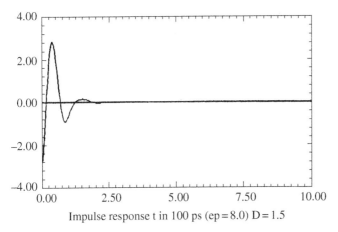

Impulse response t in 100 ps (ep = 8.0) D = 1.5

Figure 8.4 Impulse response of a conducting sphere of 1.5 inch diameter with ε_p= 8.0 and using a weighted inner product.

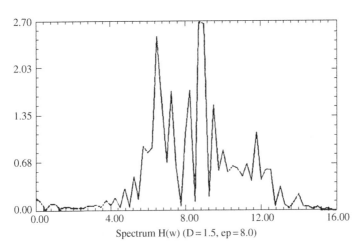

Spectrum H(w) (D = 1.5, ep = 8.0)

Figure 8.5 Spectrum of the impulse response of a conducting sphere of 1.5-in diameter with ε_p= 8.0 and using the exponentially weighted inner product.

8.3 Total Least Squares Approach Utilizing Singular Value Decomposition

8.3.1 Theory

For a linear, time invariant, causal discrete system with a zero initial state, the relation between input $x(t)$, $t = 1, 2, ..., m$, impulse response $h(t)$, $t = 1, 2, ... n$, and the output $y(t)$, $t = 1, 2, ..., m$ is given by [16];

$$y(t) = \sum_{k=1}^{n} x(t - k - 1) h(k) \tag{8.41}$$

If the impulse response $h(t)$ and input $x(t)$ of a system are known, then one can compute its response $y(t)$ from (8.41). Equation (8.41) can be put into the following matrix form:

$$\mathbf{A}\,\mathbf{h} = \mathbf{B} \tag{8.42}$$

where

$$\mathbf{B} = \begin{bmatrix} y(1) \\ y(2) \\ \vdots \\ y(m) \end{bmatrix}_{m \times 1} ; \quad \mathbf{h} = \begin{bmatrix} h(1) \\ h(2) \\ \vdots \\ h(n) \end{bmatrix}_{n \times 1} ; \quad \mathbf{A} = \begin{bmatrix} x(1) & 0 & \cdots & 0 \\ x(2) & x(1) & \cdots & 0 \\ \vdots & \vdots & \ddots & \vdots \\ x(n) & x(n-1) & \cdots & x(1) \\ \vdots & \vdots & \ddots & \vdots \\ x(m) & x(m-1) & \cdots & x(m-n-1) \end{bmatrix}_{m \times n}$$

$$\tag{8.43}$$

Deconvolution is the process of finding the impulse response vector $\mathbf{h}$ from the known values of the input $x(t)$ constituting the matrix $\mathbf{A}$ and using the output vector $\mathbf{B}$ containing the samples of the output. The discrete deconvolution problem is essentially reduced to the problem of solving a set of linear equations given by (8.42). Matrix $\mathbf{A}$ is highly ill-conditioned [6].

Conventional methods for solving (8.42) has been the method of least squares (LS). In the solution of a problem by LS, a data matrix $\mathbf{A}$ (m by n), an observation vector $\mathbf{B}$ (m by 1), and a nonsingular diagonal matrix $\mathbf{D} = \text{diag}(d_1, d_2, ..., d_m)$ are given. When there are typically more equations than unknowns, i.e., $m > n$, and so the set is overdetermined. Unless $\mathbf{B}$ belongs to the range space of $\mathbf{A}$, the overdetermined set has no exact solution and is therefore denoted by $\mathbf{Ah} = \mathbf{B}$ [6]. The unique minimal 2-norm solution to the LS problem is then given by

$$\|\mathbf{D}\,(\mathbf{B} - \mathbf{Ah})\|_2 = \min \tag{8.44}$$

where $\| \;\|_2$ indicates the Euclidean length of the vector. Solution to (8.44) is equivalent to solving

$$\mathbf{A}^T \mathbf{D}^2 \mathbf{A} \, \mathbf{h}_{L.S.} = \mathbf{A}^T \mathbf{D}^2 \, \mathbf{B} \tag{8.45}$$

or equivalently

$$\mathbf{h}_{L.S.} = (\mathbf{DA})^\dagger \mathbf{DB} \tag{8.46}$$

where † denotes the psuedoinverse of a matrix. The assumption in (8.46) is that the errors are confined only to the "observation" vector $\mathbf{B}$.

One can reformulate the ordinary LS problem as follows: Determine that $\mathbf{h}_{L.S.}$ which will satisfy

$$\mathbf{A} \, \mathbf{h}_{L.S.} = \mathbf{B} + \mathbf{T} \tag{8.47}$$

and for which

$$\text{minimize } \|\mathbf{D}\,\mathbf{T}\|_2 \\ \text{subject to } \mathbf{B} + \mathbf{T} \in \text{Range Space of } \mathbf{A} \tag{8.48}$$

The underlying assumption in the solution of the ordinary least squares problem is that the errors (equivalently noise) only occur in the observation vector $\mathbf{B}$ and that the data matrix $\mathbf{A}$ is exactly known. Often this assumption is not realistic because of sampling, modeling, or measurement errors affecting matrix $\mathbf{A}$. One way to take errors in matrix $\mathbf{A}$ into account is to introduce perturbations in $\mathbf{A}$ and solve the following problem as outlined in the next section.

8.3.2 Total Least Squares (TLS)

In the total least squares (TLS) problem [17–23] there are perturbations of both the observation vector $\mathbf{B}$ and the data matrix $\mathbf{A}$ as illustrated in Chapter 1. Consider the TLS problem as the problem of determining the solution $\hat{\mathbf{h}}$, which satisfies

$$(\mathbf{A} + \mathbf{E})\hat{\mathbf{h}} = \mathbf{B} + \mathbf{T} \qquad \text{with} \quad \mathbf{E} \in \mathbf{C}^m \tag{8.49}$$

and it is the solution to the following problem [18]

$$\text{minimize} \left\| \mathbf{D} \begin{bmatrix} \mathbf{E} \vdots \mathbf{T} \end{bmatrix} \mathbf{S} \right\|_F \text{ subject to } (\mathbf{B} + \mathbf{T}) \in \text{Range space of } (\mathbf{A} + \mathbf{E}) \tag{8.50}$$

where $[\mathbf{E} \vdots \mathbf{T}]$ represents the matrix $\mathbf{E}$ augmented by the vector $\mathbf{T}$ and the non-singular weighting matrices are known and are given by

$$\mathbf{D} = \text{diag}\,(d_1, d_2, \ldots, d_m) \quad \text{with } d_i > 0, \;\; i = 1, \ldots, m$$
$$\mathbf{S} = \text{diag}\,(s_1, s_2, \ldots, s_{n+1}) \quad \text{with } s_i > 0, \;\; i = 1, \ldots, n+1$$

and $\| \ \|_F$ denotes the Forbenius norm; i.e.,

$$\|S\|_F^2 = \sum_i \sum_j |s_{ij}|^2 \tag{8.51}$$

In the TLS problem unlike the least square problems (LS), the vector **B** or its estimation does not lie in the range space of matrix **A**. Consider the matrix $\mathbf{C}_{m \times (n+1)}$ as

$$\mathbf{C} = [\mathbf{A} \vdots \mathbf{B}] \tag{8.52}$$

such that matrices **D** and **S** are identity matrices. The SVD of matrix C can be written as [19]

$$\mathbf{C} = \mathbf{U} \sum \mathbf{V}^H$$

or

$$\mathbf{U}^H \mathbf{C} \mathbf{V} = \sum = \text{diag}\left(\sigma_1, \ldots, \sigma_p\right) \text{ with } \sigma_1 \geq \sigma_2 \geq \ldots \geq \sigma_p \geq 0$$
$$\text{and } p = \min\ (m, n+1) \tag{8.53}$$

where the superscript H denotes Hermitian (conjugate transpose) of a matrix. Both, $\mathbf{U}_{m \times m}$ and $\mathbf{V}_{(n+1) \times (n+1)}$ are unitary matrices, and $\sum$ is $m \times (n+1)$ and is "diagonal" in that $<\Sigma>_{ij} = 0$ if $i \neq j$. When $i = j$ element of the matrix then $<\Sigma>_{ij} = \sigma_i$ with σ_i real and non-negative. **V** contains the eigenvectors of $\mathbf{C}^H\mathbf{C}$, σ_i are the eigenvalues of $\mathbf{C}^H\mathbf{C}$ and **U** contains the eigenvectors of $\mathbf{C}\mathbf{C}^H$.

The decomposition of ($^{8.53}$) is called the singular value decomposition (SVD). The σ_i are called the singular values of **C**, $\mathbf{u}_i$ - the associated left singular vectors of **C**, and the $\mathbf{v}_i$ - the associated right singular vectors of **C**. The various components are related by

$$\mathbf{C}\,\mathbf{v}_i = \sigma_i\,\mathbf{u}_i \tag{8.54}$$

Furthermore,

$$\mathbf{U} = [\mathbf{u}_1, \ \mathbf{u}_2, \ldots \ldots, \mathbf{u}_m] \ \in \mathbf{C}^{m \times m} \tag{8.55}$$

$$\mathbf{V} = [\mathbf{v}_1, \ \mathbf{v}_2, \ldots \ldots, \mathbf{v}_{n+1}] \ \in \mathbf{C}^{(n+1) \times (n+1)} \tag{8.56}$$

$$\sum = \text{diag}\left(\sigma_1, \ldots, \sigma_p\right) \ \in \mathbb{R}^{m \times (n+1)} \text{ with } \sigma_1 \geq \sigma_2 \geq \ldots \geq \sigma_p \geq 0 \tag{8.57}$$

$$\mathbf{U}^H\mathbf{U} = \mathbf{I}_{m \times m} \quad \text{and} \quad \mathbf{V}^H\mathbf{V} = \mathbf{I}_{(n+1) \times (n+1)} \tag{8.58}$$

where $\mathbf{I}_{m \times m}$ and $\mathbf{I}_{(n+1) \times (n+1)}$ are identity matrices of different dimensions.

Let the rank of matrix **A** in equation (8.47) be $r < n$ (**A** is rank-deficient), the best r-dimensional subspace approximation of range $\{[\mathbf{A}\vdots\mathbf{B}]\}$ can be found by using the orthogonal projection operator $P_r^u = \sum_{i=1}^{r} u_i u_i^H$ as [23, 24] where the goal is to

$$\underset{u_1, u_2, \ldots\ldots, u_r}{\text{maximize}} \quad \sum_{i=1}^{n+1} \left\| P_r^u c_i \right\|_2^2 \quad \text{subject to} \quad u_i^H u_j = \partial_{ij} \text{ for } i,j = 1, \ldots\ldots, r \tag{8.59}$$

This is a kind of optimization problem to find the orthogonal projections of the columns c_i, $i = 1, 2, \_\ldots, n + 1$, of matrix C onto the space spanned by the r orthonormal vectors u_i, $i = 1, \ldots, r$.

Equation (8.59) can be converted to the following optimization problem:

$$\underset{u_1, u_2, \ldots\ldots, u_r}{\text{maximize}} \quad \sum_{j=1}^{n+1} u_j^H C C u_j \quad \text{subject to} \quad u_i^H u_j = \partial_{ij} \text{ for } i,j = 1, \ldots\ldots, r \tag{8.60}$$

By using the Courant-Fisher minimax characterization of the eigenvalues and eigenvectors of a symmetric matrix [25], it can be said that the "r" number of u_i using which one is trying to find the orthogonal projection operator P_r^u which are the "r" eigenvectors of CC^H associated with the "r" largest eigenvalues. These are also the "r" left singular vectors of C associated with the "r" largest singular values.

Note that

$$C = \begin{bmatrix} A \vdots B \end{bmatrix} = U \Sigma V^H = \sum_{i=1}^{n+1} \sigma_i u_i v_i^H \tag{8.61}$$

The TLS solution of equation (8.42) is any $\widehat{h}$ which satisfies

$$\left(P_r^u A \right) \widehat{h} = P_r^u B \tag{8.62}$$

or

$$P_r^u A \, \widehat{h} - P_r^u B = P_r^u \begin{bmatrix} A \vdots B \end{bmatrix} \begin{bmatrix} \widehat{h} \\ \ldots \\ -1 \end{bmatrix} = 0 \tag{8.63}$$

$P_r^u \begin{bmatrix} A \vdots B \end{bmatrix}$ can be computed as;

$$P_r^u \begin{bmatrix} A \vdots B \end{bmatrix} = \left(\sum_{i=1}^{r} u_i u_i^H \right) \left(\sum_{i=1}^{n+1} \sigma_i u_i v_i^H \right) \tag{8.64}$$

With the knowledge of $u_i^H u_j = \partial_{ij}$ and $r < n + 1$ the above equation can be simplified;

$$P_r^u [\mathbf{A} \vdots \mathbf{B}] = \left(\sum_{i=1}^{n+1} \sigma_i u_i v_i^H \right) \tag{8.65}$$

Then one may claim that the nullspace of $P_r^u [\mathbf{A} \vdots \mathbf{B}]$ which is $\begin{bmatrix} \hat{\mathbf{h}} \\ \cdots \\ -1 \end{bmatrix}$ in equa-

tion (8.63) can be spanned by $n + 1 - r$ right singular vectors of matrix $\mathbf{C}$, v_i, $i = r + 1, \ldots, n + 1$ associated with the $n + 1 - r$ smallest singular values, i.e.,

$$\begin{bmatrix} \hat{\mathbf{h}} \\ \cdots \\ -1 \end{bmatrix} = \sum_{i=r+1}^{n+1} g_i v_i = \overline{V}_r g \tag{8.66}$$

such that

$$\overline{V}_r = \begin{bmatrix} v_{r+1} \vdots \cdots \vdots v_{n+1} \end{bmatrix} \in \mathbf{C}^{(n+1) \times (n+1-r)} \tag{8.67}$$

and g is an $(n + 1 - r) \times 1$ vector.

To find the unique minimum TLS solution, one can solve the following constrained optimization problem [26, 27];

$$\begin{aligned} Min \; &\left\| \overline{V}_r g \right\|_2^2 = g^H g \\ &\text{subject to } e_{n+1}^T \overline{V}_r g = -1 \end{aligned} \tag{8.68}$$

where

$$e_{n+1} = [0 \; 0 \; \ldots \cdots 0 \; 1]^T \in \mathbb{R}^{(n+1) \times 1} \tag{8.69}$$

By applying the method of Lagrange [27], it can be proved that [23];

$$\begin{bmatrix} \mathbf{h}_{TLS} \\ \cdots \\ -1 \end{bmatrix} = \frac{-1}{e_{n+1}^T \overline{V}_r \overline{V}_r^H e_{n+1}} \overline{V}_r \overline{V}_r^H e_{n+1} \tag{8.70}$$

Manipulating equation (8.70) yields [24];

$$\mathbf{h}_{TLS} = \frac{\left\{ \sum_{i=r+1}^{n+1} v_{(n+1),i} [v_{1,i} \; v_{2,i} \cdots v_{n,i}]^H \right\}}{\sum_{i=r+1}^{n+1} v_{(n+1),i}^2} \tag{8.71}$$

or

$$\mathbf{h}_{TLS} = \frac{\left\{ \sum\limits_{i=1}^{r} V_{(n+1),i} \left[V_{1,i}\; V_{2,i} \cdots V_{n,i} \right]^{H} \right\}}{1 - \sum\limits_{i=1}^{r} v_{(n+1),i}^{2}} \tag{8.72}$$

These principles are illustrated next through experimentation using measured data.

8.3.3 Numerical Results

Consider a multiconductor transmission line with 10 ports as shown in Figures 8.6(a) and 8.6(b). To find the impulse response between the ports 1 and 10 of the transmission line (when port 1 is considered as an input and port 10 as an output). The measurement system is shown in Figure 8.7. The objective here is to illustrate that both the near-end and the far-end cross talks on a multiconductor transmission line is produced by the intramodal dispersion between the various modes propagating on the multiconductor transmission line. Since there are 5 transmission lines, it is expected that there will be 5 modes shuttling back and forth between the input and the output ports and thereby transmitting the input energy to the various ports. To understand the physics, it is then necessary to observe the impulse response between the various input and output ports of this multiconductor transmission line [28]. The actual step input applied to port 1 is shown in Figure 8.8. One can now construct a 1024 × 1024 data matrix **A**, similar to eqn. (8.43), from the 1024 digitized step voltage shown in Figure 8.8. The output is observed and recorded at port 10 as shown in Figure 8.9 through the 1024 × l observation vector **b**, similar to eqn. (8.43). The elements of the data are accurate upto 3 significant decimal digit and the SVD should consider the smallest singular value which is 1/1000 of the largest one.

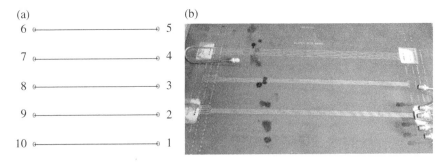

Figure 8.6 A 10 port multiconductor transmission line embedded in a multilayer dielectric, (a) Schematic diagram, (b) Actual transmission line.

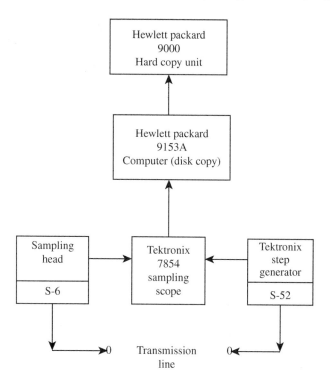

Figure 8.7 A sampling Oscilloscope is used to measure the input and the output response of the system.

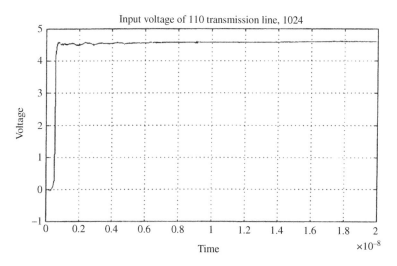

Figure 8.8 Input step voltage applied to port 1 of the multiconductor transmission line.

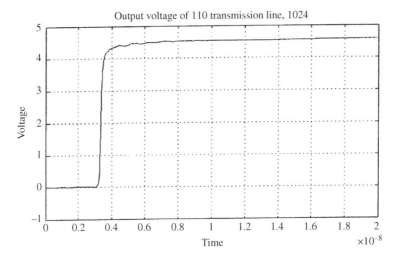

Figure 8.9 Output step response observed at port 10 of the multiconductor transmission line.

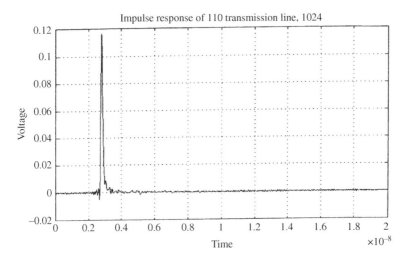

Figure 8.10 Deconvolved impulse response between ports 1 and 10 of the multiconductor transmission line.

From this output voltage (response of 1-10 transmission line to the step voltage of Figure 8.8) the deconvolved impulse response between the two ports is obtained in Figure 8.10. It is seen that the results are quite stable and they agree with the theoretical computations as illustrated in [28].

The impulse response, x_{TLS}, of the 1-10 transmission line with 307 rank for the approximating matrix is shown in Figure 8.10. The maximum and the minimum singular values of approximating matrix $\mathbf{A} \in \mathfrak{R}^{1024 \times 1024}$ is of rank 307 as $\sigma_1 = 2904.23$ and $\sigma_{302} = 2.903$ and so that the condition number is $\dfrac{\sigma_1}{\sigma_{307}} \approx 1000$. The error in this case doesn't exceed 0.4% and it is considered to be very good. The results obtained are consistent with theory of the multi-conductor transmission line as illustrated in [28].

Next, the impulse response between 1-2 transmission line is sought. The input step waveform remains the same as in Figure 8.8. The measured output response is given in Figure 8.11. Again since the data is accurate to 3 decimal places, only the first 250 singular values out of 1024 from matrix [A:b] are selected and the rest is set to zero. The maximum and minimum singular values of approximating matrix $\mathbf{A} \in \mathfrak{R}^{1024 \times 1024}$ is $\sigma_1 = 2873.04$ and $\sigma_{250} = 2.86$ and so that the condition number is $\dfrac{\sigma_1}{\sigma_{307}} \approx 1006$. The error in this case for the result obtained in Figure 8.12 for the impulse response doesn't exceed 0.2% and it is considered to be very good. The physics associated with the shape of the various pulses seen in the impulse response can be found in [28].

Next, the output voltage at port 3 is measured and is shown in Figure 8.13. It is seen that the output has a lot of noise as the measured voltage levels are quite small. The first 250 singular values out of 1024 from matrix [A:b] are selected

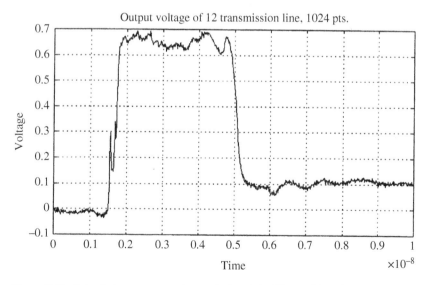

Figure 8.11 Output step response observed at port 2 of the multiconductor transmission line.

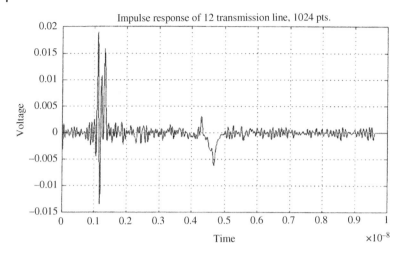

Figure 8.12 Deconvolved impulse response between ports 1 and 2 of the multiconductor transmission line.

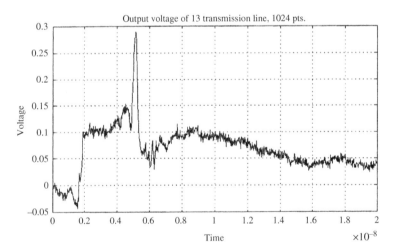

Figure 8.13 Output step response observed at port 3 of the multiconductor transmission line.

and the rest is set to zero. The maximum and minimum singular values of the approximating matrix $\mathbf{A} \in \mathfrak{R}^{1024 \times 1024}$ is $\sigma_1 = 2907.32$ and $\sigma_{250} = 2.9$ and so that the condition number is $\dfrac{\sigma_1}{\sigma_{307}} \approx 1000$. The deconvolved impulse response is shown in Figure 8.14.

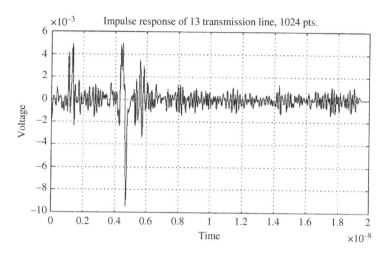

Figure 8.14 Deconvolved impulse response between ports 1 and 3 of the multiconductor transmission line.

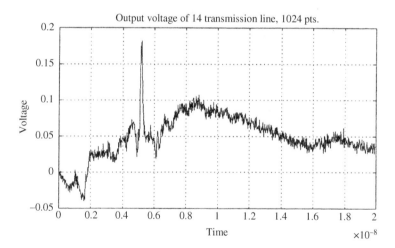

Figure 8.15 Output step response observed at port 4 of the multiconductor transmission line.

The output voltage at port 4 is measured next and is shown in Figure 8.15. The maximum and the minimum singular values of the approximating matrix $A \in \mathfrak{R}^{1024 \times 1024}$ is $\sigma_1 = 2903.912$ and $\sigma_{313} = 2.896$ and so that the condition number is $\dfrac{\sigma_1}{\sigma_{307}} \approx 1002$. The deconvolved impulse response is shown in Figure 8.16 and it has a lot of noise due to a noisy output voltage.

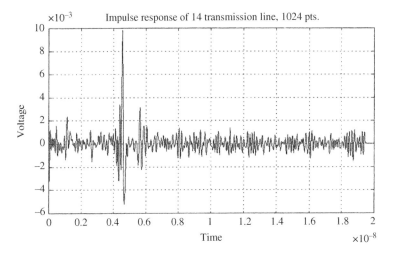

Figure 8.16 Deconvolved impulse response between ports 1 and 4 of the multiconductor transmission line.

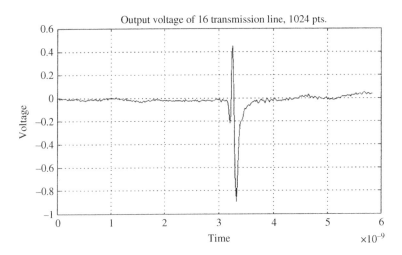

Figure 8.17 Output step response observed at port 6 of the multiconductor transmission line.

The output voltage at port 6 is measured next and is shown in Figure 8.17. The maximum and minimum singular values of the approximating matrix $A \in \mathfrak{R}^{1024 \times 1024}$ is $\sigma_1 = 2921.74$ and $\sigma_{301} = 2.918$ and so that the condition number is $\dfrac{\sigma_1}{\sigma_{301}} \approx 1001$. The deconvolved impulse response is shown in Figure 8.18.

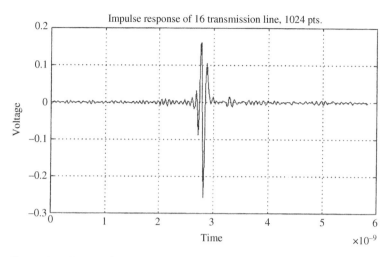

Figure 8.18 Deconvolved impulse response between ports 1 and 6 of the multiconductor transmission line.

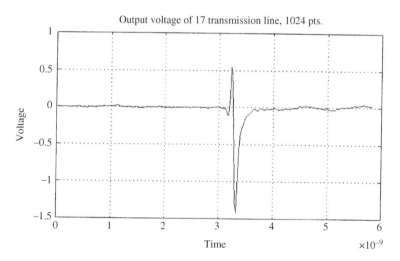

Figure 8.19 Output step response observed at port 7 of the multiconductor transmission line.

And again the computed impulsive waveforms seen in the deconvolved response makes sense as explained in [28].

The output voltage at port 7 is measured next and is shown in Figure 8.19. The maximum and the minimum singular values of the approximating matrix $\mathbf{A} \in \mathfrak{R}^{1024 \times 1024}$ is $\sigma_1 = 2921.46$ and $\sigma_{321} = 2.92$ and so that the condition number is $\dfrac{\sigma_1}{\sigma_{321}} \approx 1001$. The deconvolved impulse response is shown in Figure 8.20.

The output voltage at port 8 is measured next and is shown in Figure 8.21. The maximum and the minimum singular values of the approximating matrix $\mathbf{A} \in \mathfrak{R}^{1024 \times 1024}$ is $\sigma_1 = 2916.29$ and $\sigma_{312} = 2.90$ and so that the condition number

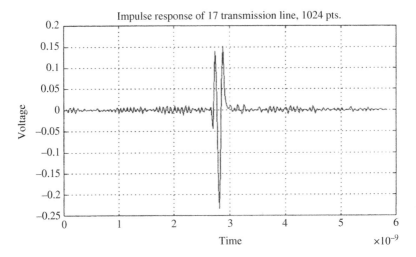

Figure 8.20 Deconvolved impulse response between ports 1 and 7 of the multiconductor transmission line.

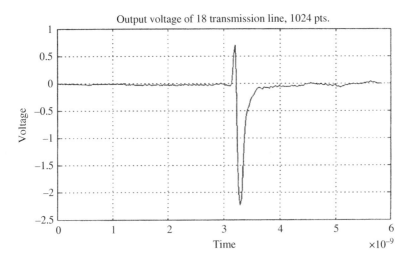

Figure 8.21 Output step response observed at port 8 of the multiconductor transmission line.

is $\dfrac{\sigma_1}{\sigma_{312}} \approx 1004$. The deconvolved impulse response is shown in Figure 8.22 and can be considered as the derivative of the output voltage.

The output voltage at port 9 is measured next and is shown in Figure 8.23. The maximum and the minimum singular values of the approximating matrix **A**

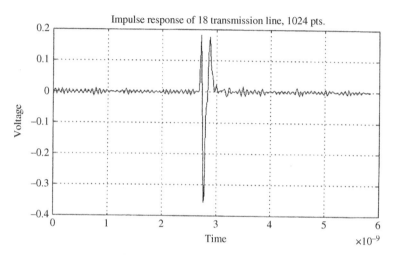

Figure 8.22 Deconvolved impulse response between ports 1 and 8 of the multiconductor transmission line.

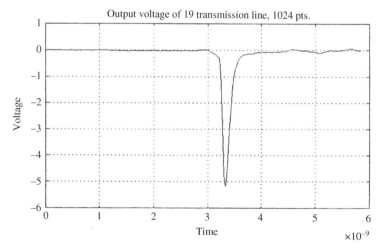

Figure 8.23 Output step response observed at port 9 of the multiconductor transmission line.

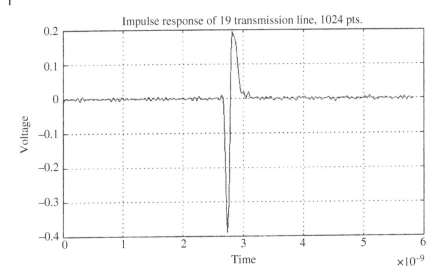

Figure 8.24 Deconvolved impulse response between ports 1 and 9 of the multiconductor transmission line.

$\in \mathfrak{R}^{1024 \times 1024}$ is $\sigma_1 = 2908.451$ and $\sigma_{302} = 2.908$ and so that the condition number is $\dfrac{\sigma_1}{\sigma_{302}} \approx 1000$. The deconvolved impulse response is shown in Figure 8.24.

8.4 Conclusion

Two methods - the conjugate gradient method with the Fast Fourier Transform Technique and the Total Least Squares based on the Singular Value Decomposition has been presented to obtain the impulse responses of some electromagnetic systems with measured input and output data. Both these methods can be automated and the truncation of the eigenvalue spectrum can be achieved by utilizing the information about the number of effective bits in the measured data. If applied properly, reasonably accurate and stable results may be obtained from these techniques.

References

1 W. E. Blass and G. W. Halsey, *Deconvolution of Absorption Spectra*, Academic Press, New York, 1981.
2 M. Mendel, *Optimal Seismic Deconvolution: An Estimation-Based Approach*, Academic Press, New York, 1983.

3 A. Ziolkowski, *Deconvolution*, IHRD, Boston, MA, 1984.

4 P. A. Jansson, *Deconvolution: With Applications in Spectroscopy*, Academic Press, New York, 1984.

5 T. K. Sarkar, D. D. Weiner, S. S. Dianat, and V. K. Jain, "Impulse Response Determination in the Time-Domain: Theory," *IEEE Transactions on Antennas and Propagation*, Vol. 30, No. 4, pp. 657–663, 1982.

6 T. Sarkar, D. Weiner, and V. Jain, "Some Mathematical Considerations in Dealing with the Inverse Problem," *IEEE Transactions on Antennas and Propagation*, Vol. 29, No. 2, pp. 373–379, 1981.

7 M. Hestenes and E. Stiefel, "Method of Conjugate Gradient for Solving Linear Systems," *Journal of Research of the National Bureau of Standards*, Vol. 49, pp. 409–436, 1952.

8 T. K. Sarkar, K. Siarkiewicz, and R. Stratton, "Survey of Numerical Methods for Solution of Large Systems of Linear Equations for Electromagnetic Field Problems," *IEEE Transactions on Antennas and Propagation*, Vol. 29, pp. 847–856, 1981.

9 T. Sarkar, E. Arvas, and S. Rao, "Application of FFT and the Conjugate Gradient Method for the Solution of Electromagnetic Radiation from Electrically Large and Small Conducting Bodies," *IEEE Transactions on Antennas and Propagation*, Vol. 34, No. 5, pp. 635–640, 1986.

10 W. Oettli, "On the Solution Set of a Linear System with Inaccurate Coefficients," *Journal of the Society for Industrial and Applied Mathematics Series B Numerical Analysis*, Vol. 2, No. 1, pp. 115–119, 1965.

11 F. I. Tseng and T. K. Sarkar, "Deconvolution of the Impulse Response of a Conducting Sphere by the Conjugate Gradient Method", *IEEE Transactions on Antennas and Propagation*, Vol. AP-35, No. 1, pp. 105–110, 1987.

12 T. K. Sarkar, F. T. Tseng, S. M. Rao, S. A. Dianat, and B. Z. Hollmann, "Deconvolution of Impulse Response from Time-Limited Input and Output: Theory and Experiment," *IEEE Transactions on Instrumentation and Measurement*, Vol. IM-34, No. 14, pp. 541–546, 1985.

13 T. K. Sarkar and X. Yang, "Accurate and Efficient Solution of Hankel Matrix Systems by FFT and the Conjugate Gradient Methods," *ICASSP '87. IEEE International Conference on Acoustics, Speech, and Signal Processing*, Dallas, TX, Vol. 12, pp. 1835–1838, 1987.

14 T. K. Sarkar and R. D. Brown, "An Ultra-Low Sidelobe Pulse Compression Technique for High Performance Radar Systems," *Proceedings of the IEEE National Radar Conference*, pp. 111–114, 1997.

15 T. K. Sarkar and R. D. Brown, "Real Time Deconvolution Using the Conjugate Gradient Algorithm," *Fifth ASSP Workshop on Spectrum Estimation and Modeling*, pp. 294–298, 1990.

16 A. V. Oppenheim and R. W. Schafer, *Discrete-Time Signal Processing*, Prentice-Hall, Englewood Cliffs, NJ, 1989.

17 G. H. Golub and C. F. Van Loan, *Matrix Computations*, Johns Hopkins University Press, Baltimore, MD, Second Edition, 1989.

18 S. Van Huffel and J. Vandewalle, *The Total Least Squares Problem, Computational Aspects and Analysis*, Frontiers in Applied Mathematics, Society for Industrial and Applied Mathematics, Philadelphia, PA, 1991.

19 G. H. Golub and C. F. Van Loan, "An Analysis of the Total Least Squares Problem," *SIAM*, Vol. 17, No. 6, pp. 883–892, 1980.

20 A. Bjorck, "Least Squares Method," in *Handbook of Numerical Analysis, Vol. I: Finite Difference Methods; Solution of Equations in R^n, P. G. Ciarlet and J. L. Lions, editors, North-Holland, Amsterdam, the Netherlands, p. 465, 1990.*

21 G. H. Golub, "Some Modified Matrix Eigenvalue Problems," *SIAM Review*, Vol. 15, pp. 318–344, 1973.

22 G. H. Golub and C. Reisch, "Singular Value Decomposition and Least Squares Solutions," *Numerische Mathematik*, Vol. 14, pp. 403–420, 1970.

23 M. D. Zoltowski, "Signal Processing Applications of the Method of Total Least Squares," in *Proceedings of 21st Asilomar Conference on Signals, Systems and Computers*, Pacific Grove, CA, pp. 290–296, 1987.

24 J. Rahman and T. K. Sarkar, "Deconvolution and Total Least Squares in Finding the Impulse Response of an Electromagnetic System from Measured Data," *IEEE Transactions on Antennas and Propagation*, Vol. 43, No. 4, pp. 416–421, 1995.

25 J. H. Wilkinson, *The Algebraic Eigenvalue Problem*, Clarendon Press, Oxford, pp. 100–101, 1965.

26 D. A. Pierre, *Optimization Theory with Applications*, Dover Publications, Inc., New York, 1986.

27 M. Aoki, *Introduction to Optimization*, The Macmillan Company, New York, 1971.

28 A. R. Djordjevic, T. K. Sarkar, and R. F. Harrington, "Time-Domain Response of Multiconductor Transmission Lines," *Proceedings of the IEEE*, Vol. 75, No. 6, pp. 743–764, 1987.

9

Performance of Different Functionals for Interpolation/Extrapolation of Near/Far-Field Data

Summary

In this chapter, first, the Chebyshev polynomials are used to extract from the data collected under non-anechoic environment a radiation pattern for the same antenna radiating in free space conditions. One of the shortcomings with this method is that some a priori information is required about the environment. Measured data has been used to illustrate the applicability of this new methodology and its improved performance over the FFT based methods. Next, the Cauchy method based on Gegenbauer polynomials is presented for antenna near-field extrapolation and the far-field estimation. Due to various physical limitations, there are often missing gaps of data in the antenna near-field measurements. However, the missing data is indispensable if we want to accurately evaluate the complete far-field pattern using the near-field to far-field transformations. To address this problem, an extrapolation method based on the Cauchy method is proposed to reconstruct the missing part of antenna near-field measurements. As the near-field data in this section is obtained on the spherical measurement surface, the far field of the antenna is calculated by the spherical near-field to far-field transformation with the extrapolated data. Some numerical results are given to demonstrate the applicability of the proposed scheme in antenna near-field extrapolation and far-field estimation. The Gegenbauer polynomials are very useful when a very large order of the polynomials is used to fit the given data. This is true for some spherical near-field measurement setup where there is a big gap in the bottom of the sphere on whose surface the measurements are made.

In addition, the performance of the Gegenbauer polynomials are compared with that of the normal Cauchy method using Polynomial expansion and the Matrix Pencil Method in filling in missing near field data on a spherical measurement range.

Modern Characterization of Electromagnetic Systems and Its Associated Metrology, First Edition.
Tapan K. Sarkar, Magdalena Salazar-Palma, Ming Da Zhu, and Heng Chen.
© 2021 John Wiley & Sons, Inc. Published 2021 by John Wiley & Sons, Inc.

9.1 Background

First, the possibility of using the Chebyshev polynomials for fitting the measured finite bandwidth frequency domain data is explored. It has been observed in previous researches [1, 2] that when scaled versions of the Bessel functions are used to approximate any causal time domain data (i.e., which exists only for $t \geq 0$), one is then fitting in the frequency domain by the analytical transforms of the scaled Bessel functions which are the Chebyshev polynomials of the first and second kinds [3]. The numerical values of the coefficients obtained by fitting the frequency domain data by the various orders of Chebyshev polynomials of the first and second kinds surprisingly follow the trend of the time domain waveforms, even though the time domain waveforms are being approximated by scaled Bessel functions [1–5]. Since, this interesting property has been observed under various different conditions, it is argued that the plot of the coefficients multiplying the scaled Bessel functions of the first kind will be able to distinguish between the various reflections and will help one to accomplish the separation between the direct wave and the various reflections present in a non-anechoic measurement.

Next, the near-field (NF) measurements of the antennas under test (AUT) are commonly used to estimate the far-field (FF) radiation patterns, as the near field (NF) to FF transformation methods are well-established [6–14]. The spherical NF scanning is widely adopted in many of the antenna measurements. The NF scan systems of the AUT are feasible to avoid the distance requirement of FF measurements, which makes it easy to measure the antenna in a small anechoic chamber. Although NF scanning reduces the requirements of measuring conditions, there still exist a great number of measurement limitations that makes it difficult to obtain the complete NF data. For example, the physical size of the scanning system limits the dimension of the AUT that could be measured. It is also difficult to rotate an electrically large aperture antenna in a small anechoic chamber, and the spherical scan surface has to be truncated. In other words, there are often missing gaps of the measurement surface in the NF measurements. However, the missing NF data is indispensable if we want to accurately evaluate the complete FF radiation pattern by the NF to FF transformations [13]. Therefore, extrapolating the available NF data can be a workable way to solve this problem. As the extrapolation problem is ill-conditioned in general, its solution is unstable and non-unique in the presence of noise. A reliable extrapolation method with some priori information is necessary in practice.

The Cauchy method [15–20] has been proposed to extrapolate frequency-domain data as explained in Chapter 3. It approximates a function by the quotient of two polynomials, while utilizing total least squares (TLS) and singular value decomposition (SVD) to obtain the coefficients of the numerator and the denominator polynomials [17]. TLS and SVD also suppress partially the effects of noise in the known data for extrapolation. With the computed coefficients, the frequency response can be extrapolated over the frequency band of interest. As the polynomials are a subset of the rational functions, the Cauchy method can extrapolate more extensive range of function shapes than least squares polynomial fitting. However, the numerator and denominator of the rational function model of the Cauchy method is defined as a sum of power functions, which is quite suitable for the numerator and denominator polynomials of small degrees. Due to the large number of samples used in the NF extrapolation and the complexity of NF distribution, the Gegenbauer polynomials [21] are adopted to model the numerator and denominator in the novel Cauchy method in this section. This newly proposed method works much better than the original one, when the missing gap of data on the measurement surface is large.

As the uncertainty of extrapolation is subject to the range of the missing data, meaningless results may appear when a very large missing gap is extrapolated. Therefore, it is necessary to introduce a reliability parameter whose value can be used to check the correctness of the near-field extrapolation results. In order to verify the reliability of the extrapolation results, we compare the rectilinear norm of the electric field strength of the missing gap with a reference value which is obtained by a simple simulation of the AUT. This verification is made under the assumption that the proportion of the field norm of a large gap would not have much difference between the measured value and simulated one, in spite of the difference of the field distribution on the measurement surface.

The performance of this method is compared with that of the original Cauchy Method using polynomials and the Matrix Pencil method when filling up the missing near-field data from a parabolic reflector antenna measured on a spherical surface.

9.2 Approximating a Frequency Domain Response by Chebyshev Polynomials

From a strictly mathematical point of view, a causal time domain response, for $t \geq 0$ cannot simultaneously be band-limited and vice versa. However, a response strictly limited in time can be assumed to be approximately band-limited, if the amplitude of the frequency response is too small (outside the region of interest) to be of any consequence.

Consider a time sequence $x(t)$ which can be expressed by an orthogonal basis $\phi_n(t, l_1)$ so that

$$
x(t) = \sum_{n=0}^{\infty} a_n \, \phi_n(t, l_1) = \sum_{n=0}^{\infty} a_n \left(\frac{t}{l_1}\right)^{-1} J_n\left(\frac{t}{l_1}\right) \tag{9.1}
$$

where a_n are the expansion coefficients, $J_n\left(\frac{t}{l_1}\right)$ is a Bessel function of the first kind and degree n [1, 2] and l_1 is a scale factor in time. The scale factor takes care of the units of time be it be nanosec, picoseconds or the like and transforms the data to its scaled version where it is easier to fit it by a limited number of basis functions. A waveform with compact time support can be expanded approximately using this orthogonal basis.

The corresponding frequency domain response $X(f)$ can be expressed by another orthogonal basis using the same set of coefficients $\{a_n\}$ [1, 2] with $j = \sqrt{-1}$ and is given by

$$X\left(\frac{f}{l_2}\right) = \sum_{n=0}^{N} \frac{2\,a_n\,(-j)^{n+1}}{n} \sqrt{1 - \left(\frac{2\pi f}{l_2}\right)^2}\, U_{n-1}\left(\frac{2\pi f}{l_2}\right); \quad |2\pi f| < l_2$$

$$= 0; \qquad\qquad\qquad |2\pi f| > l_2$$

$$(9.2)$$

The corresponding scaling parameter in the frequency domain is given by $l_2 = 1/2\pi l_1$ and $U_n(f)$ is the Chebyshev polynomial of the second kind given by $U_n(f) = \dfrac{\sin\left[(n+1)\ \cos^{-1}(f)\right]}{\sqrt{1-f^2}}$ [10, 11].

In (9.1), the causality in time is not forced whereas the signals we are dealing with in real life are causal. We also know that for a causal time domain function, the real and the imaginary parts of the transfer function must be related by the Hilbert transform. In addition, the relationship between the first and the second kind of Chebyshev polynomials are related by the Hilbert transformation, i.e.,

$$\int_{-1}^{1} \frac{\sqrt{1-y^2}\ U_{n-1}(y)}{y-x}\,dy = -\pi\,T_n(x) \tag{9.3}$$

where $T_n(x) = \cos[n\ \cos^{-1}(x)]$ is the Chebyshev polynomial of the first kind and is defined by [3, 4]. Therefore, if it is enforced that equation (9.1) must hold for a causal function then its transform will no longer be given by (9.2) but it needs to be modified. In the following presentation we now enforce the principle of causality on (9.1). When causality is enforced, then the function and its Fourier transforms are related by

$$\phi_n(t, l_1) = \left(\frac{t}{l_1}\right)^{-1} J_n\left(\frac{t}{l_1}\right) \quad \text{for}\ \ t \ge 0 \tag{9.4}$$

$$\phi_n(f, l_2) = \frac{(-j)^n}{n}\left[\left(i\sqrt{1 - \left(\frac{2\pi f}{l_2}\right)^2}\right) U_{n-1}\left(\frac{2\pi f}{l_2}\right) + T_n\left(\frac{2\pi f}{l_2}\right)\right] \quad \text{for}\ |2\pi f| < l_2$$

$$(9.5)$$

Now fit the measured data in the frequency domain by

$$X\left(\frac{f}{l_2}\right) = \sum_{n=0}^{N} \frac{(-j)^n a_n}{n} \left[\left(j\sqrt{1 - \left(\frac{2\pi f}{l_2}\right)^2}\right) U_{n-1}\left(\frac{2\pi f}{l_2}\right) + T_n\left(\frac{2\pi f}{l_2}\right)\right]$$

(9.6)

Here, the frequency scaling factor l_2 is chosen to be slightly larger than the maximum bandwidth of the available measurement data.

In this new procedure, one starts with the measured data $S_{21}(f, \phi)$ for different azimuth angles φ_i. For each azimuth angle, a set of data for different frequencies between f_1 and f_{M_2} is measured simultaneously (where M_2 is the number of the frequency domain sampled data). Correspondingly, a set of different coefficients a_n for different azimuth angles can be obtained by solving the following matrix equation using the iterative conjugate gradient method to a prespecified degree of accuracy [1–4].

$$\begin{bmatrix} \text{Re} \begin{pmatrix} \varphi_0(f_1, l_2) & \varphi_1(f_1, l_2) & \cdots & \varphi_{N-1}(f_1, l_2) \\ \varphi_0(f_2, l_2) & \varphi_1(f_2, l_2) & \cdots & \varphi_{N-1}(f_2, l_2) \\ \vdots & \vdots & \vdots & \vdots \\ \varphi_0(f_{M_2}, l_2) & \varphi_1(f_{M_2}, l_2) & \cdots & \varphi_{N-1}(f_{M_2}, l_2) \end{pmatrix} \\ \text{Im} \begin{pmatrix} \varphi_0(f_1, l_2) & \varphi_1(f_1, l_2) & \cdots & \varphi_{N-1}(f_1, l_2) \\ \varphi_0(f_2, l_2) & \phi_1(f_2, l_2) & \cdots & \varphi_{N-1}(f_2, l_2) \\ \vdots & \vdots & \vdots & \vdots \\ \varphi_0(f_{M_2}, l_2) & \varphi_1(f_{M_2}, l_2) & \cdots & \varphi_{N-1}(f_{M_2}, l_2) \end{pmatrix} \end{bmatrix}_{2M_2 \times N} \begin{bmatrix} a_0 \\ a_1 \\ \vdots \\ a_{N-1} \end{bmatrix}_{N \times 1} = \begin{bmatrix} \text{Re} \begin{pmatrix} X(f_1) \\ X(f_2) \\ \vdots \\ X(f_{M_2}) \end{pmatrix} \\ \text{Im} \begin{pmatrix} X(f_1) \\ X(f_2) \\ \vdots \\ X(f_{M_2}) \end{pmatrix} \end{bmatrix}_{2M_2 \times 1}$$

(9.7)

In summary, in this procedure one starts with the measured data $S_{21}(f, \varphi)$ which is measured as follows. The measurements are performed in a near-field/far-field spherical-range measurement system housed in an anechoic chamber of dimensions 10 m × 7.5 m × 8.5 m as shown in Figure 9.1 [5]. A simplified scheme of the whole measurement system is seen in Figure 9.2. The probe and the AUT are placed at a height of 2 *m* above the ground level separated from each other by 7.7 *m* during the whole measurement campaign. The system controlling the positioners allowed the setting of the azimuth and roll orientation of the AUT, as well as the polarization of the probe, via the software. The antenna chosen in this case are horn antennas which have wide radiation patterns. The measurement campaign was carried out in the frequency domain, measuring the $S_{21}(f)$ in the *H*-plane of a two horn transmit-receive system in the frequency band of interest for each azimuth angle ϕ from −90° to 90° at steps of 1°. Here, the bandwidth of measurements was 0.6 GHz, ranging from 3.05 GHz

Figure 9.1 The anechoic chamber with the AUT on a rollover azimuth position.

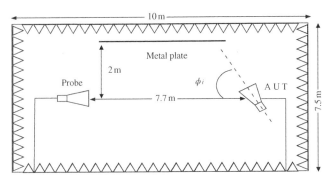

Figure 9.2 A top view of the antenna measurement system. The metal plate has been introduced into the anechoic chamber to model a non-anechoic reverberant chamber.

to 3.65 GHz. Therefore, the center frequency corresponds to the measurement at the frequency of interest of 3.35 GHz. Figure 9.3 plots the radiation pattern of the horn antenna in a true anechoic condition called the reference. Now introduce a perfectly conducting metal plate of size 5 *m* × 2 *m* centered in the same way as the antennas and located at a distance of 2 *m* from the line of the sight of the transmit-receive system. Because of the presence of the metal plate the measured radiation pattern will be different from the true one and both are plotted in Figure 9.3. The problem of interest is: can one do some processing to recover the true radiation pattern from the measured one in this non-anechoic environment? To answer this question, the following steps are carried out:

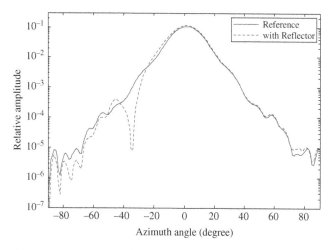

Figure 9.3 Comparison of the radiation pattern measured in the non-anechoic chamber and in the ideal chamber at the center frequency of 3.35 GHz.

First, measure the coherent (amplitude/phase) frequency response covering the bandwidth of 3.05 GHz to 3.65 GHz, i.e., $S_{21}(f, \phi)$ of the antennas in the presence of the metal plate as shown in Figure 9.4. The various reflected and the diffracted fields for the non-anechoic condition are primarily in the region covering $-20° \sim -60°$. By applying the inverse FFT to $S_{21}(f, \phi)$ one can obtain the temporal response $S_{21}(t, \phi)$ as shown in the Figure 9.5 in which the direct ray and the reflected rays can be separated as illustrated in Chapter 2.

Coming back to the approximation procedure by Chebyshev polynomials, one starts with the measurements that are performed for different azimuth angles φ_i and for each azimuth angle, a set of data for different frequencies between f_1 and f_{M_2} are measured simultaneously as shown in Figure 9.4. For each azimuth angles φ_i the matrix equation of (9.7) is solved using the conjugate gradient method (CGM) for each of the coefficients a_i. The real and the imaginary part of the coefficients a_i for each azimuth angle ϕ_i are plotted in Figure 9.6.

For the FFT-based methods as illustrated in Chapter 2, truncation is made in the time domain to eliminate the reflection and diffraction of the fields in Figure 9.5. In this new approach, the truncation is made of the Chebyshev series given by (9.6). The rationale for doing so is because from previous research [1–5] it has been observed that when a time domain function is approximated by scaled Bessel functions of the form given by (9.6), the coefficients multiplying each of the terms of the series tend to follow the variation of the time domain function that it is trying to approximate. The series is truncated when the scaled Bessel function of a certain order peaks at the time at which one would like to truncate the temporal function to prevent reflections or other secondary fields creeping in. The peak in the value of the scaled Bessel function shifts with time

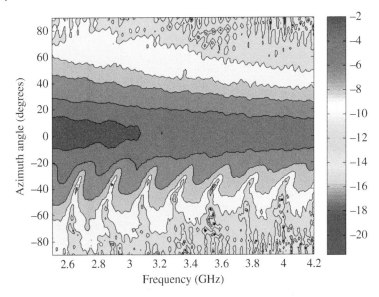

Figure 9.4 Plot of $S_{21}(f, \phi)$, the input-output frequency response for each azimuth angle.

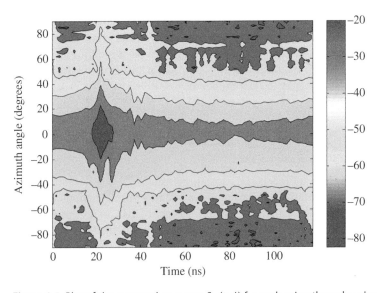

Figure 9.5 Plot of the temporal response $S_{21}(t, \phi)$ for each azimuth angle using the data from 3.05 GHz to 3.65 GHz.

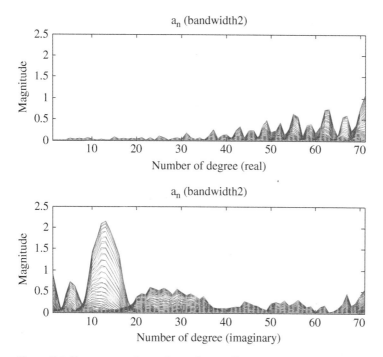

Figure 9.6 The computed set of complex coefficients a_n that fits the data for each azimuth angle.

as its order increases. In this example the waveform is truncated in the time domain around 28.92 ns. Here, $l_2 = 2 \times 10^9$. This time unit corresponds to the case when a scaled Bessel function of order 25 or 26 has the peak value in that time. Figure 9.7 shows the temporal plot of the receiving antenna response for different azimuth angles, namely the pattern, using the retained coefficients a_n up to order 25/26. Observe that the temporal response after $t = 28.92$ ns is not zero as shown in Figure 9.7, while the FFT-based method has a zero response after $t = 28.92$ ns as shown in Figure 9.8.

Finally, using the retained set of coefficients for the Chebyshev functional series, one can plot the antenna radiation pattern at the central frequency of 3.35 GHz. The amplitude response is plotted in Figure 9.9(a) along with the reference and the radiation pattern in a nonanechoic condition. By comparing Figure 9.9(a) with Figure 9.9(b) it is seen that that the current reconstruction methodology has a better performance. The mean squared error between the reference and the reconstructed response when using the FFT based methodology is approximately 0.186 when the observation angle is limited from $-90°$ to $0°$ and is 0.148 when the complete set of azimuth angle is considered form $-90°$ to $90°$. This is summarized in Table 9.1.

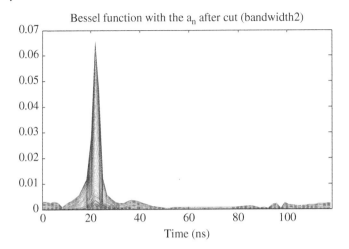

Figure 9.7 Plot of the temporal pattern using the retained set of coefficients.

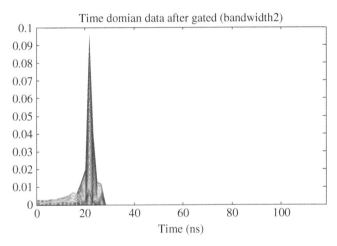

Figure 9.8 $S_{21}(t', \phi)$, the time domain data using FFT after truncating the reflected waveforms.

When using the Chebyshev polynomials to fit the measured data and then truncating the series, the corresponding mean squared error between the reference antenna pattern and the reconstructed one is approximately 0.066 when the observation angle is limited from $-90°$ to $0°$ and is 0.057 when the complete set of azimuth angle is considered form $-90°$ to $90°$. Hence, it appears that at least for this data set, the use of Chebyshev polynomials is more robust.

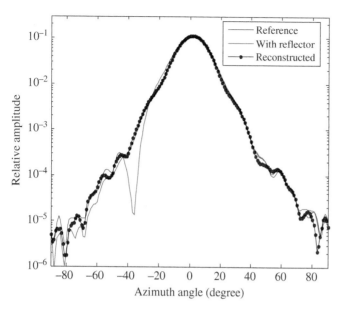

Figure 9.9(a) Amplitude pattern reconstruction using the Chebyshev Polynomials.

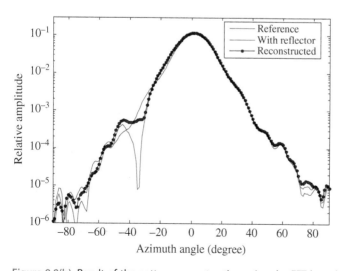

Figure 9.9(b) Result of the pattern reconstruction using the FFT-based method.

Table 9.1 The Norm of the Mean Squared Error between the Free Space Results and the Data Processed Using the Chebyshev Functions.

	Bandwidth: 0.6 GHz	
Azimuthal Angular Range	FFT	Chebyshev
$-90° \sim 0°$	0.186	0.066
$-90° \sim 90°$	0.148	0.057

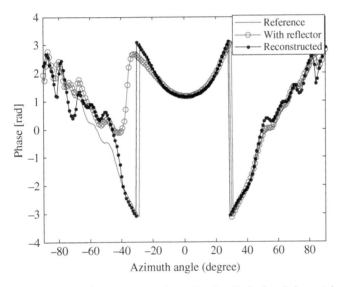

Figure 9.10(a) Phase reconstruction using the Chebyshev Polynomials.

Figure 9.10(a) plots the reconstructed phase response for this pattern restricted to $-\pi$ and $+\pi$. By comparing Figure 9.10(a) to Figure 9.10(b), generated by the FFT based method it is seen that the proposed method provides a more realistic phase response for the AUT.

Work is in progress as to what is a good way to select the appropriate scale factor in (9.1) and (9.2) [22]. In addition, it would be nice to find an automated way to truncate the series of (9.1) based on the measured data and not on the a priori knowledge of the environment. Work is in progress to address these two issues.

Next, we discuss about the use of Gegenbauer polynomials in the Cauchy method for a more stable solution procedure for some examples.

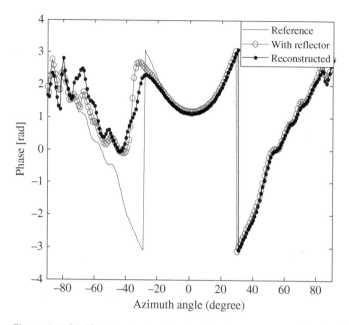

Figure 9.10(b) Phase reconstruction for the above pattern using the FFT-based method using the data between 3.05 GHz to 3.65 GHz.

9.3 The Cauchy Method Based on Gegenbauer Polynomials

The spherical NF scan system is widely used in the antenna measurements, and we adopt the spherical NF to FF transformation in this section. Consider a spherical surface of radius a over which the tangential components of the NF electric field E_θ and E_ϕ are partially measured from 0° to $(180° - \theta_g)$ in θ and from 0° to 360° in ϕ, where $(180° - \theta_g)$ to 180° is the range of the missing interval of the NF data. The extrapolating operation is performed for each ϕ_n, $n = 1$, 2, ... N_ϕ, to recover the missing NF data in the interval $[(180° - \theta_g), 180°]$ as seen in Figure 9.11. Since the uncertainty of extrapolation results is subject to the range of missing data, meaningless results could be avoided by defining a reliability functional as follows

$$\xi_r(\phi_n) = \frac{\|f(\theta_e, \phi)\|_1}{\|f(\theta, \phi)\|_1}, \phi = \phi_n \tag{9.8}$$

where $\| \ \|_1$ is the L^1 norm, $f(\theta)$, $\theta \in [0°, 180°]$, is the field component at $\phi = \phi_n$, and $f(\theta_e)$, $\theta_e \in (180° - \theta_g, 180°]$, is the field data that needs to be extrapolated. The parameter ξ_r denotes the proportion of the missing data to the complete

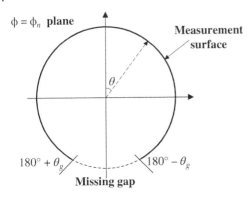

$\phi = \phi_n$ **plane**

Measurement surface

θ

$180° + \theta_g$ $180° - \theta_g$

Missing gap

Figure 9.11 The spherical NF measurement setup with a missing data in the interval $(180° - \theta_g, 180° + \theta_g)$.

field in L^1 norm, which does not contain the information of the function shape. If ξ_r is obtained approximately by a simulation of the AUT, ξ_r can provide the information to check the extrapolation results without affecting the field distribution of the simulation.

The electric field variation for the angles between ϕ_n and $(360° - \phi_n)$ can be combined together and is written in a form as follows

$$\vec{E}(\theta) = E_\theta(\theta)\hat{a}_\theta + E_\phi(\theta)\hat{a}_\phi, \ \ \theta \in [0°, 360°) \tag{9.9}$$

where $(180° - \theta_g, 180° + \theta_g)$ is the interval of the missing data. This combining operation makes $\vec{E}(\theta)$ a periodic function, and the missing interval to be extrapolated to lie in the middle of $\vec{E}(\theta)$. The reliability parameter is then given as

$$\xi = \xi_r(\phi_n) + \xi_r(360° - \phi_n) \tag{9.10}$$

After the combination of the fields between the angle ϕ_n and $(360° - \phi_n)$, we have for $\{\theta_m, \theta_m \in [0°, 180° - \theta_g] \cup [180° + \theta_g, 360°)\}$, $m = 1, 2, \dots N_s$, the measured points. The fields in the region $\{\theta_e, \theta_e \in (180° - \theta_g, 180° + \theta_g)\}$, $e = 1, 2, \dots N_e$, are the points that need to be extrapolated. As $E_\theta(\theta)$ and $E_\phi(\theta)$ has both real and imaginary parts, each part is extrapolated separately in this section. Therefore, we use $f(\theta)$ to represent the periodic function with a missing data gap that needs to be extrapolated.

In the context of the Cauchy method we define a real function $f(\theta)$ on the spherical surface for each ϕ_n, as a rational function of angle θ. In other words, $f(\theta)$ is approximated by a ratio of numerator $A(\theta)$ and denominator $B(\theta)$ polynomials as

$$f(\theta) \approx \frac{A(\theta)}{B(\theta)} = \frac{\sum_{i=0}^{N_n} a_i \theta^i}{\sum_{j=0}^{N_d} b_j \theta^j} \tag{9.11}$$

where $A(\theta)$ and $B(\theta)$ are characterized by the parameters a_i and b_j, respectively. The Cauchy method has been successfully applied to interpolate frequency domain responses [15–20], as the model of the ratio of two rational polynomials provide much more flexibility of approximation than polynomial fitting. However, if the power of the polynomials becomes too large both in the numerator $A(\theta)$ and denominator $B(\theta)$, the Cauchy method based on (9.11) will result in an ill-conditioned Vandermonde-like matrix, which is not computationally suitable for the NF extrapolation problems in this section. As a large number of sample points and high degree polynomials are needed to extrapolate the missing gap, we adopt the Gegenbauer polynomials [21] to form the numerator and denominator as follows

$$f(\theta) \approx \frac{E(\theta)}{F(\theta)} = \frac{\sum_{i=0}^{N_n} p_i C_i^{\lambda}(\theta)}{\sum_{j=0}^{N_d} q_j C_j^{\lambda}(\theta)} \tag{9.12}$$

where N_n and N_d are the polynomial degrees of the numerator and the denominator, p_i and q_j are the expansion coefficients, and $C_i^{\lambda}(\theta)$ is the Gegenbauer polynomial which is a generalization of Legendre polynomials and Chebyshev polynomials. The recurrence relations to generate the $C_i^{\lambda}(\theta)$ [21] is given by

$$C_0^{\lambda}(\theta) = 1, \qquad i = 0 \tag{9.13a}$$

$$C_1^{\lambda}(\theta) = 2\lambda\theta, \qquad i = 1 \tag{9.13b}$$

$$C_i^{\lambda}(\theta) = \frac{2\theta(\lambda + i - 1)}{i} C_{i-1}^{\lambda}(\theta) - \frac{2\lambda + i - 2}{i} C_{i-2}^{\lambda}(\theta), i \geq 2 \tag{9.13c}$$

If the function $f(\theta)$ is measured at the angular samples θ_m, for $m = 1, 2, \ldots N_s$, one can enforce equation (9.12) at the measured points, which is given by

$$E(\theta_m) - f(\theta_m)F(\theta_m) = 0 \tag{9.14}$$

Or equivalently,

$$p_0 + p_1 C_1^{\lambda}(\theta_m) + \cdots + p_{N_n} C_{N_n}^{\lambda}(\theta_m) - q_0 f(\theta_m) - q_1 f(\theta_m) C_1^{\lambda}(\theta_m) - \cdots - q_{N_d} f(\theta_m) C_{N_d}^{\lambda}(\theta_m) = 0 \tag{9.15}$$

Thus, equation (9.12) can be converted into a matrix form as follows

$$[M]_{N_s \times (N_n + N_d + 2)} \begin{bmatrix} p \\ q \end{bmatrix}_{(N_n + N_d + 2) \times 1} = 0 \tag{9.16a}$$

with

$$[M] = [E - F] \tag{9.16b}$$

$$[E] = \begin{bmatrix} 1 & C_1^\lambda(\theta_1) & \cdots & C_{N_n}^\lambda(\theta_1) \\ 1 & C_1^\lambda(\theta_2) & \cdots & C_{N_n}^\lambda(\theta_2) \\ \vdots & \vdots & \ddots & \vdots \\ 1 & C_1^\lambda(\theta_{N_s}) & \cdots & C_{N_n}^\lambda(\theta_{N_s}) \end{bmatrix} \tag{9.16c}$$

$$[F] = \begin{bmatrix} f(\theta_1) & f(\theta_1)C_1^\lambda(\theta_1) & \cdots & f(\theta_1)C_{N_d}^\lambda(\theta_1) \\ f(\theta_2) & f(\theta_2)C_1^\lambda(\theta_2) & \cdots & f(\theta_2)C_{N_d}^\lambda(\theta_2) \\ \vdots & \vdots & \ddots & \vdots \\ f(\theta_{N_s}) & f(\theta_{N_s})C_1^\lambda(\theta_{N_s}) & \cdots & f(\theta_{N_s})C_{N_d}^\lambda(\theta_{N_s}) \end{bmatrix} \tag{9.16d}$$

$$[p] = \begin{bmatrix} p_0, & p_1, & p_2, & \cdots, & p_{N_n} \end{bmatrix}^T \tag{9.16e}$$

$$[q] = \begin{bmatrix} q_0, & q_1, & q_2, & \cdots, & q_{N_d} \end{bmatrix}^T \tag{9.16f}$$

where T denotes the transpose of a matrix. If N_s is sufficiently larger than $N_n + N_d + 2$, (9.16a) is an overdetermined system of equations which makes the proposed extrapolation possible. Since there is oversampling for $f(\theta)$, it gives the upper bound of $N_n + N_d$ as follows

$$N_n + N_d \leq N_s - 2 \tag{9.17}$$

As the electric field distribution on the NF measuring surface is smooth, we need to set the degree of dominator smaller than the one of numerator, which would result in a good approximation for smooth functions. In this section, we choose the following empirical relation

$$N_d = \text{INT}(0.6N_n) \tag{9.18}$$

where INT is the operation that rounds a number to the nearest integer. In [18–20], the singular value decomposition (SVD) of $[M]$ is used to estimate the value of $N_n + N_d$. As the useful singular values for representing the data should be larger than the level of the noise, $N_n + N_d + 1$ equals to the number of nonzero singular values which is determined by a truncation parameter representing the significant decimal digits of the data. Due to the large number of the sampled points, the size of the missing gap to be extrapolated, and the complexity of different NF problems, the reliability parameter ξ introduced in (9.10) is necessary for checking and adjusting the parameter N_n and N_d.

There are two ways to solve equation (9.16a) by the total least square (TLS) method. The first way is to perform a SVD operation on $[M]$, which results in

$$[U_M][S_M][V_M]^H \begin{bmatrix} p \\ q \end{bmatrix} = 0 \tag{9.19}$$

where $[U_M]$ and $[V_M]$ are unitary matrices, and $[S_M]$ is a diagonal matrix with the singular values on the diagonal. By the TLS theory, the best solution for fitting the rational model of (9.12) to measured data is proportional to the last column of $[V_M]$, which is given by

$$[p \quad q]^T = [V_M]_{(N_n + N_d + 2)} \qquad (9.20)$$

The TLS solution for best data fitting shows an excellent performance for minimizing errors of interpolation in [15–20]. TLS can also work well for extrapolating missing data on a small interval, which is quite similar to an interpolation problem with several sample points removed. However, when the missing interval of a function becomes large, the best fitting of the existing data does not necessarily mean the best fitting of the missing part. For example, the TLS solution may allow some poles of rational function (9.14) to reside in the interval of the missing data, even though the solution (9.20) fits the model to the measured part very well. In order to solve this problem, we search the last several columns of $[V_M]$ for the best solution which avoids the unpredictably wrong result. These unwanted poles are easy to be detected by a simple finite difference operation on the extrapolated data, as the NF data is assumed to be smooth. Thus, we define a checking parameter as follows

$$s = \frac{\| f(\theta_{a+1}) - f(\theta_a) \|_\infty}{\| f(\theta_{m+1}) - f(\theta_m) \|_0} \times N_s \qquad (9.21)$$

where $\| \ \|_\infty$ and $\| \ \|_0$ denote the L^∞ and L^0 norm, respectively, and $\theta_a = \{\theta_m\} \cup \{\theta_e\}$. If $s < 10$, the extrapolation result is considered as a smooth solution in this section. Therefore, the parameter s and ξ are used to check and search the last several columns of $[V_M]$ for the best solution.

The second method for solving equation (9.16a) is to determine the coefficients of the numerator and denominator separately. As matrix $[E]$ in (9.16b) is generated by the Gegenbauer polynomials only and is not affected by the measured data, $[E]$ is free of errors and noise in the data. Thus, a QR decomposition of $[E]$ is given by

$$[E] = [Q][R] \qquad (9.22a)$$

$$[Q]^T[Q] = [I] \qquad (9.22b)$$

where $[Q]$ is a $N_s \times N_s$ orthogonal matrix, and $[R]$ is a $N_s \times (N_n + 1)$ upper triangular matrix. Next, we multiply (9.16a) with the matrix $[Q]^T$ as follows

$$[Q]^T[M] \begin{bmatrix} p \\ q \end{bmatrix} = [R - Q^T F] \begin{bmatrix} p \\ q \end{bmatrix} = 0 \qquad (9.23a)$$

$$[R] = \begin{bmatrix} P_{11} \\ 0 \end{bmatrix}_{N_s \times (N_n + 1)} , \quad [-Q^T F] = \begin{bmatrix} P_{12} \\ P_{22} \end{bmatrix}_{N_s \times (N_d + 1)} \qquad (9.23b)$$

Then, one can determine the coefficients of the denominator by performing a SVD on $[P_{22}]$ as follows

$$[P_{22}][q] = [U_P][S_P][V_P]^H[q] = 0 \tag{9.24}$$

The TLS solution for best fitting to the measured data is proportional to the last column of $[V_P]$, which is given by

$$[q] = [V_P]_{(N_d + 1)} \tag{9.25}$$

Similarly, if a large missing gap of data is to be filled in, we search the last several columns of $[V_P]$ for the best solution which avoids the unpredictably wrong result. The condition $s < 10$ is also used here. If $[q]$ is obtained, $[p]$ can be solved by the relation in (9.23a) as

$$[p] = [P_{11}]^{-1}[P_{12}][q] \tag{9.26}$$

The matrix $[P_{11}]$ might be ill-conditioned sometimes, as its condition number is affected by the sampled data. When $[P_{11}]$ is ill-conditioned, the first method based on (9.19) is used to determine the coefficients of the numerator and denominator. After $[q]$ and $[p]$ is obtained, the NF data of the missing gap can be extrapolated by the Cauchy method with Gegenbauer polynomials.

If the measured data are in the presence of noise, an iterative version of the Cauchy method is proposed to reduce the influence of noise. In each step of the iteration, the high-frequency part of the extrapolated result, which is related to noise is discarded. The main steps are summarized below:

1) *Input*: measured samples f_m.
2) *Initialization*: $k = 1, f_{input} = f_m$.
3) *Iteration*:
   i) Compute extrapolated results f_e by the Cauchy method with the input data f_{input}.
   ii) $f_k = \begin{cases} f_e, & \theta \in (180° - \theta_g, 180° + \theta_g) \\ f_m, & \theta \notin (180° - \theta_g, 180° + \theta_g) \end{cases}$.
   iii) $\widetilde{f}_k = \text{FFT}(f_k)$.
   iv) Discard high-frequency entries of $\widetilde{f}_k$.
   v) $f_u = \text{IFFT}\left(\widetilde{f}_k\right)$.
   vi) $f_{input} = f_u, \theta \notin (180° - \theta_g, 180° + \theta_g)$.
   vii) $k = k + 1$.
   Do Until $\|f_k - f_{k-1}\|_2 < \varepsilon$ or $k > K$.
4) *Output*: f_k.

In the above operations FFT stands for the fast Fourier transform and IFFT represents the Inverse Fast Fourier transform.

9.3.1 Numerical Results and Discussion

In this section, three numerical examples dealing with different antennas are presented to illustrate the applicability of the proposed extrapolation method in reconstruction of antenna NF distribution and FF pattern from incomplete samples. In these examples, the complete NF data is obtained by electromagnetic (EM) simulation package HOBBIES [23].

9.3.1.1 Example of a Horn Antenna

For the first example, a horn antenna is considered at a frequency of 1.55 GHz, as shown in Figure 9.12. The width, height, and length of the waveguide part are 0.71λ, 0.36λ, and 1.66λ, respectively. The width, height and length of the horn section are 2.84λ, 2.21λ, and 2.38λ, respectively. The distance between the center of AUT and the sampling spherical surface is 7.75λ. The horn antenna is fed by a wire pin.

For the case (a), θ_g is 30°, and the size of the NF missing hole is 60° in θ at every angle of ϕ on the measurement surface. As shown in Figure 9.12(a), the aperture of the horn is directed towards $\theta = 0°$, where the missing gap is at the bottom. Figure 9.13 shows the NF at $\phi = 90°$ extrapolated by the proposed method, which agrees well with the reference. We apply the NF to FF transformation, after the NF is completely extrapolated. Figure 9.14 and Figure 9.15 present the NF to FF transformation results by extrapolated NF and the incomplete NF data with the missing gap.

For the case (b), the aperture of the horn is directed towards $\theta = 90°$, where the missing gap is at the side of the horn antenna. The radiation pattern estimated by extrapolated NF is presented in Figures 9.16 and 9.17. As shown in Figure 9.16, the FF pattern estimated by the extrapolated NF is much better than the one obtained directly by setting the electric field to zero in the NF missing

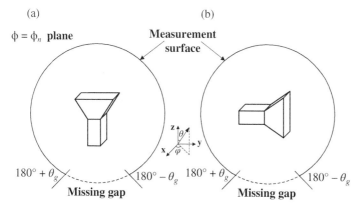

Figure 9.12 A horn antenna for the frequency 1.55 GHz. (a) Aperture of the horn directed towards q = 0°; missing gap at the bottom of the horn. (b) Aperture directed towards q = 90°; missing gap at the side of the horn.

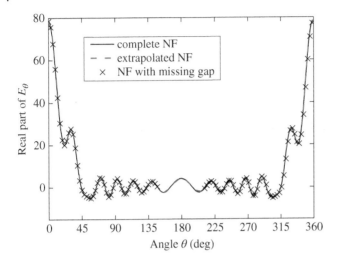

Figure 9.13 Real part of E_θ for $\phi = 90°$ on a NF measuring surface.

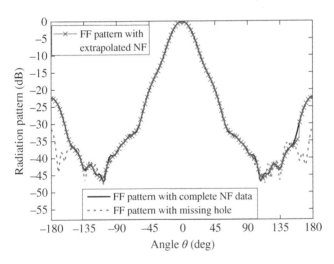

Figure 9.14 Radiation pattern of the horn antenna for the plane $\phi = 0°$.

gap. As can be seen from Figure 9.17, the radiation pattern in the horizontal plane is not affected much by the data in the missing gap, which results in a similar situation in the following examples. Therefore, the FF results in the horizontal plane are not shown in the following sections when the main beam of AUT is directed towards $\theta = 90°$.

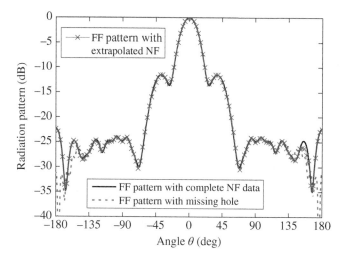

Figure 9.15 Radiation pattern of the horn antenna for the plane $\phi = 90°$.

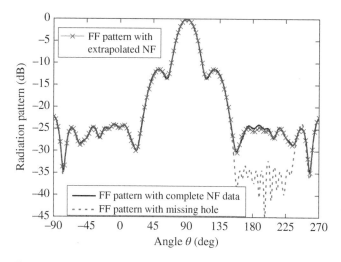

Figure 9.16 Radiation pattern of the horn antenna for the plane $\phi = 90°$.

9.3.1.2 Example of a 2-element Microstrip Patch Array

In this example, a 2-element microstrip patch array is considered at a frequency of 3 GHz, as shown in Figure 9.18. The sampling performed on a spherical surface has a radius of 1λ with the center as the AUT. The length, width, and height of the substrate are 0.5λ, 0.8λ, and 0.03λ, respectively. In this example, θ_g is $45°$, and the size of the NF missing hole is $90°$ in θ at every angle of ϕ on the

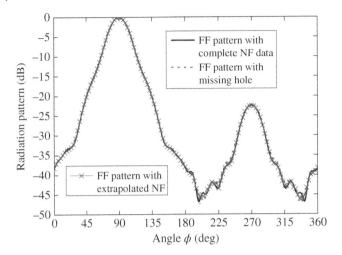

Figure 9.17 Radiation pattern of the horn antenna on the plane $\theta = 90°$.

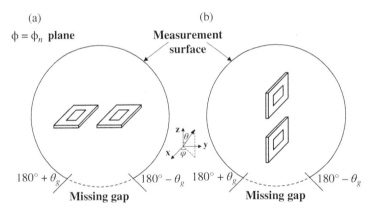

Figure 9.18 A 2-element microstrip patch array operating at 3 GHz. (a) Patch array perpendicular to the direction q=0°; missing gap at the bottom. (b) Patch array perpendicular to the direction q=90°; missing gap at the side of the patch array.

measurement surface. For the case (a) in Figure 9.18, the patch array is perpendicular to the direction $\theta = 0°$, whereas the missing gap is at the bottom of the array. For the case (b), the patch array is perpendicular to the direction $\theta = 90°$, whereas the missing gap is at the side of the array.

It is illustrated in Figure 9.19 that the extrapolated NF by the proposed method agrees well with the reference. The NF to FF transformation is applied after we obtain the extrapolated NF distribution. Figure 9.20 and Figure 9.21 show the radiation pattern of the microstrip array, which demonstrate the effectiveness of the proposed extrapolation by the NF to FF transformation results.

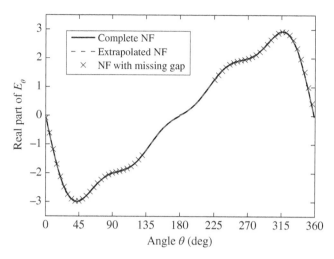

Figure 9.19 Real part of E_θ for $\phi = 90°$ on a NF measuring surface.

For the case (b), there is a gap of 90° in the data which has been removed on the side of the microstrip array as shown in Figure 9.18(b). Figure 9.22 is the NF extrapolation result for $\phi = 90°$. Figure 9.23 shows the FF pattern obtained from the extrapolated NF and incomplete NF data, respectively. As can be seen from Figure 9.23, the FF pattern estimated by the incomplete NF data has a mismatch with the reference even in the main beam, while the extrapolated NF results in a good agreement with the reference after the NF to FF transformation.

9.3.1.3 Example of a Parabolic Antenna

For the third example, a parabolic reflector antenna with 0.375 F/D (Focal Length to Diameter) ratio is considered at a frequency of 1.55 GHz. The diameter of the reflector is 6.46λ. The antenna is fed by a circular-dish backed dipole, and the dish diameter is 0.553λ with the distance of 0.3λ from the dipole. The aperture of the reflector is directed towards $\theta = 90°$, where the missing gap is at the side of the antenna as shown in Figure 9.24. In this example, θ_g is 45°, and the size of the NF missing hole is 90° in θ at every angle of ϕ on the measurement surface.

It is illustrated in Figure 9.25 and Figure 9.26 that the NF extrapolated by the proposed method agrees well with the reference. The NF to FF transformation is applied after we obtain the extrapolated NF distribution. Figure 9.27 shows the radiation pattern of the parabolic antenna. As observed, the FF pattern estimated by the incomplete NF data has a big mismatch with the reference in the side lobes. The NF to FF transformation results from the extrapolated NF data has a good agreement with the reference, which demonstrate the effectiveness of the proposed extrapolation.

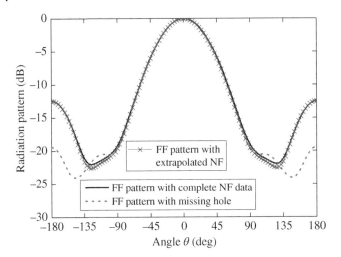

Figure 9.20 Radiation pattern of the microstrip array on the plane $\phi = 0°$.

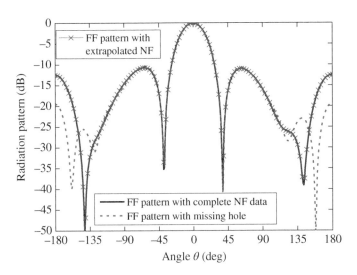

Figure 9.21 Radiation pattern of the microstrip array on the plane $\phi = 90°$.

Next we see how this method performs with respect to the Matrix pencil method and using the Cauchy method using polynomials for the interpolation of missing near field data on a spherical surface.

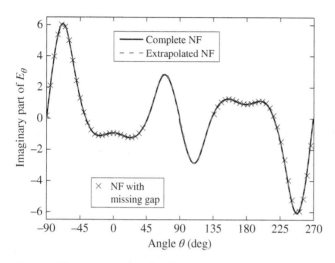

Figure 9.22 Imaginary part of E_θ for $\phi = 90°$ on a NF measuring surface.

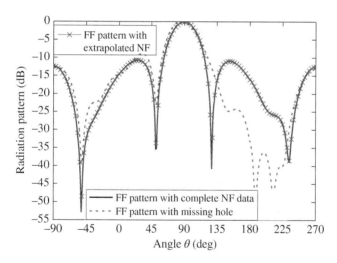

Figure 9.23 Radiation pattern of the microstrip array on the plane $\phi = 90°$.

9.4 Near-Field to Far-Field Transformation of a Zenith-Directed Parabolic Reflector Using the Ordinary Cauchy Method

The next two examples showcase the real-world applications of the Cauchy method. For the first example, the goal was to interpolate the "hole" of the near-field measurement of a parabolic reflector antenna. The "hole" that is

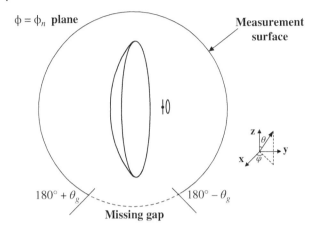

Figure 9.24 A parabolic antenna operating at a frequency 1.55 GHz.

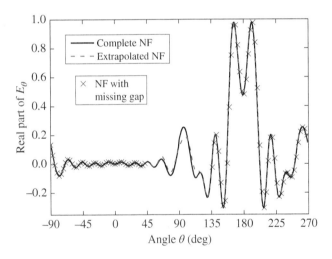

Figure 9.25 Real part of E_θ for $\phi = 110°$ on a NF measuring surface.

being referred to is a spherical area where a near-field probe is unable to reach to accurately obtain antenna measurements. The parabolic reflector antenna was simulated using HOBBIES, a Method of Moments software package. This particular example had the hole placed at 65° below the horizon (155° from zenith) and extending to 180° from zenith in θ and swept 360° in φ. Figure 9.28 illustrates the design and orientation of the zenith-directed parabolic reflector along with the spherical near-field boundaries. The blue sphere in Figure 9.28 represents the near-field boundary where simulation measurements were taken.

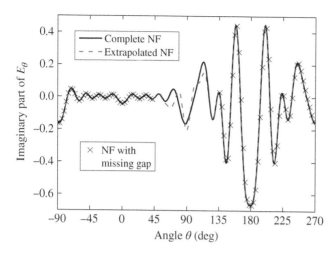

Figure 9.26 Imaginary part of E_θ for $\phi = 90°$ on a NF measuring surface.

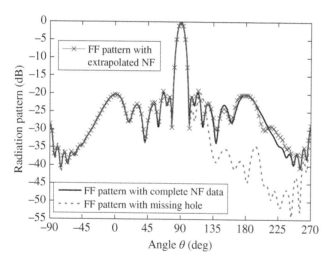

Figure 9.27 Radiation pattern of the horn antenna on the plane $\phi = 90°$.

The θ and the ϕ component of the electric near-field was sampled at $1°$ increments in θ for a total of 181 samples and at $1°$ increments in ϕ for a total of 361 samples. After simulation, the data was exported to MATLAB to be interpolated by using the Cauchy method. In MATLAB the near-field data was re-organized so that at each cut of ϕ, $360°$ of data in θ was available. For example, the near-field data at $\phi = 0°$ and $\phi = 180°$ was concatenated together to produce a full $360°$ circular cut of near-field data in θ. The data from $+155°$ to $+180°$ and $-155°$ to

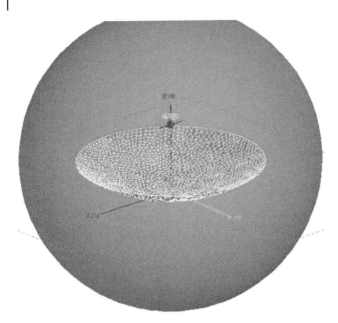

Figure 9.28 Zenith-directed parabolic reflector with a spherical near-field sampling boundary.

−180° was removed to create the hole of missing data. The data from 0° (zenith) to +154° and from 0° to −154° was sampled ($N = 310$ samples). After interpolation, the estimated near-field data was re-assembled to match the form and organization of the original data and then it was transformed to the far-field using a spherical near-field to far-field transformation code written in FORTRAN and presented in Chapter 7.

There are four components of the near-field that were interpolated, the real and imaginary parts of the θ-component and the real and imaginary parts of the φ-component. In total 720 near-field vectors were interpolated; four component near-fields with 180 cuts in φ. Due to the limited space, we will only present eight of the 720 total results. We will look at the interpolation of the real and imaginary parts of the θ and φ components of the near field for the φ cuts of φ = 0° and φ = 90°. In addition to those eight plots, we will also present the far-field that results from the spherical near-field to far-field transformation.

We will first look at the real and imaginary parts of the θ-component of the near-field (E_θ) for φ = 0°. From their singular value distributions and the error analysis code, it was determined that 45 and 65 singular values were needed to sufficiently interpolate the real and imaginary parts respectively. This leads to a polynomial order of 22 and 23 for the numerator and denominator respectively for the real part, and a numerator and denominator polynomial order of 32 and 33 for the imaginary part. Figures 9.29 and 9.30 show the result of the

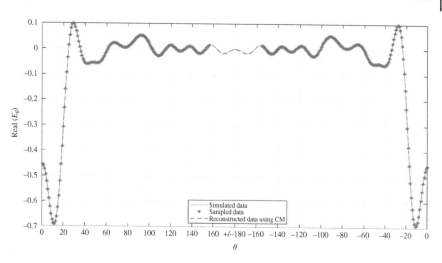

Figure 9.29 Interpolation of the real part of E_θ for $\phi = 0°$.

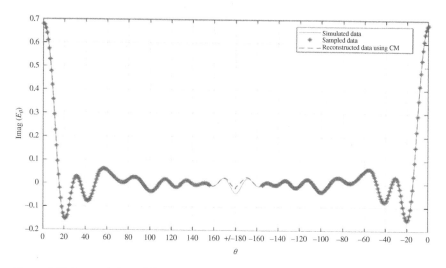

Figure 9.30 Interpolation of the imaginary part of E_θ for $\phi = 0°$.

interpolation of the real and imaginary parts of E_θ for $\phi = 0°$. The MSE of the interpolation of the real part of E_θ was calculated to be 0.0036 and for the imaginary part it was 0.0392.

Next, will look at the real and imaginary parts of E_θ for $\phi = 90°$. The number of singular values needed to accurately interpolate the real and imaginary parts was determined to be 71 and 61 respectively. Once again, these values correlate to the optimal rank of the input data matrix and are used to determine the optimal orders of the numerator and denominator polynomials. The order of the numerator and denominator polynomials for the real part of E_θ was calculated to be 35 and 36 respectively and for the imaginary part they were determined to be 30 and 31 respectively. Figures 9.31 and 9.32 show the results of the interpolation using the Cauchy method for the real and imaginary parts of E_θ for $\phi = 90°$, respectively.

By inspection, it is easy to see that the performance of the interpolation is worse for the $\phi = 90°$ cut. This is due to the fact that the amplitude of the near-field is of the order of 10^{-3} and interpolating at those levels is less accurate. Also, the fields are not symmetric. As we will see, this error does not contribute much to the final far-field result after the application of the spherical near-field to far-field transformation. The MSE of the interpolation of the real part of E_θ was calculated to be 0.1835 and for the imaginary part it was 0.2024.

The real and imaginary parts of the ϕ component of the near-field (E_φ) at $\phi = 0°$ will be presented next. Once again, the number of singular values needed to sufficiently interpolate the real and imaginary parts was determined to be 53 and 35 respectively. From these values the numerator and denominator polynomial orders for the real part of E_φ was calculated to be 26 and 27 respectively and for the imaginary part they were calculated to be 17 and 18 respectively. Figures 9.33 and 9.34 show the results of the interpolation for the real and

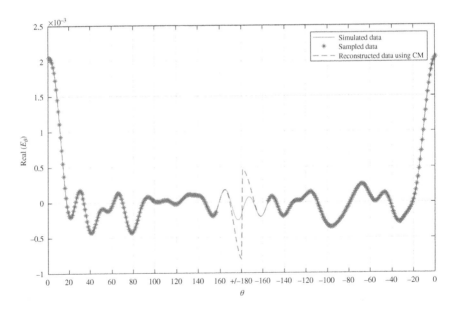

Figure 9.31 Interpolation of the real part of E_θ for $\phi = 90°$.

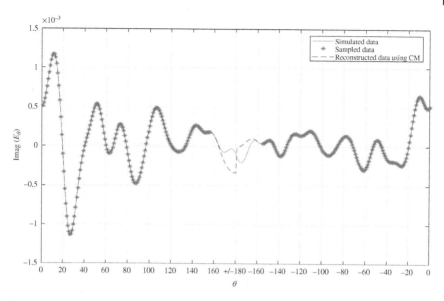

Figure 9.32 Interpolation of the imaginary part of E_θ for $\phi = 90°$.

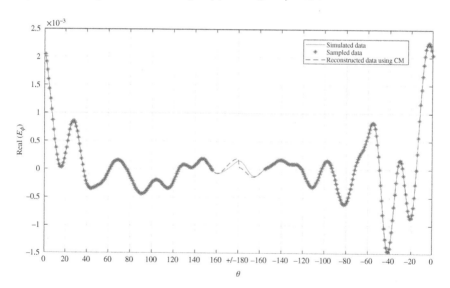

Figure 9.33 Interpolation of the real part of E_φ for $\phi = 0°$.

imaginary parts of E_φ respectively. The MSE's of the interpolation of the real and imaginary parts of E_φ at $\phi = 0°$ were found to be 0.0377 and 0.0663 respectively.

Finally, we will look at the interpolation of the real and imaginary parts of E_φ for $\phi = 90°$. The optimal number of singular values was determined to be 57 for

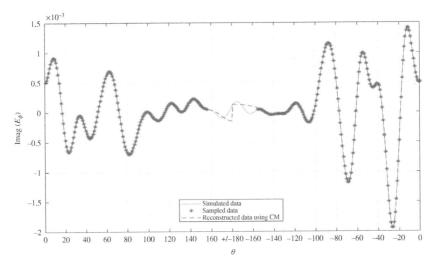

Figure 9.34 Interpolation of the imaginary part of E_φ for $\varphi = 0°$.

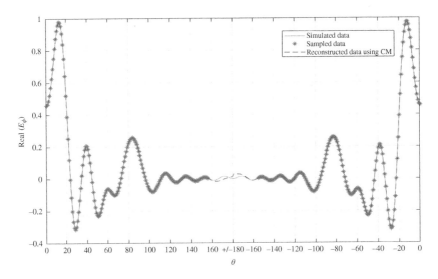

Figure 9.35 Interpolation of the real part of E_φ for $\varphi = 90°$.

the real part and 67 for the imaginary part. This led to numerator and denominator polynomial orders of 28 and 29 for the real part of E_φ, and orders of 33 and 34 for the imaginary part. Figures 9.35 and 9.36 shows the results of the interpolation for the real and the imaginary parts of E_φ, respectively.

The MSE was calculated to be 0.0161 for the interpolation of the real part and 0.0030 for the imaginary part.

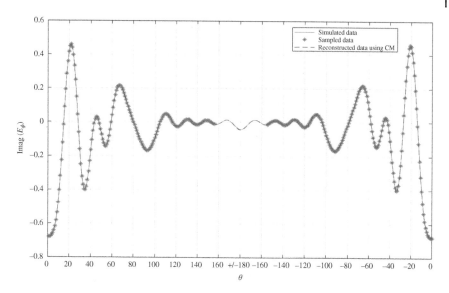

Figure 9.36 Interpolation of the imaginary part of E_φ for $\varphi = 90°$.

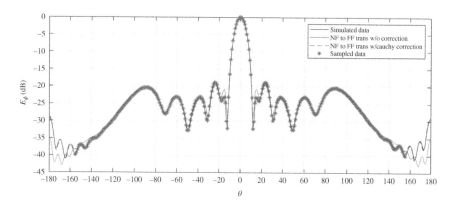

Figure 9.37 Comparison between the original far-field and interpolated far-field.

After all 720 components of the near-field were interpolated, the data was then used in a spherical near-field to far-field transformation. Figure 9.37 shows the resulting far-field ($\varphi = 0°$ cut) with the "hole" interpolated including a comparison to the simulated far-field without the hole. From Figure 9.37, one can say that it is possible to interpolate the missing data in the near-field measurements spanning 50° using the Cauchy method and then construct an accurate representation of the far-field using a spherical near-field to far-field transformation. The MSE of the resulting far-field was calculated to be 0.0070.

9.5 Near-Field to Far-Field Transformation of a Rotated Parabolic Reflector Using the Ordinary Cauchy Method

In this example, the zenith-directed reflector antenna introduced in the previous section was rotated counter-clockwise by 90°. The location of the hole of the missing data was kept the same. However, an interesting result of this antenna configuration is that the hole was able to be expanded significantly without increasing the error of the interpolation. Instead of the missing data spanning 50° like in the previous example, the hole representing the missing data was expanded to 90° with sufficient accuracy in the interpolation of the data. Figure 9.38 shows the rotated parabolic reflector antenna with the spherical near-field boundary where measurements were taken.

Eight plots will be shown just like in the previous example; the real and imaginary parts of the θ and ϕ components of the near field (E_θ and E_φ) for $\phi = 0°$ and 90°.

We will first look at the real and imaginary parts of E_θ for the cut of $\phi = 0°$. From the distribution of the singular values in conjunction with the error analysis, it was determined that the optimal rank of the matrix $[C]$ is 55 for the real part and 27 for imaginary part. This leads to a numerator polynomial order of 27 and a denominator polynomial order of 28 for the real part. For the imaginary

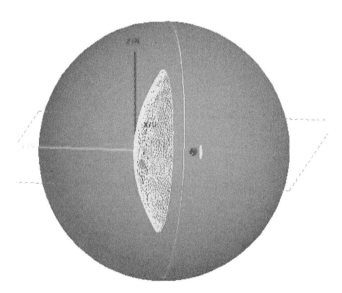

Figure 9.38 Rotated parabolic reflector antenna.

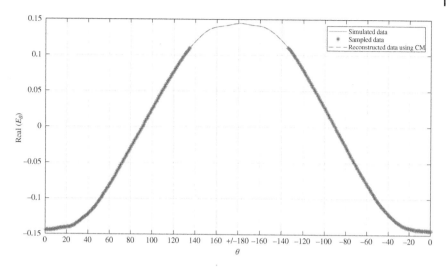

Figure 9.39 Interpolation of the real part of E_θ for $\phi = 0°$.

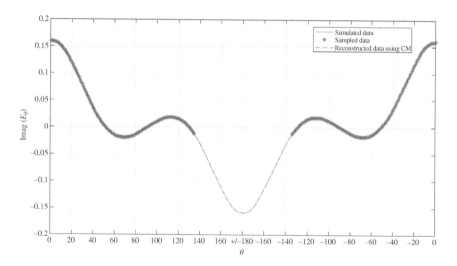

Figure 9.40 Interpolation of the imaginary part of E_θ for $\phi = 0°$.

part, the numerator polynomial order was 13 and the denominator polynomial was 14. Figures 9.39 and 9.40 show the result of the interpolation for the real and the imaginary parts of E_θ for $\phi = 0°$, respectively.

Comparing Figures 9.39 and 9.40 to Figures 9.29 and 9.30 one can observe by inspection that the error of the interpolation is much less for the rotated case versus the zenith-directed case. The near-fields of the rotated antenna at this cut of ф are not as complex, very well behaved and symmetric which improves the accuracy of interpolation greatly by the Cauchy method. The MSE of interpolating the real part of E_θ was calculated to be 0.0036 and for the imaginary part it was calculated to be 0.0079.

Next, we will look at the interpolation of the real and imaginary parts of E_θ for the ф = 90° cut. This cut showed the worst performance for the Cauchy method out of all of the cuts of ф that were interpolated. As one is dealing with very low amplitudes as well as near-field patterns that are not symmetric and for these reasons the Cauchy method did not perform well. Due to these reasons, the optimal number of singular values for the real part of E_θ was only 2. With an optimal rank of 2, this led to an interesting result for the optimal orders of the numerator and denominator polynomial. The order of the numerator polynomial was found to be 0 and the order of the denominator polynomial was found to be 1, therefore an approximate solution to the problem was not found. For the interpolation of the imaginary part which performed slightly better, the optimal number of the singular values was 65. With a rank of 65, the optimal orders of the numerator and denominator polynomial were calculated to be 32 and 33. Figures 9.41 and 9.42 show the result of interpolating the real and the imaginary parts of E_θ for ф = 90°, respectively. Due to the fact that the amplitude of the near-field for ф = 90° is so small, these errors did not contribute significantly to the overall error of the far-field after the spherical near-field to far-field transformation was applied. The MSE of the interpolation of the real part of E_θ was calculated to be 1.0748 and for the imaginary part it was calculated to be 0.6006.

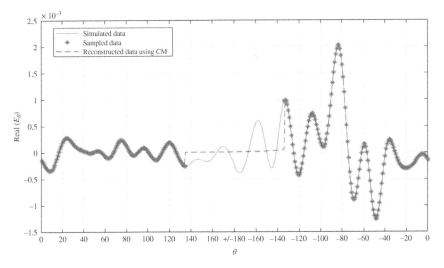

Figure 9.41 Interpolation of the real part of E_θ for ф = 90°.

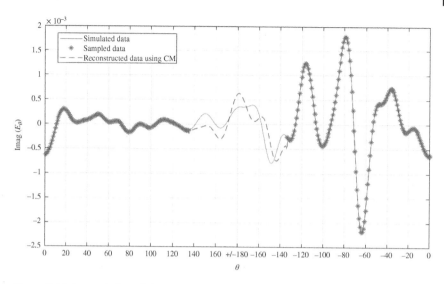

Figure 9.42 Interpolation of the imaginary part of E_θ for $\phi = 90°$.

The estimation of the real and imaginary parts of E_φ for the $\phi = 0°$ cut is shown next. The optimal rank, determined by the distribution of the singular values, was determined to be 63 for the real part of E_φ and 27 for the imaginary part. From the optimal rank of the input data matrix, the optimal order of the numerator polynomial was calculated to be 31 and 32 for the denominator polynomial for the real part of E_φ. The optimal order of the numerator and denominator polynomials for the imaginary part of E_φ was calculated to be 13 and 14 respectively. Figures 9.43 and 9.44 show the interpolation results for the real and imaginary parts of E_φ respectively.

The interpolation of the real part of E_φ had an MSE of 0.0075 and the interpolation of the imaginary part had an MSE of 0.0304.

For the last set of plots, we will look at the interpolation of the real and imaginary parts of E_φ for the ϕ cut of 90°. The optimal number of singular values and hence the optimal rank of the input data matrix was determined to be 53 for the real part and 47 for the imaginary part. With these values, the optimal orders of the numerator and denominator polynomials were calculated to be 26 and 27 respectively for the real part and 23 and 24 for the imaginary part. Figures 9.45 and 9.46 show the results of the interpolation of the real and imaginary part of E_φ for $\phi = 90°$ respectively.

The MSE of the interpolation of the real part was calculated to be 9.5767×10^{-4} and 0.0015 for the imaginary part.

After the real and imaginary parts of the θ and ϕ components of the near-field for each cut of ϕ were interpolated, the data was then used in a spherical near-field to far-field transformation. Figure 9.47 shows the resulting far-field for $\phi = 0°$.

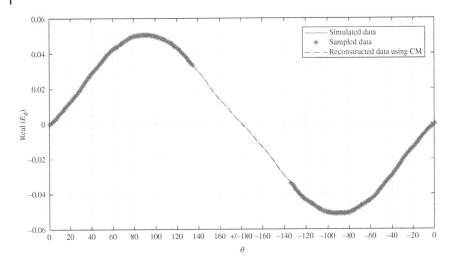

Figure 9.43 Interpolation of the real part of E_φ for $\varphi = 0°$.

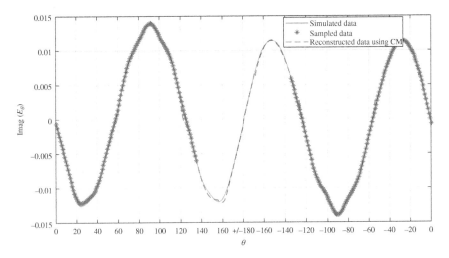

Figure 9.44 Interpolation of the imaginary part of E_φ for $\varphi = 0°$.

From Figure 9.47, the solid blue line represents the simulated far-field, the dashed red line represents the result of transforming the near-field with the hole interpolated to the far-field, and the red asterisks represent the sampled data. The solid green line shows the result of the near-field to far-field transformation if the hole of the near-field was replaced with zeroes and then transformed. It can be easily shown that interpolating the missing data of the near-field using

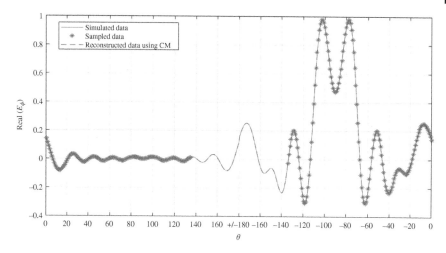

Figure 9.45 Interpolation of the real part of E_φ for $\phi = 90°$.

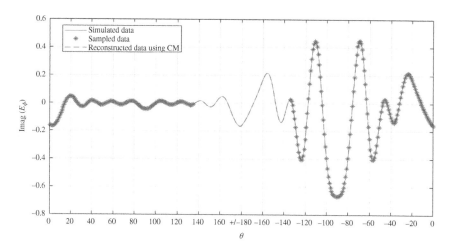

Figure 9.46 Interpolation of the imaginary part of E_φ for $\phi = 90°$.

the Cauchy method before the transformation is much more accurate than just replacing the unknown data with zeroes. The overall MSE of the far-field using the interpolated near-field data was calculated to be 0.0080. If we were to replace the hole of the missing near-field data with zeroes, then the MSE would increase to 0.3197.

Comparing the accuracy of the Cauchy method in interpolating the near-field of the zenith-directed and rotated parabolic reflector antenna's, it is obvious that it performed much better with the rotated antenna. We were even able

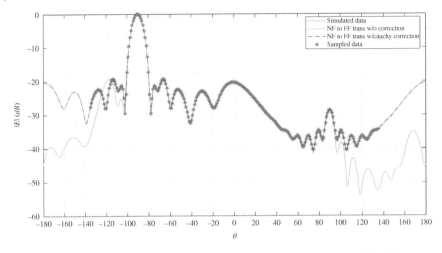

Figure 9.47 Comparison between the original far-field and the interpolated far-field.

to expand the hole of missing data from 50° to 90° in θ for the rotated reflector case. If we take a look at the complexity of the missing data in both Figures 9.37 and 9.47 one can observe that the rotated case does not have as many peaks and nulls as the zenith-directed case. The individual near-field components are also much more complex for the zenith-directed case than for the rotated case.

9.6 Near-Field to Far-Field Transformation of a Zenith-Directed Parabolic Reflector Using the Matrix Pencil Method

We are now going to revisit the zenith-directed parabolic reflector that was introduced in section 9.4 while using the Cauchy method. Similar to the previous Matrix Pencil method related sections, we will be turning each approximation into one of extrapolation instead of interpolation. The simulation parameters and antenna geometry are exactly the same as described in section 9.4 and once again was designed and simulated using HOBBIES. Please refer to Figure 9.28 for the design and orientation of the parabolic reflector antenna. In addition to changing the method of approximation from an interpolation to an extrapolation, we also must change the method of sampling to achieve the best performance with the Matrix Pencil method. For the Cauchy method, we sampled the near field from 0 degrees to +/– 155 degrees. For the matrix pencil method, we will be sampling from +/– 90 degrees to +/–135 degrees and extrapolating from +/–135 degrees to +/–180 degrees ($N = 46$ samples).

Since there are over 720 pieces of data that make up the entire complex near-field of the simulated antenna, we will not be looking at every cut for ϕ. Just like for what was shown for the Cauchy method, we will be looking at the real and imaginary parts of E_θ and $E\phi$ for the ϕ cuts of 0° and 90°. Please note that while we will be looking at the real and imaginary parts separately, the algorithm of the Matrix Pencil method actually extrapolates the complex near-field, that is the real and imaginary parts combined. For ease of comparison with the results of the Cauchy method, the real and imaginary parts of near-field were extracted from the final extrapolation result.

First, we will look at the extrapolation results for the real and imaginary parts of E_θ for $\phi = 0°$. Since the number of samples N is 46, the pencil parameter L is 22 ($L = N/2 - 1$). The optimal number of singular values M was determined to be 12 for the extrapolation of the complex near-field of E_θ. Figures 9.48 and 9.49 show the resulting real and imaginary parts of the complex near-field of E_θ for $\phi = 0°$.

The MSE of the extrapolation of the real and imaginary parts of E_θ for $\phi = 0°$ was calculated to be 0.0206 and 0.0694. While the error is higher using the Matrix Pencil method compared to the interpolation of the same fields using the Cauchy method, keep in mind that we are only using 92 samples (46 samples for each half of the near-field versus 155 samples used in the Cauchy method) and we are extrapolating a larger range in θ (45° on each side versus 25° with the Cauchy method).

Let us look at the extrapolation results for the real and imaginary parts of E_θ for $\phi = 90°$. The optimal number of singular values M was determined to be 6. Figures 9.50 and 9.51 show the resulting real and imaginary parts of the complex near-field of E_θ for $\phi = 90°$, respectively.

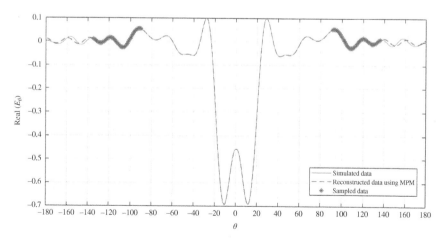

Figure 9.48 Extrapolation of the real part of E_θ for $\phi = 0°$.

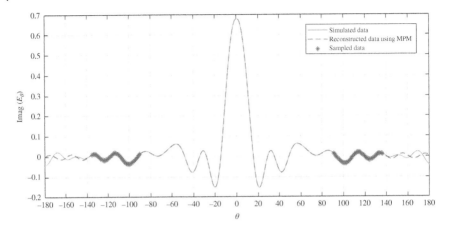

Figure 9.49 Extrapolation of the Imaginary part of E_θ for $\phi = 0°$.

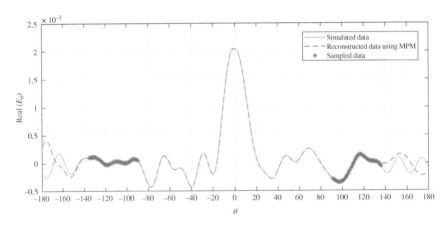

Figure 9.50 Extrapolation of the real part of E_θ for $\phi = 90°$.

The MSE's for approximating the real and imaginary near-fields as a result of extrapolating the complex near-field were found to be 0.2378 and 0.2052 respectively.

Next, we will look at the extrapolation of the real and imaginary parts of E_φ for $\phi = 0°$. The optimal number of singular values M was determined to be 6. Figures 9.52 and 9.53 show the real and imaginary parts of E_φ for $\phi = 0°$ as a result of the extrapolation of the complex near-field.

The MSE's for the extrapolation of the real and imaginary part of E_φ for $\phi = 0°$ were calculated to be 0.1474 and 0.0985 respectively.

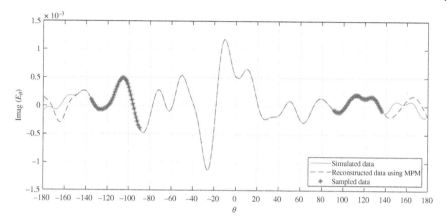

Figure 9.51 Extrapolation of the imaginary part of E_θ for $\phi = 90°$.

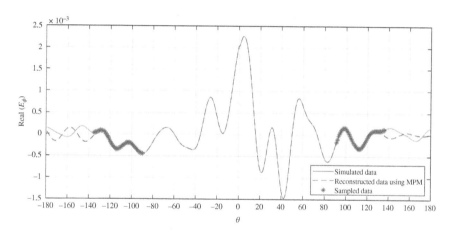

Figure 9.52 Extrapolation of the real part of E_φ for $\phi = 0°$.

For the final two plots, we will look at the real and imaginary parts of E_φ for $\phi = 90°$ as a result of the extrapolation of the complex near-field. The optimal number of singular values M was determined to be 16. Figures 9.54 and 9.55 show the extrapolation results of the real and imaginary parts of E_φ for $\phi = 90°$. The MSE's of the real and imaginary part as a result of the extrapolation of the complex near-field were calculated to be 0.0301 and 0.0268.

Once all of the near-field results were extrapolated, the total data was then exported to perform a spherical near-field to far-field transformation. The larger errors seen in the extrapolation of E_θ for $\phi = 90°$ and E_φ for $\phi = 0°$ did not

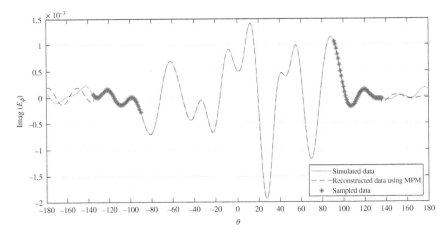

Figure 9.53 Extrapolation of the imaginary part of E_φ for $\phi = 0°$.

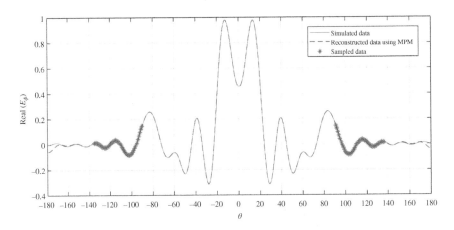

Figure 9.54 Extrapolation of the real part of E_φ for $\phi = 90°$.

have a negative impact on the final far-field result, due to the fact that those fields were much smaller in amplitudes. Figure 9.56 shows the result of the near-field to far-field transformation. In Figure 9.56, the solid blue line represents the simulated far-field, the dashed red line represents the result of transforming the extrapolated near-field to the far-field, and the red asterisks represent the sampled data. The solid green line shows the result of the near-field to far-field transformation if the hole of the near-field was replaced with zeroes and then transformed. It can be easily shown that extrapolating

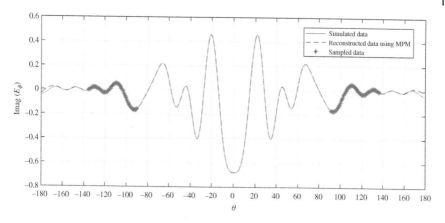

Figure 9.55 Extrapolation of the imaginary part of E_φ for $\phi = 90°$.

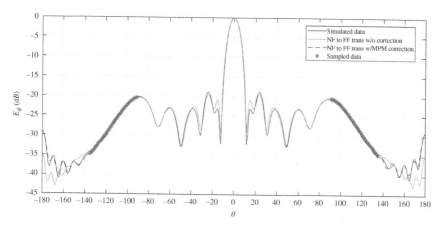

Figure 9.56 Comparison between the original far-field and the extrapolated far-field.

the missing data of the near-field using the Matrix Pencil method before filling up the missing data is much more accurate than just replacing the unknown data with zeroes. The overall MSE of the far-field using the interpolated near-field data was calculated to be 0.0104. If we were to replace the hole of missing near-field data with zeroes, then the MSE would increase to 0.0644.

While the Cauchy method had better accuracy interpolating the near-field and subsequently the far-field, the Matrix Pencil method was able to sufficiently extrapolate the near-field using less samples and it was also able to extrapolate also over a larger range in θ.

9.7   Near-Field to Far-Field Transformation of a Rotated Parabolic Reflector Using the Matrix Pencil Method

For the last example, we will be looking at the same rotated parabolic reflector antenna as was introduced in section 9.5. The rotated reflector antenna was simulated using HOBBIES and the near-field data was exported to MATLAB. For the rotated case it was difficult to get accurate results unless the hole of the missing data was reduced to 60° (30° on each side of the pattern), much like the zenith-directed case using the Cauchy method. So, the data was sampled from 90° to +/−150° (N = 61 samples) and extrapolated from +/−151 degrees to +/−180 degrees. Like in the previous section, we will be looking at the real and imaginary parts of E_θ and E_φ for the ϕ cuts of 0° and 90°.

First, we will look at the results of the extrapolation of E_θ for ϕ = 0°. Since the number of samples we are using for the input data matrix [Y] is equal to 61, the pencil parameter L is equal to 29. From the singular value distribution along with the error analysis, the optimal number of singular values M for the extrapolation of the complex near-field E_θ was determined to be 13. Using (2.17) through (2.24) one can estimate the poles and residues related to the near-field data. Figures 9.57 and 9.58 show the result of the extrapolation. The MSE of the extrapolation of the real part of E_θ was calculated to be 0.0097 and 0.0419 for the imaginary part.

Next, the extrapolation of the real and imaginary part of E_θ for ϕ = 90° is presented. From the singular values distribution, the optimal number of singular values for the extrapolation was determined to be 18. Figures 9.59 and 9.60 both show the result of the extrapolation of the real and imaginary parts of E_θ for ϕ = 90°.

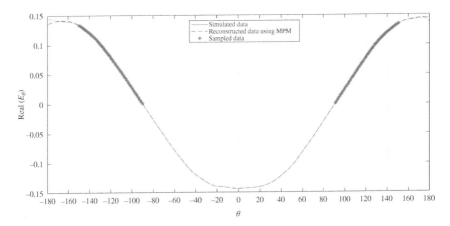

Figure 9.57 Extrapolation of the real part of E_θ for ϕ = 0°.

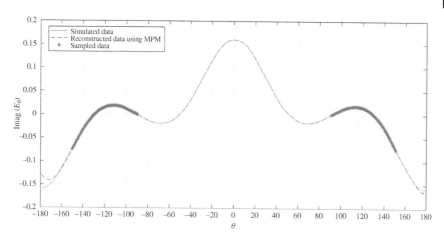

Figure 9.58 Extrapolation of the imaginary part of E_θ for $\phi = 0°$.

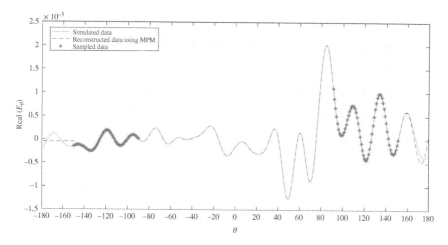

Figure 9.59 Extrapolation of the real part of E_θ for $\phi = 90°$.

The MSE for the extrapolation of the real and imaginary parts of E_θ for $90°$ was calculated to be 0.1070 and 0.0898 respectively. We are seeing higher errors here most likely due to the complexity of the near-field pattern. Notice that we are dealing with a highly unsymmetric pattern.

We will now move onto the ϕ-component of the near-field next for the ϕ cut of $0°$. The optimal number of singular values M was determined to be 11. Figures 9.61 and 9.62 show the resulting real and imaginary parts of E_φ from the extrapolation of the complex near-field. The MSE for this extrapolation was calculated to be 0.0067 for the real part and 0.0167 for the imaginary part.

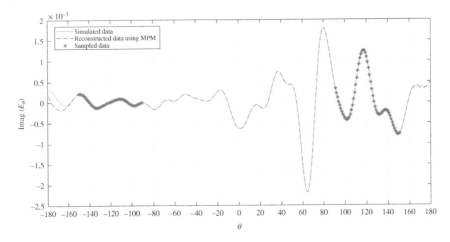

Figure 9.60 Extrapolation of the imaginary part of E_θ for $\phi = 90°$.

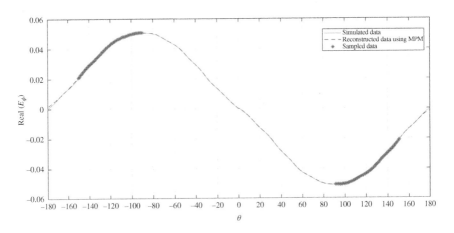

Figure 9.61 Extrapolation of the real part of E_φ for $\phi = 0°$

Finally, we will look at the result of the extrapolation for the real and imaginary parts of E_φ for $\phi = 90°$. The optimal number of singular values was determined to be 24. Figures 9.63 and 9.64 show the result of this extrapolation of the real and imaginary part of E_φ for $\phi = 90°$. The MSE of the extrapolation of the real and imaginary part of E_φ at $\phi = 90°$ was calculated to be 0.1719 and 0.2296 respectively.

Once all of the near-field patterns were extrapolated for each cuts of ϕ, the data was exported and used in a spherical near-field to far-field transformation

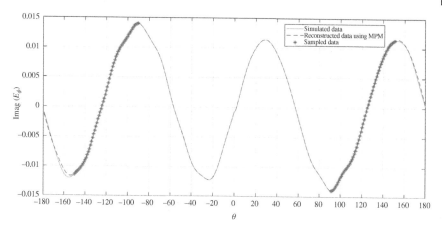

Figure 9.62 Extrapolation of the imaginary part of E_φ for $\phi = 0°$.

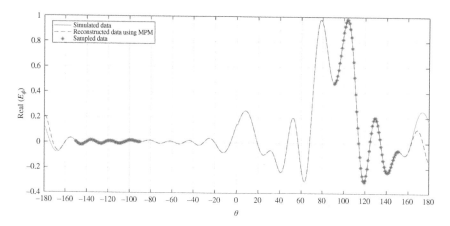

Figure 9.63 Extrapolation of the real part of E_φ for $\phi = 90°$.

code, just performed as in the previous sections. Figure 9.65 shows the normalized far-field which resulted from the transformation of the extrapolated near-field patterns. The solid blue line represents the simulated far-field, the dashed red line represents the far-field that resulted from the near-field to far-field transformation with the hole of missing data extrapolated, and the red asterisks represent the sampled data points. The solid green line is the far-field result if instead of extrapolating the missing data, zeroes were inserted. The MSE between the far-field that resulted from the near-field to far-field transformation with the missing data extrapolated and the simulated far-field was

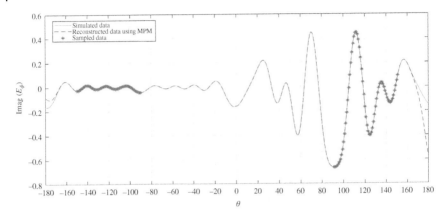

Figure 9.64 Extrapolation of the imaginary part of E_φ for $\phi = 90°$.

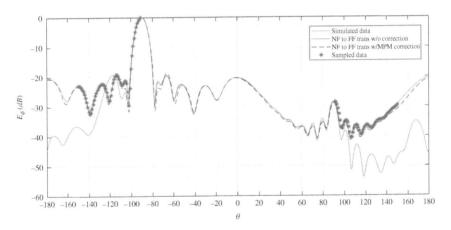

Figure 9.65 Comparison between original and extrapolated far-field.

calculated to be 0.0345. The MSE between the transformed far-field without the Matrix Pencil method correction and the simulated far-field was calculated to be 0.3197.

While the extrapolation of the near-field of the rotated parabolic reflector antenna using the Matrix Pencil method was not as accurate as the interpolation of the same near-fields using the Cauchy method, the end result after the near-field to far-field transformation is acceptable. The Matrix Pencil method also could not properly extrapolate the near-field when the gap of the missing data was increased to 90°, for which the Cauchy method performed very well. However, the accuracy of the extrapolation and resulting far-field is much better than that of the far-field without the near-fields extrapolated.

9.8   Conclusion

In this chapter, first, the Chebyshev polynomials have been used to extract from the data collected under non-anechoic environment and generate a radiation pattern for the same antenna radiating in free space conditions. The problem with this method is that some a priori information is required about the environment. Next, the Cauchy method using Gegenbauer polynomials is proposed for antenna NF extrapolation and the FF estimation particularly when the order of the approximation is large. A reliability parameter is introduced to check the correctness of the NF extrapolation results. This section illustrates the application of the proposed method in extrapolating the NF distribution of several antennas with a missing gap of the NF data. As shown in the numerical results, the radiation pattern is inaccurate if we obtain the FF directly from the incomplete NF data using a NF to FF transformation without extrapolation of the missing data. It is also demonstrated that good FF results are obtained by NF to FF transformation after extrapolating the NF data. It is worth pointing out that extrapolation can work better with a smaller range of missing data, as the error of extrapolation increases in relation to the range.

Finally, results are also presented for a parabolic reflector with missing near field data over a spherical surface using the ordinary Cauchy method using polynomials and the Matrix Pencil method.

References

1 J. Koh, T. Kim, W. Lee, and T. K. Sarkar, "Extraction of Wideband Response Using Bessel-Chebyshev Functions," *IEICE Transactions on Communications*, Vol. E85-B, No.10, pp. 2263–2272, 2002.

2 M. Yuan, J. Koh, T. K. Sarkar, W. Lee, and M. Salazar-Palma, "A Comparison of Performance of Three Orthogonal Polynomials in Extraction of Wide-Band Response Using Early Time and Low Frequency Data," *IEEE Transactions on Antennas and Propagation*, Vol. 53, No. 2, pp. 785–792, 2005.

3 M. Abramowitz and I. A. Stegun, Handbook of Mathematical Functions, National Bureau of Standards, Boulder, CO, 1970.

4 J. C. Mason and D. C. Handscomb, Chebyshev Polynomials, Champan & Hall, Boca Raton, FL, pp. 2–8, 2003.

5 Z. Du, J. I. Moon, S. Oh, J. Koh, and T. K. Sarkar, "Generation of Free Space Radiation Patterns from Non-Anechoic Measurements Using Chebyshev Polynomials," *IEEE Transactions on Antennas and Propagation*, Vol. 58, No. 8, pp. 2785–2790, 2010.

6 A. Ludwig, "Near-Field Far-Field Transformations Using Spherical-Wave Expansions," *IEEE Transactions on Antennas and Propagation*, Vol. AP-19, No. 2, pp. 214–220, 1971.

7 A. Yaghjian, "An Overview of Near-Field Antenna Measurements," *IEEE Transactions on Antennas and Propagation*, Vol. AP-34, No. 1, pp. 30–45, 1986.

8 J. J. H. Wang, "An Examination of the Theory and Practices of Planar Near-Field Measurements," *IEEE Transactions on Antennas and Propagation*, Vol. 36, No. 6, pp. 746–753, 1988.

9 O. M. Bucci and C. Gennarelli, "Use of Sampling Expansions in Near-Field-Far-Field Transformation: The Cylindrical Case," *IEEE Transactions on Antennas and Propagation*, Vol. AP-36, No. 6, pp. 830–835, 1988.

10 O. M. Bucci, C. Gennarelli, and C. Savarese, "Fast and Accurate Near-Field-Far-Field Transformation by Sampling Interpolation of Plane-Polar Measurements," *IEEE Transactions on Antennas and Propagation*, Vol. 39, No. 1, pp. 48–55, 1991.

11 P. Petre and T. K. Sarkar, "Planar Near-Field to Far Field Transformation Using an Equivalent Magnetic Current Approach," *IEEE Transactions on Antennas and Propagation*, Vol. 40, pp. 1348–1356, 1992.

12 A. Taaghol and T. K. Sarkar, "Near Field to Near/Far Field Transformation for Arbitrary Near Field Geometry Utilizing an Equivalent Magnetic Current," *IEEE Transactions on Electromagnetic Compatibility*, Vol. 38, No. 3, 536–548, 1995.

13 T. K. Sarkar, P. Petre, A. Taaghol, and R. F. Harrington, "An Alternate Spherical Near Field to Far Field Transformation," *Progress in Electromagnetics Research (PIER)*, Vol. 16, pp. 269–284, 1997.

14 T. K. Sarkar and A. Taaghol, "Near-Field to Near/Far-Field Transformation for Arbitrary Near-Field Geometry Utilizing an Equivalent Electric Current and MoM," *IEEE Transactions on Antennas and Propagation*, Vol. 47, No. 3, pp. 566–573, 1999.

15 K. Kottapalli, T. K. Sarkar, Y. Hua, E. K. Miller, and G. J. Burke, "Accurate Computation of Wide-Band Response of Electromagnetic Systems Utilizing Narrow-Band Information," *IEEE Transactions on Microwave Theory and Techniques*, Vol. 39, No. 4, pp. 682–687, 1991.

16 R. S. Adve and T. K. Sarkar, "Generation of Accurate Broadband Information from Narrowband Data Using the CM," *Microwave and Optical Technology Letters*, Vol. 6, No. 10, pp. 569–573, 1993.

17 R. S. Adve, T. K. Sarkar, S. M. Rao, E. K. Miller, and D. Pflug, "Application of the Cauchy Method for Extrapolating/Interpolating Narrow-Band System Responses," *IEEE Transactions on Microwave Theory and Techniques*, Vol. 45, No. 5, pp. 837–845, 1997.

18 J. Yang and T. K. Sarkar, "Interpolation/Extrapolation of RCS Data in Frequency Domain Using the Cauchy Method," *IEEE Transactions on Antennas and Propagation*, Vol. 55, No. 10, pp. 2844–2851, 2007.

19 W. Lee, T. K. Sarkar, H. Moon, and M. Salazar-Palma, "Computation of the Natural Poles of an Object in the Frequency Domain Using the Cauchy Method," *IEEE Antennas and Wireless Propagation Letters*, Vol. 11, pp. 1137–1140, 2012.

20 J. Yang and T. K. Sarkar, "Accurate Interpolation of Amplitude-Only Frequency Domain Response Based on an Adaptive CM," *IEEE Transactions on Antennas and Propagation*, Vol. 64, No. 3, pp. 1005–1013, 2016.

21 I. S. Gradshteyn, and I. M. Ryzhik, Table of Integrals, Series and Products, A. Jeffrey and D. Zwillinger, editors, Academic, New York, Seventh Edition, 2007.

22 Z. Mei, Y. Zhang, X. Zhao, B. H. Jung, and T. K. Sarkar, "Choice of the Scaling Factor in a Marching-on-in-Degree Time Domain Technique Based on the Associated Laguerre Functions," *IEEE Transactions on Antennas and Propagation*, Vol. 60, No. 9, pp. 4463–4467, 2012.

23 Y. Zhang, T. K. Sarkar, X. W. Zhao, D. Garcia-Donoro, W. X. Zhao, M. Salazar-Palma, and S. Ting, Higher Order Basis Based Integral Equation Solver (HOBBIES), Wiley, Hoboken, NJ, 2012.

10

Retrieval of Free Space Radiation Patterns from Measured Data in a Non-Anechoic Environment

Summary

Antenna pattern measurements are usually carried out in an anechoic chamber. However, a good anechoic chamber is very expensive to construct. Previous research has attempted to compensate for the effects of extraneous fields measured in a non-anechoic environment to obtain a free space pattern that would be measured in an anechoic chamber. This chapter illustrates a deconvolution methodology which allows the antenna measurement under a non-anechoic test environment and retrieves the free space radiation pattern of an antenna through this measured data; thus allowing for easier and more affordable antenna measurements. This is obtained by modelling the extraneous fields as the system impulse response of the test environment and utilizing a reference antenna to extract the impulse response which is used to remove the extraneous fields for a desired antenna measured under the same environment and retrieve the ideal pattern. The advantage of this process is that it does not require calculating the time delay to gate out the reflections; therefore, it is independent of the bandwidth of the antenna, and there is no requirement for prior knowledge of the test environment.

10.1 Problem Background

The anechoic chamber is a commonly used facility for the antenna far-field pattern measurement, as it provides an indoor environment and an all-weather capability. First, let us look at the normal antenna measurement carried out in the anechoic chamber (shown in Figure 10.1). Inside the chamber, the AUT (antenna under test) is mounted on the AUT tower, which provides rotation along theta (θ) and phi (φ) directions, while a probe antenna is placed at a distance away from the AUT. The walls and the floor of the room will be covered by RF absorbers to eliminate various reflected fields. Also, the mechanical

Modern Characterization of Electromagnetic Systems and Its Associated Metrology, First Edition.
Tapan K. Sarkar, Magdalena Salazar-Palma, Ming Da Zhu, and Heng Chen.
© 2021 John Wiley & Sons, Inc. Published 2021 by John Wiley & Sons, Inc.

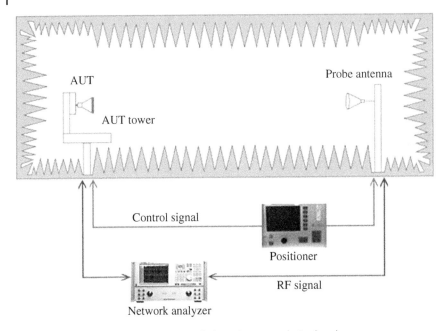

Figure 10.1 Antenna measurement carried out in an anechoic chamber.

devices present inside the chamber will be covered with RF absorbers to reduce the reflection and diffraction contributions and increase the measurement accuracy. The network analyzer is used to provide the RF signal and to measure the response received on the probe antenna. The positioner automatically controls the rotation of the AUT to generate a 3-dimentional radiation pattern.

However, it is very expensive to construct an anechoic chamber. Furthermore, to measure the radiation pattern of large antenna arrays mounted on their platforms (i.e., radar antenna array mounted on an aircraft), an indoor anechoic chamber with sufficient size would be prohibitively expensive to build. The following questions arise: How is an accurate antenna measurement carried out without using an anechoic chamber? Is it possible that one can measure the radiation pattern of the AUT in a non-anechoic environment and then do some processing to remove the artifacts of the environment? Hence, a methodology is proposed to generate a far-field pattern for the AUT that will be obtained in an anechoic environment using data measured for the AUT in a non-anechoic environment.

A large amount of research has been done to address this problem. They have used various approaches to compensate for the reflections that will occur naturally in a non-anechoic environment and generate a pattern that will approximate the free space radiation pattern. The purpose of this work is to reconstruct the free space radiation pattern using the data measured in a non-anechoic environment so that antenna pattern measurements can be carried out in any environment. As a result, the cost of the measurement will be cheaper and the measurement will be easier.

Note that, antenna measurement techniques include both far-field and near-field techniques and they perform the far-field and near-field measurement, respectively. In this chapter, only the far-field measurement is concerned as the near field techniques have been dealt with in Chapters 5-7. And the word "pattern" within the chapter indicates the far-field radiation pattern of an antenna [1–5].

10.2 Review of Pattern Reconstruction Methodologies

In the past, researchers have introduced methods for reducing the undesired reflections and diffractions of signals from the walls and objects located inside an anechoic chamber. This chapter first gives a general review of previous works on antenna pattern reconstruction. One of the methods, the FFT-based one, has already been discussed in Chapter 2 and will not be presented but methodologies related to the deconvolution method will be discussed.

Most of the existing pattern reconstruction methods can be divided into three categories based on the information that is used [6]. In the first category, the technique is to use the test-zone field for pattern correction, while in the second category the technique is to use time or frequency responses for correction. The third category's technique is to use the spatial response of the test antenna.

The Test Zone Field (TZF) compensation method [7–12] and the deconvolution method [13–16] are techniques belonging to the first category. For the TZF compensation method, the test zone field is measured over a spherical surface encompassing the test zone using a TZF probe. But this field is distorted due to extraneous fields, which are caused by reflection and diffraction responses and by the leakage from the range probe. This method provides a way to analytically remove the effect of extraneous fields in antenna pattern measurements. A spherical mode expansion (SME) of the measured test zone field is used in antenna measurements to compensate for the effects of the extraneous fields. This method basically consists of two steps. The first step is to measure the response of a reference antenna (with known radiation characteristics) in the test zone, and expand the measured TZF into spherical modes. This step is to use the measured results to calculate the coefficients for the test zone incident fields. Then in the second step, one replaces the reference antenna as the AUT and carries out the measurement again. By utilizing the measurement data and the calculated coefficients of the TZF, the radiation pattern of the AUT can be calculated. Several papers have been presented for this method, applying a matrix inversion or the FFT (Fast Fourier Transform) technique to calculate the unknown coefficients.

The deconvolution method also uses the test-zone field information and the first work was presented in 1976, but no detail information was found about it. In [14], a primary source was used to illuminate the AUT, and several secondary sources were used to imitate the environmental effects. The convolution relation between the far-field response of the AUT and the source distribution were given but without any proof. The method was verified through numerical simulations and a pilot experiment. In [15], the method was illustrated for correcting antenna measurement errors in compact antenna test ranges. The reaction theorem was applied to the AUT and the compact range antenna system to deduce the convolution equation. Measurement results of a standard horn were presented to illustrate the method. In [16], the deconvolution method was derived from the time domain convolution, and transited into the angular domain convolution by introducing the concept of the impulse response of the test environment. Numerical simulation results were presented and compared with results of the FFT-based method. All the previous work on the deconvolution method was limited to the two-dimensional case. In this chapter, derivation of the deconvolution method in the three-dimensional case will be given along with numerical simulation examples.

As the time and frequency responses of the test antenna contain similar information, different techniques utilizing either the time or the frequency domain data fall into the second category. Typical methods include the FFT-based method, the Matrix Pencil method, and equalization methods.

The FFT-based method generates the time domain response of a non-anechoic environment from its frequency response by applying the Inverse Fourier Transform [17, 18]. In the time domain, the direct signal from the transmitting antenna is detected and gated to eliminate undesired late-time echoes which are reflection and diffraction components. Then, apply the Fourier Transform to this truncated time domain response and one can obtain a cleaned up radiation pattern containing only the direct signal at the desired frequency. This method can also be used to characterize the level of reflections of the anechoic chamber [19, 20], due to the fact that the RF absorbers can reduce but not remove the reflections or diffractions.

However, a major disadvantage of this methodology is that one needs to determine the time taken by the fields to travel along a line-of-sight path (direct path) from the AUT to the probe antenna and the shortest time needed for the fields to travel through other paths besides the direct one. This can be difficult especially when the measurement site has multiple objects that are close to the direct path. Also, to have sufficient time domain resolution to perform the time gating, a large bandwidth of the measurement data is required in the frequency domain.

Another method that also applies the idea of time gating is to directly measure the far-field antenna pattern in the time domain [21]. By using the data from a single measurement in the time domain, range evaluation, pattern reconstruction and pattern error correction can be performed. However, it is difficult to carry out measurements in the time domain as a large bandwidth of the signal is required.

The Matrix Pencil method and the Oversampled Gabor Transform (OGT) essentially achieve similar goals as the FFT-based method but they require less bandwidth [22–25]. These two methods, based on the matrix-pencil and the over-sampled Gabor-transform, decompose the measured frequency response into several propagation components in the form of complex exponential functions over selected frequency intervals. By extracting only, the component contributing to the direct path of propagation and by suppressing the other components, it is possible to obtain an approximated free space radiation pattern.

P. S. H. Leather and D. Parson present an equalization technique to correct the effects of unwanted signals. A special measurement is carried out for the non-anechoic environment where an antenna is to be tested to determine the parameters of the equalizer [26–31]. By applying the idea of a matched filter, they used the adaptive equalizer to calculate the actual channel characteristics and to adjust its coefficients appropriately to approximate the free space con-dition. This method needs to have a training procedure that transmits the ideal signal to the environment, collects the responses, and records differences between the ideal signal and the received responses to calculate the coefficients. These coefficients can then be used to cancel the effects of the environment on the desired AUT.

Techniques of the third category include methods like the antenna pattern comparison (APC), novel antenna pattern comparison (NAPC) [32], and adapt-ive array strategies [33]. The APC technique was designed for measuring the reflectivity level in an anechoic chamber [34], but it can also be used to correct the measured pattern of an antenna. This technique measures the pattern of an antenna several times at different sites inside a room. Then, the recorded pat-terns are adjusted and superimposed so that the main-lobes cover each other and the corrected antenna pattern is obtained by taking the average of the meas-ured patterns. The NAPC technique requires measuring the antenna pattern twice at two different locations in the test zone. During one pattern measure-ment, the antenna location is fixed in the target zone; during the second pattern measurement, the antenna is moved as a function of the pattern angle and the corrected pattern is given by the average of the two responses. As for the adapt-ive array strategy, those spurious signals are considered as the interference sig-nals and the direction of arrival (DOA) algorithm is applied to identify and remove them.

Also, several other techniques are developed and are considered suitable in a hologram based compact antenna test range (CATR) at sub-millimeter wave-lengths (e.g., the feed scanning APC technique, the feed scanning APC tech-nique [35], the frequency shift technique [36], and the correction technique based on an adaptive array algorithm [37]).

The purpose of this chapter is to present a deconvolution-based technique and to extend the method to be used in three-dimensional environments.

10.3 Deconvolution Method for Radiation Pattern Reconstruction

Some developments have taken place in the area of antenna measurements during the recent years. And the accuracy of the measurement results is affected by factors like the Signal-to-Noise Ratio (SNR) of the measured data, the data processing algorithms, the precision of the test equipments and also the quality of the measurement environment. Large efforts have been made to improve the measurement facility but it is usually limited by the available budget. For example, use of high quality absorbing materials in the anechoic chamber is costly. This work is focused on antenna pattern reconstruction through a deconvolution method.

The concept of the deconvolution method and some results have been reported in [13–15]. In this work, the method is presented and verified from a different point of view, followed by the governing equations to implement this method and numerical examples to illustrate the process. Note that in this chapter the deconvolution method is applied to 2D test environment. Extensions of the method to 3D test environment will be given in the section 10.7.

10.3.1 Equations and Derivation

Consider making antenna measurement inside a regular room instead of an anechoic chamber. The received signal at the probe is affected by the environment (walls, floor, ceiling, and so on) and these are grouped as the environmental effects. As shown in Figure 10.2, the AUT and the probe function as the transmitting and receiving antenna, respectively. The distance between the two antennas satisfies the far field condition since the far-field pattern of an AUT is considered. The AUT will rotate during the measurement and $S_{21}(f, \phi)$ will be measured at a fixed frequency f for each rotating angle ϕ (ϕ will be the azimuth angle since the AUT rotates along the azimuth plane). The radiation pattern of the AUT is proportional to $S_{21}(f, \phi)$, as a function of the AUT rotation angle. Here, $S_{21}(f, \phi)$ contains information for both the antenna far-field pattern and the environmental effects.

Many physical processes can be represented by, and have successfully been analyzed assuming linear time-invariant (LTI) systems as models [38, 39]. And from the time domain point-of-view, the output of the system is simply the convolution of the system input and the impulse response of the system.

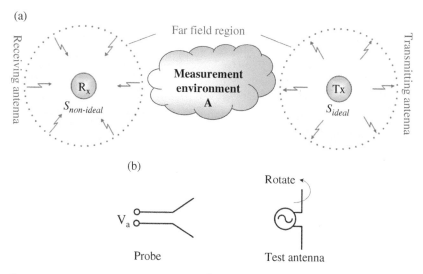

Figure 10.2 A diagram for the radiation pattern measurement system.

When the transmitter in a room generates a signal $x(t)$, it will be affected by the room and received by the receiver as $y(t)$. Suppose the impulse response of the room system is $h(t)$, then we know that:

$$y(t) = x(t) * h(t), \tag{10.1}$$

where $*$ represents a convolution in the time domain, and in the frequency domain it will be:

$$Y(f) = X(f) \times H(f), \tag{10.2}$$

where $Y(f)$, $X(f)$ and $H(f)$ are the Fourier transforms of $y(t)$, $x(t)$ and $h(t)$, respectively. Thus, for antenna measurement inside a room, the received signal at the probe will be a convolution of the transmitted signal (from the AUT) with the room impulse response. Note that, this relationship is between the time domain and the frequency domain; moreover, both the AUT and the probe are fixed in the spatial domain.

Now, let's analyze the situation when the AUT rotates during the measurement. The reflections from the room may be strong at some angles, and weak for some other angles. For each rotation angle ϕ, the AUT transmits signal along all directions and the probe also receives signal from all directions (assume that there are n directions for one cut in the plane). The signal which is transmitted along direction i travels along a certain path and is received from direction j. We define this path as ji. And the corresponding environmental effect along this path is defined as h_{ji}. So the total signals received at the probe along direction

1 will be a summation of the signals from the transmitter radiating in all directions and through paths h_{11}, h_{12}, ..., h_{1n}. Then, one will have:

$$R_x^1 = h_{11}T_x^1 + h_{12}T_x^2 + \ldots + h_{1i}T_x^i + \ldots + h_{1n}T_x^n, \qquad (10.3)$$

where

R_x^1 is the signal received at the probe along direction 1
T_x^i is the signal transmitted to the AUT along direction i
h_{1i} is the environmental effect along the path $1i$
n is the number of directions that are considered.

Similarly, there will be a signal received at the probe from direction 2, 3, until n. And there will be corresponding equations for R_x^2, R_x^3,..., R_x^n. This results in

$$
\begin{aligned}
R_x^1 &= h_{11}T_x^1 + h_{12}T_x^2 + \ldots + h_{1i}T_x^i + \ldots + h_{1n}T_x^n \\
R_x^2 &= h_{21}T_x^1 + h_{22}T_x^2 + \ldots + h_{2i}T_x^i + \ldots + h_{2n}T_x^n \\
&\vdots \\
R_x^n &= h_{n1}T_x^1 + h_{n2}T_x^2 + \ldots + h_{ni}T_x^i + \ldots + h_{nn}T_x^n
\end{aligned}
\qquad (10.4)
$$

and one can rewrite them in the matrix form as:

$$
\begin{bmatrix} R_x^1 \\ \vdots \\ R_x^n \end{bmatrix} =
\begin{bmatrix} h_{11} & \cdots & h_{1n} \\ \vdots & \ddots & \vdots \\ h_{n1} & \cdots & h_{nn} \end{bmatrix} \cdot
\begin{bmatrix} T_x^1 \\ \vdots \\ T_x^n \end{bmatrix}
\qquad (10.5)
$$

where $R_x^1 \cdots R_x^n$ represent the signals received at the receiver from all n directions; $T_x^1 \cdots T_x^n$ stand for the signals sent by the transmitter along all n directions; h_{ji} stands for the environmental effect for the path that the signal is transmitted along direction i and is received along direction j.

So for each azimuth (rotation) angle ϕ_1, $S_{21}(\phi_1)$ is a summation of the received signals as $R_x^1 + R_x^2 + \cdots + R_x^n$ (at ϕ_1). By using h_1 to represent the summation of h_{11}, h_{21}, until h_{n1}, and also other h_{ji} terms, one can write $S_{21}(\phi_1)$ as a summation of $T_x^1 h_1 + T_x^2 h_2 + \cdots + T_x^n h_n$:

$$
\begin{aligned}
S_{21}(\phi_1) = R_x^1 + R_x^2 + \cdots + R_x^n \text{ (at } \phi_1) = sum
\begin{cases}
T_x^1(h_{11} + h_{21} + \cdots + h_{n1}) & h_1 \\
T_x^2(h_{12} + h_{22} + \cdots + h_{n2}) & h_2 \\
\vdots \\
T_x^n(h_{1n} + h_{2n} + \cdots + h_{nn}) & h_n
\end{cases} \\[6pt]
= T_x^1 h_1 + T_x^2 h_2 + \cdots + T_x^n h_n
\end{aligned}
\qquad (10.6)
$$

where,

$$h_1 = h_{11} + h_{21} + \cdots + h_{n1}$$
$$h_2 = h_{12} + h_{22} + \cdots + h_{n2}$$
$$\vdots$$
$$h_n = h_{1n} + h_{2n} + \cdots + h_{nn}$$

$$(10.7)$$

Then, the AUT is rotated every 1 degree, the angle ϕ changes from ϕ_1 to ϕ_2. The corresponding field component transmitted along direction 1 will then be transmitted along direction 2. And we can get a similar equation for $S_{21}(\phi_2)$ as presented earlier:

$$S_{21}(\phi_2) = sum \begin{cases} h_{12}T_x^1 + h_{13}T_x^2 + \ldots + h_{1n}T_x^{n-1} + h_{11}T_x^n \\ h_{22}T_x^1 + h_{23}T_x^2 + \ldots + h_{2n}T_x^{n-1} + h_{21}T_x^n \\ \vdots \\ h_{n2}T_x^1 + h_{n3}T_x^2 + \ldots + h_{nn}T_x^{n-1} + h_{n1}T_x^n \end{cases}$$

$$(10.8)$$

$$= T_x^1 h_2 + T_x^2 h_3 + \cdots + T_x^n h_1$$

Similarly, we can get such expressions for every single ϕ and write them in the matrix form as:

$$\begin{bmatrix} S_{21}(\phi_1) \\ \vdots \\ S_{21}(\phi_n) \end{bmatrix} = \begin{bmatrix} h_1 & h_2 & \cdots & h_n \\ h_2 & h_3 & \cdots & h_1 \\ \vdots & & & \vdots \\ h_n & h_1 & \cdots & h_{n-1} \end{bmatrix} \cdot \begin{bmatrix} T_x^1 \\ \vdots \\ T_x^n \end{bmatrix} = (h_1, h_2, \ldots, h_n) \otimes_\phi (T_x^1, T_x^2, \ldots, T_x^n)$$

$$(10.9)$$

where $\otimes_\phi$ represents a convolution in the angular domain. It is easy to observe that the $[h]$ matrix is shifted by one element for each row in Eq. (10.9). So the multiplication of matrix $[h]$ and vector of $[T_x]$ will be a convolution of two vectors. Since, $(T_x^1, T_x^2, \ldots, T_x^n)$, is the signal transmitted by the AUT, and forms the free space radiation pattern of the AUT; while $(S_{21}(\phi_1), S_{21}(\phi_2), \ldots, S_{21}(\phi_n))$, which is the actual signal received by the probe in the presence of the environment, forms the non-ideal radiation pattern. One can then conclude that the measured non-ideal signal can be represented as an angular convolution between the ideal signal and the environmental responses at the frequency f, given by

$$P_{non\text{-}ideal\ AUT}(\phi, f) = P_{ideal\ AUT}(\phi, f) \otimes_\phi A(\phi, f)$$

$$(10.10)$$

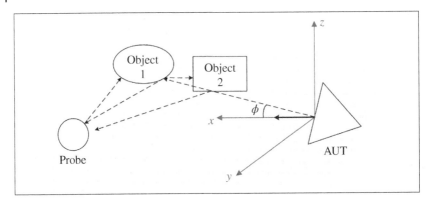

Figure 10.3 Multiple reflections exist between the AUT and the probe in the azimuth plane.

The above derivation of equation (10.10) is not very rigorous while the following derivation is more rigorous, and it also can be found in [16]. First assume that an AUT can generate an ideal pencil beam pattern (the radiated signal will be along one direction in the far field). As shown in Figure 10.3, the signal which is transmitted along ϕ experiences multiple reflections from Object 1 and Object 2 in the azimuth plane. It is assumed that the objects do not change size or positions in time (time invariant). The received time domain signal at the probe, when the AUT is rotated through ϕ, is unique with respect to other angles. Therefore the time domain response along angle ϕ can be described as an impulse response along ϕ or a spatial signature of ϕ. The reason for making the pencil beam assumption here is that the environment affects the radiated signal differently along different angles, and one must first consider the environmental effects in one direction only.

The measured signal at the probe, which contains various reflections, can be represented by a time convolution of the ideal signal (without any reflection) and the impulse response along ϕ_L, can be written as:

$$P_{non\text{-}ideal}(\phi_L, t) = P_{ideal}(\phi_L, t) \otimes A(\phi_L, t) \tag{10.11}$$

or

$$P_{non\text{-}ideal}(\phi_L, f) = P_{ideal}(\phi_L, f) \times A(\phi_L, f) \tag{10.12}$$

where $\otimes$ denotes a time convolution, and

$P_{non\text{-}ideal}(\phi_L, t)$ is the non-ideal time domain signal at the probe in the presence of the reflections for the angle ϕ_L;

$P_{non\text{-}ideal}(\phi_L, f)$ is the non-ideal frequency domain signal at the probe in the presence of the reflections for the angle ϕ_L;

$P_{ideal}(\phi_L, t)$ is the ideal time domain signal without any reflection for the angle ϕ_L;

$P_{ideal}(\phi_L, f)$ is the ideal frequency domain signal without any reflection for the angle ϕ_L;

$A(\phi_L, t)$ is the impulse response of the environment with objects present when the AUT has a pencil beam pointing along the angle ϕ_L;

$A(\phi_L, f)$ is the frequency domain response of the environment with objects present when the AUT has a pencil beam pointing along the angle ϕ_L.

Note that $A(\phi_L, t)$ represents the contribution of various reflections from the environment along the angle ϕ_L and is independent of the particular AUT.

In a real situation, the AUT will radiate towards every direction in the spatial domain and cannot have an ideal pencil beam pattern. One can define the impulse response of the environment along the rotation angle ϕ_L as $\hat{A}(\phi_L, f)$ when the AUT does not have an ideal pencil beam pattern. Then, using (3.12) one has:

$$P_{non\text{-}ideal}(\phi_L, f) = P_{ideal}(\phi_L, f) \times \hat{A}(\phi_L, f) \tag{10.13}$$

where, $\hat{A}(\phi_L, f)$ is the frequency domain impulse response along the angle ϕ_L when the AUT does not have an ideal pencil beam pattern, which is more practical.

Now, $\hat{A}(\phi_L, f)$ contains information of the beam pattern of the AUT as well as the environmental effects at ϕ_L. However, one needs the true impulse response $A(\phi_L, f)$ or $A(\phi_L, t)$, which is independent of the beam pattern of the AUT. Then $\hat{A}(\phi_L, f)$ can be considered as a convolution in the angular domain of the normalized beam pattern and the true impulse response. That is,

$$\hat{A}(\phi_L, f) = \frac{P_{ideal}(\phi, f)}{P_{ideal}(\phi_L, f)} \otimes_\phi A(\phi, f) \Big|_{\phi = \phi_L} \tag{10.14}$$

Here, $P_{ideal}(\phi, f)$ is the ideal pattern of an antenna without any reflection from the environment at the frequency f, and $\otimes_\phi$ is the convolution operator in the angular domain. When the AUT doesn't have an ideal pencil beam pattern, it radiates the wave towards different directions, and is affected differently by the environment. And the combination of such effects is equal to the term $\hat{A}(\phi_L, f)$. By substituting (10.14) into (10.13) one have,

$$P_{non\text{-}ideal}(\phi_L, f) = P_{ideal}(\phi, f) \otimes_\phi A(\phi, f) \Big|_{\phi = \phi_L} \tag{10.15}$$

For a general angle of ϕ, one have:

$$P_{non\text{-}ideal}(\phi, f) = P_{ideal}(\phi, f) \otimes_\phi A(\phi, f) \tag{10.16}$$

which leads to the same conclusion as shown in (10.10).

Therefore, the beam pattern of the AUT in the presence of reflections can be considered as a convolution in the angular domain between the ideal beam pattern of the AUT and the impulse response of the environment. And one knows that by taking the FFT will transform the convolution in the angular domain to the multiplication in the transform domain, which one can name as the angle-frequency domain. Thus the impulse response $A(\phi, f)$ can easily be calculated by taking the IFFT of (10.16) using the measurement data of a reference antenna whose ideal pattern is known. Once $A(\phi, f)$ is calculated, the ideal pattern $P_{ideal}(\phi, f)$ of the AUT can be obtained for any antenna measured in the same environment through (10.16). This requires two assumptions: First, the environment is unchanged during measurements for the reference antenna and the AUT. Second, because both the probe and the AUT are considered as part of the environment, sizes of the reference antenna and the AUT need to be similar. In this way, the change of an antenna will not cause a sudden change of the environment.

Also, one condition needs to be mentioned is that the AUT inside the environment should radiate as if it is located in the free space, and in other words, the current distribution on the AUT should remain close to the ideal one. The reason is that one is reconstructing the radiation pattern of the AUT to approximate the free space radiation pattern. If the current distribution on the antenna has been dramatically changed by the environment, then the radiation pattern will also be changed. And after the reconstruction, the reconstructed pattern will be a pattern that has been affected by the environment. To satisfy this condition, the AUT should be positioned at a "safe" distance away from the environment, to minimize the changes in the current on the AUT due to the environment. A reference for this "safe" distance can be the free space far-field distance, which is commonly taken as $2D^2/\lambda$, where D is the maximum overall dimension of the antenna and λ is the wavelength of operation [1].

10.3.2 Steps Required to Implement the Proposed Methodology

After obtaining (10.16), one can design a procedure to demonstrate this procedure through simulation. HOBBIES [38] will be used as the EM simulation tool for all the examples presented in this work. In the software, the radiation pattern measurement model consists of an AUT, a probe antenna, and PEC plates which serve as the environment. The simulation is carried out first in a non-anechoic environment using two reference (standard) antennas, whose ideal patterns are known, and a deconvolution is performed to compute the environmental response A. Then, the simulation for the AUT in the same environment is carried out and an estimate of its free space radiation pattern using the environmental response A that has been extracted from the reference antenna. The entire procedure can be summarized by the 4 steps:

1) At a fixed frequency, measure the reference (standard) antenna response $P_{non\text{-}ideal}(\phi, f)$ in a non-anechoic environment. Also the reference antenna response $P_{ideal}(\phi, f)$ is known.

2) Calculate $A(\phi, f)$ using the equation $P_{non\text{-}ideal}(\phi, f) = P_{ideal}(\phi, f) \otimes_\phi A(\phi, f)$.

3) At the same frequency, replace the reference antenna with the AUT and keep the rest of the environment unchanged, measure the AUT in the same way as in step (1), and let $P_{non\text{-}ideal\,AUT}(\phi, f)$ be the result.

4) Obtain the ideal response of the AUT, $P_{ideal\,AUT}(\phi, f)$, through deconvolution using:

$$P_{non\text{-}ideal\,AUT}(\phi, f) = P_{ideal\,AUT}(\phi, f) \otimes_\phi A(\phi, f).$$

From the above procedures, one can observe that the deconvolution method only requires a single frequency measurement, and it is independent of the bandwidth of the antenna; while the FFT based approach require broadband characteristics for the antenna. The deconvolution method requires no prior knowledge of the system or the test environment. Most importantly, it doesn't require the antenna radiation pattern measurement to be carried out in an anechoic chamber.

10.3.3 Processing of the Data

Specifically, the following procedures are the rules of thumb for setting up simulation models and processing the simulated data generated through numerical electromagnetics code:

1) In this work, all simulation examples will be carried out using HOBBIES to perform the full wave EM simulation. First, build the simulation model for the antenna radiation pattern measurement system which consists of an AUT (transmitting antenna), a probe (receiving antenna) and reflectors. The reflectors will reflect the radiated fields from the AUT and can be modeled by the PEC plates around the antennas.

2) In HOBBIES, set the operation mode as "ANTENNA (one generator at a time)", and set an excitation port for both the AUT and the probe. During the simulation, the AUT and the probe will be the transmitter and the receiver, respectively. To simulate the antenna measurement process, the AUT rotates along itself in the azimuth plane with a step of 1°, and for each rotation angle the model is simulated and the data for S_{21} is collected.

3) The S_{21} data along each azimuth angle ϕ forms the radiation pattern of the AUT. When the AUT rotates one complete rotation in the azimuth plane, there will be 360 data points, i.e. $-180°$, $-179°, -178°,...,$ $179°$, and then the measured data will repeat this sequence. Therefore, the ideal pattern $P_{ideal}(\phi, f)$ and the non-ideal pattern $P_{non\text{-}ideal}(\phi, f)$ as well as the environmental effects $A(\phi, f)$, are all periodic sequences of period 360° in the angular

ϕ domain. Note that if the model is symmetrical (both the environment and the antennas are symmetrical) in the azimuth plane, one can then reduce the number of simulation points by half.

4) As described in Section 10.3.2, first use a reference antenna as the AUT and simulate the antenna within a non-anechoic environment to obtain the non-ideal radiation pattern of the reference antenna $P_{non\text{-}ideal-Ref}(\phi,f)$. Also, one can obtain the ideal radiation pattern of the reference antenna $P_{ideal\text{-}Ref}(\phi,f)$ by simulating the model without the environment (the antenna would be like in the free space without the non-anechoic environment). In reality, one can obtain the ideal pattern of the reference antenna through its datasheet. Similarly, by replacing the reference antenna with the desired AUT and carry out the simulation, one can obtain the ideal pattern $P_{ideal\text{-}AUT}(\phi,f)$ and the non-ideal pattern $P_{non\text{-}ideal-AUT}(\phi,f)$ for the desired AUT. The ideal pattern $P_{ideal\text{-}AUT}(\phi,f)$ will be the goal of the reconstructed pattern.

5) Now it is necessary to apply (10.16) to reconstruct the radiation pattern of the AUT. According to Section 10.3.2.4 of [40], the convolution of two periodic sequences is the multiplication of the corresponding discrete Fourier series. Let $\tilde{x}_1(n)$ and $\tilde{x}_2(n)$ be the two periodic sequences of period N with the discrete Fourier series denoted by $\tilde{X}_1(k)$ and $\tilde{X}_2(k)$, respectively. It can be written as:

$$\tilde{x}_3(n) = \sum_{m=0}^{N-1} \tilde{x}_1(m)\tilde{x}_2(n-m) \tag{10.17}$$

$$\tilde{X}_3(k) = \tilde{X}_1(k)\tilde{X}_2(k) \tag{10.18}$$

6) From (10.16) one has the following equations:

$$P_{non\text{-}ideal-Ref}(\phi,f) = P_{ideal\text{-}Ref}(\phi,f) \otimes_\phi A(\phi,f) \tag{10.19}$$

$$P_{non\text{-}ideal-AUT}(\phi,f) = P_{ideal\text{-}AUT}(\phi,f) \otimes_\phi A(\phi,f) \tag{10.20}$$

Therefore, by taking the FFT of both sides of (10.19), the angular convolution operator will become the multiplication operator. One can derive the environment effects $A(\phi, f)$ as:

$$A(\phi,f) = ifft\left(\frac{fft(P_{non\text{-}ideal-Ref}(\phi,f))}{fft(P_{ideal\text{-}Ref}(\phi,f))}\right) \tag{10.21}$$

where *ifft* represents the inverse Fourier transform.

7) Take the FFT of both sides of (10.20) and substitute the environmental effects $A(\phi, f)$ into the equation, to obtain:

$$P_{ideal\text{-}AUT}(\phi,f) = ifft\left(\frac{fft(P_{non\text{-}ideal-AUT}(\phi,f))}{fft(A(\phi,f))}\right)$$

$$= ifft\left(\frac{fft(P_{non\text{-}ideal-AUT}(\phi,f)).fft(P_{ideal\text{-}Ref}(\phi,f))}{fft(P_{non\text{-}ideal-Ref}(\phi,f))}\right) \tag{10.22}$$

Note that the division and multiplication in (10.21) and (10.22) are element by element operations on the vectors. All data processing can be performed off-line using a commercial software package (MATLAB 7, The MathWorks Inc., Natick, MA, 2000). In MATLAB, the function Y=fft (X) returns the Discrete Fourier Transform (DFT) of vector X, computed with a Fast Fourier Transform (FFT) algorithm. Similarly, the function Y=ifft(X) returns the Inverse Discrete Fourier Transform (IDFT) of vector X.

8) Compare the reconstructed pattern of the AUT from (10.22) with the simulated result $P_{ideal-AUT}(\phi, f)$.

10.3.4 Simulation Examples

10.3.4.1 Example I: One PEC Plate Serves as a Reflector

To verify the idea just presented, numerical examples are presented using different antennas, a horn, a helix, and a Yagi antenna. The horn antenna is set to be the reference antenna and as the probe antenna for all the examples. It is interesting to observe that in this methodology, the probe antenna need not be small as its response will be factored out. The goal is to remove the effects of the extraneous fields due to the presence of PEC reflectors and retrieve the free space radiation pattern of the helical antenna and the Yagi antenna. For all the examples, it is assumed that the environment does not vary with time.

The simulation model is shown in Figure 10.4. This includes two antennas; on the right hand side is a horn antenna which is the probe, while the antenna on the left hand side is the AUT. There are two different AUTs, Figures 10.5 and 10.6 provide the dimensions of the helical antenna and the 6-element Yagi antenna. Figure 10.7 shows the dimensions of the horn antenna (the reference antenna and also the same antenna as the probe antenna). The PEC plate serves as the reflector, and is 1.25 meter away from the two antennas. One condition however needs to be satisfied in this procedure which is that the current distribution on the feed dipole should remain close to the ideal one. And an approximation for the "safe" distance would be the free space far-field distance $2D^2/\lambda$,

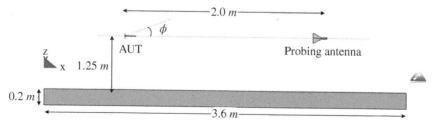

Figure 10.4 Model of the measurement system with one PEC plate as the reflector.

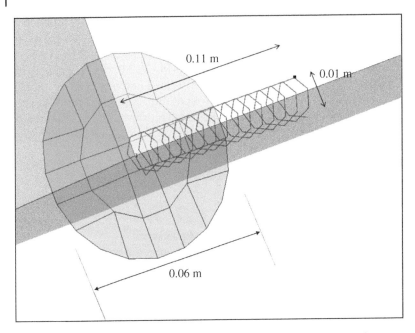

Figure 10.5 Dimensions of the helical antenna with a reflecting plate (AUT).

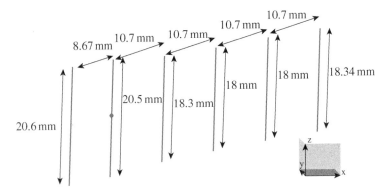

Figure 10.6 Dimensions of the 6-element Yagi antenna (AUT).

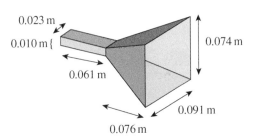

Figure 10.7 Dimensions of the horn antenna which is used as a probe.

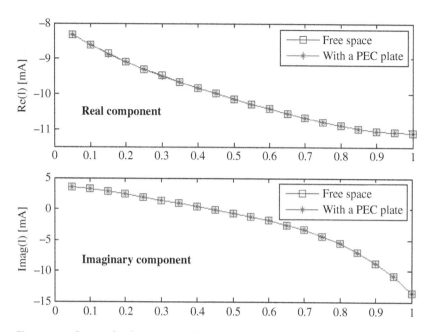

Figure 10.8 Current distribution on the feed dipole of the helical antenna with one PEC plate.

which is 1.12 m (D is 0.155 m as the largest diagonal size of the antennas). So the distance between the PEC plate and the antenna is selected to be 1.25 m. Figure 10.8 provides the real and imaginary parts of the current distribution on the feed dipole of the helical antenna, and it shows that they are not affected by the PEC plate (the current distribution on the other antenna is omitted due to limited space).

The numerical simulations and the processing of the followed the procedures as described in Section 10.3. The azimuth angle ϕ of the AUT was varied from −180° to 179° with a 1° step. The S_{21} data was collected at each azimuth angle ϕ to form the plot of the radiation pattern. Equation (10.22) was used to calculate the reconstructed pattern for the AUT. And the reconstructed results were compared with the simulated ideal pattern of the AUT to illustrate the performance of the deconvolution method.

Note that use of the PEC plate instead of a dielectric one as the reflector is to increase the level of reflections and make the environmental effects as strong as possible. Therefore, the presence of PEC plates will greatly distort the radiation pattern of the AUT and the performance of the deconvolution method on pattern reconstruction can be illustrated more clearly.

Figure 10.9 shows the comparison of the amplitude pattern for the horn antenna with and without the PEC reflector at 7 GHz; while Figure 10.10 shows the phase component of the radiation pattern. The blue dashed line termed

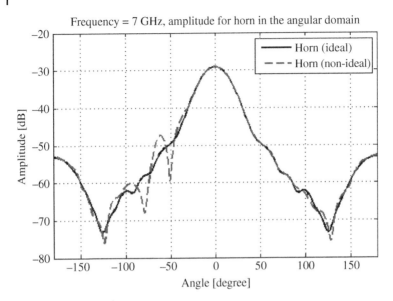

Figure 10.9 Amplitude pattern for the horn antenna with one PEC plate as the reflector.

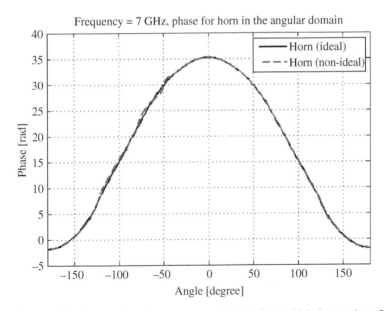

Figure 10.10 Phase pattern for the horn antenna with one PEC plate as the reflector.

non-ideal is the result in the presence of the PEC reflector, while the black line is for the free space radiation pattern. It is easy to observe the difference between the two curves due to the reflection between the azimuth angles $-40°$ and $-130°$, where the main beam of the horn antenna is pointing at the PEC plate. The environmental response $A(\phi, f)$ extracted from the model is shown in Figure 10.11.

Figure 10.12 shows the comparison of the amplitude pattern for the helical antenna with and without the PEC reflector at 7 GHz; while Figure 10.13 shows the phase component. The blue dashed line termed *non-ideal* is the result in the presence of the PEC reflector while the reconstructed pattern is indicated by the red line termed *reconstructed*. The goal is to reconstruct the pattern from this non-ideal data, and by factoring in the environmental response data, one can obtain a clean pattern for the helical antenna. One can see that the reconstructed pattern is very close to the ideal pattern of the helical antenna. Therefore, it indicates that the reflections and diffractions caused by the PEC plate have been factored out by using the deconvolution method. Compared with the result of FFT-based method as shown in [41], the deconvolution method generates better results.

Similarly, one can also apply the deconvolution method to reconstruct a clean radiation pattern for the 6-element Yagi antenna using data measured in a non-anechoic environment. For the same environment, one replaces the helical

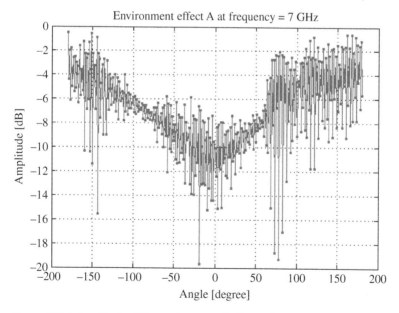

Figure 10.11 Amplitude of the environmental effects when one PEC plate is used as the reflector.

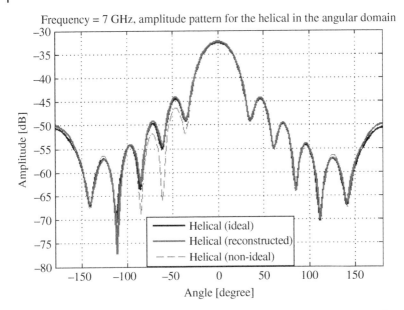

Figure 10.12 Amplitude pattern for the helical antenna with one PEC plate as the reflector.

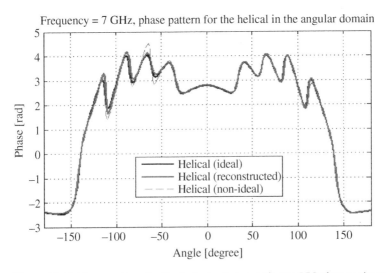

Figure 10.13 Phase pattern for the helical antenna with one PEC plate as the reflector.

antenna with the 6-element Yagi antenna. And by using the environmental response data, one can similarly obtain a clean pattern for the yagi. Figure 10.14 shows the comparison of the amplitude pattern for the Yagi antenna with and without the PEC reflector at 7 GHz; while Figure 10.15 shows the phase component. Again, one can observe that the reconstructed pattern has greatly reduced the presence of the undesired reflections from the plates

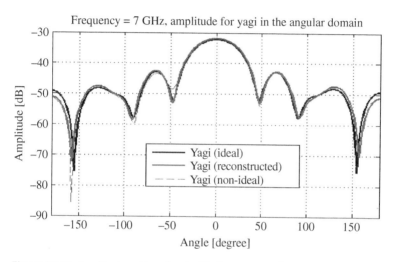

Figure 10.14 Amplitude pattern for the Yagi antenna with one PEC plate as the reflector.

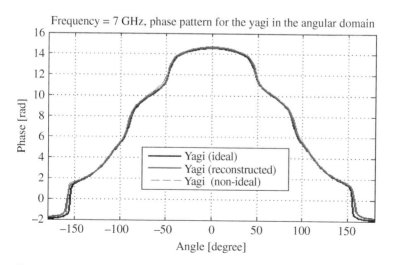

Figure 10.15 Phase pattern for the yagi antenna with one PEC plate as the reflector.

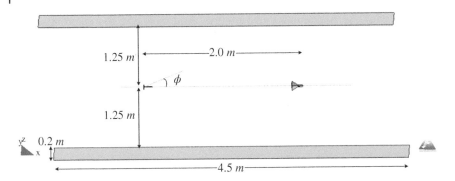

Figure 10.16 Model of the measurement system with two PEC plates as the reflector.

in the measurements, and the processed result is very close to the ideal pattern of the Yagi antenna.

10.3.4.2 Example II: Two PEC Plates Now Serve as Reflectors

The previous example has shown the pattern reconstruction for the helical antenna and the 6-element Yagi antenna using the data generated in a non-anechoic environment. That example presented a simple case with the presence of only one PEC plate, which is a very simple environment. To evaluate the performance of the deconvolution method using more complicated environments, a second example is presented with two PEC plates as reflectors. Figure 10.16 shows the schematic diagram of the model, two PEC plates have the same size and are located at symmetric positions towards the antennas (the symmetry is applied to reduce the number of simulation points by half). The same procedure is applied as in Section 10.3.4.1 to reconstruct the radiation pattern for the helical antenna and the 6-element Yagi antenna. Again, we need to check the current distribution on the feed dipole of the helical antenna. Figure 10.17 indicates that the current distribution is not affected by the PEC plates.

Figure 10.18 shows the comparison of the amplitude pattern for the horn antenna with and without the PEC reflector at 7 GHz; while Figure 10.19 shows the phase component of the pattern. The blue dashed line indicates the result in the presence of the PEC reflectors. It is easy to observe that there are large reflections between the azimuth angles range [−40°, −130°] and [40°, 130°], where the main beam of the horn antenna is pointing at the PEC plates. The environmental response $A(\phi, f)$ extracted from the model is shown in Figure 10.20.

Figure 10.21 shows the comparison of the amplitude pattern for the helical antenna with and without the PEC reflector at 7 GHz; while Figure 10.22 shows the phase component of the pattern. The blue dashed line termed *non-ideal*

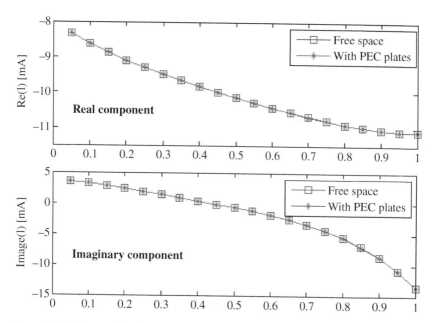

Figure 10.17 Current distribution on the feed dipole of the helical antenna with two PEC plates.

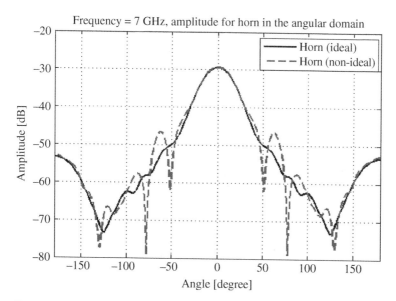

Figure 10.18 Amplitude pattern for the horn antenna with two PEC plates as the reflector.

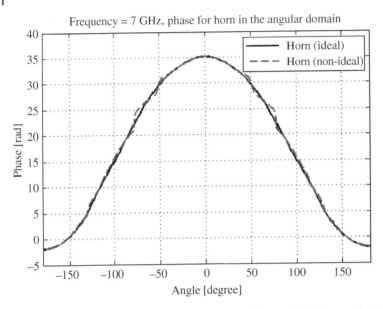

Figure 10.19 Phase pattern for the horn antenna with two PEC plates as the reflector.

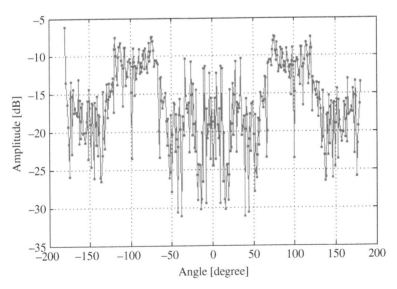

Figure 10.20 Amplitude of the environmental effects when two PEC plates are used as the reflector.

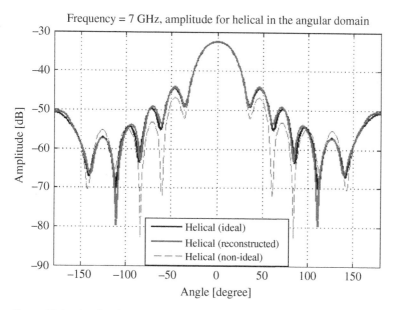

Figure 10.21 Amplitude pattern for the helical antenna with two PEC plates as the reflector.

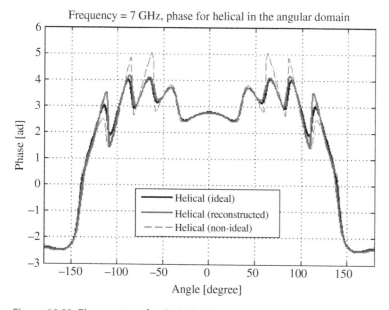

Figure 10.22 Phase pattern for the helical antenna with two PEC plates as the reflector.

indicates the result in the presence of the PEC reflector, the reconstructed pattern is indicated by the red line termed *reconstructed*. Compared with the example in Section 10.3.4.1, the reflected response for this example is stronger. Still, one can observe that the reconstructed pattern is very close to the ideal pattern. This indicates that the echoes caused by the PEC plates have been successfully compensated for by using the deconvolution method.

Similarly, the deconvolution method is applied to reconstruct a clean radiation pattern for the 6-element Yagi antenna. Figure 10.23 shows the comparison of the amplitude pattern for the yagi antenna with and without the PEC reflector at 7 GHz; while Figure 10.24 shows the phase component of the pattern. Again, one can observe that the reconstructed pattern is very close to the ideal pattern of the Yagi antenna.

10.3.4.3 Example III: Four Connected PEC Plates Serve as Reflectors

This third example presents a more complicated case, to fully evaluate the deconvolution method for the pattern reconstruction from the data measured in a non-anechoic environment. This example uses four connected PEC plates as the reflectors around the antenna, as shown in Figure 10.25. Four PEC plates form a rectangular cylinder enclosing the AUT and the probe antenna. The entire model is 4.5 *m* long and 2.5 *m* wide. Similarly, it is necessary to check the current distribution on the feed dipole of the helical antenna.

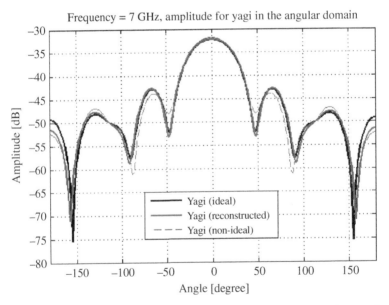

Figure 10.23 Amplitude pattern for the yagi antenna with two PEC plates as the reflector.

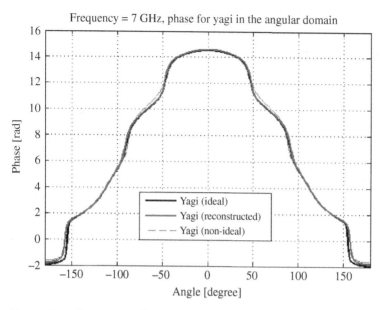

Figure 10.24 Phase pattern for the yagi antenna with two PEC plates as the reflector.

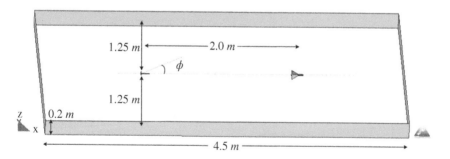

Figure 10.25 Model of the measurement system with four PEC plates as the reflector.

Figure 10.26 shows that the current distribution is slightly affected by the PEC plates.

The same procedure was carried out to reconstruct the radiation pattern for the helical antenna and the 6-element yagi antenna. Figure 10.27 shows the comparison of the amplitude pattern for the horn antenna with and without the PEC reflector at 7 GHz; while Figure 10.28 shows the phase component of the pattern. The blue dashed line termed *non-ideal* is the result in the presence of the PEC reflectors. It is easy to observe that the reflection from the plates affect the radiation pattern of the antenna for all azimuth angles; especially the

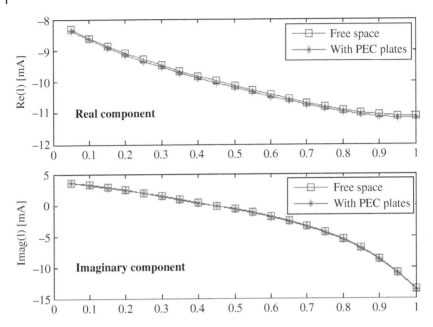

Figure 10.26 Current distribution on the feed dipole of the helical antenna with four PEC plates.

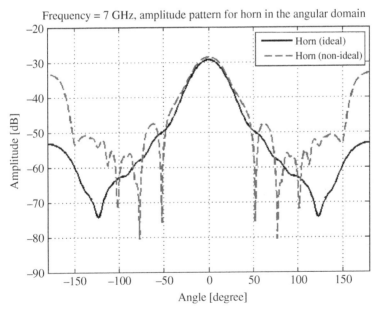

Figure 10.27 Amplitude pattern for the horn antenna with four PEC plates as the reflector.

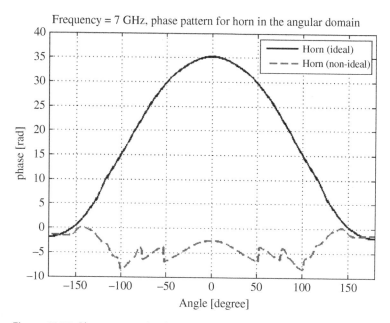

Figure 10.28 Phase pattern for the horn antenna with four PEC plates as the reflector.

back lobe level of the pattern has been greatly increased due to the PEC plates. The environmental response $A(\phi, f)$ extracted from the model is shown in Figure 10.29.

The reconstructed pattern for the helical antenna is given in Figures 10.30 and 10.31. Figure 10.30 shows the comparison of the amplitude pattern with and without the PEC reflectors at 7 GHz; while Figure 10.31 shows the phase component. The blue dashed line termed *non-ideal* is the result in the presence of the PEC reflector, while the reconstructed pattern is indicated by the red line termed *reconstructed*. Compared with the example in Section 10.3.4.2, the reflected responses from the PEC plates for this example are much stronger at the back side. However, the deconvolution method still obtained a very good reconstructed pattern which is close to the ideal pattern of the helical antenna. This indicates that the reflection and diffraction contributions caused by the PEC plates have been successfully compensated for by using the deconvolution method.

The deconvolution method is also applied to reconstruct the pattern for the 6-element Yagi antenna from its non-anechoic measured responses. Figure 10.32 shows the comparison of the amplitude pattern for the Yagi antenna with and without the PEC reflectors at 7 GHz; while Figure 10.33 shows the phase component of the pattern. Again, it is seen that the reconstructed pattern is very good.

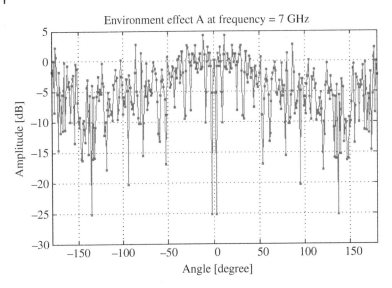

Figure 10.29 Amplitude of the environmental effects when four PEC plates are used as the reflector.

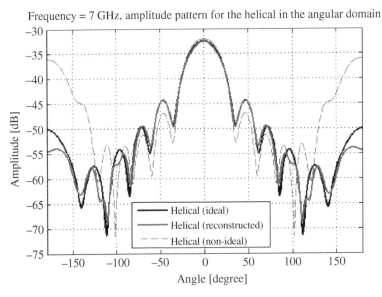

Figure 10.30 Amplitude pattern for the helical antenna with four PEC plates as the reflector.

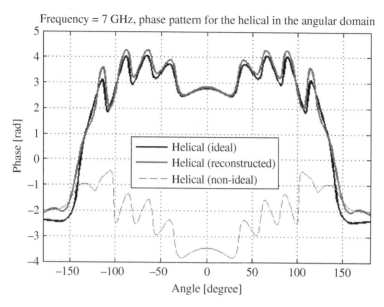

Figure 10.31 Phase pattern for the helical antenna with four PEC plates as the reflector.

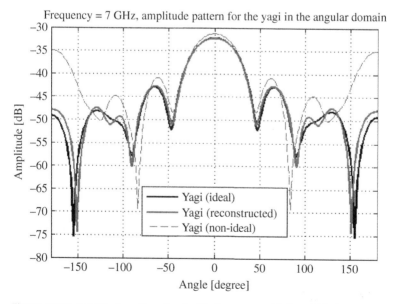

Figure 10.32 Amplitude pattern for the Yagi antenna with four PEC plates as the reflector.

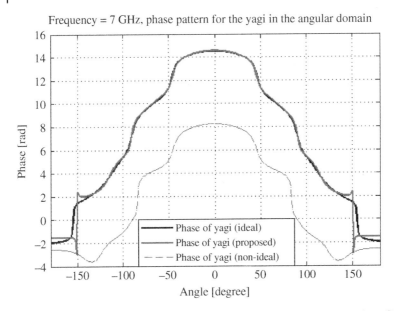

Figure 10.33 Phase pattern for the Yagi antenna with four PEC plates as the reflector.

10.3.4.4 Example IV: Use of a Parabolic Reflector Antenna as the AUT

Previous examples have shown that the deconvolution method works under three different environmental settings to reconstruct the pattern of a helical antenna and a yagi antenna. To better illustrate the applicability of the deconvolution method, it is necessary to introduce a parabolic reflector antenna to be the AUT as an example for the pattern reconstruction. Figure 10.34 shows the model of a parabolic reflector antenna with its feeding network. The dish diameter D was chosen to be 0.16 m, while the depth d was 0.04 m, and so the focal length f was calculated as $f = D^2/16d = 0.04$ m. The feeding element is a dipole antenna positioned at the focal point inside a waveguide.

The AUT is chosen as the parabolic reflector antenna, and the same procedure is carried out to retrieve the radiation pattern of the AUT. Figure 10.35 shows the comparison of the amplitude pattern for the parabolic reflector antenna with and without the PEC plates at 7 GHz; while Figure 10.36 shows the phase component of the radiation pattern. The blue dashed line termed *non-ideal* is the result in the presence of PEC plates, the reconstructed pattern is indicated by the red line termed *reconstructed*. It is seen that the reconstructed pattern using the deconvolution method is close to the ideal pattern for the main lobe but does not fit well for the side lobes and the back lobe.

To find out the reason, it is necessary to first look at the far-field distance for this parabolic reflector antenna which is $2D^2/\lambda = 1.20$ meter (D is 0.16 m as the

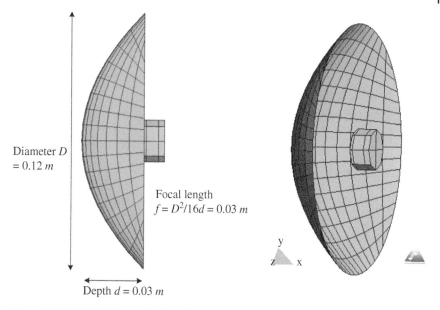

Diameter D = 0.12 m

Focal length $f = D^2/16d = 0.03$ m

y

z   x

Depth $d = 0.03$ m

Figure 10.34 Model of a parabolic reflector antenna with its feeding structure.

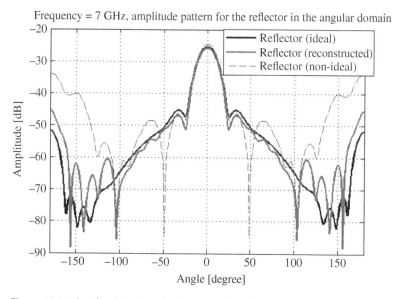

Figure 10.35 Amplitude pattern for the parabolic reflector antenna.

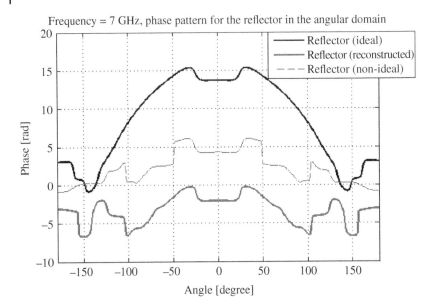

Figure 10.36 Phase pattern for the parabolic reflector antenna.

diagonal size of the antenna). And the distance between the reflector and PEC plates is $1.25 - 0.16/2 = 1.17$ m, which is smaller than the free space far-field distance in this model. This means the PEC plates are not in the free space far-field region of the parabolic reflector antenna, and this may distort the radiation pattern.

Thus, the parabolic reflector antenna was re-designed and reduced in size to be of diameter $D = 0.12$ m, depth $d = 0.03$ m, and focal length as 0.03 m. so that the free space far-field distance is reduced to be $2D^2/\lambda = 0.67$ m (D is 0.12 m as the diagonal size of the antenna). Then, the distance between the reflector antenna and PEC plates is $1.25 - 0.12/2 = 1.19$ m, which is larger than the free space far-field distance in the new model.

The same pattern reconstruction procedure was carried out for the new parabolic reflector antenna. Figure 10.37 shows the comparison of the amplitude pattern for the new parabolic reflector antenna with and without the PEC plates at 7 GHz; while Figure 10.38 shows the phase component of the radiation pattern. The blue dashed line termed *non-ideal* is the result in the presence of PEC plates, the reconstructed pattern is indicated by the red line termed *reconstructed*. One can observe that the reconstructed pattern using the deconvolution method has been greatly improved for both the side lobes and the back lobe when compared with the previous parabolic reflector antenna model.

Frequency = 7 GHz, amplitude pattern for the reflector in the angular domain

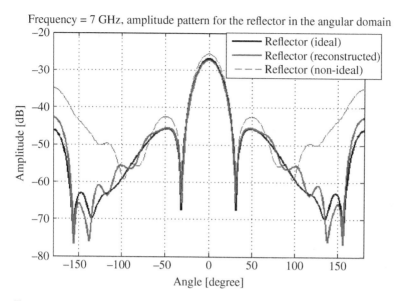

Figure 10.37 Amplitude pattern for the new parabolic reflector antenna.

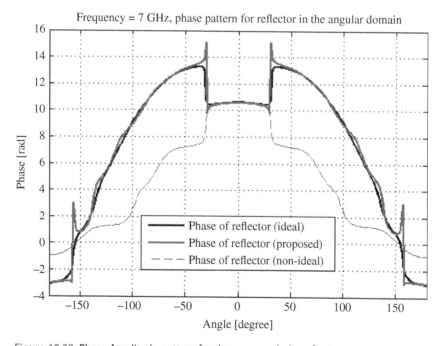

Figure 10.38 Phase Amplitude pattern for the new parabolic reflector antenna.

10.3.5 Discussions on the Deconvolution Method for Radiation Pattern Reconstruction

The previous sections have introduced the deconvolution method to reconstruct the free space radiation pattern of an antenna using data measured in a non-anechoic environment. Numerical examples have been presented to show that under different environmental setup, the deconvolution method can be successfully used to reconstruct the radiation pattern for a helical antenna, a 6-element Yagi antenna, and a parabolic reflector antenna. These examples demonstrate the general idea of the deconvolution method and give one the confidence to carry out more comprehensive analysis on it.

The deconvolution method is a general methodology for pattern reconstruction. In this chapter, more examples will be presented to analyze the applicability of the method and find out what factors would limit its performance. First, the use of different types probe antennas will be discussed, and then different sizes of antennas will be used to evaluate the method. After that, the environmental effects will be changed by varying the size of the PEC plates and compare the reconstructed results.

10.4 Effect of Different Types of Probe Antennas

Previously, numerical examples have been presented using different AUTs to test the deconvolution method. For all those examples, a standard gain horn antenna as the probe has been used. Here, different types of probe antennas will be used to change the effect of the probe and the simulations will be repeated to verify the applicability of this method.

10.4.1 Numerical Examples

10.4.1.1 Example I: Use of a Yagi Antenna as the Probe

For this example, the environmental setting using 4 PEC plates is kept the same, like that in the example of Section 10.3.4.3, but the probe antenna is replaced with the Yagi antenna, as shown in Figure 10.39. The reconstruction procedure is the same as before. First, a standard gain horn is used as the AUT. With the presence of four PEC plates, one will obtain a distorted radiation pattern of the horn in the simulation and can therefore extract the environmental effects based on its ideal pattern. Then, the horn is replaced by the helical/yagi antenna as the AUT, and the distorted radiation pattern of the helical/yagi antenna is measured. By substituting the environmental effects, one can derive the free space radiation pattern of the helical/yagi antenna. The size for the yagi and the helical antenna are as shown in Figures 10.6 and 10.5, respectively (for simplicity, the same Yagi antenna has been used both as the probe and the AUT).

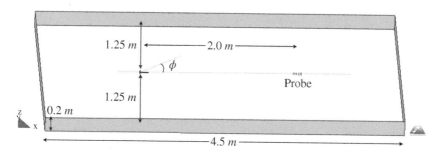

Figure 10.39 Model of the measurement system with a Yagi as the probe.

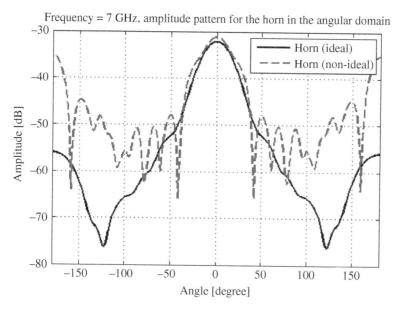

Figure 10.40 Amplitude pattern for the horn antenna system with a yagi as the probe.

Figures 10.40 and 10.41 present the received pattern for the horn antenna. Figure 10.40 shows the comparison of the amplitude pattern for the horn antenna with and without the four PEC plates; while Figure 10.42 shows the phase component. The reconstructed results for the helical antenna are shown in Figures 10.43 and 10.44. Figure 10.42 shows the comparison of the amplitude pattern; while Figure 10.43 shows the phase component.

The deconvolution method is now applied to reconstruct a clean radiation pattern for the Yagi antenna. The comparison of the amplitude pattern for the Yagi antenna with and without the PEC plates is given in Figure 10.44; while Figure 10.45 shows the phase component. When compared with examples in

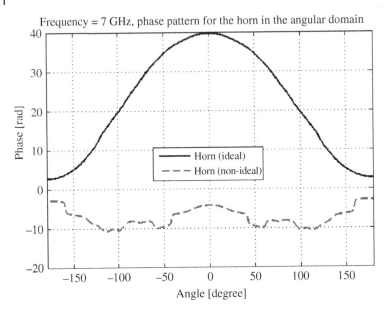

Figure 10.41 Phase pattern for the horn antenna system with a Yagi as the probe.

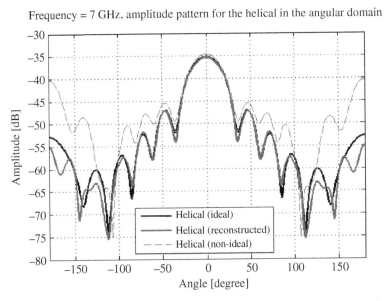

Figure 10.42 Amplitude pattern for the helical antenna with a Yagi as the probe.

Frequency = 7 GHz, phase pattern for the helical in the angular domain

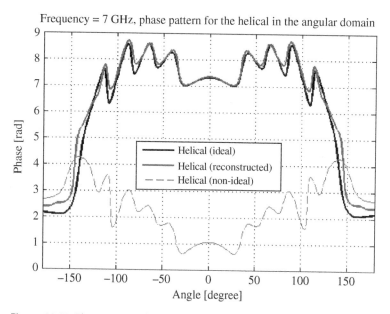

Figure 10.43 Phase pattern for the helical antenna with a Yagi as the probe.

Frequency = 7 GHz, amplitude pattern for the yagi in the angular domain

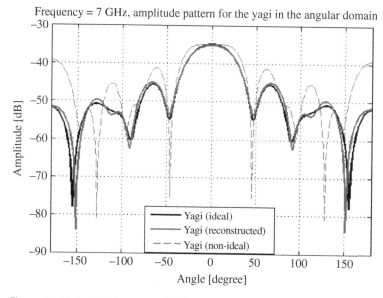

Figure 10.44 Amplitude pattern for the Yagi antenna with a Yagi as the probe.

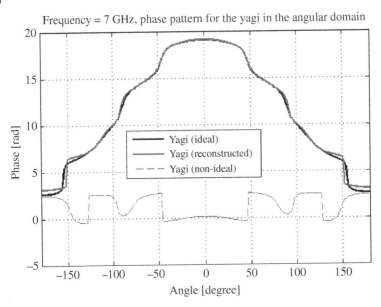

Figure 10.45 Phase pattern for the Yagi antenna with a Yagi as the probe.

Section 10.3.4.3, where the horn antenna is used as the probe, the reconstructed results by using the Yagi antenna as the probe also obtained a pattern well approximated to the ideal pattern of the AUT.

10.4.1.2   Example II: Use of a Parabolic Reflector Antenna as the Probe

The previous example has shown that choosing the Yagi antenna as the probe does not affect the reconstructed results. For this example, we'd like to use a parabolic reflector antenna as the probe, and observe if this would make any difference for the deconvolution method. The size of the parabolic reflector antenna is shown in Figure 10.34. The model for the simulation is shown in Figure 10.46. And the AUT for this example is also a parabolic reflector antenna with the same size as the probe, for simplicity.

The same procedure was applied to reconstruct the radiation pattern for the parabolic reflector antenna. Since the procedures for deconvolution has been described before, here only the reconstructed result for the parabolic reflector antenna as the AUT is presented. The received responses at the probe are shown in Figures 10.47 and 10.48. The reconstructed results for the parabolic reflector antenna are shown in Figures 10.49 and 10.50 for the amplitude and the phase component of the radiation pattern, respectively. From those figures, one can observe that when the parabolic reflector antenna is used as the probe, the reconstructed pattern is still well approximated to the ideal antenna pattern.

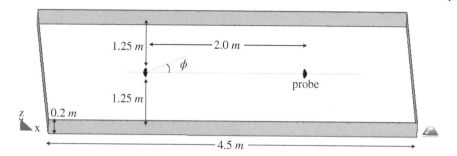

Figure 10.46 Model of the measurement system with a parabolic reflector as the probe.

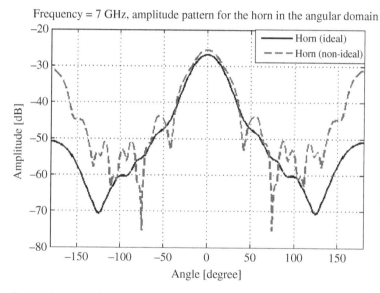

Figure 10.47 Amplitude pattern for the horn antenna with a parabolic reflector as the probe.

10.4.1.3 Example III: Use of a Dipole Antenna as the Probe

The previous two examples have well supported the deconvolution method for pattern reconstruction. However, not all types of antennas may be suitable for use as a probe antenna.

For this example, we'd like to use a dipole antenna as the probe, and observe the reconstructed results. The simulation model is shown in Figure 10.51. The dipole antenna is a half-wave dipole with the length of 2 *cm* (simulation frequency is 7 GHz) and the radius of 0.18 *mm*. The AUTs of this example are the same helical antenna and the Yagi antenna as used in previous examples. The standard gain horn antenna is used as the reference antenna to derive the

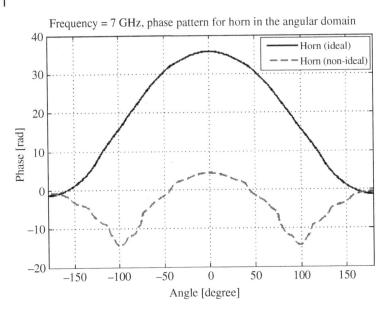

Figure 10.48 Phase pattern for the horn antenna with a parabolic reflector as the probe.

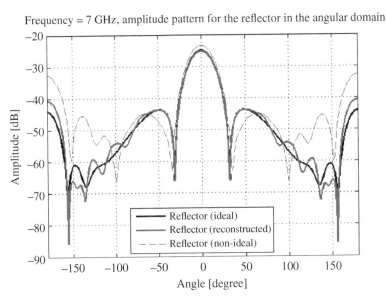

Figure 10.49 Amplitude pattern for the parabolic reflector antenna with a parabolic reflector as the probe.

Frequency = 7 GHz, phase pattern for the reflector in the angular domain

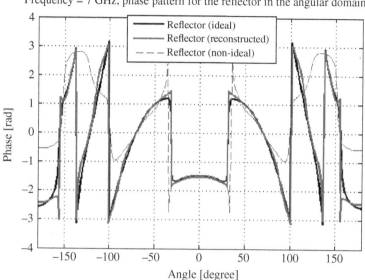

Figure 10.50 Phase pattern for the parabolic reflector antenna with a parabolic reflector as the probe.

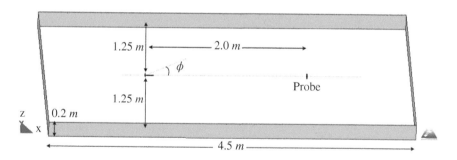

Figure 10.51 Model of the measurement system with a dipole as the probe.

environmental effects. The reconstruction procedure is the same as described in the previous examples.

Figure 10.52 shows the comparison of the amplitude pattern for the horn antenna with and without the four PEC plates; while Figure 10.53 shows the phase component. The reconstructed results for the helical antenna and the yagi antenna are shown in Figures 10.54 to 10.57. The comparison of the amplitude pattern for the helical / yagi antenna is shown in Figure 10.54 and 10.56; while Figures 10.55 and 10.57 shows the phase component.

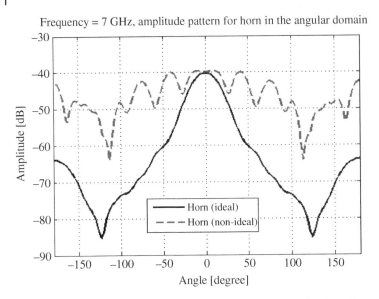

Figure 10.52 Amplitude pattern for the horn antenna with a dipole as the probe.

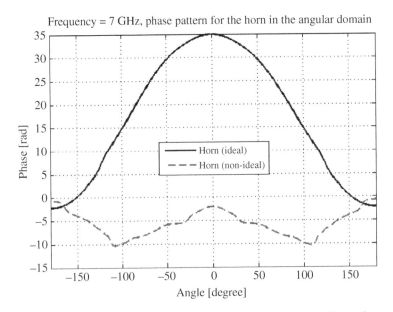

Figure 10.53 Phase pattern for the horn antenna with a dipole as the probe.

Frequency = 7GHz, amplitude pattern for the helical in the angular domain

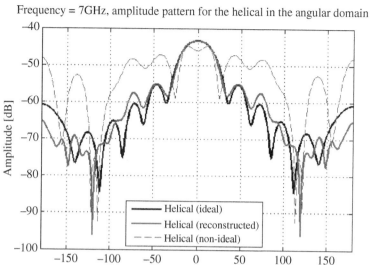

Figure 10.54 Amplitude pattern for the helical antenna with a dipole as the probe.

Frequency = 7GHz, phase pattern for the helical in the angular domain

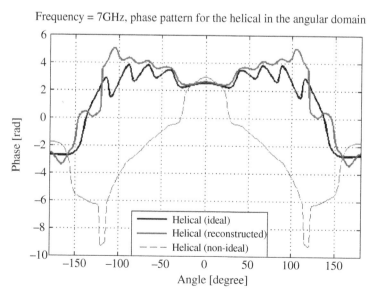

Figure 10.55 Phase pattern for the helical antenna with a dipole as the probe.

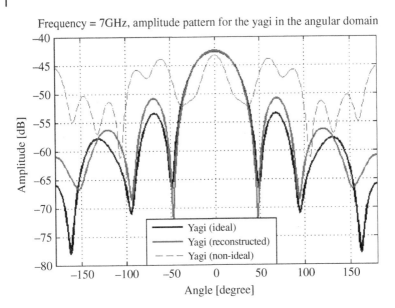

Figure 10.56 Amplitude pattern for the yagi antenna with a dipole as the probe.

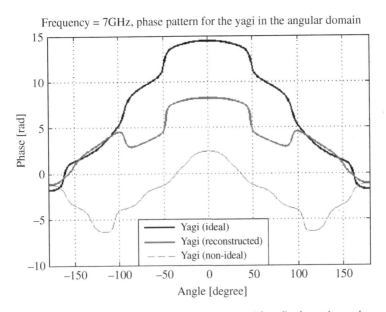

Figure 10.57 Phase pattern for the yagi antenna with a dipole as the probe.

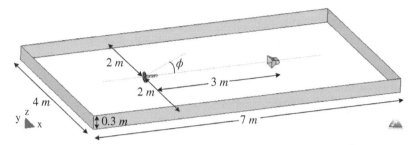

Figure 10.58 The new model of the measurement system with four PEC plates.

From these figures, one can observe that the distorted patterns are totally different from the free space radiation patterns. The reconstructed results are not as good as in the previous two examples, and only the main lobe of the pattern is well reconstructed. One can realize that when a dipole is used as the probe, due to its omni-directional property along the plane, it receives responses from the environment and the AUT equally. When the reflection responses are large enough, the direct response from the AUT will be overwhelmed by the reflection responses, and thus generating distorted reconstructed results. Therefore, the probe antenna required for the pattern reconstruction for the deconvolution method needs to be an antenna with a proper front-to-back ratio.

The two examples in Section 10.4.1.1 and Section 10.4.1.2 show that with a different probe the effect of the probe antenna to the received signal is different. As shown in Figure 10.40 and Figure 10.47, under the same environment but with a different probe the received non-ideal signals are different due to the effect of the probe. However, this effect will be cancelled and will not affect the reconstructed result. The deconvolution method requires one to perform two tests, one for a reference antenna and the other for the AUT. During the two tests, the same probe will be used to collect the signal, thus resulting in the same probe effect for two tests, and this effect can be cancelled out.

Note that the reference antenna used for extracting the environmental effects should not be any omni-directional antenna either. If the reference antenna generates an omni-directional radiation pattern, the received signal at the probe will be a constant value when the reference antenna rotates in the azimuth plane. No matter how the environment changes, the received pattern will not reveal the change of the environment.

10.5 Effect of Different Antenna Size

In this Section, we'd like to present examples using different simulation frequencies, and using different electrical sizes for the antennas to demonstrate that the deconvolution method is a general method and works for different sizes of antennas. Previous examples are all under the operating frequency of 7 GHz, the current example will change the frequency to be 1.5 GHz. And all the antennas used in the simulation are redesigned for them to work at this new frequency.

The new model for the measurement system also has four PEC plates as the reflectors around the antenna, which is shown in Figure 10.25. Four PEC plates form a rectangular cylinder enclosing the probe and the AUT. A horn antenna is still used as the probe, and the AUTs are a 6-element Yagi and a helical antenna with a back plate. New models for the horn, helix, and the Yagi antenna are shown in Figures 10.59, 10.60, and 10.61, respectively. For the operation frequency of 1.5 GHz, those antennas now have different electrical sizes. In previous examples, the horn, the helix, and the Yagi antenna models have the largest electrical sizes as 3.6 λ, 2.7 λ, and 1.3 λ, respectively; while the new models now have the largest electrical sizes as 1.5 λ, 1.6 λ, and 1.3 λ, respectively. And their radiation properties are different from the previous models (except the Yagi, whose electrical size is almost the same). In this model, both antennas are kept under a 2 m distance, which is 10 λ, away from the four PEC plates; while in the old model, the distance was 29 λ.

For this new models used in the measurement system, it is also necessary to satisfy the condition that the current distribution on the feed dipole of the AUT should not change dramatically with the environment. Figure 10.62 shows that the current distribution on the feed dipole of the helical antenna is barely affected by the PEC plates (the current distribution on other antennas is omitted). The same pattern reconstruction procedure was carried out for this example. Figure 10.63 shows the comparison of the amplitude pattern for the horn

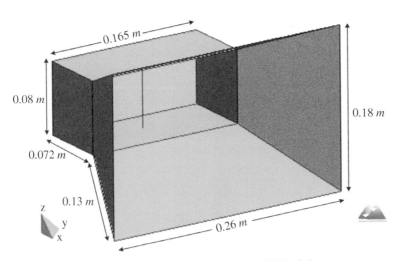

Figure 10.59 Dimensions of a new horn antenna model (Probe).

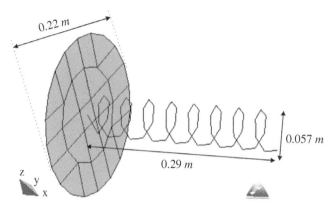

Figure 10.60 Dimensions of a new helical antenna model with a reflecting plate (AUT).

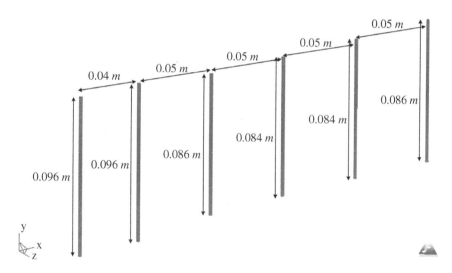

Figure 10.61 Dimensions of a new 6-element Yagi antenna model (AUT).

antenna with and without the PEC reflector at 1.5 GHz; while Figure 10.64 shows the phase component of the radiation pattern. The extracted environmental response $A(\phi, f)$ is shown in Figure 10.65.

The reconstructed results are shown through the following figures. Figure 10.66 gives the comparison of the amplitude pattern for the helical antenna; while Figure 10.67 shows the phase component of the radiation pattern. The comparison of the amplitude pattern for the yagi antenna is shown in Figure 10.68; while the phase component is given in Figure 10.69. When compared with the examples in the previous chapter, under this new

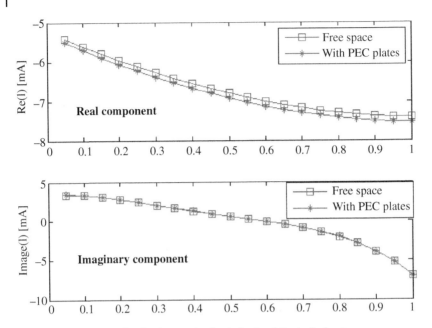

Figure 10.62 Current distribution on the feed dipole of the helical antenna.

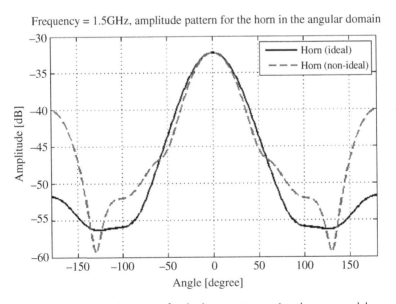

Figure 10.63 Amplitude pattern for the horn antenna using the new model.

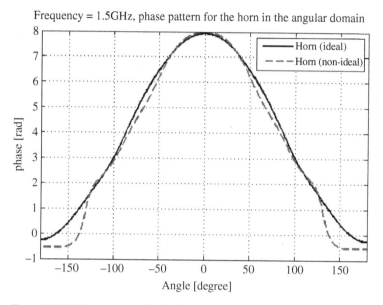

Figure 10.64 Phase pattern for the horn antenna using the new model.

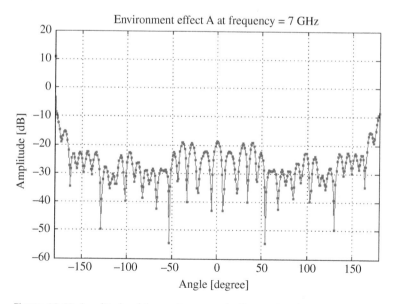

Figure 10.65 Amplitude of the environmental effects using the new model.

Frequency = 1.5GHz, amplitude pattern for the helical pattern in the angular domain

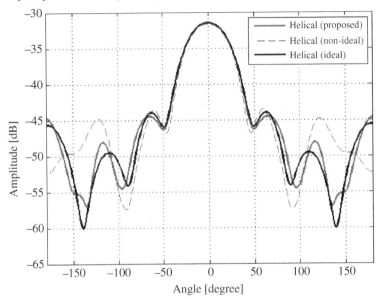

Figure 10.66 Amplitude pattern for the helical antenna using the new model.

Frequency = 1.5GHz, phase pattern for the helical pattern in the angular domain

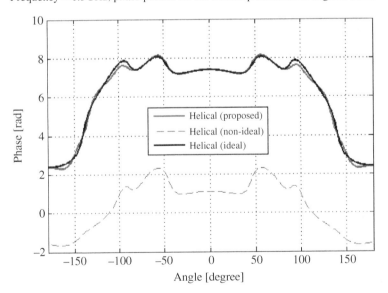

Figure 10.67 Phase pattern for the helical antenna using the new model.

Frequency = 1.5GHz, amplitude pattern for the yagi in the angular domain

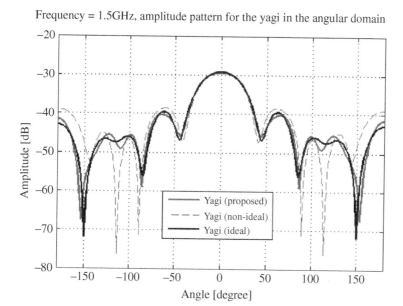

Figure 10.68 Amplitude pattern for the yagi antenna using the new model.

Frequency = 1.5GHz, phase pattern for the yagi in the angular domain

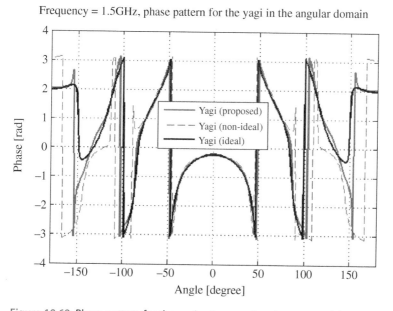

Figure 10.69 Phase pattern for the yagi antenna using the new model.

environment, the reconstructed patterns using the deconvolution method are still well approximated to the ideal patterns of the AUTs. This example indicates that the deconvolution method can successfully reconstruct the radiation pattern for different electrical sizes of antennas operating at different frequencies.

10.6 Effect of Using Different Sizes of PEC Plates

The purpose of the deconvolution method in this section, is to retrieve the ideal radiation pattern using the data measured in a non-anechoic environment. An important question would be: **How would the environment itself affect the performance of the pattern reconstruction?** In previous examples, one started from the simplest example with only one PEC plate, and increased to two plates, and then to four plates. During this process, the applicability of the deconvolution method on pattern reconstruction regarding the effect of different environments has been illustrated. Figures 10.70 and 10.71 give comparisons of the reconstructed results for the helical antenna and the yagi antenna when the number of the PEC plates increase. The left side show the amplitude patterns while the right side show the phase patterns. The units and axes are the same as in the previous examples. Note that these two figures are results summarized from previous examples described earlier.

It is shown that when there is only one PEC plate, the reconstructed results are always the best for both AUTs; while the reconstructed results are always the worst when there are four PEC plates. It indicates that when the number of PEC plates increase, the environment gets more complicated, there are more reflections and diffractions, and the reconstructed results will get worse.

To delve deeper into the question that how different environments affect the pattern reconstruction quality, one need to further change the environmental effects. One way is to add more PEC objects between the antennas and the PEC plates. However, if the object is set too close to the antennas, this may change the current distribution on the AUT and affect the reconstructed results. Another way is to increase the width of the PEC plates and generate stronger reflections.

The following example is presented to illustrate how the pattern reconstruction is affected by the width of the PEC plates. For comparison purposes, the model in Section 10.3.4 can be used and use that result as a reference. In Section 10.3.4, the AUTs and the probe antenna are newly designed and the simulation frequency is 1.5 GHz. Four PEC plates form a contour around the AUT and the probe. The helical antenna and the yagi antenna are the AUTs; while the horn antenna is the probe. We will keep those settings in the following example, but change the width of the PEC plates from 0.3 *m*

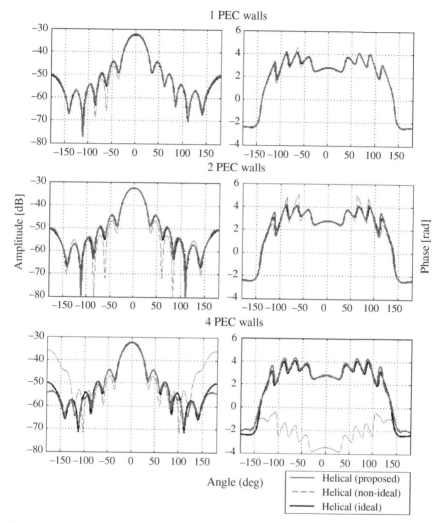

Figure 10.70 Comparison of the reconstructed patterns for the helical antenna under different number of PEC plates.

to be 0.1 m and 0.5 m, respectively. Figure 10.72 shows the model using different widths of PEC plates. It is seen that when the plate width is 0.1 m, it is of a relatively narrow strip compared to the antenna size; when the width is 0.5 m, it is relatively wide.

The pattern reconstruction procedure is the same as described in the previous examples. And the reconstructed results are shown in Figures 10.73 and 10.74. The two figures list the comparisons of the reconstructed results under different

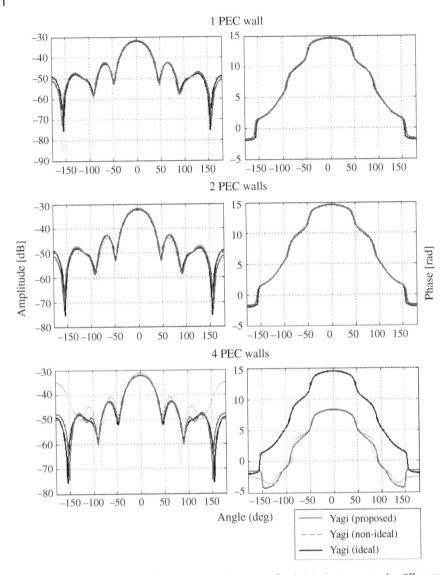

Figure 10.71 Comparison of the reconstructed patterns for the Yagi antenna under different number of PEC plates.

width settings when the AUTs are the helical antenna and the yagi antenna, respectively. And they clearly show that the reflections gets stronger and the reconstructed results get worse when the PEC plate width gets increased. We also calculated and compared the error between the ideal pattern and the

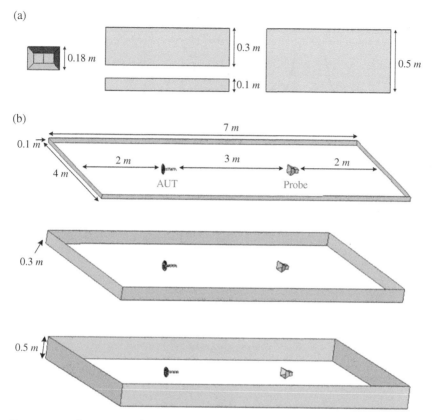

Figure 10.72 Comparison of the models with different PEC plate widths.

reconstructed pattern with respect to the rotation angle ϕ under the three width settings. And the error is defined as:

$$error(\phi) = |S_{21}(\phi) - S'_{21}(\phi)| \tag{10.23}$$

where $S_{21}(\phi)$ is the ideal pattern of the AUT and $S'_{21}(\phi)$ is the reconstructed pattern of the AUT. The results for the helical antenna and the yagi antenna are shown in Table 10.1 and Table 10.2, respectively. The Mean / STD / Maximum values of the error term represent for the average / standard deviation / largest values of error with respect to angle ϕ. And the values in the tables have a multiplier of the order of 10^{-4}.

Based on the above examples, one can conclude that when the number of PEC plates increases or the width of the PEC plates gets larger, the environment becomes more complicated and there will be stronger reflections, and the reconstructed results get worse. One can claim that the effectiveness of the pattern reconstruction is inversely proportional to the complexity of the environment.

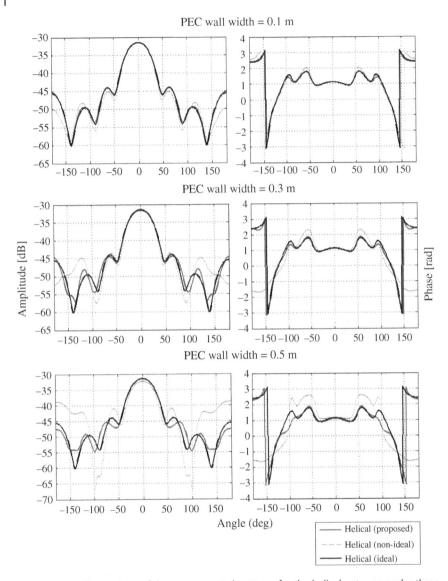

Figure 10.73 Comparison of the reconstructed patterns for the helical antenna under three PEC plate width settings.

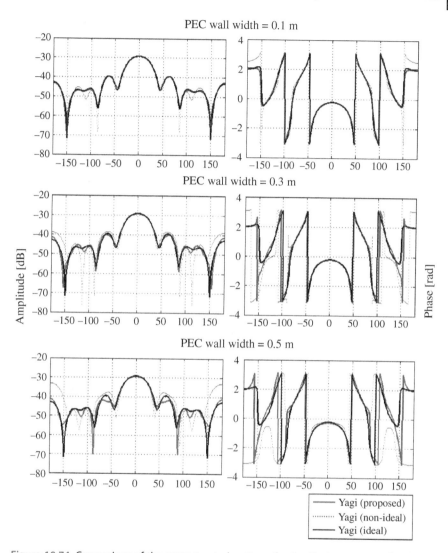

Figure 10.74 Comparison of the reconstructed patterns for the Yagi antenna under three PEC plate width settings.

Table 10.1 Reconstruction error for the helical antenna in the angular domain at 1.5 GHz.

Error (×10⁻⁴)	Width = 0.1 m	Width = 0.3 m	Width = 0.5 m
Mean Error level	1.06	3.82	6.55
STD Error level	0.68	2.91	4.80
Maximum Error level	2.55	10.26	18.92

Table 10.2 Reconstruction error for the yagi antenna in the angular domain at 1.5 GHz.

Error ($\times 10^{-4}$)	Width = 0.1 m	Width = 0.3 m	Width = 0.5 m
Mean Error level	1.59	5.45	8.99
STD Error level	0.97	3.48	7.07
Maximum Error level	4.28	12.35	31.21

10.7 Extension of the Deconvolution Method to Three-Dimensional Pattern Reconstruction

In the last sections, models of different sizes are simulated and discussed. Examples in Section 10.6 show that reconstructed results of the model with the plate width of 0.5 *m* are not as good as for the other examples. And we conclude that the effectiveness of the pattern reconstruction is inversely proportional to the complexity of the environment. So the question comes to our mind naturally: how would the deconvolution method handle the case when the plate width is much larger than the antenna size?

For all previous examples, we constitute the environment such that the PEC plates are around the probe and the AUT. And the PEC plates, whether they have one or two or four plates, whether they are wide or narrow, are always located at the azimuth plane of the antennas. But this would not be the practical case in a real measurement environment. If the measurement is carried out inside a room, there would be concrete plates around the room, the floor, and the ceiling. Therefore, there would be much more reflections from all directions inside the room compared with the examples described earlier. And it would have more practical meaning if the deconvolution method can address these complex cases. This section presents the theoretical derivation and numerical examples for a realistic environment in three dimensions.

10.7.1 Mathematical Characterization of the Methodology

In Section 10.3.1, we start by considering an AUT which can generate an ideal pencil beam pattern. As shown in Figure 10.2, the AUT, the probe and two objects are located on the same plane, and the received response at the probe is a function of the rotation angle ϕ. So the time domain response along angle ϕ can be described as an impulse response along ϕ or a spatial signature of ϕ. And the equation (10.16), which is repeated here

$$P_{non\text{-}ideal\ AUT}(\phi,f) = P_{ideal\ AUT}(\phi,f) \otimes_\phi A(\phi,f)$$

reveals that the measured non-ideal signal can be represented as an angular convolution between the ideal signal and the environmental responses. The convolution is in the azimuth angle ϕ domain, while the pattern measurements are considered in 2D (along the azimuth angle only), even though the numerical simulations are carried out in 3D.

In a complicated environment, when objects and antennas are not in the same plane, the impulse response of the environment is not only related to the azimuth angle, but also related to the elevation angle. Similarly, we first assume an AUT generates an ideal pencil beam pattern (the radiated signal will be along one direction in the far field). As shown in Figure 10.75, the gray plane is the azimuth plane where the AUT, the probe and Object 1 are located; while Object 2 is above the plane. It is easy to know that reflections occur in both ϕ and θ angles. The received time domain signal at the probe, when the AUT is at the rotation angles (θ, ϕ), is unique with respect to other angles and will not be related to the response of the AUT along other angles. Therefore, the received response at the probe will be a function of both the azimuth angle ϕ and the elevation angle θ.

The measured signal at the probe, which contains various reflections, can be represented as a convolution in time by the ideal signal (without any reflection) and the impulse response of the environment along the rotation angle (θ_I, ϕ_I). It can be written as (similar to previous case):

$$P_{non\text{-}ideal}(\theta_L, \phi_I, t) = P_{ideal}(\theta_L, \phi_I, t) \otimes A(\theta_L, \phi_I, t) \tag{10.24}$$

or

$$P_{non\text{-}ideal}(\theta_L, \phi_I, f) = P_{ideal}(\theta_L, \phi_I, f) \cdot A(\theta_L, \phi_I, f) \tag{10.25}$$

where $\otimes$ denotes a time convolution, and

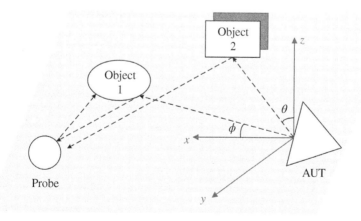

Figure 10.75 Multiple reflections exist in the whole spatial domain.

$P_{non\text{-}ideal}(\theta_L, \phi_I, t)$ is the non-ideal time domain signal at the probe in the presence of the reflections for the rotation angle (θ_L, ϕ_I);

$P_{non\text{-}ideal}(\theta_L, \phi_I, f)$ is the non-ideal frequency domain signal at the probe in the presence of the reflections for the rotation angle (θ_L, ϕ_I);

$P_{ideal}(\theta_L, \phi_I, t)$ is the ideal time domain signal without any reflection for the rotation angle (θ_L, ϕ_I);

$P_{ideal}(\theta_L, \phi_I, f)$ is the ideal frequency domain signal without any reflection for the rotation angle (θ_L, ϕ_I);

$A(\theta_L, \phi_I, t)$ is the impulse response of the environment when the AUT has a ideal pencil beam pattern pointing along the angle (θ_L, ϕ_I);

$A(\theta_L, \phi_I, f)$ is the frequency domain response of the environment when the AUT has a ideal pencil beam pattern pointing along the angle (θ_L, ϕ_I).

Note that $A(\theta_L, \phi_I, t)$ represents the reflection contributions from the environment at the rotation angle (θ_L, ϕ_I) and is independent of the particular AUT. It is the spatial signature of the environment.

In a real situation, the AUT will radiate towards every direction in the spatial domain and cannot have an ideal pencil beam pattern. One can define the impulse response of the environment along the angle (θ_L, ϕ_I) as $\hat{A}(\theta_L, \phi_I, f)$ when the AUT does not have an ideal pencil beam pattern. Then, by using (10.25) we have:

$$P_{non\text{-}ideal}(\theta_L, \phi_I, f) = P_{ideal}(\theta_L, \phi_I, f) \cdot \hat{A}(\theta_L, \phi_I, f) \tag{10.26}$$

Here, $\hat{A}(\theta_L, \phi_I, f)$ is the impulse response of the environment along the angle (θ_L, ϕ_I) when the AUT does not have a ideal pencil beam pattern. It is not the true impulse response of the environment but also contains the pattern information of the AUT. For the deconvolution method, we need to know the true impulse response $A(\theta_L, \phi_I, f)$ of the test environment. Then $\hat{A}(\phi_L, f)$ can be considered as a convolution in the angular domain of the normalized beam pattern and the true impulse response. That is,

$$\hat{A}(\theta_L, \phi_I, f) = \frac{P_{ideal}(\theta, \phi, f)}{P_{ideal}(\theta_L, \phi_I, f)} \otimes_{\phi,\theta} A(\theta, \phi, f) \Big|_{\theta = \theta_L, \phi = \phi_I} \tag{10.27}$$

Here, $P_{ideal}(\theta, \phi, f)$ is the ideal pattern without any reflection from the environment at the frequency f. And $\otimes_{\phi,\theta}$ is the 2D convolution operator for both the azimuth angle ϕ and the elevation angle θ, in other words, this operator is a two-dimensional operator works on each row and column of the matrix. By substituting (5.4) into (5.3), we have:

$$P_{non\text{-}ideal}(\theta_L, \phi_I, f) = P_{ideal}(\theta, \phi, f) \otimes_{\phi,\theta} A(\theta, \phi, f) \Big|_{\theta = \theta_L, \phi = \phi_I} \tag{10.28}$$

For a general angle of (θ, ϕ), we have:

$$P_{non\text{-}ideal}(\theta, \phi, f) = P_{ideal}(\theta, \phi, f) \otimes_{\phi,\theta} A(\theta, \phi, f) \tag{10.29}$$

which is an extension of the equation (10.16). Therefore, the radiation pattern of the AUT in the presence of reflections can be considered as two-dimensional convolution in the angular domain between the ideal pattern of the AUT and the impulse response of the environment. Note that the convolution is now carried out in both the azimuth angle ϕ domain and the elevation angle θ domain, thus the 3D pattern measurements (in both azimuth angle and elevation angle) are carried out to form the matrix of $P_{non\text{-}ideal}(\theta, \phi, f)$, $P_{ideal}(\theta, \phi, f)$ and $A(\theta, \phi, f)$. And we know that taking the FFT will transform the convolution in the angular domain to the multiplication in the other domain. Thus the impulse response of the environment $A(\theta, \phi, f)$ can be extracted by taking the Inverse Fourier Transform of (10.29) after a reference antenna is measured in the environment. And the ideal radiation pattern of the AUT $P_{ideal}(\theta, \phi, f)$ can then be obtained for any AUT measured in the same environment using (10.29).

10.7.2 Steps Summarizing for the Methodology

The procedure to carry out the deconvolution method in 3D environments is very similar to the procedure described in Section 10.3. First, carry out the measurement in a 3D non-anechoic environment using two reference antennas, whose ideal patterns are known, and then perform a deconvolution to estimate the environmental response A. Then, use the AUT as the transmitter and carry out the measurement in the same environment and estimate the ideal radiation pattern of the AUT. The entire procedure consists of 4 steps:

1) At a fixed frequency, measure the reference antenna in a non-anechoic environment and obtain the received response $P_{non\text{-}ideal}(\theta, \phi, f)$. Also the ideal response for the reference antenna $P_{ideal}(\theta, \phi, f)$ is known.
2) Calculate the environmental effects $A(\theta, \phi, f)$ using the equation:

$$P_{non\text{-}ideal}(\theta, \phi, f) = P_{ideal}(\theta, \phi, f) \otimes_{\phi, \theta} A(\theta, \phi, f)$$

3) At the same frequency, use the AUT as the transmitter at the same position and measure the AUT in the same non-anechoic environment as described in step (1), and let $P_{non\text{-}ideal\,AUT}(\theta, \phi, f)$ be the result.
4) By substituting the environmental effects $A(\theta, \phi, f)$ into the equation one can obtain the ideal response of the AUT $P_{ideal\,AUT}(\theta, \phi, f)$ through deconvolution:

$$P_{non\text{-}ideal\,AUT}(\theta, \phi, f) = P_{ideal\,AUT}(\theta, \phi, f) \otimes_{\phi, \theta} A(\theta, \phi, f)$$

Again, one can see that the deconvolution method only requires a single frequency measurement, and does not need any prior knowledge of the test environment.

10.7.3 Processing the Data

Section 10.3 gives the data processing procedures for 2D radiation pattern reconstruction. This section presents the following procedures for processing the simulated data for 3D radiation pattern reconstruction. The procedures will be similar to that for 2D pattern reconstruction but rotate the AUT in three-dimensions.

1) First, create the simulation model for the 3D pattern measurement system. For both the reference antenna and the AUT, each antenna needs to have two simulation models. One model only has antennas (the probe and the AUT), and there is no PEC plates as reflectors. Such a model is to simulate the free space condition so as to obtain the ideal pattern of the antenna. The other model has both antennas and PEC plates, and the objective is to simulate the non-anechoic environment to obtain the non-ideal pattern of the antenna.

2) For each model, rotate the AUT along itself for a step of $10°$ for both the azimuth angle and the elevation angle (θ, ϕ), and for each rotation angle the model is simulated and the S_{21} data is collected. The S_{21} data along each spatial angle (θ, ϕ) forms the radiation pattern of the AUT. There are 36 data points along the elevation angle θ, i.e. $0°$, $10°$,..., $350°$. For each elevation angle, the AUT rotates one loop in the azimuth plane and there will be 36 data points in one loop, i.e. $-180°$, $-170°$,..., $170°$. One can imagine this as the AUT rotates in a step of $10°$ in the azimuth plane and changes the elevation angle in a step of $10°$. In total, there will be 36*36 data points forming a period in the angular domain (θ, ϕ), i.e. $(0°, -180°)$, $(0°, -170°)$,..., $(0°, 170°)$, $(10°, -180°)$, $(10°, -170°)$,..., $(350°, 160°)$, $(350°, 170°)$ in a two dimensional matrix form, and then the data will repeat this sequence. The data at the spatial angle of $(360°, 180°)$ is equal to that of $(0°, -180°)$ and belongs to the next period. Therefore, the non-ideal pattern $P_{non\text{-}ideal\text{-}AUT}$ (θ, ϕ, f), the ideal pattern $P_{ideal\text{-}AUT}(\theta, \phi, f)$ and the environmental effects A (θ, ϕ, f) all are two-dimensional periodic sequences in the angular domain (θ, ϕ). Note that in the normal spherical coordinate system, the elevation angle θ varies between $[0°, 180°]$. Here, the θ angle will change its value between $[0°, 360°]$ to form a period in the spatial angular domain. However, we only need to simulate the model when θ varies between $[0°, 180°]$, from which the data set of θ between $[180°, 360°]$ can be derived. Since there is no explicit definition for the data set of θ between $[180°, 360°]$ in the normal spherical coordinate system, we need to derive that data through using the conversion between the spherical coordinate system and the Cartesian coordinate system, as shown in the Appendix 10A. Then, both (θ, ϕ) change its value within 2π range and form a periodic 2D matrix.

3) Simulate the model for the reference antenna in the free space and the desired non-anechoic environment to obtain the ideal and non-ideal

patterns of the reference antenna $P_{ideal-\text{Re}\,f}(\theta, \phi, f)$ and $P_{non-ideal-\text{Re}\,f}(\theta, \phi, f)$, respectively. Similarly, by replacing the reference antenna with the desired AUT and carry out the simulation, one can obtain the ideal pattern $P_{ideal-\text{AUT}}(\theta, \phi, f)$ and the non-ideal pattern $P_{non-ideal-\text{AUT}}(\theta, \phi, f)$ for the desired AUT. The ideal pattern of the AUT $P_{ideal-\text{AUT}}(\theta, \phi, f)$ will be the goal of the reconstructed pattern.

4) Now we need to apply (10.29) to reconstruct the 3D radiation pattern of the AUT. According to Section 3.9 of [40], the convolution of two 2D periodic sequences is the multiplication of the corresponding 2D matrix of discrete Fourier series. Let $\tilde{x}_1(m, n)$ and $\tilde{x}_2(m, n)$ be two periodic sequences of period N^*M with the 2D discrete Fourier series denoted by $\tilde{X}_1(k, l)$ and $\tilde{X}_2(k, l)$, respectively. It can be written as:

$$\tilde{x}_3(m, n) = \sum_{q=0}^{M-1}\sum_{r=0}^{N-1} \tilde{x}_1(q, r)\tilde{x}_2(m - q, n - r) \tag{10.30}$$

$$\tilde{X}_3(k, l) = \tilde{X}_1(k, l)\tilde{X}_2(k, l) \tag{10.31}$$

5) From (10.28) and (10.29) we have the following equations:

$$P_{non-ideal-\text{Re}f}(\theta, \phi, f) = P_{ideal-\text{Re}f}(\theta, \phi, f) \otimes_{\theta,\phi} A(\theta, \phi, f) \tag{10.32}$$

$$P_{non-ideal-\text{AUT}}(\theta, \phi, f) = P_{ideal-\text{AUT}}(\theta, \phi, f) \otimes_{\theta,\phi} A(\theta, \phi, f) \tag{10.33}$$

Therefore, by taking the 2D-FFT of both sides of (10.25), the angular convolution operator will become the multiplication operator. One can derive the environment effects $A(\phi, f)$ as:

$$A(\theta, \phi, f) = \mathit{ifft}_2\left(\frac{\mathit{fft}_2(P_{non-ideal-\text{Ref}}(\theta, \phi, f))}{\mathit{fft}_2(P_{ideal-\text{Ref}}(\theta, \phi, f))}\right) \tag{10.34}$$

where fft_2 and ifft_2 denote the 2D-FFT and 2D-IFFT operator, respectively.

6) Take the 2D-FFT of both sides of (10.33) and substitute the environment effects $A(\theta, \phi, f)$ into the equation, to obtain:

$$\begin{aligned} P_{ideal-\text{AUT}}(\theta, \phi, f) &= \mathit{ifft}_2\left(\frac{\mathit{fft}_2(P_{non-ideal-\text{AUT}}(\theta, \phi, f))}{\mathit{fft}_2(A(\theta, \phi, f))}\right) \\ &= \mathit{ifft}_2\left(\frac{\mathit{fft}_2(P_{non-ideal-\text{AUT}}(\theta, \phi, f)) \cdot \mathit{fft}_2(P_{ideal-\text{Ref}}(\theta, \phi, f))}{\mathit{fft}_2(P_{non-ideal-\text{Ref}}(\theta, \phi, f))}\right) \end{aligned} \tag{10.35}$$

Note that the division and multiplication in (10.34) and (10.35) are element by element operations for 2D matrices. All data processing will be performed off-line using a commercial software package (MATLAB 7, The MathWorks Inc., Natick, MA, 2000). In MATLAB, the function Y = fft2(X) returns the two-dimensional Discrete Fourier Transform of matrix

X, computed with a Fast Fourier Transform algorithm. Similarly, the function $Y = \text{ifft2}(X)$ returns the two-dimensional Inverse Discrete Fourier Transform of matrix X. All the data used in (10.35) need to be processed through step 2 first. Note that when the environment setting is symmetrical along the elevation angle $\theta = 90°$, some elements of the 2D-FFT of the 2D matrices generate zero values, so in (10.35) there exists $0 \cdot (0/0)$ value, which should be zero. However, N/A value was generated in MATLAB due to the numerical error. To conduct the 2D-IFFT of the matrix, we manually set those N/A values to be zero, which they should be, as shown in Appendix 10B.

7) Compare the reconstructed pattern of the AUT from (10.35) with the simulated result $P_{ideal-\text{AUT}}(\theta, \phi, f)$.

10.7.4 Results for Simulation Examples

In this section, examples for 3D pattern reconstruction under five different environmental settings are presented by using the Yagi antenna and the parabolic reflector antenna as the AUT. The five environmental settings include: four wide PEC plates around the antennas (shown in Section 10.7.4.1), four PEC plates as well as the PEC ground (shown in Section 10.7.4.2), six PEC plates form an unclosed contour around the antennas (shown in Section 10.7.4.3), six PEC plates forming a closed contour (shown in Section 10.7.4.4), and six dielectric plates forming a closed contour (shown in Section 10.7.4.5). Those five environmental settings present a full picture of how the deconvolution method extracts the ideal pattern from the non-ideal signal under 3D environments. For the examples, illustrated next, a 6-element Yagi antenna and a new parabolic reflector antenna will be the AUTs. Again, a horn antenna is set to be both the reference antenna and the probe. Dimensions of the horn antenna and the Yagi antenna are given in Figures 10.59 and 10.61, respectively. Dimensions of the new parabolic reflector antenna are shown in Figure 10.76. The numerical simulations and the processing of the simulation data will follow the procedures as described in Section 10.7.3. The simulation model in this example is similar to the one shown in Figure 10.72. The difference is that the PEC plates have a width of 2 m and are 2 m away from the antennas in this example, as shown in Figure 10.77.

The S_{21} data was collected for each spatial angle (θ, ϕ) to form the 3D radiation pattern. Equation (10.33) was used to calculate the reconstructed 3D pattern for the AUT. And the reconstructed results were compared with the simulated ideal pattern of the AUT to illustrate the performance of the deconvolution method. HOBBIES was used to perform the full wave EM simulation and the operation frequency was 1.5 GHz.

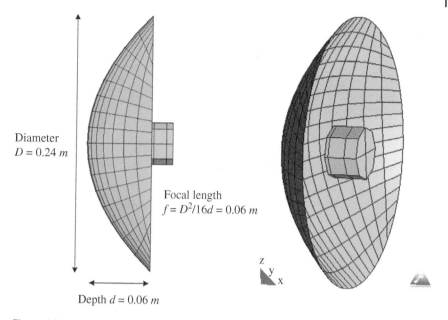

Diameter
$D = 0.24\ m$

Focal length
$f = D^2/16d = 0.06\ m$

Depth $d = 0.06\ m$

Figure 10.76 Model of a new parabolic reflector antenna with its feeding structure.

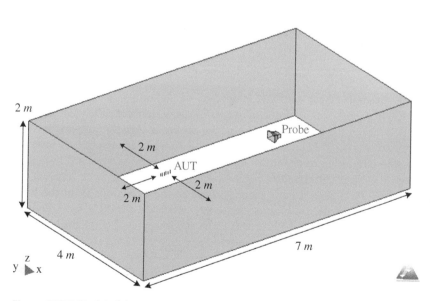

Figure 10.77 Model of the measurement system with four very wide PEC plates.

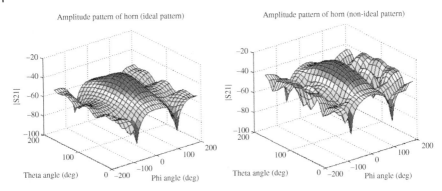

Figure 10.78 Three-dimensional plot of the ideal and non-ideal radiation patterns of the horn antenna.

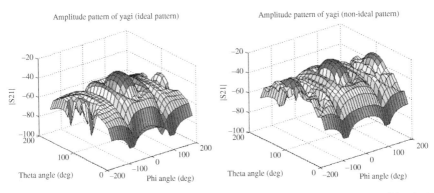

Figure 10.79 Three-dimensional plot of the ideal and non-ideal radiation patterns of the Yagi antenna.

10.7.4.1 Example I: Four Wide PEC Plates Serve as Reflectors

The ideal (free space) and non-ideal (under the presence of the PEC plates) radiation patterns of the horn, the Yagi, and the parabolic reflector antenna for the 3D case are shown in Figures 10.78, 10.79 and 10.80, respectively. The vertical axis is the amplitude of S_{21} in dB scale. It is easy to observe the differences between the ideal patterns and the non-ideal patterns. The back lobes in all three figures have been greatly increased due to the reflections from the PEC plates. Note that at the rotation angle $(\theta, \phi) = (90°, 0°)$, the AUT faces toward the probe; while at $(\theta, \phi) = (90°, 180°)$, the AUT rotates 180° in the azimuth plane and faces back to the probe. And the back lobe is located around the angle of $\phi = 180°$.

As mentioned in Section 10.5, one condition needs to be satisfied is that the current distribution on the feed dipole of the AUT should remain close to the ideal one. Figures 10.81 and 10.82 give both the real and imaginary part of the current distribution on the feed dipole of the two AUTs, the Yagi antenna and the parabolic reflector antenna, respectively. It shows that the current distribution on the feed dipole of the parabolic reflector is slightly changed (since the feed dipole is located inside the waveguide); while the current distribution of

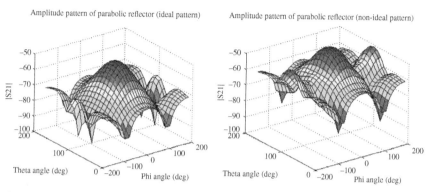

Figure 10.80 Three-dimensional plot of the ideal and non-ideal radiation patterns of the parabolic reflector antenna.

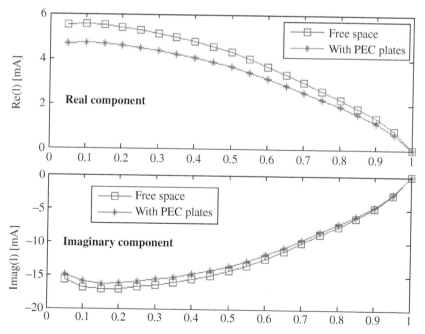

Figure 10.81 Current distribution on the feed dipole on the Yagi antenna with four very wide PEC plates.

the Yagi has been partly affected by the PEC plates. To mitigate this change of the current distribution due to the PEC plates, it is necessary to increase the distance between the plates and the AUT. However, this will dramatically increase the computational size of the problem and make the simulation

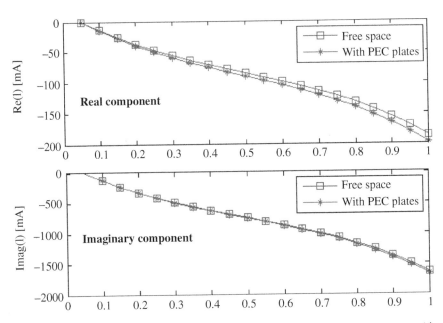

Figure 10.82 Current distribution on the feed dipole of the parabolic reflector antenna with four wide PEC plates as the reflector.

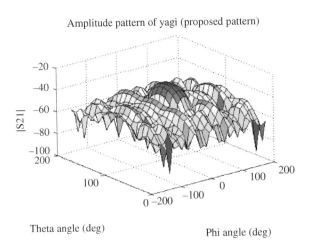

Figure 10.83 Three-dimensional plot of the reconstructed pattern for the Yagi antenna.

difficult to carry out using modest computational resources. Therefore, we will keep the current model settings, but we should expect some level of differences between the ideal pattern and the reconstructed results.

The pattern reconstruction procedure in 3D environment follows the steps as described in Section 10.7.2. The 3D reconstructed pattern of the Yagi antenna is shown in Figure 10.83. To better illustrate the reconstructed pattern along different spatial angles, the 2D cuts of different phi angles and theta angles are shown in Figures 10.84 and 10.85, respectively. The reconstructed pattern is indicated by the red line while the ideal pattern is indicated by the black line. The blue dash line is for the non-ideal pattern. In Figure 10.84 the x-axis is the elevation angle θ, and the y-axis is the amplitude of S_{21} in dB scale. In Figure 10.85 the x-axis is the azimuth angle ϕ, and the y-axis is the amplitude of S_{21} in dB scale. Note that the 2D cuts at $\phi = 0°$ and $\theta = 90°$ give the patterns of the principal planes for the AUT, and they are more representative in illustrating the performance of the 3D pattern reconstruction.

Similarly, for the pattern reconstruction of the parabolic reflector antenna, the 3D reconstructed pattern is shown in Figure 10.86; while the 2D cuts of different phi angles and theta angles are shown in Figures 10.87 and 10.88, respectively. As shown in the figures, the results of the 3D pattern reconstruction for the parabolic reflector antenna are of engineering accuracy. The reconstructed pattern is very close to the ideal pattern. While for the Yagi antenna the results are not as good as previous 2D reconstruction examples. The pattern shapes for some 2D cuts are a little different from the ideal ones. But the reconstructed pattern basically follows the trend of the ideal pattern and the undesired reflections have been greatly compensated by using the deconvolution method. One reason for the difference in the patterns is that the current distribution on the feed dipole of the AUT has changed and this leads to the change of the free space radiation pattern of the AUT. The other possible reason may be due to the sparse sampling of the data in the angular domain. In the previous examples, for the 2D reconstruction a 1° step in ϕ was used; while for the 3D reconstruction example a 10° step was chosen for sampling in both θ and ϕ. Even though the total number of data points (equals 36*36) is much larger than that for the 2D examples, the density of the data points in the angular domain is much lower. And this sparse sampling of the data points may not sufficiently characterize the environmental effects, especially when the environments are complicated or the reflections are strong.

10.7.4.2 Example II: Four PEC Plates and the Ground Serve as Reflectors

The simulation model of this example includes the antenna, four PEC plates and the PEC ground, as shown in Figure 10.89. The PEC plates and the ground serve as the reflectors and reflect the fields from the AUT in all directions. Those four PEC plates all have the same size of 3 m by 4 m and are connected with the PEC ground plane (the blue plane in Figure 10.89), just like the four plates and the

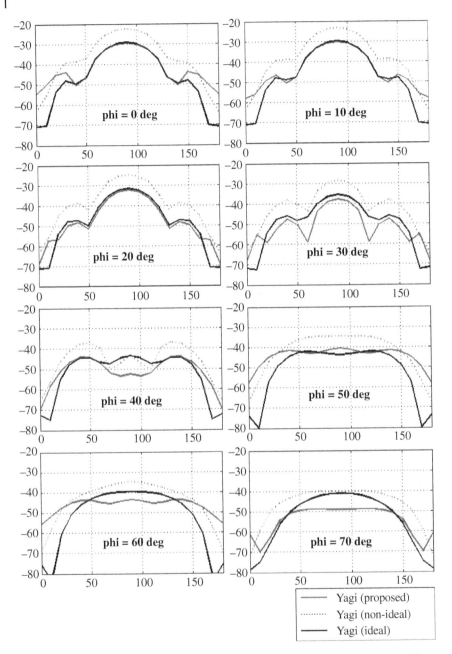

Figure 10.84 Comparison of the reconstructed patterns for the Yagi antenna along different φ angles with four wide PEC plates as the reflector.

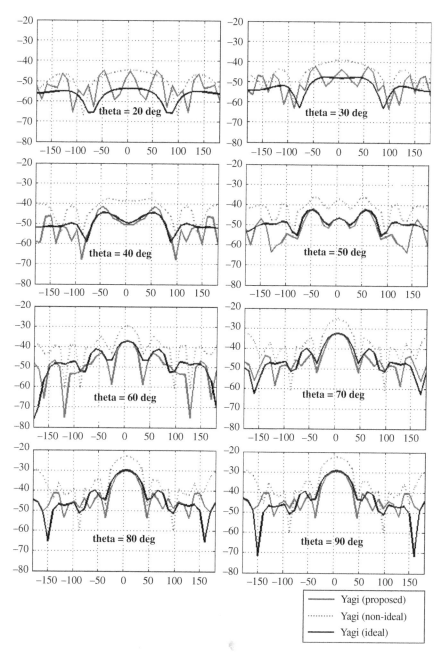

Figure 10.85 Comparison of the reconstructed patterns for the Yagi antenna along different θ angles with four wide PEC plates as the reflector.

Amplitude pattern of parabolic reflector (proposed pattern)

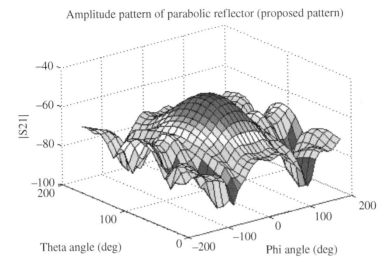

Figure 10.86 Three-dimensional plot of the reconstructed pattern for the parabolic reflector antenna.

floor of a room. The AUT and the probe are 2.5 meters away from the PEC ground. The PEC ground is modeled with an infinite large PEC plane. Due to the property of an infinite large PEC plane, the PEC ground can be substituted by adding the image of the model with respect to the PEC ground, as shown by the red circle in Figure 10.89.

The ideal and non-ideal (under the presence of PEC plates) radiation patterns of the horn antenna, the Yagi antenna and the parabolic reflector antenna in 3D are shown in Figures 10.90, 10.91 and 10.92, respectively. The vertical axis is the amplitude of S_{21} in dB scale. It is easy to observe the differences between the ideal patterns and the non-ideal patterns. Especially the back lobes of all the three figures are much stronger due to the reflections from the PEC plates and the ground.

Now let's look at the current distribution on the feed dipole on the Yagi antenna, which is shown in Figure 10.93. It shows that the real part of the current distribution on the yagi has been partly affected by the PEC plates, which means the reconstructed pattern in this environment will be somewhat different from the free space ideal pattern. And we should expect this in the reconstructed results.

The pattern reconstruction procedures follow the steps as described in Section 10.7. And the reconstructed 3D pattern for the yagi antenna is shown in Figure 10.94. To better display the reconstructed pattern along different spatial angles, Figures 10.95 and 10.96 present the 2D cuts for different ϕ and θ, respectively. Since the model is not symmetrical in θ domain, the θ cuts take

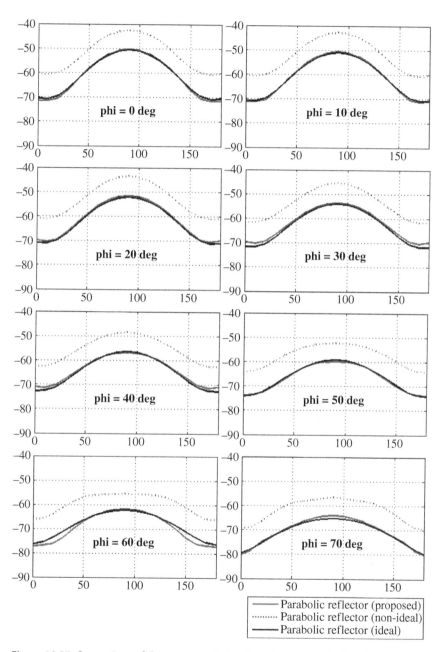

Figure 10.87 Comparison of the reconstructed patterns for the parabolic reflector antenna along different φ with four wide PEC plates as the reflector.

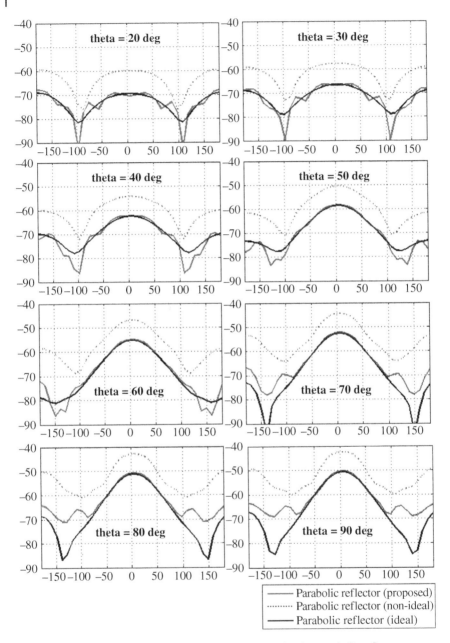

Figure 10.88 Comparison of the reconstructed patterns for the parabolic reflector antenna along different θ angles with four wide PEC plates as the reflector.

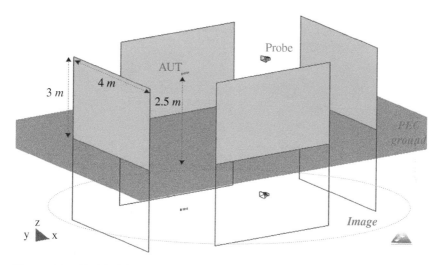

Figure 10.89 Model of the measurement system with four PEC plates and the ground as the reflector.

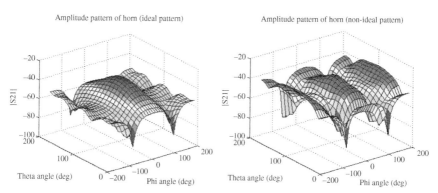

Figure 10.90 Three-dimensional plot of the ideal and non-ideal radiation patterns of the horn antenna.

several cut planes around the principal plane ($\theta = 90°$). The reconstructed pattern is indicated by the red line while the ideal pattern is indicated by the black line. The blue dash line is for the non-ideal pattern. In Figure 10.95 the x-axis is the elevation angle θ, the y-axis is the amplitude of S_{21} in dB scale. In Figure 10.96 the x-axis is the azimuth angle ϕ, the y-axis is the amplitude of S_{21} in dB scale.

Similarly, for the pattern reconstruction of the parabolic reflector antenna, the 3D reconstructed pattern is shown in Figure 10.97; while the 2D cuts of different ϕ and θ are shown in Figures 10.98 and 10.99, respectively. The

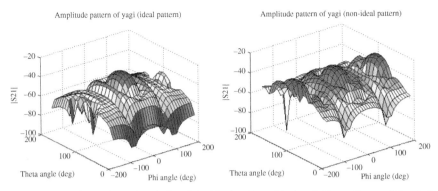

Figure 10.91 Three-dimensional plot of the ideal and non-ideal radiation patterns of the Yagi antenna.

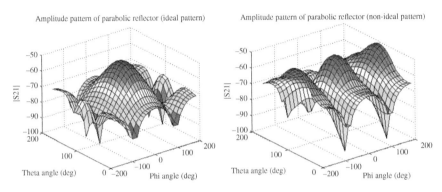

Figure 10.92 Three-dimensional plot of the ideal and non-ideal radiation patterns of a parabolic reflector antenna.

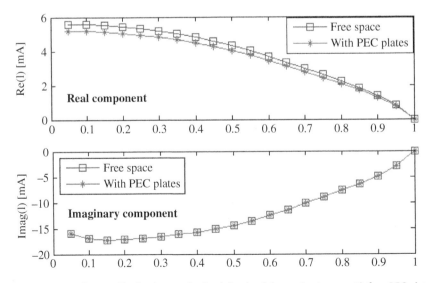

Figure 10.93 Current distribution on the feed dipole of the yagi antenna with four PEC plates and the ground as the reflector.

Amplitude pattern of yagi (proposed pattern)

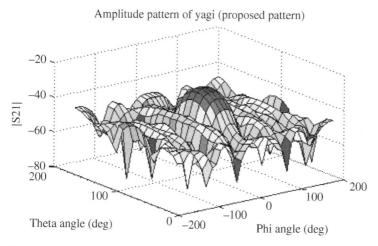

Figure 10.94 Three-dimensional plot of the reconstructed pattern for the yagi antenna.

reconstructed results for the Yagi antenna are not as good as the results shown in Example I (Section 10.7.4.1). The plots for the phi-cuts show that along θ equals 0° and 180° the reconstructed patterns have even stronger reflections than the non-ideal patterns. But the plots for the theta-cuts still show that the reconstructed pattern is improved through the deconvolution method. For the parabolic reflector antenna, the reconstructed results are much better. Figures 10.98 and 10.99 clearly illustrate that reflections of the non-ideal patterns have been greatly compensated, and nulls of the reconstructed pattern have been shifted to the right positions.

10.7.4.3 Example III: Six Plates Forming an Unclosed Contour Serve as Reflectors

The simulation model with six PEC plates forming an unclosed contour is shown in Figure 10.100, which provides a side and a top-down view of the model. The PEC plates are set in a symmetrical way and have the same size of the four plates at the side. The PEC plates on the bottom and the top also have the same size. These six PEC plates will serve as the reflectors and reflect the fields of the AUT from all directions. The antenna measurement simulation is carried out inside the contour.

The ideal and non-ideal (under the presence of PEC plates) radiation patterns of the horn antenna, the Yagi antenna and the parabolic reflector antenna in 3D are shown in Figures 10.101, 10.102 and 10.103, respectively. The vertical axis is the amplitude of S_{21} in dB scale. It is clear that there are visual differences between the ideal patterns and the non-ideal patterns. Especially the back lobes for all the three figures are much stronger than the ideal ones due to the reflections from the PEC plates.

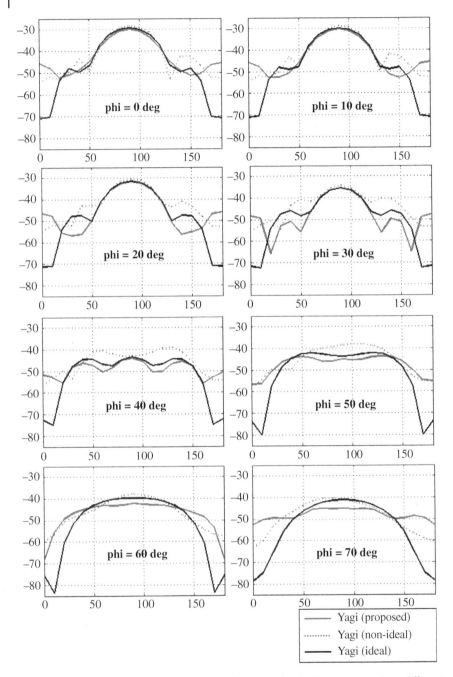

Figure 10.95 Comparison of the reconstructed patterns for the Yagi antenna along different φ with four PEC plates and the ground as the reflector.

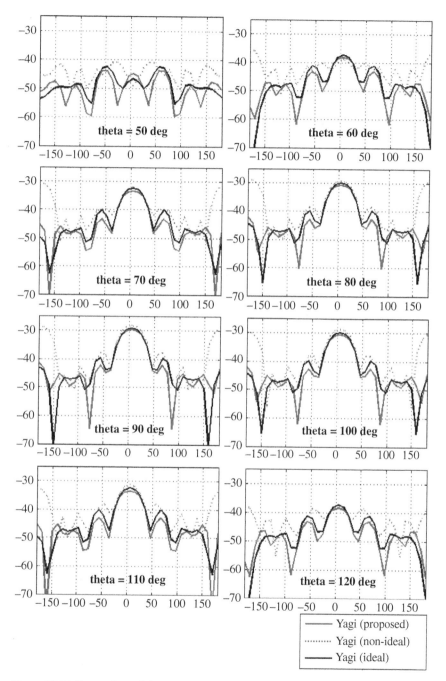

Figure 10.96 Comparison of the reconstructed patterns for the yagi antenna along different θ with four PEC plates and the ground as the reflector.

Amplitude pattern of parabolic reflector (proposed pattern)

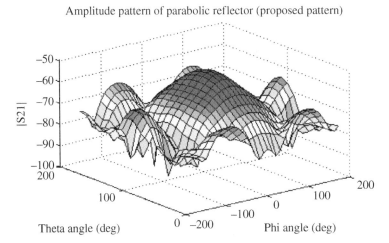

Figure 10.97 Three-dimensional plot of the reconstructed pattern for the parabolic reflector antenna.

Now let's look at the current distribution on the feed dipole of the Yagi antenna, which is shown in Figure 10.104. It shows that the current distribution on the Yagi has been partly affected by the PEC plates.

The pattern reconstruction procedures in 3D environments follow the steps as described in Section 10.3. And the reconstructed 3D pattern for the Yagi antenna is shown in Figure 10.105. To better demonstrate the reconstructed results along different spatial angles, Figures 10.106 and 10.107 give the 2D cuts for different ϕ and θ, respectively. The reconstructed pattern is indicated by the red line while the ideal pattern is indicated by the black line. The blue dash line is for the non-ideal pattern. In Figure 10.106 the x-axis is the elevation angle θ, the y-axis is the amplitude of S_{21} in dB scale. In Figure 10.107 the x-axis is the azimuth angle ϕ, the y-axis is the amplitude of S_{21} in dB scale.

Similarly, for the pattern reconstruction of the parabolic reflector antenna, the 3D reconstructed pattern is shown in Figure 10.108; while the 2D cuts of different ϕ and θ are shown in Figures 10.109 and 10.110, respectively.

As shown in the figures, the reconstructed results are not as good as the results shown in Example I in Section 10.7.4.1. The reconstructed pattern can basically follow the trend of the ideal pattern and compensate for the reflections for the main lobe; but the reconstructed results for the side lobes are not good. Except the effect of the changes in current distribution shown in 10.104, the two PEC plates at the top and the bottom of the antennas also add to the complexity of reflections and thus distorting the reconstructed patterns.

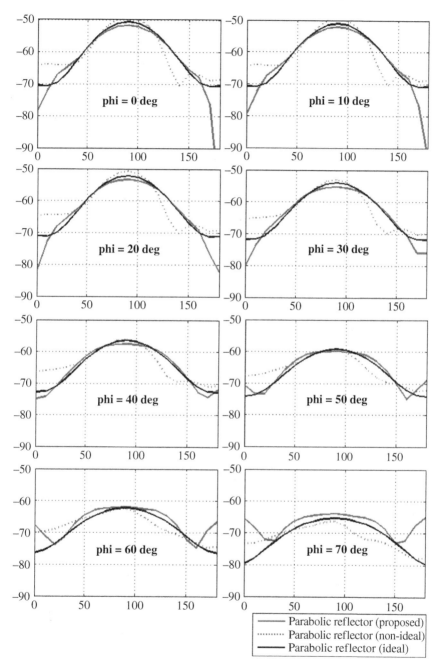

Figure 10.98 Comparison of the reconstructed patterns for the parabolic reflector antenna along different φ with four PEC plates and the ground as the reflector.

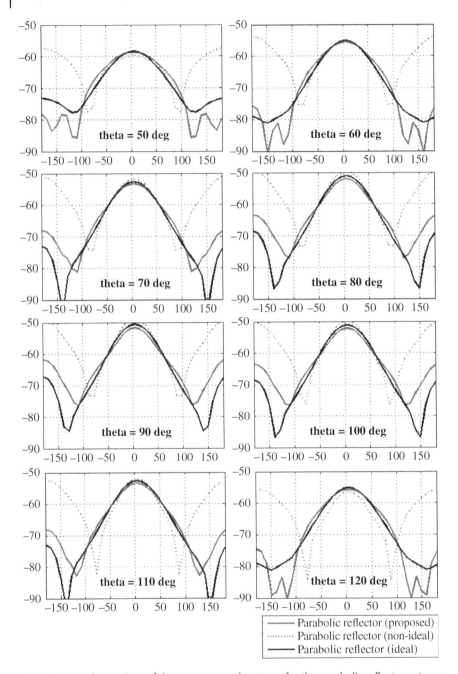

Figure 10.99 Comparison of the reconstructed patterns for the parabolic reflector antenna along different θ with four PEC plates and the ground as the reflector.

(a)

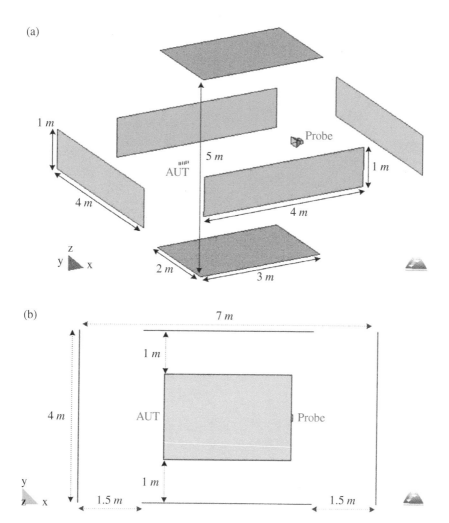

1 m

5 m

Probe

AUT

1 m

4 m

4 m

z

y x

2 m

3 m

(b)

7 m

1 m

4 m

AUT

Probe

y

z x

1 m

1.5 m

1.5 m

Figure 10.100 Model of the measurement system with six PEC plates forming an unclosed contour.

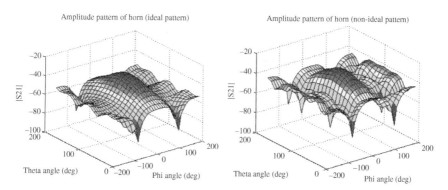

Amplitude pattern of horn (ideal pattern)

Amplitude pattern of horn (non-ideal pattern)

Figure 10.101 Three-dimensional plot of the ideal and non-ideal radiation patterns of the horn antenna.

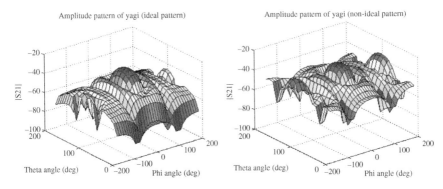

Figure 10.102 Three-dimensional plot of the ideal and non-ideal radiation patterns of the Yagi antenna.

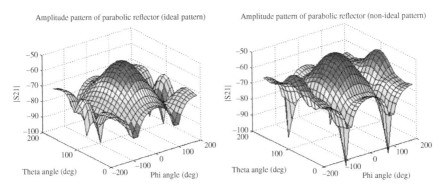

Figure 10.103 Three-dimensional plot of the ideal and non-ideal radiation patterns of the parabolic reflector antenna.

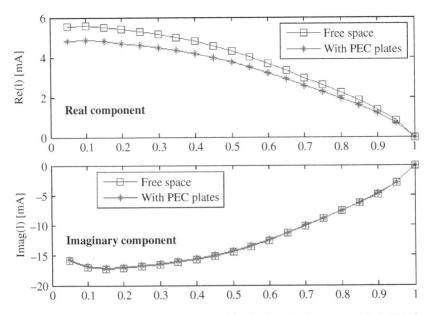

Figure 10.104 Current distribution on the feed dipole of the Yagi antenna with six PEC plates as the reflector.

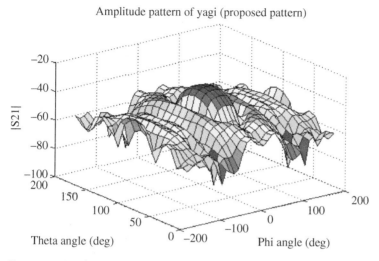

Amplitude pattern of yagi (proposed pattern)

Figure 10.105 Three-dimensional plot of the reconstructed pattern for the Yagi antenna.

10.7.4.4 Example IV: Antenna Measurement in a Closed PEC Box

The previous three examples have shown the pattern reconstruction under the environments of open (unclosed) PEC contour. This example is to illustrate the case in which the antenna measurement is carried out in a closed PEC box, where the deconvolution method fails. **This is because an antenna does not radiate when placed inside a PEC box.**

The model for this example is shown in Figure 10.111 (displayed in the transparent mode). The PEC box has the size of 4 *m* by 4 *m* by 7 *m*; while the AUT and the probe antenna have a distance of 3 *m* between them and they are 2 *m* away from the surrounding PEC plates. Again, the 6-element Yagi antenna and the parabolic reflector antenna are used as the AUT while the horn is used as the probe antenna. The sampling of data points in the angular domain is every 10 degree as a step in both θ angle and ϕ angle.

The ideal and non-ideal (under the presence of PEC plates) radiation patterns of the horn antenna, the Yagi antenna and the parabolic reflector antenna in 3D plot are shown in Figures 10.112, 10.113 and 10.114, respectively. The vertical axis is the amplitude of S_{21} in dB scale. It is seen that there are tremendous differences between the ideal patterns and the non-ideal patterns. The non-ideal patterns don't have a regular shape of radiation patterns with the main lobe and the back lobe. They are more like random values. And the average level of the non-ideal patterns is much higher than the ideal patterns. This will be explained later.

Let's also look at the current distribution on the feed dipole of the Yagi antenna as shown in Figure 10.115. As illustrated in the figure, the imaginary part of the current distribution within the PEC box is much larger than the ideal one; while the real part of the current distribution within the PEC box is

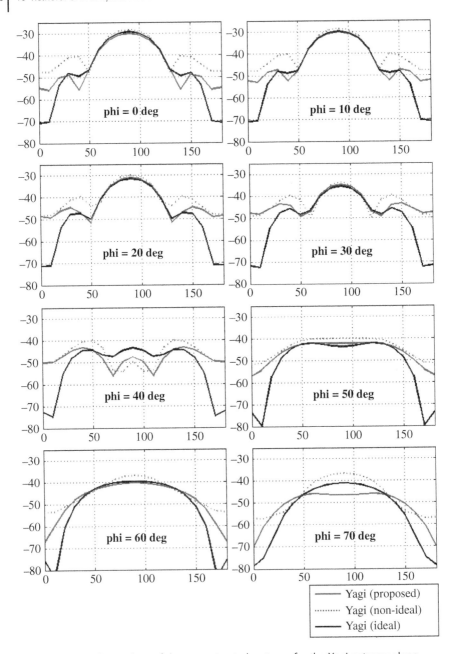

Figure 10.106 Comparison of the reconstructed patterns for the Yagi antenna along different ϕ with six PEC plates as the reflector.

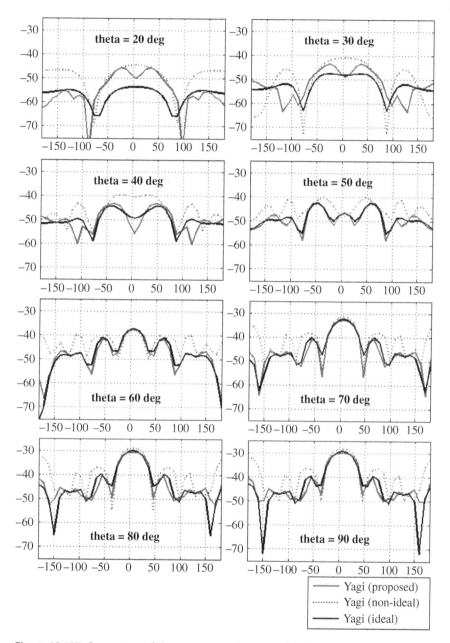

Figure 10.107 Comparison of the reconstructed patterns for the yagi antenna along different θ with six PEC plates as the reflector.

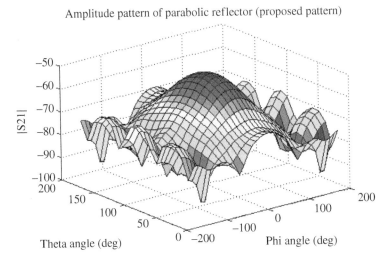

Figure 10.108 Three-dimensional plot of the reconstructed pattern of the parabolic reflector antenna.

practically of zero value. It is known that inside a closed PEC box, the energy generated by the AUT is purely imaginary, which is induced by the imaginary part of the current. So the field inside the closed PEC box would be the near-field only, and there would be no far-field inside or outside the PEC box, which is why a PEC box is commonly used as the shield box in the Wheeler-Cap method for measuring the radiation efficiency of an antenna. Therefore, the measured pattern under such environment is not the far-field pattern. This also explains why the average level of the non-ideal radiation pattern is much larger than the ideal pattern in Figures 10.112–10.114. Because all the energy is conserved inside the PEC box as the near-field. Therefore, we do not dwell on this example any further and deal with the case if the antenna is placed inside a dielectric cube which in reality would be a room.

10.7.4.5 Example V: Six Dielectric Plates Forming a Closed Contour Simulating a Room

The previous four examples have used PEC plates to model walls of a room, but this is not realistic in a real measurement as in reality the rooms are mainly consist of dielectric walls mimicking bricks. The idea of using the PEC plates is to enhance the environmental effects and make it easier to be analyzed. On the other hand, that greatly increases the difficulty of the reconstruction problem, since the reflection and diffraction components are strong. Actually, they are too strong so that the current distribution of the AUT changes a lot under these environmental settings. This example models the environment as a room formed by six dielectric plates, which is much more realistic.

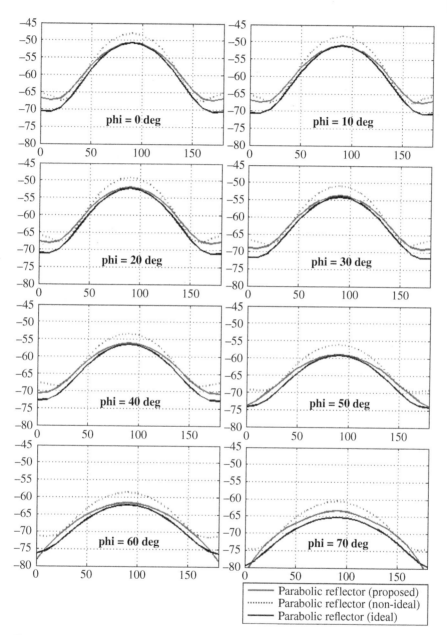

Figure 10.109 Comparison of the reconstructed patterns for the parabolic reflector antenna along different φ with six PEC plates as the reflector.

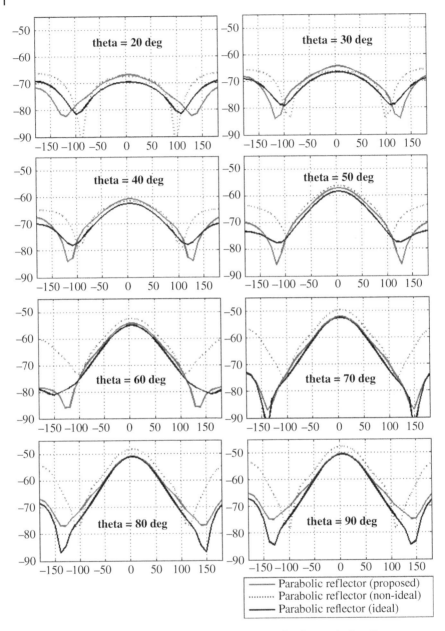

Figure 10.110 Comparison of the reconstructed patterns for the parabolic reflector antenna along different θ with six PEC plates as the reflector.

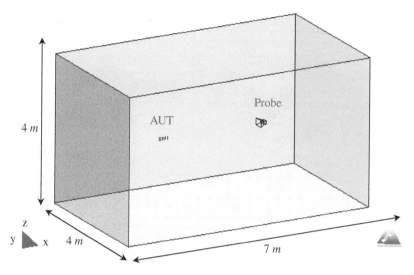

Figure 10.111 Model of the measurement system within a closed PEC box.

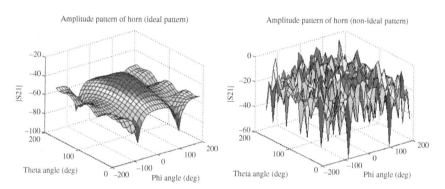

Figure 10.112 Three-dimensional plot of the ideal and non-ideal radiation patterns of the horn antenna (antenna in a closed PEC box).

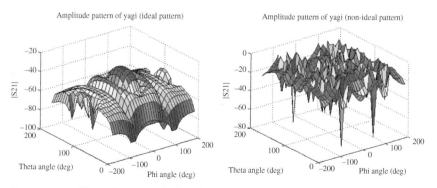

Figure 10.113 Three-dimensional plot of the ideal and non-ideal radiation patterns of the Yagi antenna (antenna in a closed PEC box).

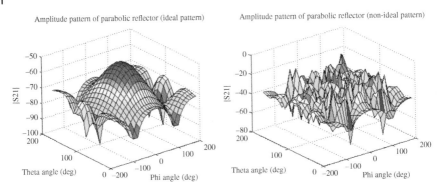

Figure 10.114 Three-dimensional plot of the ideal and non-ideal radiation patterns of the parabolic reflector antenna (antenna in a closed PEC box).

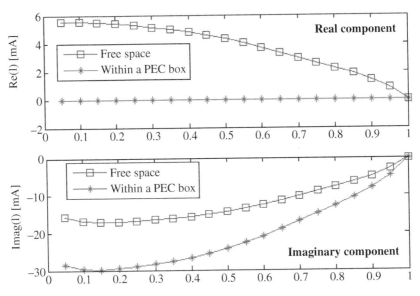

Figure 10.115 Current distribution on the feed dipole of the yagi antenna. (antenna within a closed PEC box).

The simulation model of this example includes the AUT, the probe and six dielectric plates around the antennas, as shown in Figure 10.116. The antennas are inside the plated box while the dielectric material starts from the plates and extends to infinity. Therefore, the dielectric plates around the antennas will be infinitely thick. This is a simplified model of a regular room where the plates are made of concrete blocks with a finite thickness of around 1 feet. The radiated fields from the AUT will reflect, diffract and refract from the dielectric plates.

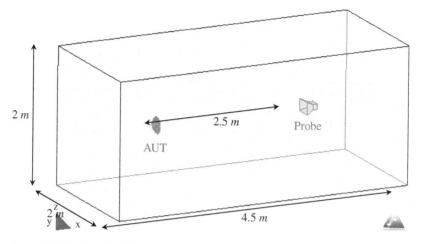

Figure 10.116 Model of the measurement system within a closed box with 6 dielectric plates.

The dielectric box has a size of 2 *m* by 2 *m* by 4 *m*, and the six plates separate the air and the dielectric material, which is concrete ($\varepsilon_r = 2.2$, $\tan\delta = 0.011$). The AUT and the probe are 2.5 meters away from each other and are 1 meter away from plates around.

For this model, the differences between the ideal and non-ideal (under the presence of dielectric plates) radiation patterns are much smaller. We will not show the 3D plot here since they appear very close in the 3D plots, but the comparison will be illustrated for the patterns in 2D cuts.

First let's check the current distribution on the feed dipole of the Yagi antenna, which is shown in Figure 10.117. It shows that the current distribution on the feed dipole of the Yagi is very slightly changed by the dielectric plates, which means the reflection and diffraction components are much smaller under this environment compared with the previous examples. And we should expect this in the reconstructed results. Based on the experience gained in simulating the previous examples we expect that the reconstructed pattern should be closer to the ideal pattern.

We followed the procedures as described in Section 10.7.2 and reconstructed the patterns for the Yagi antenna and the parabolic reflector antenna. To better display the differences between the ideal pattern, non-ideal pattern, and the reconstructed pattern, 2D cuts along different spatial angles are given in the following figures.

For the Yagi antenna, 2D cuts along different ϕ and θ are shown in Figures 10.118 and 10.119, respectively. The reconstructed pattern is indicated by the red line while the ideal pattern is indicated by the black line. The blue dash line is for the non-ideal pattern. In Figure 10.118 the *x*-axis is the elevation angle θ, the *y*-axis is the amplitude of S_{21} in dB scale. In Figure 10.119 the *x*-axis is the azimuth angle ϕ, the *y*-axis is the amplitude of S_{21} in dB scale.

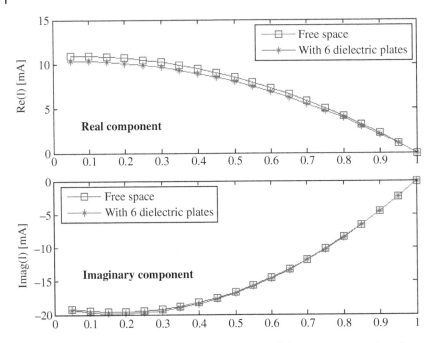

Figure 10.117 Current distribution on the feed dipole of the Yagi antenna when the test environment is a closed box with 6 dielectric plates.

Similarly, for the pattern reconstruction of the parabolic reflector antenna, the 2D cuts along different ϕ and θ are shown in Figures 10.120 and 10.121, respectively.

The results for both the Yagi and the parabolic reflector antenna show that the reconstructed patterns agree very well to the ideal patterns. The differences between the ideal patterns and the reconstructed patterns are mainly located in the regions $\theta \in [0°, 50°]$, $\theta \in [130°, 180°]$, $\phi \in [-180°, -100°]$, and $\phi \in [100°, 180°]$, which are mainly in the domain of the side lobes and back lobes.

This example also indicates that under a more realistic environment setting, when the current distribution of the AUT is not much affected by the test environment, the deconvolution method can achieve a very good reconstructed result. For previous four examples, those are the worst scenarios where PEC plates are used as the reflector. The PEC plates generate much more reflections and diffractions than the dielectric plates and affect the current distribution on the AUT a lot.

The previous 2D pattern reconstruction has been extended to 3D environments. The five examples shown above discussed the pattern reconstruction in different 3D environments. The reconstructed results can greatly mitigate

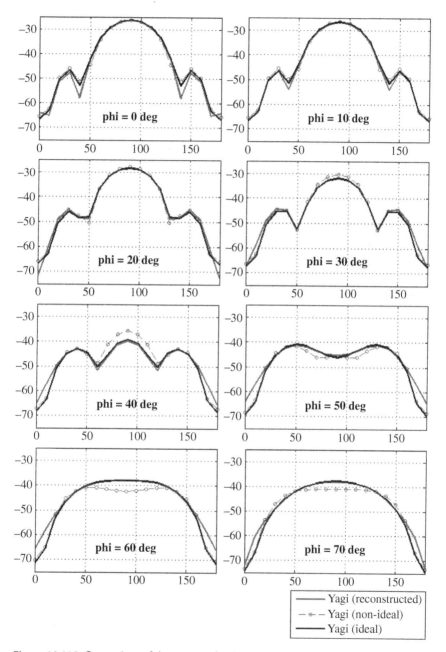

Figure 10.118 Comparison of the patterns for the Yagi antenna along different φ when the test environment is a closed box with 6 dielectric plates.

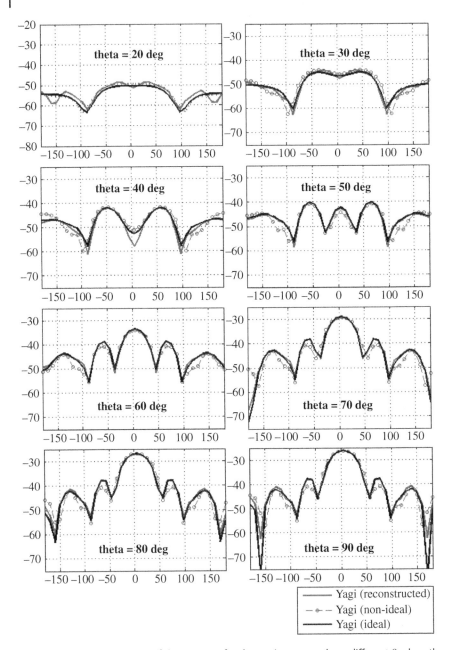

Figure 10.119 Comparison of the patterns for the yagi antenna along different θ when the test environment is a closed box with 6 dielectric plates.

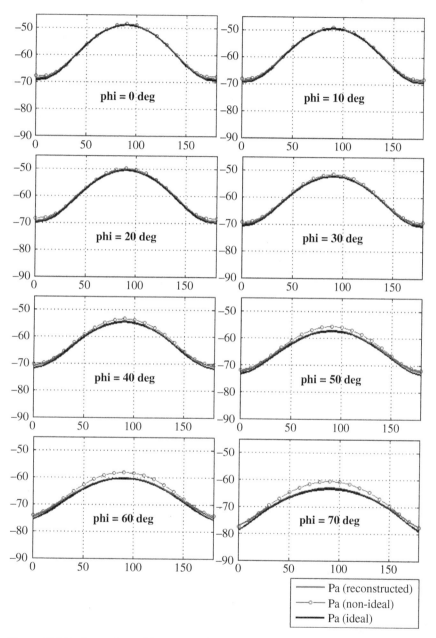

Figure 10.120 Comparison of the patterns for the parabolic reflector antenna along different φ when the test environment is a closed box with 6 dielectric plates.

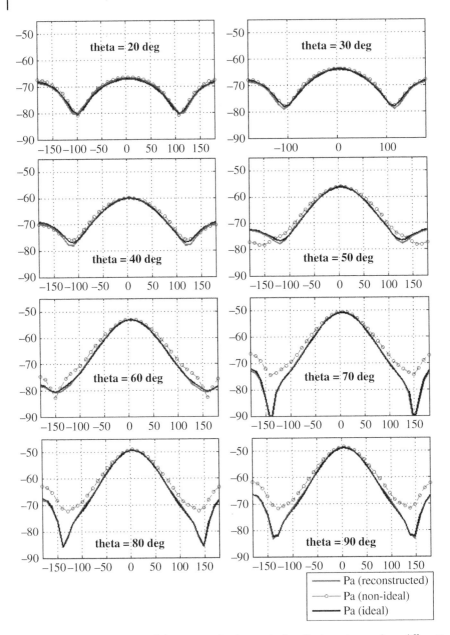

Figure 10.121 Comparison of the patterns for the parabolic reflector antenna along different θ when the test environment is a closed box with 6 dielectric plates.

the undesired reflections and approximate the ideal patterns. One major reason for the differences between the ideal patterns and the reconstructed patterns is due to the change of the current distribution on the AUT.

10.8 Conclusion

This presentation has focused on illustrating and extending a deconvolution algorithm to retrieve the free space far-field pattern of an AUT from its measured radiation pattern in a non-anechoic environment. Extensive numerical examples are given for illustration.

The environmental effects has been modelled as an impulse response of the test environment in the spatial domain. And the measured non-ideal pattern under the environment is an angular convolution between the ideal pattern (free space far-field pattern) of the AUT and the environmental responses. Therefore, under any test environment, one can obtain the free space far-field pattern for an unknown antenna with two antenna measurements. One measurement is for the calibration of the environment using a reference antenna whose pattern is known, and the other is for the measurement of an AUT in the same test environment. This requires two assumptions that the environment is unchanged during measurements for the two antennas and sizes of the reference antenna as well as the AUT need to be similar.

Numerical examples indicate that the method is not limited to the sizes of the antennas or the frequencies under test, as long as the current distribution on the AUT is not much affected by the environments. However, the method does have a limitation to the choice of the probe antenna. The failure of the dipole antenna example shows that the probe needs to be a directional antenna. And the performance of the method is inversely proportional to the complexity of the environments. The complexity of the environments shows up in two ways in our simplified models, the number of PEC objects and the width of the PEC plates. The deconvolution method is extended to three-dimensional environments to reconstruct the 3D radiation pattern for an AUT, in both azimuth and elevation angles. Numerical examples are given to evaluate the 3D pattern reconstruction under five different environmental settings. Those settings cover different 3D environments and present a full picture of how the deconvolution method retrieves the ideal pattern from the non-ideal pattern. Results indicate that with the PEC plates as the reflector, the reconstructed patterns are roughly approximate to the ideal patterns, but are not as good as the results shown for the examples presented for 2D. And for the environment of a closed PEC box, the reconstruction will fail since there will be no far-field formed inside a closed PEC box. While for the example with dielectric plates as the reflector, reconstructed results are much improved and are very well approximated to the ideal pattern of an AUT.

We find that one major reason for the differences between the ideal patterns and the reconstructed patterns is due to the change of the current distribution on the AUT. When PEC plates are used as the reflector, they cause much more reflections and diffractions than the dielectric plates and affect the current distribution on the AUT a lot. The current distribution change directly affects the radiation pattern of the AUT. However, an environment with dielectric plates is more realistic than an environment with large PEC plates. Therefore, one can expect a better reconstructed result from the deconvolution method under realistic environment settings.

So far, multiple numerical examples have been presented to model the antenna radiation pattern measurement system and test the effectiveness of the deconvolution method. Those examples have shown the method can achieve a good approximation for the free space radiation pattern of the AUT from the data measured in non-anechoic environments. From the small samples of presented examples, one can observe some good features or advantages of the deconvolution method as compared to other pattern reconstruction methods available in the literature. First, this method is independent of the bandwidth of an antenna and there is no requirement of prior knowledge of the system or the test environment. Also, this method not only generates an approximated free space radiation pattern, but also provides the knowledge of the phase component of the pattern. And this phase knowledge is necessary for some applications.

Still, there are lots of work that needs to be done to improve the deconvolution method:

1) Besides the measurement of a single AUT, the deconvolution method is also useful and promising for measurements of a large target, especially for antennas mounted on large platforms. For those large targets, indoor measurements inside an anechoic chamber are very difficult. The deconvolution method provides a way to obtain a quick and relative accurate estimation of the radiation pattern for such large targets. Due to the limited time and computational resources, we are currently unable to simulate the pattern reconstruction for antennas mounted on large platforms. This situation is suggested to be evaluated in real measurements.

2) The deconvolution method is not only useful in antenna pattern measurements in non-anechoic environments, but also could be applied in characterizing the reflection level of an anechoic chamber. It will be very interesting to model an anechoic chamber with absorbing materials and simulate the antenna measurements in such environment. And the deconvolution method can be used to extract the reflections and diffractions within the anechoic chamber.

3) As shown previously, we choose PEC plates and dielectric plates to represent the reflector. For realistic measurements in a large room, the radiated fields

would reflect from wooden tables, concrete walls around the room, the floor, and the ceiling of the room. Therefore, a more realistic numerical simulation needs to add these realistic objects into the model. And the reconstructed results would be more convincing.

4) So far, we have evaluated deconvolution method through numerical simulations. A better way is to test the method would be in real measurements. This requires one to choose an environment and carry out a real antenna pattern measurement. Then, follow procedures of the deconvolution method to measure the received response at the probe and extract the environmental effects from measured result of a reference antenna. And use it in a subsequent measurement for an AUT to extract its ideal pattern, and compare the reconstructed pattern with that measured in an anechoic chamber.

We hope the deconvolution method could be applied into the real antenna measurements, and can be used to save the expenses on building an anechoic chamber. So the antenna pattern measurement could be more affordable and flexible.

Appendix A: Data Mapping Using the Conversion between the Spherical Coordinate System and the Cartesian Coordinate System

As mentioned in Section 10.3, the FFT transforms the convolution of 2D periodic sequences into the multiplication of their corresponding discrete Fourier series. And the sequences that we select should be a period of the 2D periodic sequences. For example, if we take a step of $10°$ along the azimuth angle ϕ, the data at $\phi = -180°, -170°, \ldots, 160°, 170°$ should be a period of the periodic sequences, as shown below:

$$\phi = -\infty, \ldots, -180°, -170°, \ldots, 170°, 180°, 190°, \ldots, 530°, \ldots, +\infty$$

$$\underbrace{\hspace{3cm}}_{\text{One period}} \quad \underbrace{\hspace{3cm}}_{\text{One period}}$$

Also note that the end point of one period should be continuous to the start point of the next period, like the transition from $\phi = 170°$ to $180°$. These two properties can be visualized as: we rotate an object along the azimuth angle and the object will return to its original starting place after rotating one loop and then it starts to rotate for the next loop. This is the case for our 1D pattern reconstruction.

These two properties also apply to the 2D situation. For example, if one changes both ϕ and θ in a step of $10°$, a 2D matrix is formed by listing the (θ, ϕ) in a plane, as shown below (θ varies along the column and ϕ varies along the row):

$$
\begin{bmatrix}
(0°, -180°) & (0°, -170°) & \cdots & (0°, 170°) \\
(10°, -180°) & (10°, -170°) & \cdots & (10°, 170°) \\
\vdots & \vdots & \ddots & \vdots \\
(170°, -180°) & (170°, -170°) & \cdots & (170°, 170°) \\
(180°, -180°) & (180°, -170°) & \cdots & (180°, 170°)
\end{bmatrix}_{19 \times 36}^{\theta \times \phi}
$$

We would think this is also a period of the 2D periodic sequences, and it should be continuously transited to the next period along either ϕ or θ. However, it turns out to be not true.

If we duplicate this matrix and pad it at its side as shown, we see that the two adjacent data (θ, ϕ) of $(180°, -180°)$ and $(0°, -180°)$ should have a step difference of $10°$, which is not. Therefore, the original 2D matrix cannot satisfy our requirement.

$$
(\theta, \phi)
\begin{bmatrix}
(0°, -180°) & (0°, -170°) & \cdots & (0°, 170°) \\
(10°, -180°) & (10°, -170°) & \cdots & (10°, 170°) \\
\vdots & \vdots & \ddots & \vdots \\
(170°, -180°) & (170°, -170°) & \cdots & (170°, 170°) \\
(180°, -180°) & (180°, -170°) & \cdots & (180°, 170°)
\end{bmatrix}_{19 \times 36}
$$

$$
\begin{bmatrix}
(0°, -180°) & (0°, -170°) & \cdots & (0°, 170°) \\
(10°, -180°) & (10°, -170°) & \cdots & (10°, 170°) \\
\vdots & \vdots & \ddots & \vdots \\
(170°, -180°) & (170°, -170°) & \cdots & (170°, 170°) \\
(180°, -180°) & (180°, -170°) & \cdots & (180°, 170°)
\end{bmatrix}_{19 \times 36}
$$

However, if we complete the data matrix by padding the data in the elevation angle θ, specifically adding θ between $[180°, 360°)$, we can guarantee that the above two properties can be satisfied. Since there is no explicit definition for the data set of θ between $[180°, 360°]$ in the normal spherical coordinate system, we need to derive that data by using the conversion between the spherical coordinate system and the Cartesian coordinate system, as shown below:

$$
\begin{cases}
x = r \sin \theta \cos \phi \\
y = r \sin \theta \sin \phi \\
z = r \cos \theta
\end{cases}
\tag{10.36}
$$

One can verify that the data in the region $\theta \in [0°, 180°]$ and $\phi \in [-180°, 180°]$ can be fully mapped to the data in the region $\theta \in [180°, 360°]$ and $\phi \in [-180°, 180°]$ by using the angular conversion:

$$
(r, \theta, \phi) = (r, 2\pi - \theta, \phi \pm \pi)
\tag{10.37}
$$

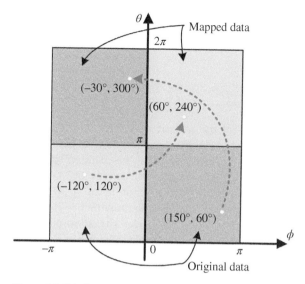

Figure 10.122 Data mapping illustration.

It is easy to see that the Cartesian coordinates generated by the two sides of the (10.37) are the same. Therefore, the data set of θ between $[180°, 360°]$ can be derived. Then, both θ and ϕ change its value within a range of 2π and form a periodic 2D matrix.

This data mapping process is shown in Figure 10.122. The lower layer data blocks are the original data when θ is between $[0°,180°]$ while the upper layer data blocks are the mapped data when θ is between $[180°,360°]$. The data blocks with the same color indicate the mapping location for the spatial angle (θ, ϕ). And the dashed lines show two examples of data mapping using (10.37), i.e., the spatial angle (θ, ϕ) at $(-120°,120°)$ should be mapped to $(60°,240°)$; while the angle at $(150°,60°)$ should be mapped to $(-30°,300°)$.

Appendix B: Description of the 2D-FFT during the Data Processing

As mentioned in Section 10.3, we need to perform a 2D-FFT on matrices and transform the angular convolution into the multiplication in the other domain. According to (10.35), the ideal pattern of the AUT can be calculated as

$$fft_2(P_{ideal-\text{AUT}}(\theta, \phi, f)) = \frac{fft_2(P_{non-ideal-\text{AUT}}(\theta, \phi, f)) \cdot fft_2(P_{ideal-\text{Ref}}(\theta, \phi, f))}{fft_2(P_{non-ideal-\text{Ref}}(\theta, \phi, f))}$$

$$(10.38)$$

and the 2D-FFT is defined as:

$$X(m,n) = \sum_{q=1}^{M}\sum_{r=1}^{N} x(q,r) W_M^{(m-1)(q-1)} W_N^{(n-1)(r-1)} \quad (m = 1, ..., M; n = 1, ..., N)$$

$$W_N = e^{-j(2\pi/N)}$$

$$(10.39)$$

where M and N are the dimensions of the matrix, and $X(m, n)$ denotes the discrete Fourier transform (DFT) of $x(q, r)$. In our examples, we take a step of 10° in the spatial domain, and the matrix will be padded to be a 36 by 36 square matrix (as shown in Appendix 10.9), i.e., $M = N = 36$. And both $x(q, r)$ and $X(m, n)$ should be a matrix with the dimension of 36 by 36.

When the environment is symmetrical along the plane with the elevation angle $\theta = 90°$, and the dimension N is even, it can be proved that some elements of $X(m, n)$ should be zero when m is even and n is odd. The proof is shown as below.

If we set $m = 2k, n = 2t - 1, (k, t = 1, 2, ..., N/2)$, then m is even and n is odd. So we have:

$$X(2k, 2t - 1) = \sum_{q=1}^{N}\sum_{r=1}^{N} x(q,r) W_N^{(2k-1)(q-1)} W_N^{(2t-2)(r-1)} \quad (k, t = 1, ...N/2)$$

$$(10.40)$$

By separating the summation operator, we have:

$$X(2k, 2t - 1) = \sum_{q=1}^{N} W_N^{(2k-1)(q-1)} \sum_{r=1}^{N} x(q,r) W_N^{(2t-2)(r-1)}$$

$$(10.41)$$

We can separate (10.41) as:

$$X(2k, 2t - 1) = \sum_{q=1}^{N} W_N^{(2k-1)(q-1)} C_q$$

$$(10.42)$$

$$C_q = \sum_{r=1}^{N} x(q,r) W_N^{(2t-2)(r-1)}$$

$$(10.43)$$

For a fixed value of q, (10.43) is a weighted summation of row elements at row q, which can be viewed as a constant. Also, the terms $W_N^k = e^{-j(2\pi/N)k}$ are the complex roots on the unit circle, as shown in Figure 10.123.

If C_q have the value of 1, we will have $X(2k, 2t - 1) = 0$ from the summation lemma:

$$\sum_{q=1}^{N} W_N^{-(k-1)(q-1)} = 0, \quad (k = 1, 2, ..., N)$$

$$(10.44)$$

Figure 10.123 Points of complex roots on the unit circle.

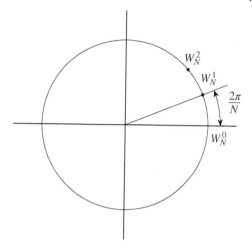

Due to the symmetrical environment setting, for any $r = 1, 2, ..., N$, we have:

$$x(1,r) = x(19,r), x(2,r) = x(18,r), ..., x(9,r) = x(11,r) \tag{10.45}$$

Therefore, C_q have the following values

at the upper unit circle: $C_1 = C_{19}, C_2 = C_{18}, ..., C_9 = C_{11}$ (10.46)

at the lower unit circle: $C_{20} = C_{36}, ..., C_{27} = C_{29}$ (10.47)

C_{10} and C_{28} are on the positive and negative axis, respectively (10.48)

And (10.42) can be expanded as the following summation of 8 terms as:

$$X(2k, 2t-1) = sum \begin{cases} \displaystyle\sum_{q=2}^{9} W_N^{(2k-1)(q-1)} C_q + \sum_{q=11}^{18} W_N^{(2k-1)(q-1)} C_q \\[2ex] W_N^{(2k-1)(1-1)} C_1 + W_N^{(2k-1)(19-1)} C_{19} \\[1ex] W_N^{(2k-1)(10-1)} C_{10} + W_N^{(2k-1)(28-1)} C_{28} \\[2ex] \displaystyle\sum_{q=20}^{27} W_N^{(2k-1)(q-1)} C_q + \sum_{q=29}^{36} W_N^{(2k-1)(q-1)} C_q \end{cases}$$

$$(10.49)$$

The 4 terms in the center cancel out to be zero; while the first line and the 4th line also cancel out to be zero. Therefore $X(2k, 2t-1)$ will becomes zero value. Thus $fft_2(P_{ideal-\text{AUT}}(\theta, \phi, f))$ in (10.38) will be $0 \cdot (0/0)$ in value, which should be zero. But N/A value was generated in MATLAB due to the numerical errors encountered in the computations. To conduct the 2D-IFFT of the matrix, we manually set those N/A values to be zero, which they should be.

References

1 M. N. Abdallah, T. K. Sarkar, M. Salazar-Palma, and V. Monebhurrun, "Where Does the Far Field of an Antenna Start? [Stand on Standards]," *IEEE Antennas and Propagation Magazine*, Vol. 58, No. 5, pp. 115–124, 2016.

2 P. W. Arnold, "The Slant Antenna Range," *IEEE Transactions on Antennas and Propagation*, Vol. AP-14, No. 5, pp. 658–659, 1966.

3 R. C. Johnson and R. J. Poinsett, *Compact Antenna Range Techniques*, Technical Report, 1966.

4 E. B. Joy, W. M. Leach, Jr., and G. P. Rodrigue, "Applications of Probe-Compensated Near-Field Measurements," *IEEE Transactions on Antennas and Propagation*, Vol. 26, No. 3, pp. 379–389, 1978.

5 W. L. Stutzman and G. A. Thiele, *Antenna Theory and Design*, John Wiley & Sons, New York, Second Edition, 1998.

6 V. Viikari, J. Mallat, J. Ala-Laurinaho, J. Mallat, and A. V. Raisanen, "New Pattern Correction Techniques for Submm-Wave CATRs," *Proceedings of The European Conference on Antennas and Propagat.(EuCAP)*, 2006.

7 D. N. Black, E. B. Joy, M. G. Guler, and R. E. Wilson, "Range Field Compensation," *Proceedings Antenna Measurement Techniques Association Symposium*, pp. 3B-19–3B-24, 1991.

8 D. N. Black and E. B. Joy, "Test Zone Field Compensation," *IEEE Transactions on Antennas and Propagation*, Vol. 43, No.4, pp. 362–368, 1995.

9 J. T. Toivanen, T. A. Laitinen, S. Pivnenko, and L. Nyberg, "Calibration of Multi-Probe Antenna Measurement System Using Test Zone Field Compensation," *3rd European Conference on Antennas and Propagation (EuCAP'09)*, Berlin, Germany, pp. 2916–2920, 2009.

10 J. T. Toivanen, T. A. Laitinen, and P. Vainikainen, "Modified Test Zone Field Compensation for Small Antenna Measurements," *IEEE Transactions on Antennas and Propagation*, Vol. 58, No. 11, pp. 3471–3479, 2010.

11 R. Pogorzelski, "Extended Probe Instrument Calibration (EPIC) for Accurate Spherical Near-Field Antenna Measurements," *IEEE Transactions on Antennas and Propagation*, Vol. 57, No. 10, pp. 3366–3371, 2009.

12 R. Pogorzelski, "Experimental Demonstration of the Extended Probe Instrument Calibration (EPIC) Technique," *IEEE Transactions on Antennas and Propagation*, Vol. 58, No. 6, pp. 2093–2097, 2010.

13 S. Raz and R. Kastner, "Pattern Reconstruction from Distorted Incident Wave Measurements," *Antennas and Propagation Society International Symposium*, Vol. 14, p. 280, 1976.

14 J. C. Bennet and A. Griziotis, "Removal of Environmental Effects from Antenna Radiation Patterns by Deconvolution Processing," *Proceedings of the IEE Conference*, Pub. 219, Pt. 1, pp. 224–228, 1983.

15 P. L. Garcia Muller, J. L. Cano, and R. Torres, "A Deconvolution Method for Correcting Antenna Measurement Errors in Compact Antenna Test Ranges,"

Proceedings of the 17th Annual Antenna Measurement Techniques Association (AMTA) Meeting & Symposium, USA, pp. 509–514, November 1995.

16 J. Koh, A. De, T. K. Sarkar, H. Moon, W. Zhao, and M. Salazar-Palma, "Free Space Radiation Pattern Reconstruction from Non-Anechoic Measurements Using an Impulse Response of the Environment," *IEEE Transactions on Antennas and Propagation*, Vol. 60, No. 2, pp. 821–831, 2012.

17 S. Loredo, M. R. Pino, F. Las-Heras, and T. K. Sarkar, "Echo Identification and Cancellation Techniques for Antenna Measurement in Non-Anechoic Test Sites," *IEEE Antennas and Propagation Magazine*, Vol. 46, No. 1, pp. 100–107, 2004.

18 S. Loredo, G. Leon, S. Zapatero, and F. Las-Heras, "Measurement of Low-Gain Antennas in Non-Anechoic Test Sites Through Wideband Channel Characterization and Echo Cancellation," *IEEE Antennas and Propagation Magazine*, Vol. 51, No. 1, pp. 128–135, 2009.

19 A. V. Kalinin, "Anechoic Chamber Wideband Antenna Measurements," *IEEE Aerospace and Electronic Systems Magazine*, Vol. 21, No. 1, pp. 21–24, 2006.

20 M. E. Hines and H. E. Stinehelfer, "Time-Domain Oscillographic Microwave Network Analysis Using Frequency-Domain Data," *IEEE Transactions on Microwave Theory and Techniques*, Vol. 22, No. 3, pp. 276–282, 1974.

21 J. Marti-Canales and L. P. Ligthart, "Modeling and Pattern Error Correction of Time Domain Far-Field Antenna Measurements," *IEE Proceedings on Microwaves, Antennas Propagation*, Vol. 148, No. 2, pp. 133–136, 2001.

22 B. Fourestie, Z. Altman, and M. Kanda, "Anechoic Chamber Evaluation Using the Matrix Pencil Method," *IEEE Transactions on Electromagnetic Compatibility*, Vol. 41, No. 3, pp. 169–174, 1999.

23 B. Fourestie, Z. Altman, J. Wiart, and A. Azoulay, "On the Use of the Matrix-Pencil Method to Correlate Measurements at Different Test Sites," *IEEE Transactions on Antennas and Propagation*, Vol. 47, No. 10, pp. 1569–1573, 1999.

24 B. Fourestie, Z. Altman, and M. Kanda, "Efficient Detection of Resonances in Anechoic Chambers Using the Matrix Pencil Method," *IEEE Transactions on Electromagnetic Compatibility*, Vol. 42, No. 1, pp. 1–5, 2000.

25 B. Fourestie and Z. Altman, "Gabor Schemes for Analyzing Antenna Measurements," *IEEE Transactions on Antennas and Propagation*, Vol. 49, No. 9, pp. 1245–1253, 2001.

26 P. S. H. Leather and D. Parson, "Equalization for Antenna-Pattern Measurements: Established Technique – New Application," *IEEE Antennas and Propagation Magazine*, Vol. 45, No. 2, pp. 154–161, 2003.

27 P. S. H. Leather, D. Parsons, J. Romeu, S. Blanch, and A. Aguasca, "Correlation Techniques Applied to Antenna Pattern Measurement," *Electronics Letters*, Vol. 40, No. 10, pp. 572–574, 2004.

28 P. S. H. Leather and J. D. Parsons, "Plane Wave Spectra, Test-Zone Fields and Simulation of Antenna-Pattern Measurements," *Electronics Letters*, Vol. 39, No. 25, pp. 1780–1782, 2003.

29 P. S. H. Leather and D. Parsons, "Improved Antenna Pattern Measurements Using Equalization," *IEE Antennas and Propagation Newsletter*, pp. 4–7, October 2002.

30 P. S. H. Leather and D. Parsons, "Equalization: A Technique to Improve the Accuracy of Antenna Radiation Pattern Measurements," *IEE Conference Publication*, Vol. 1, pp. 102–106, 2003.

31 P. S. H. Leather, D. Parsons, and J. Romeu, "Signal Processing Techniques Improve Antenna Pattern Measurement," *IEE AMS 2004*, pp. 97–100, 2004.

32 W. D. Burnside and I. J. Gupta, "A Method to Reduce Stray Signal Errors in Antenna Pattern Measurements," *IEEE Transactions on Antennas and Propagation*, Vol. 42, No. 3, pp. 399–405, 1994.

33 M. D. Migliore, "Filtering Environmental Reflections in Far-Field Antenna Measurement in Semi-Anechoic Chambers by an Adaptive Pattern Strategy," *IEEE Transactions on Antennas and Propagation*, Vol. 52, No. 4, pp. 1112–1115, 2004.

34 J. Appel-Hansen, "Reflectivity Level of Radio Anechoic Chambers," *IEEE Transactions on Antennas and Propagation*, Vol. 21, No. 4, pp. 490–498, 1973.

35 V. Viikari, J. Mallat, J. Ala-Laurinaho, J. Mallat, and A.V. Raisanen, "A Feed Scanning Based APC Technique for Compact Antenna Test Ranges," *IEEE Transactions on Antennas and Propagation*, Vol. 53, No. 10, pp. 3160–3165, 2005.

36 V. Viikari, J. Mallat, J. Ala-Laurinaho, J. Hakli, and A. Raisanen, "A Frequency Shift Technique for Pattern Correction in Hologram-Based CATRs," *IEEE Transactions on Antennas and Propagation*, Vol. 54, No. 10, pp. 2963–2968, 2006.

37 V. Viikari, V.-M. Kolmonen, J. Salo, and A. V. Raisanen, "Antenna Pattern Correction Technique Based on an Adaptive Array Algorithm," *IEEE Transactions on Antennas and Propagation*, Vol. 55, No. 8, pp. 2194–2199, 2007.

38 Y. Zhang, T. K. Sarkar, X. Zhao, D. Garcia-Donoro, W. Zhao, M. Salazar-Palma, and S. Ting, *Higher Order Basis Based Integral Equation Solver (HOBBIES)*, John Wiley & Sons, Hoboken, NJ, 2012.

39 A. V. Oppenheim, A. S. Willsky, and I. T. Young, *Signals and Systems*, Prentice Hall, Englewood Cliffs, NJ, 1983.

40 A. V. Oppenheim and R. W. Schafer, *Digital Signal Processing*, Prentice Hall, Englewood Cliffs, NJ, 1975.

41 W. Zhao, *Retrieval of Free Space Radiation Patterns Through Measured Data in a Non-Anechoic Environment*, Ph.D. Thesis, Syracuse University, Syracuse, New York, 2013.

42 Z. Du, J. I. Moon, S. Oh, J. Koh, and T. K. Sarkar, "Generation of Free Space Radiation Patterns From Non-Anechoic Measurements Using Chebyshev Polynomials," *IEEE Transactions on Antennas and Propagation*, Vol. 58, No. 8, pp. 2785–2790, 2010.

Index

a

adaptive array strategy 577
adaptive Cauchy interpolation
  algorithm 159
  advantage 171–172
  flowchart 160, 161
  hollow conducting box 163–168
  microstrip patch antenna 162–164
  short dual patch antenna array 168–171
additive noise case, Cauchy
  method 131–136
amplitude-only data 352, 356
  array of probes 320
  calculating far-field 428–442
  direct optimization approach (see direct
    optimization approach)
  interpolatory Cauchy method for
    high resolution wideband response
    generation 138–148
  non-minimum phase response
    generation 148–158
  method of moments 351
  near-field to far-field
    transformation 320, 322, 352
  normalization component 346, 347
analytical spherical near-field to far-field
  transformation 453–463
  experimental data 465–468
  features 463–464
  numerical simulations 464–468
  patch microstrip array
    antennas 465–468

synthetic data, four dipole
  array 464–465
anechoic chambers 520, 523, 524
  antenna radiation pattern
    measurements 57–60, 573, 574
antenna diagnostic methods 236
antenna pattern comparison (APC)
  technique 577
antenna radiation pattern
  measurements 527
  in anechoic chambers 57–60
antenna under test
  azimuth radiation pattern 60
  FFT based method 58, 61–64, 67–74
  MPM technique 58, 64–74
  on rollover azimuth positioner 58, 59
  computational methodology
    363–365
  development 363
  integral equation formulation 365–367
  integro-differential equations 367–369
  Rayleigh limit 58
  sample numerical results 369–384
antenna under test (AUT) 235
  anechoic chamber, radiation patterns
    in 573, 574
  estimation error 320
  far-field pattern 384, 468
  horn antenna 369
  magnetic currents 319, 352
  method of moments approach
    362, 407

Modern Characterization of Electromagnetic Systems and Its Associated Metrology, First Edition.
Tapan K. Sarkar, Magdalena Salazar-Palma, Ming Da Zhu, and Heng Chen.
© 2021 John Wiley & Sons, Inc. Published 2021 by John Wiley & Sons, Inc.

antenna under test (AUT) (*cont'd*)
  near-field to far-field
    transformation   520
  pattern reconstruction, using
    deconvolution method
    576, 608
    data processing   585–587
    equations and derivation
    578–584
    HOBBIES, 584
    improvements required in
    674–675
    parabolic reflector antenna   604–607
    PEC plates as reflectors   587–604
    probe antenna types, effect
    of   608–619
    procedures   584–585
    three-dimensional   632–673
    using different electrical antenna
    sizes   619–626
    using different PEC plate size
    effects   626–632
  radiating structure   352
  radiation pattern measurements
    azimuth radiation pattern   60
    FFT based method   58, 61–64, 67–74
    MPM technique   58, 64–74
    on rollover azimuth positioner   58, 59
    source reconstruction method   362–364
    TZF compensation method   575

b

base station antenna
  near-field to far-field
    transformation   304–307
  reference volume for   310, 312–313
base terminal stations (BTS)   298–300
basis functions   2, 241, 246, 277, 279, 280,
    304, 307, 321, 325, 326, 355, 367, 445,
    446, 471, 473, 522
Beatty standard, MPM
  frequency-domain response of   49
  impedance step discontinuities   44
  impulse response of   45, 46, 52, 53

  magnitude and phase response of   44, 46,
    47, 49, 53, 55
  terminated with a short   51–57
  terminated with matched load
    44, 45
  time domain impulse response   47, 48,
    50, 51, 54, 56, 57
Bessel function   454, 520, 522, 525, 527, 528
bias-variance tradeoff, in reduced rank
    modelling   3–6
block Toeplitz matrix   319
  CGFFT, 329, 334, 442, 443
  fast Fourier transform   327
boundary value problems   1
bounded input and bounded output (BIBO)
    response   2
broadband device characterization, using
    Cauchy method   127, 129, 130

c

Cauchy method (CM)   6, 521
  adaptive interpolation algorithm   159
  advantage   171–172
  flowchart   160, 161
  hollow conducting box   163–168
  microstrip patch antenna   162–164
  short dual patch antenna
    array   168–171
  broadband device characterization   127,
    129, 130
  E-plane waveguide filter
    characterization   172–175
  for filter analysis   125–128
  on Gegenbauer polynomials
    531–543
  high resolution wideband response
    generation   138–148
  MATLAB codes   181–187
  MoM program   120–122
  near-field to far-field
    transformation   543–551
  noise contaminated data
    additive noise component,
    perturbation   131–136

Gaussian random variables 134–137
  numerical example 136–138
  perturbation of invariant
    subspaces 130–131
  non-minimum phase response
    generation 148–158
  non-uniform sampling 158
  object's external resonant frequency
    extraction 176–180
  optical computations, interpolating results
    for 123–125
  origin 107
  system response estimation
    examples of 112–120
    interpolation/extrapolation
      procedure 112
    uniqueness of 108
causal systems 192
CG-FFT *see* conjugate gradient-fast Fourier
    transformation (CG-FFT) method
characteristic impedance, MPM, 32
  definitions 33
  of microstrip line 33
    at fixed frequency 35–36
    quasi static solution 35
    voltage and current distribution 34
  for static case in vacuum 33
Chebyshev polynomials 520, 533, 569
  frequency domain response by 521–531
chessboard-like magnetic current
    357–359
commercial antennas, radiation pattern
    evaluation
  electromagnetic analysis 299
  far field pattern 297
  human exposure recommendations/
    regulations 298
  near-field to far-field transformation
    base station antenna 304–307
    pyramidal horn antenna 307–311
  source reconstruction method
    EqMC distribution 301–303
    goal/objective 300
    problem formulation 301–304

conjugate gradient-fast Fourier
    transformation (CG-FFT)
    method 321, 368
  block Toeplitz structure 329,
    334, 442
  fast Fourier transform 368
  magnetic currents 368, 380
  magnetic dipoles 347
  2D-FFT, 349
conjugate gradient method (CGM) 319,
    363, 368, 369
  with fast Fourier transform
    advanced convolution operator 495
    numerical computation 496–498
    numerical results 498–500
  plane conducting sheet
    impulse response 498–500
    transient waveform scattered
      from 498, 499
cost function 18, 353
  optimization of 356–357
Cramer–Rao bound (CRB) 81
  vs. signal-to-noise ratio (SNR) 82, 83

d

data mapping process
  2D-FFT, 677–679
  using spherical *vs.* cartesian coordinate
    system conversion 675–677
deconvolution method 491, 494, 501
  radiation pattern
    reconstruction 576, 608
    data processing 585–587
    equations and derivation 578–584
    HOBBIES, 584
    improvements required in 674–675
    parabolic reflector antenna 604–607
    PEC plates as reflectors 587–604
    probe antenna types, effect
      of 608–619
    procedures 584–585
    three-dimensional 632–673
    using different electrical antenna
      sizes 619–626

deconvolution method (*cont'd*)
  using different PEC plate size
    effects   626–632
dipole antenna
  input impedance of, interpolating missing
    data   210, 213, 214
  as probe   613, 619
    helical antenna   615, 617
    horn antenna   615, 616
    Yagi antenna   615, 618
dipole arrays   443, 447
dipole probes
  antenna pattern
    measurements   361–384
  array measurement   407–428
direction of arrival (DOA) estimation,
    MPM, 81–85
direct optimization approach   321
  cost function   356–357
  equivalent current
    representation   354–356
  experimental data   358–359
  numerical simulation   357–358
discrete-time Fourier transform (DTFT)   4

e

Eckart–Young–Mirsky theorem   18
electric current distribution   445
electric field integral equation (EFIE)   34,
    37, 241, 276, 277, 319, 321, 380, 442,
    443, 445
electromagnetic (EM) transient
    waveform   21
elliptic partial differential equation   1
E-plane waveguide filter
    characterization   172–175
equalization technique   577
equivalence principle   239, 241, 277
equivalent magnetic current approach, NF to
    FF transformation   240
  description   241–245
  electric field integral equation   241
  equivalence principle   241
  integral equation, solution of   245–249
  moment method procedure   241

patch microstrip array, co-polarization
    characteristics   249–276
equivalent magnetic current (EMC)
    distribution   301–305,
    307, 310, 352–355, 364,
    469–472
equivalent magnetic dipole array
    approximation   321, 328–334,
    348–350
extrapolation   107
  frequency domain data
    samples   200–203
  noise contaminated data   26
  system response estimation using Cauchy
    method   112

f

fast Fourier transform (FFT)   319, 343, 352,
    356, 362, 368, 385, 428
feature selection   3
FFT-based method   576
  antenna radiation pattern
    measurements   58, 61–64, 67–74
filter analysis, using Cauchy
    method   125–128
fixed probe array
  methodology   429–430
  sample numerical results   430–441
Fortran based computer program   483–488
Fourier transform   340–347, 350,
    377, 406
  fast Fourier transform   319, 343, 352,
    356, 362, 368, 385, 428
  and Hilbert transform   195–196
    for analog and discrete cases   199–200
free space radiation pattern retrieval *see*
    radiation pattern reconstruction
frequency domain data samples
  extrapolation/interpolation
    technique   200–203
  interpolating missing data   203–215
    Cauchy method *vs.* Hilbert transform
      method   203–208
    dipole antenna, input impedance
      of   210, 213, 214

microstrip bandpass filter   203, 205, 208–211
microstrip notch filter   213, 215
frequency domain response, by Chebyshev polynomials   521–531
Frobenius norm   17

g

Galerkin's type testing   442, 443, 446
Gaussian random variables   134–137
Gegenbauer polynomials   519
   Cauchy method on   531–543
Gram-Schmidt orthogonalization procedure   216
Green's function   324, 463
   fast Fourier transform   356
   integral equation approach   343
   invariant properties of   355
   three-dimensional free space   365
   two dimensional Fourier transform of   342

h

Hankel matrix   66
Hanning window   201–203
helical antenna
   amplitude pattern   621, 624
      with dipole antenna as probe   615, 617
      with double PEC plate reflector   594, 597
      with four PEC plate reflector   601, 602
      with single PEC plate reflector   591, 592
      with Yagi probe antenna   609, 610
   current distribution on feed dipole   622
      with double PEC plate reflector   594, 595
      with four PEC plate reflector   599, 600
      with single PEC plate reflector   589
   dimensions   587, 588, 621
   pattern retrieval using different PEC plate size effects   626–632
   phase pattern   621, 624
      with dipole antenna as probe   615, 617

      with double PEC plate reflector   594, 597
      with four PEC plate reflector   601, 603
      with single PEC plate reflector   591, 592
      with Yagi probe antenna   609, 611
higher order basis based integral equation solver (HOBBIES)   91, 92, 176–179, 367, 378, 406, 430, 564
Hilbert inner product   493
Hilbert-Schmidt convolution operator   494
Hilbert transform   191
   for analytic signal   197
   and causality   194–195
   convolution product   197
   and Fourier transform   195–196
      for analog and discrete cases   199–200
   inverse of   199
   minimum phase systems   193
   properties   195–199
   for real valued function   193
   self-mapping property   197
   for spectrum computation   213, 221–225
Hilbert transformation   522
horn antenna   537–539, 638
   amplitude pattern   620, 622
      with dipole antenna as probe   615, 616
      with double PEC plate reflector   594, 595
      with four PEC plate reflector   599, 600
      with parabolic reflector as probe   612, 613
      with single PEC plate reflector   589, 590
      with Yagi probe antenna   609
   antenna under test   369
   dimensions   587, 588, 620
   pattern retrieval using different PEC plate size effects   626–632
   phase pattern   621, 623
      with dipole antenna as probe   615, 616
      with double PEC plate reflector   594, 596
      with four PEC plate reflector   599, 601

horn antenna (*cont'd*)
  with parabolic reflector as
      probe   612, 614
  with single PEC plate
      reflector   589, 590
  with Yagi probe antenna   609, 610
  phase response generation
      dimensions of   152
      input Gaussian pulse   152
      phase functions   152, 153
      power spectrum density   152, 153
      schematic representation   151
      time domain responses   154–155
  pyramidal, near-field to far-field
      transformation   307–311
  three-dimensional radiation pattern
      reconstruction   640, 649, 651, 657,
      659, 665
Huygen's principle   341
hyperbolic partial differential equation   2

i

ill-posed deconvolution problem
      491–492
independent and identically distributed (IID)
      random process   3
initial value problems   2
inner product   196, 495, 498
integral equations
  approach   341–344, 350
  formulation   323–325, 365–367
  mathematical formulations   239–240
integro-differential equations   367–369
invariant subspaces, perturbation
      of   130–131
inverse discrete Fourier transform   201,
      202, 327
iterative conjugate gradient (CG)
      method   235, 238
iterative nonparametric technique *see*
      Hilbert transform

k

Karhunen-Loeve expansion   4
Kramers–Kronig relationship   195

l

least squares (LS) method
  minimal 2-norm solution
      501–502
  reformulation of   502
Legendre polynomials   533
linear time-invariant (LTI) system   2, 6, 21,
      108, 149, 194
Lomb periodogram approach   192,
      216–217, 222, 227

m

mathematical physics   1
MATLAB codes
  Cauchy method   181–187
  matrix pencil method
      implementation   96–99
matrix method formulation   469
matrix pencil method (MPM)   6, 21, 23, 577
  antenna radiation pattern
      measurements   57–74
  for characteristic impedance evaluation, of
      transmission line   32–36
  direction of arrival (DOA)
      estimation   81–85
  MATLAB codes   96–99
  miscellaneous applications of   95
  multiple objects identification, in free
      space   91–95
  network analyzer measurements   44–57
  for noise contaminated data   22
      estimated mean squared error   29, 31
      numerical data illustrations   26–32
      singular value decomposition
      24, 25
      system response, interpolation/
      extrapolation   26
      white Gaussian noise   29, 30
  oscillatory functional variation, in
      Sommerfeld integrals   85–91
  rotated parabolic reflector   564–568
  SEM poles, single estimate for
      74–81
  for S-parameters computation   37–44

zenith-directed parabolic
reflector   558–563
Maxwell's equations   2, 319
method of moments (MoM) procedure
antenna under test   428
E-field integral equations   367
electric field integral equation   319
equivalent magnetic currents   321, 362,
385, 407
formulation   325–328
matrix equation   321
method of moments (MoM)
program   120–122
microstrip filter, interpolating missing data
bandpass filter   203, 205, 208–211
notch filter   213, 215
microstrip patch antenna
adaptive interpolation
algorithm   162–164
phase response generation
154–157
modal expansion method   339–341,
345, 469
moment method procedure   241, 277
MoM program *see* method of moments
(MoM) program
Moore-Penrose pseudo-inverse   24
MPM *see* matrix pencil method (MPM)
multiple objects identification, in free
space   91–95

n

near-field (NF) to far-field (FF)
transformation   236, 237
amplitude-only data   320, 322, 352
antennas under test   520
for arbitrary near-field geometry using
equivalent electric current
analytic continuity concept   276, 277
description   278–281
equivalence principle   277
moment method procedure   277
patch microstrip array, co-polarization
characteristics   281–297

Cauchy method   543–551
dipole probe array
measurement   406–428
EMC distribution   470–472
amplitude and phase   473, 474
two-dimensional   473–475
integral equation formulation   323–325,
341–344
linear patch array   475–476
matrix equation formulation
325–328
matrix method formulation   469
matrix pencil method   558–568
microstrip patch array   477–479
modal expansion method   339–341, 469
numerical examples   344–351
for planar scanning   319, 321
radial electric field retrieval
algorithm   472–473
radiation patterns   520
reflector antenna   477–482
of rotated parabolic reflector   552–558,
564–568
sample numerical results   329–337
spherical (*see* spherical near-field to far-
field transformation)
use of magnetic dipole array   328
using equivalent magnetic current
approach   240
description   241–245
electric field integral equation   241
equivalence principle   241
integral equation, solution
of   245–249
moment method procedure   241
patch microstrip array, co-polarization
characteristics   249–276
of zenith-directed parabolic
reflector   543–551, 558–563
network analyzer measurements,
MPM, 44–57
noise contaminated data, MPM for   22
estimated mean squared error
29, 31

noise contaminated data, MPM for (*cont'd*)
  numerical data illustrations   26–32
  singular value decomposition   24, 25
  system response, interpolation/
    extrapolation   26
  white Gaussian noise   29, 30
non-anechoic environments   520, 524, 525
  antenna radiation pattern
    measurements   58–74
  free space radiation pattern retrieval (*see*
    radiation pattern reconstruction)
non ionizing radiation (NIR), of
    electromagnetic fields   298
novel antenna pattern comparison (NAPC)
    technique   577
Nyquist sampling   221, 227, 229, 362, 377,
    428, 470

o
object's external resonant frequency
    extraction   176–180
oversampled Gabor transform (OGT)   577

p
Paley–Wiener criterion   192
parabolic partial differential equation   2
parabolic reflector antenna   541–543
  amplitude pattern   604, 605, 607
  model with feeding structure
    604, 605
  phase pattern   604, 606, 607
  as probe antenna   612–615
  three-dimensional radiation pattern
    reconstruction   638, 639, 641–643,
    646–651, 654–656, 658, 659,
    662–664, 666–668, 671, 672
PEC plate reflector
  amplitude of environmental effects
    with double PEC plate   594, 596
    with four PEC plate   602
    with single PEC plate   591
  helical antenna (*see* helical antenna)
  horn antenna (*see* horn antenna)
  simulation model of measurement system
    with double PEC plate   594

    with four PEC plate   599
    with single PEC plate   587
  Yagi antenna (*see* Yagi (Yagi-Uda)
    antenna)
perfect electric conductors (PEC)   322
  high resolution amplitude
    response   142, 143
  phase response generation   156–158
  spheres model   92–95
periodogram approach   192, 216–217,
    222, 227
planar modal expansion method   338
planar near-field antenna
    measurements   442, 444, 447
power spectral density (psd)   4
principle of causality   194, 195
probe correction
  amplitude only data without   359
  formulation   446
  methodology   443–446
  sample numerical results   447–448
pseudo Green's functions   463
pseudo-morphic high electron mobility
    transistor (PHEMT)
    characterization   127, 129, 130

q
QR decomposition of matrix   111

r
radar cross section (RCS)
    calculation   122, 123
radial electric field retrieval
    algorithm   472–473
radiation pattern reconstruction
  deconvolution method   576, 608
    data processing   585–587
    equations and derivation   578–584
    HOBBIES, 584
    improvements required in   674–675
    parabolic reflector antenna   604–607
    PEC plates as reflectors   587–604
    probe antenna types, effect
      of   608–619
    procedures   584–585

three-dimensional 632–673
using different electrical antenna
sizes 619–626
using different PEC plate size
effects 626–632
equalization technique 577
FFT-based method 576
matrix pencil method 577
TZF compensation method 575
random process 3–4
Rayleigh limit 58, 350
rectangular probe array 385
methodology 385–387
sample numerical results 387–406
reduced-rank modelling 3–6
RFID tags 429
rotated parabolic reflector
matrix pencil method 564–568
ordinary Cauchy method 552–558

S

short dual patch antenna (SDPA) array
adaptive interpolation
algorithm 168–171
elements 168
radiation patterns, using adaptive Cauchy
method 168–171
shorted Beatty standard, MPM, 51–57
SI evaluation *see* Sommerfeld integrals (SI)
evaluation
signal-to-noise ratio (SNR) 82–86
signum function 195
singularity expansion method (SEM) 177
singularity expansion method (SEM)
poles 74–81
singular value decomposition (SVD) 6–15,
491, 494, 503
of matrix 109
noise level 534
TLS implemented through
501–516
total least squares (TLS) and 520
Sommerfeld integrals (SI) evaluation
using MPM, 85–91

using weighted-average method 88–91
source reconstruction method
(SRM) 235, 362
advantages 236
commercial antennas, radiation pattern
evaluation
EqMC distribution 301–303
goal/objective 300
problem formulation 301–304
commercial antennas, radiation pattern
evaluation of 297–313
drawback 238
overview 238–239
S-parameters computation, using
MPM, 37–44
specific absorption rate (SAR) limits, for
devices 298
spectral analysis, for nonuniformly
spaced data
Hilbert transform relationship 221, 223
least squares methodology 213,
217–221
magnitude estimation 223–228
periodogram approach 216–217,
222, 227
spherical near-field to far-field
transformation
analytical technique 453–463
experimental data 465–468
features 463–464
numerical simulations 464–468
patch microstrip array
antennas 465–468
synthetic data, four dipole
array 464–465
Fortran based computer
program 483–488
strict sense stationary (SSS)
process 3, 4

t

Taylor's theorem 107, 108
test zone field (TZF) compensation
method 575

three-dimensional radiation pattern
    reconstruction, deconvolution
    method
  closed PEC box   659, 662, 665, 666
  data processing   636–638
  dielectric plates forming closed
    contour   662, 666–673
  four PEC plate and ground as
    reflectors   638, 643–651
  four wide PEC plate as reflectors   638,
    640–643
  mathematical characterization
    632–635
  PEC plates forming an unclosed
    contour   638, 651–661
  procedure   635
time-difference-of-arrival (TDOA)
    technique   92
time domain function   525
total least squares (TLS) method   15–19,
    110, 111, 502–506
  multiconductor transmission line with
    10 ports
    deconvolved impulse
      response   508–516
    input step voltage   506, 507
    output step voltage   506, 509–515
    schematic diagram   506
  and singular value decomposition   520
transmission line characteristic impedance
    evaluation   32–36
truncation error   338, 341, 343, 345,
    346, 349
2-element microstrip patch array
    539–541

w

wave equation   2
weighted-average method (WAM), SI
    evaluation   88–91
white Gaussian noise   29, 30, 81, 82,
    118–120
wide-sense stationary (WSS) process   4

y

Yagi (Yagi-Uda) antenna   377–379, 397,
    398, 401, 419, 437, 638
  amplitude pattern   621, 625
    with dipole antenna as probe   615, 618
    with double PEC plate reflector   598
    with four PEC plate reflector   601, 603
    with single PEC plate reflector   593
  dimensions   587, 588, 621
  phase pattern   621, 625
    with dipole antenna as probe   615, 618
    with double PEC plate
      reflector   598, 599
    with four PEC plate reflector   601, 604
    with single PEC plate reflector   593
  as probe   608–612
  three-dimensional radiation pattern
    reconstruction   640, 642, 646, 650,
    651, 653, 658, 659, 661, 665, 666, 670
  using different PEC plate size
    effects   626–632

z

zenith-directed parabolic reflector
  matrix pencil method   558–563
  ordinary Cauchy method   543–551